Carbon Dioxide and Environmental Stress

This is a volume in the

PHYSIOLOGICAL ECOLOGY series
Edited by Harold A. Mooney

A complete list of books in this series appears at the end of the volume.

Carbon Dioxide and Environmental Stress

Edited by

Yiqi Luo
Biological Sciences Center
Desert Research Institute
Reno, Nevada

Harold A. Mooney
Department of Biological Sciences
Stanford University
Stanford, California

Academic Press
San Diego London Boston New York Sydney Tokyo Toronto

Cover photo: Open top chambers for CO_2 enrichment and UV fluorescent tubes. For more information see Chapter 6 by J. Rozema et al.

This book is printed on acid-free paper. ♾

Academic Press
a division of Harcourt Brace & Company
525 B Street, Suite 1900, San Diego, California 92101-4495, USA
http://www.apnet.com

Academic Press
24-28 Oval Road, London NW1 7DX, UK
http://www.hbuk.co.uk/ap/

Library of Congress Catalog Card Number: 99-60087

International Standard Book Number: 0-12-460370-X

PRINTED IN THE UNITED STATES OF AMERICA
99 00 01 02 03 04 BB 9 8 7 6 5 4 3 2 1

Contents

14. Ecosystem Modeling of the CO_2 Response of Forests on Sites Limited by Nitrogen and Water

Ross E. McMurtrie and Roderick C. Dewar

Part III
Synthesis and Summary

15. Diverse Controls on Carbon Storage under Elevated CO_2: Toward a Synthesis

Christopher B. Field

16. Interactive Effects of Carbon Dioxide and Environmental Stress on Plants and Ecosystems: A Synthesis

Yiqi Luo, Josep Canadell, and Harold A. Mooney

Contributors

Numbers in parentheses indicate the pages on which the authors' contributions begin.

Göran I. Ågren (333), Department of Ecology and Environmental Research, Swedish University of Agricultural Sciences, SE-750 07 Uppsala, Sweden

Jeffrey S. Amthor (33), Environmental Sciences Division, Oak Ridge National Laboratory, Oak Ridge, Tennessee 37831

Marilyn C. Ball (139), Ecosystem Dynamics Group, Research School of Biological Sciences, Australian National University, Canberra, ACT, Australia

Martyn Caldwell (169), Department of Rangeland Resources, Utah State University, Logan, Utah 84326

Josep Canadell (393), CSIRO Wildlife and Ecology, Lynehem ACT 2602, Australia

Weixin Cheng[1] (245), Biological Sciences Center, Desert Research Institute, Reno, Nevada 89506

Sharon A. Cowling (289), Department of Botany, University of Toronto, Toronto, Ontario, Canada M5S 3B2

Grant R. Cramer (139), Department of Biochemistry, University of Nevada, Reno, Nevada 89557

R. M. M. Crawford (61), Plant Science Laboratory, St. Andrews University, St. Andrews KY16 95H, United Kingdom

Roderick C. Dewar[2] (347), School of Biological Science, University of New South Wales, Sydney NSW 2052, Australia

Christopher B. Field (373), Department of Plant Biology, Carnegie Institution of Washington, Stanford, California 94305

Erik P. Hamerlynck (107), Department of Biological Sciences, Rutgers University, Newark, New Jersey 07102

Theodore C. Hsiao (3), Department of Land, Air, and Water Resources, Hydrology Program, University of California at Davis, Davis, California 95616

[1] Present Address: Department of Biological Sciences, Louisiana State University, Baton Rouge, Louisiana 70803.

[2] Present Address: Unité de Bioclimatologie, INRA Centre de Bordeaux, 33883 Villenave d'Ornon, France.

Bruce A. Hungate[3] (265), Smithsonian Environmental Research Center, Edgewater, Maryland 21037

Robert B. Jackson[4] (3), Department of Botany, University of Texas at Austin, Austin, Texas 78713

Dean N. Jordan (107), Department of Biological Sciences, University of Nevada at Las Vegas, Las Vegas, Nevada 89154

Yiqi Luo[5] (309, 393), Biological Sciences Center, Desert Research Institute, Reno, Nevada 89512

Harold A. Mooney (393), Department of Biological Sciences, Stanford University, Stanford, California 94305

Ross E. McMurtrie (347), School of Biological Science, University of New South Wales, Sydney NSW 2052, Australia

Rana Munns (139), CSIRO Plant Industry, Canberra ACT 2601, Australia

Eva J. Pell (193), Department of Plant Pathology, Pennsylvania State University, University Park, Pennsylvania 16802

Andrea Polle (193), Forstbotanisches Institut, Georg-August-Universität Göttingen, D-37077 Göttingen, Germany

Stephen A. Prior (215), National Soil Dynamics Laboratory, ARS-USDA, Auburn, Alabama 36849

Edward B. Rastetter (333), The Ecosystem Center, Marine Biological Laboratory, Woods Hole, Massachusetts 02543

Hugo H. Rogers (215), National Soil Dynamics Laboratory, ARS-USDA, Auburn, Alabama 36831

Jelte Rozema (169), Systems Ecology and Plant Ecophysiology, Department of Biology, Vrije University, 1081 HV Amsterdam, The Netherlands

G. Brett Runion (215), School of Forestry, Auburn University, Auburn, Alabama 36849

Rowan F. Sage (289), Department of Botany, University of Toronto, Toronto, Ontario, Canada M5S 3B2

Gaius R. Shaver (333), The Ecosystem Center, Marine Biological Laboratory, Woods Hole, Massachusetts 02543

Stanley D. Smith (107), Department of Biological Sciences, University of Nevada at Las Vegas, Las Vegas, Nevada 89154

Alan Teramura (169), College of Natural Sciences, University of Hawaii, Honolulu, Hawaii 96822

H. Allen Torbert (215), Grassland, Soil, and Water Research Laboratory, ARS-USDA, Temple, Texas 76502

D. W. Wolfe (61), Cornell University, Ithaca, New York 14850

[3] Present Address: Department of Biological Sciences, Northern Arizona University, Flagstaff, Arizona 86011.

[4] Present Address: Department of Botany, Duke University, Durham, North Carolina 27703.

[5] Present Address: Department of Botany and Microbiology, University of Oklahoma, Norman, Oklahoma 73019.

Preface

Of all the global changes that are occurring on the planet, the increase in the carbon dioxide (CO_2) concentration of the atmosphere is the most well documented and the most troublesome because it has both direct and indirect effects on the operation of the earth system. Since the industrial revolution, the CO_2 concentration in the atmosphere has increased from about 275 ppm to the current level of 365 ppm. This concentration is expected to continue to increase, doubling from the current level during the next century. These changes will have profound effects both on the climate system and on the earth's primary productivity because CO_2 is the major greenhouse gas as well as a substrate for the production of biomass. There is a vast literature on the direct effects of enhanced CO_2 on plants (Lemon, 1983) and to a lesser extent on the effects on total ecosystem processes (Koch and Mooney, 1996). Basically, it has been found that increasing CO_2 will increase plant production, but only to the extent that other resources, such as water and nutrients, are not limiting. Thus, in many ecosystems the effects will be small. However, the impact of CO_2 in increasing the water use efficiency of plants can be profound and can change the water balance of a site and hence the population dynamics of an ecosystem.

Although we do have a considerable amount of information on the direct effects of CO_2 on plants and ecosystems, we cannot be complacent about our ability to predict what the future will bring. The effects of CO_2 that we will see will be moderated by other global changes that influence plant productivity such as drought, salinity, nutrients, temperature, and atmospheric pollutants.

This volume presents an initial discussion of these important interactions. The objectives of this book are twofold: (i) to explore and summarize our current understanding of how CO_2 interacts with other environmental stressors and (ii) through this review process to stimulate future explicit experimentation on these interactions, particularly at the ecosystem level. Experiments on factor interactions are difficult, particularly when performed at the ecosystem level. However, despite these difficulties this is where we must put our efforts if we are going to be able to predict what our future world will look and act like.

Chapters in this book are organized into three parts: (I) CO_2 and stress interactions (Chapters 1–10); (II) evolutionary, scaling, and modeling studies of CO_2 and stress interactions (Chapters 11–14); and (III) summary and synthesis (Chapters 15 and 16). Each chapter in Part I summarizes up-to-date knowledge on and speculates, where the knowledge is lacking, about interactive effects of CO_2 with water (Chapters 1 and 2), temperature (Chapters 3 and 4), salinity (Chapter 5), UB-B (Chapter 6), ozone (Chapter 7), and nutrients (Chapters 8–10) on plants and ecosystems. Part II offers broad perspectives beyond experimental measurements of plant and ecosystem ecophysiology to facilitate our comprehension of the complexity of CO_2 and stress interactions. Part III highlights knowns and unknowns of the interactions and discusses strategies for future research.

Acknowledgments

This book resulted from an initial consideration of these ideas during a meeting at the Granlibakken Conference Center, Lake Tahoe, California, sponsored by the National Science Foundation's EPSCoR Program to the State of Nevada and by the Electric Power Research Institute. The Desert Research Institute of the University of Nevada served as the host institution. Special thanks go to Drs. Jeff Seemann and Tim Ball, who led the effort of organizing the Granlibakken meeting. We also thank Roger Kreidberg for indexing the book. This project is an activity of the Global Change and Terrestrial Ecosystems of the International Biosphere Geosphere Program.

H. A. MOONEY AND Y. LUO

References

Koch, G. W., and H. A. Mooney, eds. (1996). "Carbon Dioxide and Terrestrial Ecosystems." Academic Press, San Diego.

Lemon, E. R., ed. (1983). "CO_2 and Plants: The Response of Plants to Rising Levels of Atmospheric Carbon Dioxide." Westview Press, Boulder.

I

Interactions of CO_2 with Water, Temperature, Salinity, UV-B, Ozone, and Nutrients

1

Interactive Effects of Water Stress and Elevated CO_2 on Growth, Photosynthesis, and Water Use Efficiency

Theodore C. Hsiao and Robert B. Jackson

I. Introduction

Of all the physical stresses in the global environment, water deficit is probably the most important in determining plant growth and productivity worldwide. At the same time, plant water use and growth are strongly influenced by climatic conditions and CO_2 concentration in the atmosphere. Of particular interest is the fact that as the level of CO_2 is raised above the present ambient level, photosynthesis is commonly enhanced and transpiration is often reduced, resulting in a higher efficiency of water use, and plant growth and productivity are generally increased. Limited data also show that elevated levels of CO_2 may facilitate plants' adjustment to drought. The rise in atmospheric CO_2 due to fossil fuel burning and other anthropogenic activities will continue for decades and centuries to come, although the extent of the rise is uncertain and a matter of debate. The broad consensus is that this rise will result in hotter and drier environments in many parts of the world, which would also affect plant productivity in addition to the effects of rising CO_2. How water deficits and elevated CO_2 interact to impact plant productivity and water use efficiency (WUE) is a pivotal question in the consideration of future changes of natural and managed terrestrial ecosystems. For natural communities, differences in WUE under elevated CO_2 may determine the success in adaptation and competition of plant species, and ultimately in community succession, in environments of generally warmer temperature and more frequent drought (Ehleringer and Cerling, 1995). For managed communities, where water is the major limiting factor, productivity is determined by the WUE of the crop or tree, or of individual species making up the community, and the amount of water available. This productivity in turn affects population dynamics at higher trophic levels. Conversely, since the standing biomass of plants is a major sink for atmospheric CO_2, the growth and succession of plant communities in turn play a role in modulating the future rise in CO_2.

More than 90% of the dry matter (biomass) produced by the plant comes from assimilated CO_2. Hence, productivity depends on the capture of photosynthetically active radiation (PAR) by the plant and the use of the radiation for photosynthesis. This chapter focuses on three critical aspects of crop productivity as affected by water deficit and elevated CO_2. The first is expansive growth of leaves and roots. Leaf growth underlies canopy development and hence PAR capture. Next comes photosynthesis and its adjustment to the environment. The last aspect discussed is WUE.

II. Expansive Growth, Water Stress, and Elevated CO_2

Because productivity is dependent on radiation capture, the growth of leaves and the enlargement of the canopy to maximize PAR interception play a critical role in plant productivity. Leaf growth by cell enlargement is extremely sensitive to water stress (Boyer, 1970; Acevedo *et al.*, 1971; Hsiao and Jing, 1987). Even mild water stress reduces the rate of leaf area development, leading to a smaller foliage canopy (Bradford and Hsiao, 1982). At the cellular and organ level, the effects of water deficit may be examined in terms of water uptake, turgor pressure and osmotic adjustment, and the ability of the cell wall to expand under a given force supplied by turgor pressure (Fig. 1).

Because most of the cell volume is occupied by water, water uptake (*WT* in Fig. 1) is closely linked to cell expansion. The water potential (Ψ) of the growth zone of an organ must be lower by a certain amount relative to its surrounding to sustain water uptake and growth (Boyer, 1985). The low water potential of the growth zone is maintained by the continuous uptake (plus internal generation, if any) of solutes (*ST* in Fig. 1). Growth is also underlain by the irreversible expansion (plastic deformation) of the cell wall and the accretion of components of the cell wall and of the protoplasm. Expansive growth is manifested, however, only when turgor pressure (Fig. 1, Ψ_p) provides the force necessary to stretch and increase the cell wall area irreversibly (Ray *et al.*, 1972). There is a minimal turgor, known as the *yield threshold,* below which irreversible volume expansion will not occur. Above the yield threshold, the rate of expansive growth is taken to be proportional to the amount of turgor pressure that is above the yield threshold value, in accordance with the equation of Lockhart (1965). Since

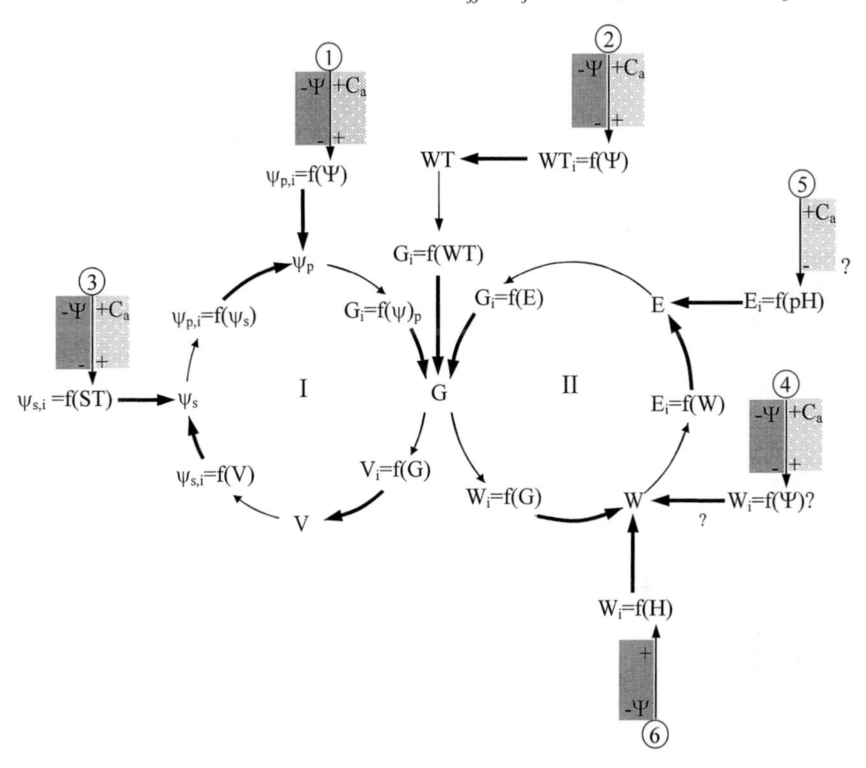

Figure 1 Dependence of expansive growth on the underlying interactive parameters, and the possible points of impact by water deficit ($-\Psi$) and elevated CO_2 ($+C_a$). G is the rate of expansive growth; V is cell volume; ψ_s is solute potential of the growing cell; ψ_p is turgor or pressure potential; Ψ is water potential; ST is rate of solute transport to the growing cell; WT is rate of water transport (water uptake); W is cell wall synthesis and metabolism; E is cell wall extending ability (volumetric extensibility and yield threshold turgor); H is the effect of negative hormonal signals; and pH is cell wall pH. Each parameter or process is depicted as a function of one or more other parameters. For example, $G = f(\psi_p, WT, E)$. Partial or component functions are denoted by the subscript i. Converging heavy arrows indicate the summing of the partials to make the whole. Light arrows indicate that the whole in turn serves as the variable on which the partial of another function depends. Points of impact by water deficit ($-\Psi$) and elevated CO_2 (+Ca) are labeled by circled numbers (e.g., ②, with the algebraic sign (±) at the head of the arrow indicating whether the effect is positive (raising) or negative (lowering). A question mark (?) indicates that the impact or causal relationship is uncertain or speculative. Loops are indicative of the interlocking nature of the processes or parameters and are labeled I and II for easy reference.

turgor is a function of the water potential and osmotic potential of the cell, reduction in water potential due to water stress (Fig. 1, Impact 1) may reduce turgor and hence the rate of expansive growth. That effect is easily seen when turgor of growing tissue is reduced by the sudden imposition of water stress (Green, 1968; Acevedo *et al.*, 1971; Hsiao and Jing, 1987; Pardossi *et al.*, 1994; Frensch and Hsiao, 1995). Growth can recover at least partially, however, through osmotic adjustment to restore turgor, or through enhancement of the ability of the cell wall to expand at a given turgor.

Normally water uptake and solute uptake by growing cells occur in concert. The rate of solute accumulation (Fig. 1, *ST*) of growing cells matches their rate of volume expansion (Fig. 1, *V*) so that solute potential (Sharp *et al.*, 1990) and turgor pressure of growing cells (Frensch and Hsiao, 1994) remain relatively constant with time. When growth is slowed or stopped by a reduction in turgor, the simplest mechanism one can visualize for osmotic adjustment is for solute importation to continue unabated while water uptake and volume expansion are restricted. Consequently, solutes become more concentrated in the cell, lowering the cell water potential and allowing a recovery in water uptake, leading to a higher turgor and recovery in growth. The growth rate under water stress, however, would be slower compared to that without the imposed stress. The available data on the kinetics of turgor and growth recovery (Frensch and Hsiao, 1995) are consistent with this view. There is no evidence for an acceleration in solute accumulation in response to the onset of water stress. In fact, in a portion of the growth zone, solute accumulation is slowed by water stress lasting many hours as evinced by its slower net deposition rate of solutes (after accounting for dilution by volume expansion) (Sharp *et al.*, 1990; Walker and Hsiao, 1993). Also, indirect evidence (Frensch and Hsiao, 1995; Frensch, 1997) based on the rate of turgor recovery indicates that the rate of solute transport into the growth zone may be reduced within minutes after the onset of water stress. How such a rapid response in solute transport is brought about by water stress is a matter of speculation. In any event, the reduction in expansive growth under water stress may be viewed as the combined results of more concentrated solutes being necessary to maintain a particular turgor at a lower level of water potential and the likely slower solute transport into the growth zone. These effects are generally depicted on the left side (I) of Fig. 1.

The other major perturbation caused by water stress is in the ability of the cell wall to expand at a given turgor pressure, which reflects directly or indirectly all metabolic and hormonal effects on the wall [see right side (II) of Fig. 1]. Recent studies have shown clearly that the yield threshold turgor decreases in maize roots within a few minutes in response to osmotic or water stress. That enables the root to grow at the same (Hsiao and Jing, 1987) or slightly slower rate (Frensch and Hsiao, 1994) in spite of reduced turgor. On the other hand, in earlier studies where turgor was calculated as the difference between water potential and osmotic potential, growth of leaves was often shown to be slower in plants subjected for some time to water stress in spite of the apparent maintenance of turgor (Michelena and Boyer, 1982; Matthews *et al.*, 1984; Van Volkenburgh and Boyer, 1985; Hsiao and Jing, 1987), indicating an apparent loss in the ability of the cell wall to expand. These earlier results cannot, however, be easily compared to the more recent data obtained by measuring cell turgor directly with a pressure microprobe under short-term water stress (minutes to a few hours). It is now known that water stress shortens the growth zone of roots (Sharp *et al.*, 1988) and leaves (Walker and Hsiao, 1993) over periods of many hours or even within an hour (Frensch and Hsiao, 1995). So the reduced total rate of growth at a given turgor might be the result of reducing the portion of the organ that is actually growing, and not the consequence of a reduction in the intrinsic capacity of the cell wall to expand.

Plants grown under elevated CO_2 often have larger leaves (Fig. 2; Prior *et al.*, 1991; Lawlor and Mitchell, 1991) and the leaf area expansion rate is faster (Morison and Gifford, 1984; Cure *et al.*, 1989). Studies of the underlying processes are only beginning (Ferris and Taylor, 1994; Ranasinghe and Taylor, 1996), and the results up to now are not at all clear. The several possible mechanisms that could explain the faster and better growth are summarized in a general way as impact points in Fig. 1. One obvious possibility is the enhanced assimilate supply under high CO_2, which, generally speaking, should lead to a higher growth rate. More specifically, the enhanced assimilation rate presumably will result in higher rates of solute importation into the growing cells (Fig. 1, Impact 3), enabling the cells to maintain a high turgor in spite of a high expansion rate. That is consistent with the reported more negative solute potential in the growth zone of roots under elevated CO_2 (Ferris and Taylor, 1994), but is not supported by data obtained on leaves in the same laboratory (Ranasinghe and Taylor, 1996). A better assimilate supply also may enhance the expanding ability of the growing cell wall, as reported by the same authors. The supporting data, however, are not convincing. Additional studies need to be done with more definitive techniques to determine if the instantaneous yield threshold is indeed lower when growing under elevated CO_2 and the volumetric extensibility higher. Another possibility is that the faster leaf growth is the result of improved plant water status under elevated CO_2 (Fig. 1, Impacts 1, 2, and 4). Because leaf growth is sensitive to even very mild water stress, any improvement in water status should lead to faster leaf expansion. There is also the likelihood that the faster growth is promoted by a more acid cell wall at elevated CO_2 (Fig. 1, Impact 5), according to the acid growth theory (e.g., Cleland, 1980). The equilibration between air CO_2 and carbonic acid in the wall solution dictates a lower wall pH, other things being equal. Air CO_2 at a concentration of 3% has been shown to stimulate the growth of oat coleoptile in darkness (Nishizawa and Suge,

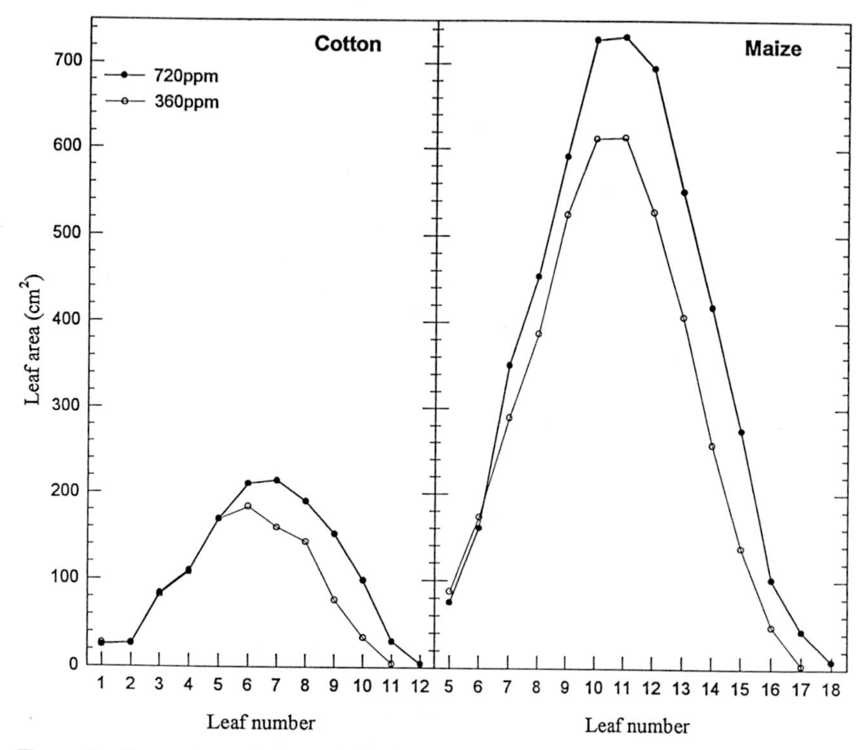

Figure 2 Comparison of areas of individual leaves of cotton and maize plants grown in growth chambers at either normal (360 μmol mol^{-1}) or elevated (720 μmol mol^{-1}) CO_2. Leaf numbers are in the order of leaf emergence and only main stem leaves were measured for cotton. Note that the plants were still highly vegetative with younger leaves (higher in leaf number) above the leaves of peak leaf area still actively enlarging. Growth was at day/night temperatures of 27/20°C and day/night relative humidities of 40/80%. PAR was 770 μmol m^{-2} s^{-1} for the day time. (From original data of T. C. Hsiao.)

1995). There appears to be no thorough study of the effects of CO_2 in the range of expected future air CO_2 concentration in relation to acid growth.

Regardless of the underlying mechanisms, the fact that leaf growth is accelerated under elevated CO_2 has important implications. During the early vegetative stage when the canopy is small and incomplete, faster leaf growth will increase the amount of PAR intercepted by the canopy, leading to more CO_2 assimilated per plant or per unit land area. This effect is compounded with time (Bradford and Hsiao, 1982) and can result in a much larger plant and higher productivity, provided that radiation capture remains limiting and other resources such as mineral nutrients and water are not. This point is further discussed in a subsequent section.

III. Stomata, Photosynthesis, and Water Stress

For a given leaf area or canopy size of a plant, and hence the amount of radiation it captures, the amount of CO_2 assimilated photosynthetically by the plant is dependent on photosynthetic capacity, intercellular CO_2 concentration, and epidermal (mostly stomatal) conductance of the leaves (ignoring boundary layer/canopy resistances discussed later). It has been known for several decades that epidermal conductance and photosynthesis are reduced by water stresses that are sufficiently severe (for a review, see Boyer, 1976). Initially, attention was largely directed at stress-induced partial stomatal closure and the consequent restriction of CO_2 diffusion into the intercellular space as the mechanism causing slower photosynthesis. On closer examination, however, the early data also showed that often there was a nonstomatal inhibitory effect accompanying stomatal closure (Boyer, 1971, 1976; for a review, see Hsiao, 1973). Some earlier studies also indicated a concerted response to water stress of the stomatal and nonstomatal components of photosynthesis (Redshaw and Meidner, 1972; see review by Hsiao, 1973). Studies showing a linear relationship between leaf photosynthesis and epidermal conductance, with intercellular CO_2 (C_i) remaining constant at a given air CO_2 concentration at different levels of water stress (Wong *et al.*, 1979), are likely a reflection of this coordination. Later studies showed that a number of metabolic steps or enzymes could be affected by water stress of a particular level, leading to a reduction in photosynthetic capacity. In isolated chloroplasts, photosynthetic processes are resistant to mild or moderate water stress (Kaiser, 1987). Generally speaking, when water stress develops rapidly, photosynthetic capacity appears not to be reduced if the stress is not very severe (Bradford and Hsiao, 1982; Kirschbaum, 1987; Sharkey, 1987). When similar levels of stress develop more slowly, however, photosynthetic capacity could be reduced (Sharkey and Seemann, 1989), as it would be under severe water stress (Kirschbaum, 1987). That reduction may in fact be adaptive (Sharkey, 1987) and possibly represents a coordinated modulation of the photosynthesis system (Bradford and Hsiao, 1982) and probably other plant processes. Clearly, the sequence of changes in the different parts of photosynthesis complex upon the onset of water stress are important, but very few studies provide such information. Summarized here are some of the more interesting changes observed in the last decade. The apparent conflicts in findings need further resolution.

Earlier *in vitro* studies (Younis *et al.*, 1979) pointed to impaired photophosphorylation under severe water stress with a conformational change in the coupling factor as the possible cause. A later reexamination (Ortiz-Lopez *et al.*, 1991) using current techniques to assess coupling factor activity *in vivo,* however, indicated no significant impairment and concluded that nonstomatal limitation of photosynthesis under water stress cannot be explained by impaired photophosphorylation. For bean plants (*Phaseolus vulgaris*) under a moderate water stress developed over a 4-d period, Sharkey and Seemann (1989) found that assimilation and C_i were reduced substantially, indicating effects largely due to stomatal closure, while the A/C_i curve also showed a reduction in photosynthetic capacity. The ribulose bisphosphate (RuBP) content of the leaf was unaffected by the moderate water stress while phosphoglyceric acid (PGA) content was reduced markedly. These changes in metabolite contents suggest that ribulose bisphosphate carboxylase/oxygenase (Rubisco) activity *in vivo* might be reduced (Sharkey, 1987), in spite of a lack of detectable *in vitro* change in the activity or state of the enzyme. As water stress continued and became more severe, assimilation was reduced to nearly zero and C_i became higher than in the well-watered control. PGA content decreased further and RuBP content also declined. In contrast, RuBP was found by Tezara and Lawlor (1995) to decline more markedly than assimilation as water stress developed over days while C_i remained essentially the same in sunflower plants grown under apparently low irradiance. Their C_i data were not that convincing because assimilation did not appear to be linearly related to conductance. There was no clear indication of a decline in Rubisco activity measured *in vitro* until stress was severe and assimilation reduced to nearly zero. Tezara and Lawlor (1995) concluded that a problem with RuBP regeneration is the likely cause of reduced assimilation under water stress. This viewpoint is consistent at least in part with the earlier assessment of water stress effects on photosynthesis by Farquhar *et al.* (1987), who concluded that inhibition of RuBP regeneration follows an initial decline in C_i as water stress develops.

In a number of studies (von Caemmerer and Farquhar, 1984; Sharkey, 1985; Sharkey and Seemann, 1989), photosynthesis of water-stressed plants was found to respond only minimally or not at all to increases in C_i above the normal level for well-watered controls under normal ambient CO_2. Sharkey (1987, 1990) interpreted this lack of stimulation by CO_2 as an indication of an end-product limitation, that is, photosynthesis impaired by reduced rates of sucrose or starch synthesis (Vassey and Sharkey, 1989), which presumably came from a down-regulation of some key enzymes in sucrose and starch anabolism in response to the reduction in C_i brought about initially by stomatal closure (Sharkey, 1990).

To complete the discussion on photosynthesis and stomata in relation to water stress, several other aspects need to be mentioned. One is nonuniform or patchy stomatal closure (Daley *et al.,* 1989; Terashima, 1992) in leaves induced by stress leading to misleading values of calculated C_i. The value of C_i could be quite low in the patches where stomata are closed but the calculated value would be dominated by the C_i of the patches where stomata are open. This may give rise to C_i values that appear to be unaffected by water stress if the closure in the closed patches is nearly complete and lateral diffusion of CO_2 between the patches is minimal. On the other hand, less extreme patchy behavior should not lead to large errors in calculating C_i, especially if the patches are small in size and not separated by relatively impervious boundaries formed by veins. In the studies cited, patchy closure, visualized by autoradiography of leaves fed $^{13}CO_2$, was clearly seen in one case (Sharkey and Seemann, 1989), relatively minor in another (Ortiz-Lopez *et al.*, 1991), and not seen in a third case (Tezara and Lawlor, 1995). The patchiness problem was exaggerated for a few years following its discovery and probably is minimal or unimportant for most species if water stress develops not suddenly but at a relatively slow speed as is common in nature (Wise *et al.*, 1992).

Another aspect is the possible photoinhibition of photosynthesis in stressed leaves if C_i is reduced too much by stomatal closure. Here again the problem was initially given excessive attention after its discovery, but is now thought to be minimal (Kirschbaum, 1987) or uncommon (Sharkey and Seeman, 1989) for mesophytic species. It is, however, an important issue for xerophytes, which withstand long periods of completely closed stomata during drought (Osmond *et al.*, 1987). The last aspect is the potential role of abscisic acid (ABA). The fact that ABA accumulates in leaves under water stress and that applied ABA inhibits stomatal opening has long been recognized (Hsiao, 1973; Bradford and Hsiao, 1982). Effects of applied ABA on C_i and certain parts of photosynthetic carbon metabolism are similar (Seemann *et al.*, 1987; Raschke, 1987). ABA export from roots to leaves is also known to be promoted by drying soil (e.g., Davies and Zhang, 1991). It is not yet clear, however, if ABA affects photosynthesis mostly through its effect on stomata, or if nonstomatal effects of ABA on reactions in the carbon reduction cycle are equally important (Seemann *et al.*, 1987; Raschke, 1987).

IV. Stomata, Photosynthesis, and Elevated CO_2

From the beginning of research on the physiologic effects of atmospheric CO_2, reductions in leaf epidermal (including stomatal) conductance (g_e) and transpiration per unit leaf area (T) and increases in CO_2 assimilation (A) have been observed as CO_2 increases (Gaastra, 1959; van Bavel, 1974). In the last decade studies have focused on plants *grown* at subambient and elevated CO_2, rather than exposing a leaf or plant to a temporary CO_2 change. Here we discuss first the response of stomata and g_e to changes in CO_2. In general, g_e is substantially reduced as the growth CO_2 concentration (C_a) increases, though there are some apparent exceptions. The decrease in g_e occurs whether the comparison is from subambient to ambient CO_2 (e.g., Polley *et al.*, 1992) or ambient to elevated CO_2 (e.g., Kimball and Idso, 1983). It also holds for both C_3 and C_4 species, and for trees as well as herbs (though see later discussion). Morison and Gifford (1984) found that g_e decreased 36% on average for 11 crop and herb species grown in pots at ambient and twice-ambient CO_2. Numerous field experiments have also shown typical reductions of 20–60% (e.g., Knapp *et al.*, 1993; Jackson *et al.*, 1994; Hileman *et al.*, 1994).

While it is clear that substantial reductions in g_e are common, there are large differences among studies and species. Field *et al.* (1995) examined data for 23 tree species in ambient and elevated CO_2 (typically 35 vs. 70 Pa) and found an average reduction in g_e of 23%. For plants grown in growth chambers, the average reduction was 31%. When only the data from open-top chambers were examined, the average reduction in g_e was 17% for plants grown in pots, and 4% for plants grown in the ground. Results from the first free-air CO_2 enrichment (FACE) experiment in a forest system showed reductions in g_e to be minimal for *Pinus taeda* in elevated CO_2 (Ellsworth *et al.*, 1995). Curtis (1996) reviewed the gas-exchange responses of 41 tree species to elevated CO_2. Increased CO_2 led to moderate reductions in g_e for unstressed tree species, but g_e was not changed significantly for the few studies on stressed trees. Trees grown for more than 100 days in elevated CO_2 showed the largest reductions. Meta-analysis was used to allow statistical comparison of diverse studies by different groups and a number of variables were taken into account (Curtis, 1996). To understand the variation in responses and deduce causality, however, it is necessary to examine possible metabolic adjustments to elevated CO_2, as is done in a following section.

In spite of the reductions in g_e, photosynthesis of C_3 species is generally enhanced as C_a is increased above the current ambient level (Gaastra, 1959), at least for hours to days. CO_2 concentration in the intercellular space (C_i) rises approximately in proportion to the increase in C_a. This enhances carboxylation and suppresses oxygenation activity of the photosynthetic enzyme, RuBP carboxylase/oxygenase (Rubisco). Over longer terms, elevations in CO_2 often reduce A (used here routinely on a basis of per unit leaf area, unless specified otherwise). To separate stomatal and nonstomatal components and to evaluate the different parts of the photosynthesis processes in their response to CO_2 and water stress, it is advantageous (Sage, 1994) to use curves of net CO_2 assimilation vs. C_i (A/C_i curves, Fig. 3) obtained by gas exchange carried out at a range of C_a. The curves show A to be zero at the CO_2 compensation point C_i, and to increase with increases in C_i until assimilation is saturated with respect to CO_2. The value of C_i is dependent on the CO_2 concentration external to the leaf as well as the photosynthetic capacity and g_e. In nearly all studies of C_3 species,

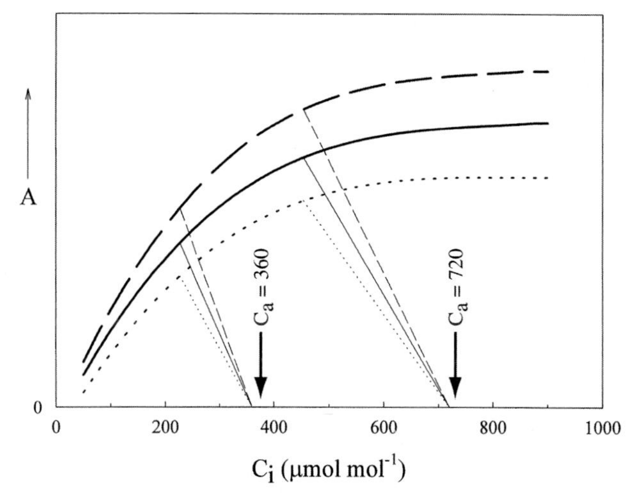

Figure 3 Curves of photosynthetic CO_2 assimilation (A) vs. intercellular CO_2 concentration (C_i) showing either a hypothetical downward (--------) or upward (———) adjustment in photosynthetic capacity of the leaf after long-term growth under elevated CO_2, as compared to the curve for the leaf grown under normal CO_2 (———). Normal C_a is taken as 360 μmol mol^{-1} and elevated C_a is taken as 720 μmol mol^{-1}. Straight lines are drawn to connect the operational point and the point on the C_i axis that equals the value of C_a used for growth. The operational point is defined as the point where the A value is that measured under the given C_a and the calculated corresponding C_i. The slope of the straight line is the negative value of epidermal conductance g_e. Note that as drawn, g_e is higher under the lower C_a, and g_e adjusts with the downward or upward adjustment in photosynthetic capacity.

except when there is extreme stress, the slope of the A/C_i curve at the C_i level corresponding to the current C_a is positive. That is, A increases with an increase in C_a above the present ambient. What complicates the matter is the fact that in most cases the A/C_i curve changes after the plant is exposed to and grown at a different C_a. Two possible changes are schematically depicted in Fig. 3. These changes, reflecting changes in photosynthetic capacity independent of the stomatal response to CO_2, have been reviewed (Sage, 1994), and are discussed in some detail in the next section. The curves in Fig. 3 also depict possible changes in g_e associated with changes in C_a or in photosynthetic capacity. The value of g_e is the negative slope of the straight line connecting C_a and the corresponding C_i at the assimilation operational point.

A well-known difference among species is that C_4 plants generally respond much less in growth to increases in CO_2 than do C_3 plants (Wong, 1979; Morison and Gifford, 1984; Kirkham *et al.*, 1991), although substantial positive responses have been observed in some C_4 species (Drake and Leadley, 1991; Bowes, 1993). The lower response in photosynthesis is the result of C_4 species having the mechanism to "concentrate" CO_2 in the bundle sheath cells, the site of final carboxylation. Thus, photosynthesis is already operating at a point near CO_2 saturation in the present ambient CO_2 concentration. Rising atmospheric CO_2 will probably benefit C_3 species over C_4 species (Henderson *et al.*, 1994).

V. Adjustment in Photosynthetic Capacity under Elevated CO_2

The long-term impact of elevated CO_2 on photosynthesis is hard to predict because of adjustments in photosynthetic capacity of the plants growing under high CO_2, which result in changes in the A/C_i curves. The adjustment in photosynthetic capacity had been mostly downward (Fig. 3, dotted curve) in earlier studies, often referred to as down-regulation or negative acclimation, but no adjustment or upward adjustment (Fig. 3, dashed curve) is now also known to occur (Campbell *et al.*, 1988; Sage *et al.*, 1989; Arp and Drake, 1991; Habash *et al.*, 1995). In reviews, Stitt (1991) and Gifford (1992) concluded that the adjustment may be dependent on the ability of the plant to initiate new sinks for extra assimilates when photosynthesis is enhanced by high CO_2. Arp (1991) related the adjustment to the size of the pot used to grow the plants, with less tendency for downward adjustment in pots larger than 3500 cm^3. Consistent with this is the lack of significant downward adjustment in plants growing in the field under elevated CO_2 (Radin *et al.*, 1987; Campbell *et al.*, 1988, 1990; Arp and Drake, 1991), or only moderate downward adjustment (Jacob *et al.*, 1995, after 8 years of growth under high CO_2!). The conclusion is also supported by more direct data. The transfer of downwardly adjusted cotton plants under elevated CO_2 from extremely small pots to larger pots allowed their photosynthetic capacity to recover within 4 d (Thomas and Strain, 1991). Pot size was also found to be positively correlated with the extent of increase in root-to-shoot ratio under high CO_2 (Arp, 1991), fitting the hypothesis that a shortage of sinks (root growth) for assimilates is at least one potential cause for downward adjustment.

Rubisco concentrations have been shown to be reduced in downwardly adjusted plants (Wong, 1979; Stitt, 1991; Riviere-Rolland and Betsche, 1996). Stitt (1991) examined possible metabolic mechanisms underlying the adjustments of photosynthesis and concluded that decreases in photosynthetic capacity arose from decreases in key photosynthetic enzymes, including Rubisco. This may be the result of shifting resources to grow more assimilate sinks so that source-to-sink ratios are in balance (Sage *et al.*, 1989; Stitt and Schulze, 1994). The evidence for inhibition of photosynthesis by chloroplast distortion due to starch grains accumulated under high CO_2 was considered by Stitt (1991) to be inconclusive.

The resources that may become more limiting under high CO_2 are mineral nutrients. So far attention has been focused primarily on nitrogen and phosphorus. Generally for a given environment A is highly correlated with leaf nitrogen content (Evans, 1989; Wong, 1990). Plants low in nitrogen showed more downward adjustment under high CO_2 in a number of studies (e.g., Wong, 1979; Jacob *et al.*, 1995) but not in all. In a review, Pettersson and McDonald (1994) concluded that more studies with better monitoring of plant nitrogen status are needed before the interaction of nitrogen nutrition with adjustment in photosynthesis under high CO_2 can be clarified. That view may be too pessimistic in light of the more recent work. A reevaluation (Sage, 1994) of literature gas-exchange data for plants grown under different levels of CO_2 yielded considerable insight on the various patterns of allocation of resources (largely nitrogen based) to the components of photosynthesis to accommodate shifts in growth and nutrition. Evidence supporting the hypothesis of shifting resources also came from studies on photosynthesis of transgenic plants differing widely in Rubisco content and grown at different nitrogen levels (Stitt and Schulze, 1994), but without CO_2 as a variable. Additional evidence came from Van Oosten and Besford (1994), who showed that the level of mRNA coding for the small subunit of the Rubisco protein was lower in plants growing under high CO_2 and that feeding of sucrose or glucose to excised tomato leaves reduced the mRNA level. This suggests that one cause of downward adjustment may be the repression of the genes coding for Rubisco by the photosynthetic end products (sugars), which accumulate in plants under high CO_2 when sink capacity is insufficient.

Inorganic phosphate is used in the phosphorylation of intermediates in carbohydrate metabolism and is particularly important as a part of the shuttle that moves the principal photosynthetic products (as triose phosphates) out of the chloroplasts into the cytoplasm (Geiger and Giaquinta, 1982). Photosynthesis has been shown to be limited when there is a sudden and strong inhibition of sucrose or starch synthesis or when there is a sudden and marked increase in photosynthetic CO_2 fixation, tying up the free phosphate (Stitt, 1991). Evidence for phosphate limitation under less extreme and long-term conditions came from the recent work of Riviere-Rolland and Betsche (1996). They found that pea chloroplast phosphate translocator increased under high CO_2 and the increase was accentuated by growing the plant under low phosphate nutrition. The increase is considered to be a way to alleviate low phosphate in the chloroplast and improve the balance in carbon partitioning between starch and soluble carbohydrates. These recent studies appeared to have provided a metabolic conceptual base to better analyze the diverse range of adjustments in photosynthesis to elevated CO_2.

Assuming that the A/C_i curve of a leaf acclimated to elevated CO_2 is known, there is still the question of how much stomatal conductance of acclimated leaves would be reduced by the increase in C_a and the corresponding change in C_i. The problem would be greatly simplified, however, if the C_i/C_a ratio remained conservative in the acclimated plants. As discussed in a later section, that fortunately appears to be the case.

VI. Interactions between Effects of Elevated CO_2 and Water Stress on Photosynthesis

The preceding discussions brought out the obvious ameliorating effects of elevated CO_2 on water stress. The lower g_e induced by increasing CO_2 usually would result in reduced transpiration. This would not only stretch out the limited water supply, but would also raise the leaf water status slightly for a given situation (Prior *et al.*, 1991; Roden and Ball, 1996) because a smaller difference in Ψ between the soil and the leaves would be required to drive the reduced rate of water uptake. Because higher C_a usually means a higher C_i, the reduction in C_i due to stomatal closure under water stress, postulated to lead to reduced photosynthetic capacity (Farquhar *et al.*, 1987; Sharkey, 1990), should be ameliorated to some extent by elevations in CO_2. On the other hand, water stress, by reducing or limiting the increase in C_i under elevated CO_2, may delay the down-regulation of photosynthesis under high CO_2 (Roden and Ball, 1996). In the case of severe water stress, however, the beneficial effect of elevated CO_2 may not materialize because photosynthesis may be CO_2 insensitive (Sharkey, 1987). Although not yet compared in detail in the same experiment, the downward adjustment in photosynthesis under elevated CO_2 appears to have much in common with the reduction in photosynthetic capacity effected by water stress, at least in terms of end-product inhibition. There are still less direct interactions, such as changes in leaf area, in water use efficiency, or in assimilates available, which may affect photosynthetic behavior or photosynthesis integrated over the plant life cycle. These are elaborated on in Section X. Water use efficiency, however, is a topic by itself and is taken up next.

VII. General Aspects of Plant Water Use Efficiency

With enhanced photosynthesis and reduced g_e, the efficiency of water use for the production of assimilates may be expected to improve under elevated CO_2, which is indeed the case. But first, we provide a brief review of the general aspects of plant water use efficiency.

In economics terms, water use efficiency (WUE) is the ratio of output (product) to input (water). Since the output and input can be defined or measured in a number of ways, WUE can be expressed differently with different implications (Fischer and Turner, 1978; Eamus, 1991; Hsiao, 1993; Morison, 1993). The most basic physiologically and important in terms of primary productivity is the amount of CO_2 assimilated through photosynthesis relative to the water lost through transpiration, here termed photosynthetic WUE (Hsiao, 1993). When measured for short time intervals photosynthetic WUE (WUE_{at}) is the ratio of A to transpiration rate T, a ratio also termed *transpiration efficiency* in the literature. WUE_{at} can also refer to the long-term outcome at an integrated level of organization such as the whole plant and a canopy made up of a population of plants. In that case, the loss of CO_2 as respiration of nonphotosynthetic organs and by the whole plant at night needs to be taken into account. Unfortunately, much of the data reported so far in the literature have been of the near instant or short-term nature, often only for individual leaves, and are of very limited value in deducing WUE for the whole plant in the field over a significant duration. Although some data on canopy WUE_{at} for time intervals of days have been published in the last few years (e.g., Baldocchi, 1994), WUE over longer terms is most commonly evaluated as the ratio of biomass produced to the total amount of water consumed in evapotranspiration. Both biomass transpirational WUE (WUE_{mt}) and biomass consumptive WUE (WUE_{me}) are intimately tied to photosynthetic WUE, but differ from the latter because the conversion of CO_2 to biomass is not one to one. Also, in the case of WUE_{me}, soil evaporation as well as transpiration are included in the amount of water consumed.

Numerous studies on biomass WUE have been published, with much of the data from the beginning of this century analyzed and interpreted by de Wit (1958). Later results were summarized by Hanks (1983) and other authors in a volume on crop water use efficiency (Taylor *et al.*, 1983). Most of the data showed a nearly constant WUE_{me} for a given species under the ambient CO_2 level when normalized for the evaporative demand of the atmosphere (de Wit, 1958; Tanner and Sinclair, 1983) regardless of the amount of water supply. This conservative behavior led to the suggestion of estimating plant productivity from WUE_{me} and the amount of available water (Fischer and Turner, 1978; Anonymous, 1982), which is now often used in modeling productivity. The near constancy of WUE_{me} for a given evaporative demand has been interpreted largely in terms of the common pathway across the air boundary layer and stomata shared by the CO_2 assimilation and transpiration process (Hsiao and Acevedo, 1974; Fischer and Turner, 1978; Tanner and Sinclair, 1983). That commonality, however, is only part of the explanation. Equally important is the fact that A and T also have in common the reliance on radiation absorption as the energy source to drive the processes, as pointed out by Hsiao and Bradford (1983) and Hsiao (1990). In spite of the fact that assimilation uses only the visible part of the solar radiation spectrum whereas transpiration uses energy provided by radiation of all wavelengths, changes in leaf area and display and in plant spacing affect the energy supply for the two processes in a nearly identical manner. Thus, A and T on a per unit land area basis are closely linked and go hand in hand, leading to a relatively constant WUE_{me} for given species in a given climate (Hsiao, 1990, 1993). There is reason to believe that the same conservative behavior of biomass WUE would hold for conditions of elevated CO_2.

The conclusion that WUE_{me} remains nearly constant regardless of the degree of water deficit is in sharp contrast to many conclusions drawn in studies on single leaves in gas-exchange chambers. It may be argued that these studies are valid only for the condition of the chamber, where the principle of energy balance as it operates in the field is circumvented by controlling the leaf temperature (see Section IX).

VIII. Water Use Efficiency and Elevated CO_2

Because increases in CO_2 generally enhance photosynthesis and reduce stomatal conductance (Morison, 1987), WUE is expected to increase under elevated CO_2, as is almost universally observed (Morison, 1993; Jackson *et al.*, 1994). The increase, however, is also highly variable among different studies according to several surveys of the literature (Eamus, 1991; Enoch and Honour, 1993; Morison, 1993), whether WUE is expressed as the ratio of A to T or as the ratio of biomass produced per unit of evapotranspiration. The more recent publications reporting observed WUE showed similar variability in results and did not shed much light on the problem. In the case of the FACE experiment on cotton growing in an open field, WUE_{me} was found to be improved 28% in one year and 39% in the next year for the well-watered treatment, and 19% and 37% for the dry treatment, when CO_2 was elevated from ambient to 550 ppm (Mauney *et al.*, 1994). The improvement was essentially due to improved biomass production as ET was similar or even higher (Kimball *et al.*, 1994) under elevated CO_2. On the other hand, canopy WUE_{at} of the same FACE study, over three periods of 5–9 d each, was found (Hileman *et al.*, 1994) to be statistically not different between the two CO_2 levels, although there was a tendency for it to be higher at the 550-ppm level. In a study also on cotton but in sunlit chambers at controlled temperature, Reddy *et al.* (1995) found that a doubling of growth CO_2 improved WUE_{at} by an average of 50% for three growth temperatures, at least for the one daily cycle presented. In an experiment in the Canberra sunlit phytotron, two wheat cultivars, Matong and Quarrion, were found to show improvement in WUE_{mt} of 60% and 78%, respectively, for the well-watered treatment and 66% and 61%, respectively, for the dry treatment (Samarakoon *et al.*, 1995). The apparently conflicting results among the different authors are not surprising considering some of the uncontrolled variables affecting growth and water use and the diverse experimental methods used. More importantly, rarely are the ET or transpiration data normalized for the evaporative demand of the growth environment. In fact, except for a few rare cases (e.g., Samarakoon *et al.*, 1995), these studies did not provide sufficient information to make such normalizations. Evaporative demand is determined by weather and affects transpiration markedly but often has little impact on photosynthesis. Only when differences in evaporative conditions are accounted for can data on WUE be transferrable and not site specific (de Wit, 1958). By analyzing the extensive literature, Tanner and Sinclair (1983) arrived at relatively simple equations for the analysis of WUE under different climatic conditions.

In a review on WUE as affected by rising CO_2, Eamus (1991) discussed many of the complicated mechanisms or interactive processes that have a final impact on WUE. These included the variable response of photosynthesis to high CO_2, downward photosynthetic adjustment, stomatal behavior as influenced by hormones, the controversial topic of the mechanism of stomatal response to CO_2 (Hsiao, 1976; Mansfield and Atkinson, 1990), integrated long-term change vs. short-term change, the interaction of CO_2 with temperature, the impact of stomata on transpiration as related to the degree of coupling of the canopy with the atmosphere, and the problems of scaling up from leaves and plants to regions and continents (McNaughton and Jarvis, 1991). Eamus (1991) touched on the complication introduced by faster leaf area development under high CO_2, which is the likely reason for the often reported similar water use per plant or per unit land area under elevated CO_2 (e.g., Kimball *et al.*, 1994) in spite of reduced stomatal opening. Gifford (1992) has been concerned with this complication for some time and recently published an interesting study of the effect (Samarakoon *et al.*, 1995). In terms of WUE, however, the concern may be unnecessary because leaf area or the degree of canopy cover of the land has nearly the same impact on assimilation as well as transpiration, as already pointed out earlier. Similarly, a number of the complications discussed by Eamus (1991) may be overstated since there exist some unifying principles and integrative parameters which can serve as a basis for the systematic approach in quantifying WUE under elevated CO_2 and environmental stresses, as developed in the following section.

IX. Framework for Response of Photosynthesis and Water Use Efficiency to Elevated CO_2

In view of the varied responses of photosynthesis and WUE to increases in CO_2, a systematic approach is needed to assess responses quantitatively. Hsiao (1993) proposed a conceptual framework for such a purpose, based on the gas exchange of single leaves, but dealing only with the gaseous phase and avoiding the complicating effects of the aqueous phase where metabolic reactions occur. Making the usual assumption of steady-state conditions, the rate of assimilation (A) is equated to the rate of CO_2 transport from the bulk air to the intercellular space, driven by the difference in CO_2 concentration (ΔC) between the bulk air and the intercellular space:

$$A = g_{ae}(C_a - C_i) = g_{ae}(\Delta C), \tag{1}$$

where g is the conductance for CO_2, with the transport pathway segment across the air boundary layer and the leaf epidermis denoted by the subscript ae. The advantage of Eq. (1) is that assimilation is considered only in terms of CO_2 transport in the gaseous phase, and hence a direct treatment of the complex metabolic changes and liquid-phase transport associated with changes in environmental factors is avoided. Instead, these changes are reflected in the value of C_i relative to C_a, a ratio termed α. Only physical processes are involved in the gaseous phase, so all terms in Eq. (1) are well defined and can be experimentally determined.

For easy assessments of the impact of elevated CO_2 or stress on leaf assimilation and transpiration based on theory, Hsiao (1993) advocated using relative changes. Starting with Eq. (1) and designating the original situation with the subscript o and the new situation with the subscript n, the relative change in A, A_n/A_o, is given by

$$\frac{A_n}{A_o} = \frac{g_{ae,n}\,(C_{a,n} - C_{i,n})}{g_{ae,o}\,(C_{a,o} - C_{i,o})} \tag{2a}$$

$$= \frac{g_{ae,n}(1 - \alpha_n)\,C_{a,n}}{g_{ae,o}(1 - \alpha_o)\,C_{a,o}}. \tag{2b}$$

Equation (2b) is derived by recognizing that $\alpha = C_i/C_a$ and $\alpha C_a = C_i$. Should α remain the same under the original and new conditions, Eq. (2b) reduces to

$$\frac{A_n}{A_o} = \frac{g_{ae,n}\,C_{a,n}}{g_{ae,o}\,C_{a,o}}. \tag{3}$$

Starting with the work of Wong *et al.* (1979) and Ball and Berry (1982), evidence is accumulating that generally there is a strong tendency for C_i

to be proportional to C_a; that is, for α to be approximately constant in diverse species under a range of environmental conditions conducive for photosynthesis (Morison, 1987; Hsiao, 1993). However, α does change in some situations. It increases slightly when the difference in water vapor concentration between the leaf intercellular space and the outside air is reduced in several species (Morison, 1987), and more so when PAR is reduced to a very low level (Morison, 1987; Bolaños and Hsiao, 1991). The ratio decreased under water stress in some studies [e.g., von Caemmerer and Farquhar (1984) but not in others (Hsiao, 1993)].

For the simplest case of constant α when Eq. (3) is applicable, the increase in assimilation is proportional to the relative increase in C_a modulated by the reduction in relative gaseous-phase conductance. For example, the equation predicts that if C_a is doubled but gaseous phase conductance is reduced to 80% of the original, then the new assimilation rate would be 60% higher than the original. Should α not remain the same under elevated CO_2, then the slightly more complicated equation, Eq. (2b), will have to be used to predict the outcome. In all cases, knowledge of constancy or variability of α is critical. Equation (3) is useful in evaluating data on photosynthetic response to elevated CO_2. Data showing a response greater than the relative increase in C_a may be questioned since g_{ae} is invariably reduced under a higher C_a, unless α is shown to be increased to compensate for the decrease in g_{ae}.

To consider WUE, an equation analogous to Eq. (1) is written for transpiration:

$$T = g'_{ae}(W_i - W_a) = g'_{ae}(\Delta W), \tag{4}$$

where g_{ae} denotes the conductance of the air boundary layer and leaf epidermis for water vapor, with the prime differentiating it from the conductance for CO_2, which has no prime. The driving force for transpiration is the difference in water vapor concentration (ΔW) between the intercellular space (W_i) and the bulk air (W_a). Due to its lighter molecular mass, water vapor diffuses faster than CO_2 and $g'_{ae} = 1.6 g_{ae}$ (Farquhar and Sharkey, 1982).

To obtain photosynthetic WUE, the ratio of A to T, we combine Eqs. (1) and (4) while recognizing that $g_{ae} = 0.625 g'_{ae}$,

$$\mathrm{WUE_{at}} = \frac{g_{ae}}{g'_{ae}} \frac{\Delta C}{\Delta W} = 0.625 \frac{\Delta C}{\Delta W}. \tag{5}$$

This equation highlights the importance of knowing ΔC and ΔW when evaluating WUE, and holds regardless of whether the leaf is under high or present level of CO_2, is well watered and under optimal conditions, or is affected by other stresses.

To evaluate the impact of elevated CO_2 or water stress on WUE, again the comparison of the original situation (designated by the subscript o) to the new situation (designated by the subscript n) in relative terms (Hsiao, 1993) has a considerable advantage. The changes to be evaluated, from the original to the new, may be in C_a, in the level of water stress, or for that matter, in the level of any other stress. Consequently, the approach is general and not restricted to only some environmental parameters. Based on Eq. (5), the relative WUE_{at} is

$$\frac{WUE_{at,n}}{WUE_{at,o}} = \frac{\Delta C_n \Delta W_o}{\Delta C_o \Delta W_n} = \frac{(1 - \alpha_n) C_{a,n} \Delta W_o}{(1 - \alpha_o) C_{a,o} \Delta W_n}. \tag{6}$$

The second part of Eq. (6) is derived by recognizing again that $\alpha C_a = C_i$. As mentioned, α tends to be conservative in the face of variations of a number of environmental factors, that is, $\alpha_o = \alpha_n$. In that case, Eq. (6) simplifies to

$$\frac{WUE_{at,n}}{WUE_{at,o}} = \frac{C_{a,n} \Delta W_o}{C_{a,o} \Delta W_n}. \tag{7}$$

Accordingly, for situations where the CO_2 level stays the same but water stress is imposed, Eq. (7) indicates that the ratio of WUE_{at} with and without stress would simply be $\Delta W_o / \Delta W_n$. For the same atmospheric humidity, ΔW would depend on W_i, which in turn is determined by leaf temperature, as the intercellular space is virtually saturated with water vapor and saturation vapor concentration is a function of temperature. When stomatal opening and transpiration are reduced by water stress, leaf temperature rises as dictated by the principle of energy balance, leading to a higher W_i. Consequently, $\Delta W_o / \Delta W_n < 1$ and $WUE_{at,n} < WUE_{at,o}$ as long as C_i remains the same. How much ΔW changes under water stress would depend on stomatal characteristics as well as atmospheric conditions affecting leaf energy balance. Other things being equal, the more severe the stress and the greater the stomatal closure, the more WUE_{at} would be reduced.

The conclusion that WUE_{at} is reduced under water stress appears to be inconsistent with the conservative nature of WUE_{me}, and is also in apparent conflict with many single leaf studies in the literature which suggest that WUE_{at} is improved by water stress. It should be realized that the situations described above are not those occurring in the typical gas-exchange chambers used to determine WUE of isolated leaves. There, leaf temperature is controlled or largely dictated by the chamber temperature, the principle of energy balance as it operates in the open field is largely circumvented, and ΔW is kept nearly constant. Hence, WUE_{at} would remain the same if C_i does not change with water stress, and it is likely to increase if C_i declines. As for the tendency of WUE_{me} of a given species under a given climate to remain constant instead of declining with water stress, there are reasons that this might occur. One is that C_i (and hence α) is often reduced under water stress, which may be sufficient to counteract the decrease in the ratio of $\Delta W_o/\Delta W_n$ [Eq. (6)], keeping WUE_{at} unchanged. Another reason is that as water stress develops and intensifies, the plant's first line of defense is a restriction of leaf and canopy area growth while keeping stomata open (Bradford and Hsiao, 1982). So the bulk of the biomass is produced when neither α nor ΔW is altered. Only when stress becomes severe enough is g_e reduced and ΔW raised, with the possibility of reducing WUE_{at}. The amount of biomass produced under the more severe stress, however, is likely to be small because of reduced assimilation. Hence the overall WUE_{me} usually appears to be relatively unaffected by water stress when normalized for the evaporative demand of the climate.

For the case of changes in CO_2 level in the absence of significant water stress, Eq. (7) indicates that the change in WUE_{at} would be less than proportional to the change in C_a, whether the change is positive or negative, and regardless of species. How much less is dependent on how sensitive stomatal response is to CO_2. This is because ΔW is negatively related to g_e and g_e is negatively related to C_a, other things being equal.

Because the potential changes in WUE caused by water stress are likely to be minor or insignificant, the combined impact of CO_2 elevation and water stress is likely largely due to the increase in C_a, consistent with the common observation that WUE is increased under elevated CO_2 with or without water stress.

The framework outlined here is based on the gas exchange of single and isolated leaves. Scaling up to canopy and ecosystem levels would involve additional complications (Jarvis and McNaughton, 1986; Baldocchi *et al.*, 1991; Kim and Verma, 1991; Jackson *et al.*, 1991, 1998). Nonetheless, for a canopy that is short and covering most of the soil, indications are that it acts essentially as a big leaf and obeys the Penman-Monteith combination equation (Monteith, 1973; Choudhury *et al.*, 1986). In fact, in experiments conducted by Hsiao and colleagues (unpublished) in controlled environmental chambers on several crop species with detailed monitoring of ΔW, WUE_{me} over growth periods of a few to many weeks appears to obey, at least approximately, the prediction of Eq. (7). At the same time, scaling up to canopies of tall trees and forests is likely not to be simple (Jarvis *et al.*, 1976).

Strictly speaking, α is conservative (Ball and Berry, 1982) only when it represents the ratio of CO_2 concentration in the intercellular space to CO_2 concentration of the air at the leaf epidermis just outside the stomatal pore (C_e). Implicit in the procedure in arriving at Eq. (6) is the approximation that $C_i/C_e \cong C_i/C_a$. How good the approximation is depends on the conductance of the air boundary layer (g_a) relative to conductance of the leaf epidermis (mostly stomatal). So Eq. (6) should be good for relatively isolated leaves under turbulent conditions, but becomes less and less accurate as turbulence decreases and leaves become more crowded in a canopy.

X. Overview and Conclusion

The effects and interactions of water deficit and elevated CO_2 on plants have been discussed in terms of growth by cell enlargement, stomatal behavior, CO_2 assimilation, adjustments in photosynthetic capacity in relation to sink capacity and nitrogen and phosphorus nutrition, and water use efficiency. All of these processes or parameters are pivotal in determining plant productivity. In terms of productivity, the impact on these processes due to water stress and rising CO_2 may be summarized by a diagram. In Fig. 4, assimilation per unit land area, which cumulatively, after the deduction of respiratory CO_2 loss, determines productivity, is shown to be dependent on the amount of PAR absorbed by the canopy and on the assimilation per unit of absorbed PAR (PAR use efficiency). Absorbed PAR, in turn, is jointly determined by solar radiation and by effective leaf area. Assimilation per unit of absorbed PAR is jointly determined by photosynthetic capacity, C_i, and g_e of the leaves. Effective leaf area is a function of the rate of expansive growth of leaves, as long as canopy cover of the ground is incomplete. The impact of water deficit and elevation of C_a on growth and effective leaf area is shown to be counteractive. That is, water deficit has a negative effect while elevated CO_2 has a positive effect. The impact of water deficit and C_a elevation on C_i is similarly counteractive. On the other hand, the impact of water deficit and C_a elevation on g_e is mutually reinforcing, both causing the closing of stomata. The resultant reduction in g_e, by slowing down transpiration, has a negative feedback effect on the development of water deficit. Similarly, water deficit development is linked to effective leaf area via negative feedback. These feedback links tend to modulate marked changes and stabilize the system. Impacts of water deficit and C_a elevation on photosynthetic capacity can also be mutually reinforcing, or can be minimal. Because of the counteractive impact of water deficit and C_a elevation on at least two of the parameters in the top row depicted in Fig. 4, the overall impact on assimilation per unit land area and productivity is generally counteractive. That is, elevated CO_2 tends to alleviate the negative effect of water deficit, and water deficit tends to reduce the beneficial effects of elevated CO_2.

This chapter is of limited scope and focuses only on selected topics; other important processes or parameters impacted by water deficit and CO_2 interactions were omitted. These include respiration reduction under elevated CO_2 and water deficit, variable and sometimes complex effects of water deficit and generally promotive effects of elevated CO_2 on the

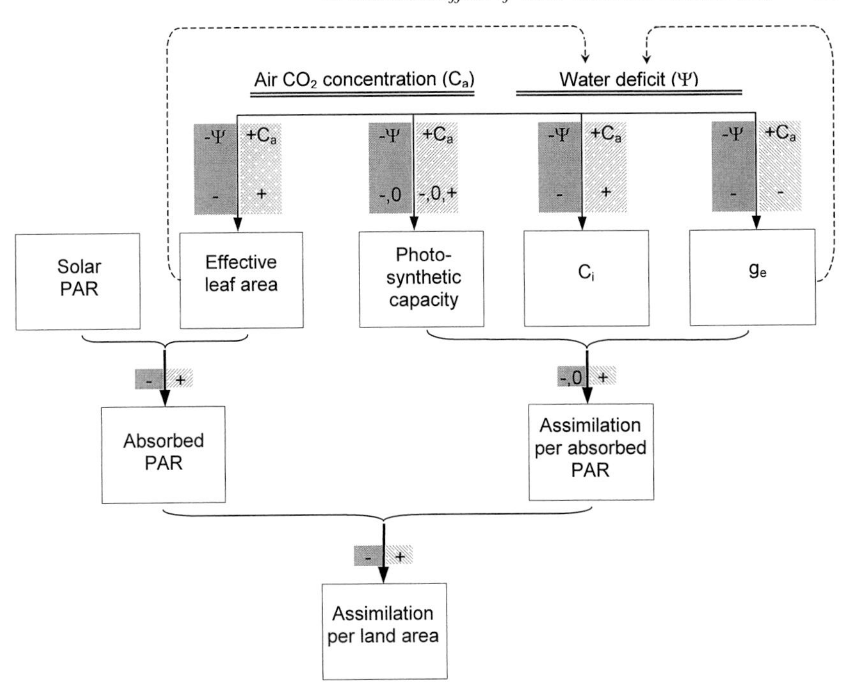

Figure 4 Counteractive and mutually reinforcing effects of elevated CO_2 ($+C_a$, cross-hatched) and water deficit ($-\Psi$, gray shade) on parameters underlying canopy CO_2 assimilation. Canopy assimilation per unit land area (or per plant) is depicted as the product of absorbed PAR per unit land area (or per plant) and radiation use efficiency (assimilation per unit of absorbed PAR). In turn, absorbed PAR is a function of solar radiation and effective leaf area per unit land area (or per plant), and radiation use efficiency is a function of photosynthetic capacity, intercellular CO_2 concentration, and epidermal conductance of the foliage. Effects of air CO_2 concentration and water deficit are shown as thin arrows with the impact being either positive, zero, or negative, as indicated by the algebraic signs at the head of the arrow. For photosynthetic capacity, the impact of water deficit could be either negative or zero, and the impact of elevated CO_2 could be either negative, zero, or positive. The overall effects on assimilation per land area are counteractive as indicated by the negative sign for water deficit and positive sign for elevated CO_2. Dashed lines and arrows represent negative feedback effects.

partitioning of assimilates to fruits and storage-roots, and a number of other less well delineated effects such as those on organ development or senescence. Clearly our knowledge base is deficient and many challenges lie ahead. Some recent developments, such as the new insights gained on the adjustment in photosynthetic capacity in response to assimilates and sinks, lend encouragement. Much more forbidding is the challenge for societal action in modulating the speed of global change and its impact.

Acknowledgments

The authors wish to thank the organizers of the current volume and the original conference at which some of these ideas were presented. The work of T.C.H. is partly supported by DOE grant DE-FG03-93ER6187 and through WESTGEC/NIGEC. R.B.J. acknowledges NIGEC/DOE and the Advanced Research Program of the state of Texas for support of CO_2-related research.

References

Acevedo, E., Hsiao, T. C., and Henderson, D. W. (1971). Immediate and subsequent growth responses of maize leaves to changes in water status. *Plant Physiol.* **48,** 631–636.

Anonymous. (1982). Drought resistance in crops with emphasis on rice. p. 404. International Rice Research Institute, Los Baños, Philippines.

Arp, W. J. (1991). Effects of source-sink relations on photosynthetic acclimation to elevated CO_2. *Plant Cell Environ.* **14,** 869–875.

Arp, W. J., and Drake, B. G. (1991). Increased photosynthetic capacity of *Scirpus olneyi* after 4 years of exposure to elevated atmospheric CO_2. *Plant Cell Environ.* **14,** 1004–1008.

Baldocchi, D. D. (1994). A comparative study of mass and energy exchange over a closed C_3 (wheat) and open C_4 (corn) canopy: II. CO_2 exchange and water use efficiency. *Agric. For. Meteorol.* **67,** 291–321.

Baldocchi, D. D., Luxmore, R., Hatfield, J. (1991). Discerning the forest from the trees: An essay on scaling canopy stomatal conductance. *Agric. For. Meteorol.* **54,** 197–226.

Ball, J. T., and Berry, J. A. (1982). The C_i/C_s ratio: a basis for predicting stomatal control of photosynthesis. *Carnegie Institution of Washington Yearbook,* pp. 88–92.

Bolaños, J. A., and Hsiao, T. C. (1991). Photosynthetic and respiratory characterization of field grown tomato. *Photosynth. Res.* **28,** 21–32.

Bowes, G. (1993). Facing the inevitable: Plants and increasing atmospheric CO_2. *Annu. Rev. Plant Physiol. Plant Mol. Biol.* **44,** 309–332.

Boyer, J. S. (1970). Leaf enlargement and metabolic rates in corn, sorghum, and sunflower at various leaf water potentials. *Plant Physiol.* **46,** 233–235.

Boyer, J. S. (1971). Nonstomatal inhibition of photosynthesis in sunflower at low leaf water potentials and high light intensities. *Plant Physiol.* **48,** 532–536.

Boyer, J. S. (1976). Photosynthesis at low water potentials. *Phil. Trans. R. Soc. London Ser. B* **273,** 501–512.

Boyer, J. S. (1985). Water transport. *Annu. Rev. Plant Physiol.* **36,** 473–516.

Bradford, K. J., and Hsiao, T. C. (1982). Physiological responses to moderate water stress. *In* "Encyclopedia of Plant Physiology, N.S. Physiological Plant Ecology II. Water Relations and Carbon Assimilation" (O. L. Lange *et al.,* eds.), pp. 263–324. Springer-Verlag, Berlin.

Campbell, W. J., Allen, L. H., Jr., and Bowes, G. (1988). Effects of CO_2 concentration on Rubisco activity, amount and photosynthesis in soybean leaves. *Plant Physiol.* **88,** 1310–1316.

Campbell, W. J., Allen, L. H., Jr., and Bowes, G. (1990). Response of soybean canopy photosynthesis to CO_2 concentration, light and temperature. *J. Exp. Bot.* **41,** 427–433.

Choudhury, B. J., Raginato, R. J., and Idso, S. B. (1986). An analysis of infrared temperature observations over wheat and calculation of latent heat flux. *Agric. For. Meteorol.* **37**(1), 75–88.

Cleland, R. E. (1980). Auxin and H^+-excretion: The state of our knowledge. *In* "Plant Growth Substances" (F. Skoog, ed.), pp. 71–78. Academic Press, New York.

Cure, J. D., Rufty, T. W., Jr., and Israel, D. W. (1989). Alterations in soybean leaf development and photosynthesis in a CO_2-enriched atmosphere. *Bot. Gaz.* **150,** 337–345.

Curtis, P. S. (1996). A meta-analysis of leaf gas exchange and nitrogen in trees grown under elevated carbon dioxide. *Plant Cell Environ.* **19,** 127–137.

Daley, P. F., Raschke, K., Ball, J. T., and Berry, J. A. (1989). Topography of photosynthetic activity of leaves obtained from video images of chlorophyll fluorescence. *Plant Physiol.* **90,** 1233–1238.

Davies, W., and Zhang, J. (1991). Root signals and the regulation of growth and development of plants in drying soil. *Ann. Rev. Plant Physiol. Molec. Biol.* **42,** 55–76.

de Wit, C. T. (1958). Transpiration and Crop Yields. *Versl. Landbouwkd. Onderz.* (*Purdoc*) 64.6.

Drake, B. G., and Leadley, P. W. (1991). Canopy photosynthesis of crops and native plant communities exposed to long-term elevated CO_2. *Plant Cell Environ.* **14,** 853–860.

Eamus, D. (1991). The interaction of rising CO_2 and temperatures with water use efficiency. *Plant Cell Environ.* **14,** 843–852.

Ehleringer, J. R., and Cerling, T. E. (1995). Atmospheric CO_2 and the ratio of intercellular to ambient CO_2 concentrations in plants. *Tree Physiol.* **15,** 105–111.

Ellsworth, D. S., Oren, R., Huang, C., Philips, N., and Hendrey, D. R. (1995). Leaf and canopy responses to elevated CO_2 in a pine forest under free-air CO_2 enrichment. *Oecologia* **104,** 139–146.

Enoch, H. Z., and Honour, S. J. (1993). Significance of increasing ambient CO_2 for plant growth and survival, and interactions with air pollution. *In* "Interacting Stresses on Plants in a Changing Climate" (M. B. Jackson and C. R. Black, eds.), pp. 51–75. Springer-Verlag, Berlin.

Evans, J. R. (1989). Photosynthesis and nitrogen relationships in leaves of C_3 plants. *Oecologia* **78,** 9–19.

Farquhar, G. D., and Sharkey, T. D. (1982). Stomatal conductance and photosynthesis. *Ann. Rev. Plant Physiol.* **33,** 317–345.

Farquhar, G., Masle, J., and Hubrick, K. (1987). Effects of drought, salinity and soil compaction on photosynthesis, transportation and carbon isotope composition of plants. *Curr. Topics Plant Biochem. Physiol.* **6,** 147–155.

Ferris, R., and Taylor, G. (1994). Elevated CO_2, water relations and biophysics of leaf extension in four chalk grassland herbs. *New Phytol.* **127,** 297–307.

Field, C. B., Jackson, R. B., and Mooney, H. A. (1995). Stomatal responses to increased CO_2: Implications from the plant to the global scale. *Plant Cell Environ.* **18,** 1214–1225.

Fischer, R. A., and Turner, N. C. (1978). Plant productivity in the arid and semiarid zones. *Annu. Rev. Plant Physiol.* **29,** 277–317.

Frensch, J. (1997). Primary responses of root and leaf elongation to water deficits in the atmosphere and in the soil solution. *J. Exp. Bot.* **48,** 985–999.

Frensch, J., and Hsiao, T. C. (1994). Transient responses of cell turgor and growth of maize roots as affected by changes in water potential. *Plant Physiol.* **104,** 247–254.

Frensch, J., and Hsiao, T. C. (1995). Rapid response of the yield threshold and turgor regulation during adjustment of root growth to water stress in *Zea mays*. *Plant Physiol.* **108,** 303–312.

Gaastra, P. (1959). Photosynthesis of crop plants as influenced by light, carbon dioxide, temperature and stomatal diffusive resistance. *Mededel. Landbouwbogeschool. Wageningen* **59,** 1–68.

Geiger, D. R., and Giaquinta, R. T. (1982). Translocation of photosynthate. *In* "Photosynthesis" (Govindjee, ed.), pp. 345–386. Academic Press, New York.

Gifford, R. M. (1992). Interaction of carbon dioxide with growth-limiting environmental factors in vegetation productivity: Implications for the global carbon cycle. *Adv. Bioclimatol.* **1,** 24–58.

Green, P. B. (1968). Growth physics in Nitella: A method for continuous *in vivo* analysis of extensibility based on a micro-manometer technique for turgor pressure. *Plant Physiol.* **43,** 1169–1184.

Habash, D. Z., Paul, M. J., Parry, M. A. J., Keys, A. J., and Lawlor, D. W. (1995). Increased capacity for photosynthesis in wheat grown at elevated CO_2: The relationship between electron transport and carbon metabolism. *Planta* **197,** 482–489.

Hanks, R. J. (1983). Yield and water-use relationships: An overview. *In* "Limitations to Efficient Water Use in Crop Production" (H. M. Taylor, W. R. Jordan, and T. R. Sinclair, eds.). Am. Soc. Agron., Madison, WI.

Henderson, S., Hattersley, P., von Caemmerer, C. B., and Osmond, C. B. (1994). Are C_4 pathway plants threatened by global climatic change. *In* "Ecophysiology of Photosynthesis" (Ernst-D. Schulze and M. Caldwell, eds.), pp. 529–549. Springer-Verlag, Berlin.

Hileman, D. R., Huluka, G., Kenjige, P. K., Sinha, N., Bhattacharya, N. C., Biswas, P. K., Lewin, K. F., Nagy, J., and Hendrey, G. R. (1994). Canopy photosynthesis and transpiration of field-grown cotton exposed to free-air CO_2 enrichment (FACE) and differential irrigation. *Agric. For. Meteorol.* **70,** 189–207.

Hsiao, T. C. (1973). Plant responses to water stress. *Annu. Rev. Plant Physiol.* **24,** 519–570.

Hsiao, T. C. (1976). Stomatal ion transport. *In* "Encyclopedia of Plant Physiology. New Series, Vol. 2. Transport in Plants II. Part B. Tissues and Organs" (U. Lüttge and M. G. Pitman, eds.), pp. 195–221. Springer-Verlag, Berlin.

Hsiao, T. C. (1990). Crop water requirement and productivity. *In* "Proceedings of the 5th International Conference on Irrigation" pp. 5–18. Hidekel Printing House Ltd., Tel-Aviv.

Hsiao, T. C. (1993). Effects of drought and elevated CO_2 on plant water use efficiency and productivity. *In* "Interacting Stresses on Plants in a Changing Climate" (M. B. Jackson and C. R. Black, eds.), pp. 435–465. Springer-Verlag, Berlin.

Hsiao, T. C., and Acevedo, E. (1975). Plant responses to water deficits, water-use efficiency, and drought resistance. *In* "Plant Modification for More Efficient Water Use" (J. F. Stone, ed.), pp. 59–84. Elsevier Sci. Publ. Co., Amsterdam.

Hsiao, T. C., and Bradford, K. J. (1983). Physiological consequences of cellular water deficits. *In* "Limitations to Efficient Water Use in Crop Production" (H. M. Taylor, W. R. Jordan, and T. R. Sinclair, eds.). Am. Soc. Agron., Madison, WI.

Hsiao, T. C., and Jing, J. (1987). Leaf growth in response to water deficits. *In* "Physiology of Cell Expansion During Plant Growth" (D. J. Cosgrove and D. P. Knievel, eds.). American Society of Plant Physiology, Rockville, MD.

Jackson, R. B., Sala, O. E., Field, C. B., and Mooney, H. A. (1994). CO_2 alters water use, carbon gain, and yield for the dominant species in a natural grassland. *Oecologia* **98,** 257–262.

Jackson, R. B., Woodrow, I. E., and Mott, K. A. (1991). Nonsteady-state photosynthesis following an increase in photon flux density (PFD): Effects of magnitude and duration of inital PFD. *Plant Physiol.* **95,** 498–503.

Jackson, R. B., Sala, O. E., Paruelo, J. M., and Mooney, H. A. (1998). Ecosystem water fluxes for two grasslands in elevated CO_2: A modeling analysis. *Oecologia* **113,** 537–546.

Jacob, J., Greitner, C., and Drake, B. G. (1995). Acclimation of photosynthesis in relation to Rubisco and nonstructural carbohydrate contents and *in situ* carboxylase activity in *Scirpus olneyi* grown at elevated CO_2 in the field. *Plant Cell Environ.* **18,** 875–884.

Jarvis, P. G., and McNaughton, K. G. (1986). Stomatal control of transpiration: Scaling up from leaf to region. *Adv. Ecol. Res.* **15,** 1–49.

Jarvis, P. G., James, G. B., and Landberg, J. J. (1976). Coniferous forest. *In* "Vegetation and the Atmosphere. Case study" (J. L. Monteith, ed.), pp. 171–240. Academic Press, London.

Kaiser, W. M. (1987). Effects of water deficit on photosynthetic capacity. *Physiol. Plant.* **71,** 142–149.

Kim, J., and Verma, S. (1991). Modeling canopy photosynthesis: Scaling up from a leaf to canopy in a temperate grassland ecosystem. *Agric. For. Meteorol.* **57,** 187–208.

Kimball, B. A., and Idso, S. B. (1983). Increasing atmospheric CO_2: Effects on crop yield, water use and climate. *Agric. Water Mgmt.* **7,** 55–72.

Kimball, B. A., LaMorte, R. L., Seay, R. S., Pinter, P. J., Rokey, R. R., Hunsaker, D. J., Dugas, W. A., Heuer, M. L., Mauney, J. R., Hendrey, G. R., Lewis, K. F., and Nagy, J. (1994). Effects of free-air CO_2 enrichment on energy balance and evapotranspiration of cotton. *Agric. For. Meteorol.* **70,** 259–278.

Kirkham, M. B., He, H., Bolger, T. P., Lawlor, D. J., and Kanemasu, E. T. (1991). Leaf photosynthesis and water use of big bluestem under elevated carbon dioxide. *Crop. Sci.* **31,** 1589–1594.

Kirschbaum, M. (1987). Water stress in *Eucaluptus pauciflora:* Comparison of effects on stomatal conductance with effects on the mesophyll capacity for photosynthesis and investigation of a possible involvement of photoinhibition. *Planta* **171,** 466–473.

Knapp, A. K., Hamerlynck, E. P., and Owensby, C. E. (1993). Photosynthetic and water relations responses to elevated CO_2 in the C_4 grass *Andropogon gerardii. Int. J. Plant Sci.* **154,** 459–466.

Lawlor, D. W., and Mitchell, R. A. C. (1991). The effects of increasing CO_2 on crop photosynthesis and productivity: A review of field studies. *Plant Cell Environ.* **14,** 807–818.

Lockhart, J. A. (1965). Cell extension. *In* "Plant Biochemistry" (J. Bonner and J. E. Varner, eds.), pp. 826–849. Academic Press, New York.

Mansfield, T. A., and Atkinson, C. J. (1990). Stomatal behavior in water stressed plants. *In* "Stress Responses in Plants: Adaptation and Acclimation Mechanisms" (R. G. Alscher and N. S. Allen, eds.), pp. 241–264. Wiley-Liss, New York.

Matthews, M. A., Van Volkenburgh, E., and Boyer, J. S. (1984). Acclimation of leaf growth to low leaf water potentials in sunflower. *Plant Cell Environ.* **7,** 199–206.

Mauney, J. R., Kimball, B. A., Pinter, P. J., and Lamorte, R. L. (1994). Growth and yield of cotton in response to a free-air carbon dioxide enrichment environment. *Agric. For. Meteorol.* **70,** 49–67.

McNaughton, K., and Jarvis, P. G. (1991). Effects of spatial scale of stomatal control of transpiration. *Agric. For. Meteorol.* **54,** 279–301.

Michelena, V. A., and Boyer, J. S. (1982). Complete turgor maintenance at low water potentials in the elongating region of maize leaves. *Plant Physiol.* **69,** 1145–1149.

Monteith, J. L. (1973). "Principles of Environmental Physics." Edward Arnold Ltd., London.

Morison, J. I. L. (1987). Intercellular CO_2 concentration and stomatal response to CO_2. *In* "Stomatal Function" (E. Zeiger, G. D. Farquhar, and I. R. Cowan, eds.), pp. 229–251. Stanford University Press, Stanford, CA.

Morison, J. I. L. (1993). Response of plants to CO_2 under water limited conditions. *Vegetatio* **104/105,** 193–209.

Morison, J. I. L., and Gifford, R. M. (1984). Plant growth and water use with limited water supply in high CO_2 concentrations. I. Leaf area, water use and transpiration. *Aust. J. Plant Physiol.* **11,** 361–374.

Nishizawa, T., and Suge, H. (1995). Ethylene and carbon dioxide: Regulation of oat mesocotyl growth. *Plant Cell Environ.* **18,** 197–203.

Ortiz-Lopez, A., Ort, D. R., and Boyer, J. S. (1991). Photophosphorylation in attached leaves of *Helianthus annuus* at low water potentials. *Plant Physiol.* **96,** 1018–1025.

Osmond, C. B., Austin, M. P., Berry, J. A., Billings, W. D., Boyer, J. S., Dacey, J. W. H., Nobel, P. S., Smith, S. D., and Winner, W. E. (1987). Stress physiology and the distribution of plants. *BioScience* **37,** 38–47.

Pardossi, A., Pritchard, J., and Tomos, A. D. (1994). Leaf illumination and root cooling inhibit bean leaf expansion by decreasing turgor pressure. *J. Exp. Bot.* **45,** 415–422.

Pettersson, R., and McDonald, A. J. S. (1994). Effects of nitrogen supply on the acclimation of photosynthesis to elevated CO_2. *Photosynth. Res.* **39,** 389–400.

Polley, H. W., Johnson, H. B., and Mayeux, H. S. (1992). Carbon dioxide and water fluxes of C_3 annuals and C_3 and C_4 perennials at subambient CO_2. *Funct. Ecol.* **6,** 693–703.

Prior, S. A., Rogers, H. H., Sionit, N., and Patterson, R. P. (1991). Effects of elevated atmospheric CO_2 on water relations of soya bean. *Agric. Ecosystems Environ.* **35,** 13–25.

Radin, J. W., Kimbal, B. A., Hendrix, D. L., and Mauney, J. R. (1987). Photosynthesis of cotton plants exposed to elevated levels of carbon dioxide in the field. *Photosynth. Res.* **12,** 191–203.

Ranasinghe, S., and Taylor, G. (1996). Mechanism for increased leaf growth in elevated CO_2. *J. Exp. Bot.* **47**(296), 349–358.

Raschke, K. (1987). Metabolic changes during declines of photosynthesis: Responses to dry air and to abscisic acid. *Curr. Topics Plant Biochem. Physiol.* **6,** 104–118.

Ray, P. M., Green, P. B., and Cleland, R. E. (1972). Role of turgor in plant cell growth. *Nature* **239,** 163–164.

Reddy, V. R., Reddy, K. R., and Hodges, H. F. (1995). Carbon dioxide enrichment and temperature effects on cotton canopy photosynthesis, transpiration, and water-use efficiency. *Field Crops Res.* **41,** 13–23.

Redshaw, A. J., and Meidner, H. (1972). Effects of water stress on the resistance to uptake of carbon dioxide in tobacco. *J. Exp. Bot.* **23,** 229–240.

Riviere-Rolland, H., and Betsche, T. (1996). Adaptation of pea to elevated atmospheric CO_2: Rubisco, phosphoenolpyruvate carboxylase and chloroplast phosphate translocator at different levels of nitrogen and phosphorus nutrition. *Plant Cell Environ.* **19,** 109–117.

Roden, J. S., and Ball, M. C. (1996). The effect of elevated CO_2 on the growth and photosynthesis of two eucalyptus species exposed to high temperatures and water deficits. *Plant Physiol.* **111,** 909–919.

Sage, R. F. (1994). Acclimation of photosynthesis to increasing CO_2: The gas exchange perspective. *Photosynth. Res.* **39,** 351–368.

Sage, R. F., Sharkey, T. D., and Seemann, J. R. (1989). Acclimation of photosynthesis to elevated CO_2 in five C_3 species. *Plant Physiol.* **89,** 590–596.

Samarakoon, A. B., Muller, W. J., and Gifford, R. M. (1995). Transpiration and leaf area under elevated CO_2—effects of soil water status and genotype in wheat. *Aust. J. Plant Physiol.* **22,** 33–44.

Seemann, J. R., Sharkey, T. D., Wang, J. L., and Osmond, C. B. (1987). Environmental effects on photosynthesis, nitrogen-use efficiency, and metabolite pools in leaves of sun and shade plants. *Plant Physiol.* **84,** 796–802.

Sharkey, T. D. (1985). Photosynthesis in intact leaves of C_3 plants: Physics, physiology and rate limitation. *Bot. Rev.* **51,** 53.

Sharkey, T. D. (1987). Carbon reduction cycle intermediates in water stressed, intact leaves. *Curr. Topics Plant Biochem. Physiol.* **6,** 88–103.

Sharkey, T. D. (1990). Water stress effects on photosynthesis. *Photosynthetica* **24,** 651.

Sharkey, T. D., and Seemann, J. (1989). Mild water stress effects on carbon-reduction-cycle intermediates, ribulose biphosphate carboxylase activity, and spatial homogeneity of photosynthesis in intact leaves. *Plant Physiol.* **89,** 1060–1065.

Sharp, R. E., and Davies, W. J. (1989). Regulation of growth and development of plants growing with a restricted supply of water. *In* "Plants Under Stress" (H. G. Jones, T. L. Flowers, and M. B. Jones, eds.), pp. 72–93. Cambridge University Press, London.

Sharp, R. E., Silk, W. K., and Hsiao, T. C. (1988). Growth of the maize primary root at low water potentials. I. Spatial distribution of expansive growth. *Plant Physiol.* **87,** 50–57.

Sharp, R. E., Hsiao, T. C., and Silk, W. K. (1990). Growth of the maize primary root at low water potential. II. Role of growth and deposition of hexose and potassium in osmotic adjustment. *Plant Physiol.* **93,** 1337–1346.

Stitt, M. (1991). Rising CO_2 levels and their potential significance for carbon flow in photosynthetic cells. *Plant Cell Environ.* **14,** 741–762.

Stitt, M., and Schulze, D. (1994). Does Rubisco control the rate of photosynthesis and plant growth? An exercise in molecular ecophysiology. *Plant Cell Environ.* **17,** 465–487.

Tanner, C. B., and Sinclair, T. R. (1983). Efficient water use in crop production: Research or Re-search? *In* "Limitations to Efficient Water Use in Crop Production" (H. M. Taylor, W. R. Jordan, and T. R. Sinclair, eds.). American Society of Agronomy, Madison, WI.

Taylor, H. M., Jordan, W. R., and Sinclair, T. R. (1983). "Limitations to Efficient Water Use in Crop Production." American Society of Agronomy, Madison, WI.

Terashima, I. (1992). Anatomy of non-uniform leaf photosynthesis. *Photosynth. Res.* **31,** 195–212.

Tezara, W., and Lawlor, D. W. (1995). Effects of water stress on the biochemistry and physiology of photosynthesis in sunflower. *In* "Photosynthesis: From Light to Biosphere" (P. Mathis ed.), pp. 625–628. Kluwer Academic Publishers, Rothamsted.

Thomas, R. B., and Strain, B. R. (1991). Root restriction as a factor in photosynthetic acclimation of cotton seedlings grown in elevated carbon dioxide. *Plant Physiol.* **96,** 627–634.

van Bavel, C. H. M. (1974). Antitranspirant action of carbon dioxide on intact sorghum plants. *Crop. Sci.* **14,** 208–212.

Van Oosten, J. J., and Besford, R. T. (1994). Sugar feeding mimics effect of acclimation to high CO_2–rapid down regulation of rubisco small subunit transcripts but not of the large subunit transcripts. *J. Plant Physiol.* **143,** 306–312.

Van Volkenburgh, E., and Boyer, J. S. (1985). Inhibitory effects of water deficit on maize leaf elongation. *Plant Physiol.* **77,** 190–194.

Vassey, T. L., and Sharkey, T. D. (1989). Mild water stress of *Phaseolus vulgaris* plants leads to reduced starch synthesis and extractable sucrose phosphate synthase activity. *Plant Physiol.* **89,** 1066–1070.

von Caemmerer, S., and Farquhar, G. D. (1984). Effects of partial defoliation, changes of irradiance during growth, short-term water stress and growth at enhances $p(CO_2)$ on the photosynthetic capacity of leaves of *Phaseolus vulgaris* L. *Planta* **160,** 320–329.

Walker, S., and Hsiao, T. (1993). Osmoticum, water and dry matter deposition rates in the growth zone of sorghum leaves under water deficit conditions. *Environ. Exp. Bot.* **33,** 447–456.

Wise, R. R., Ortiz-Lopez, A., and Ort, D. R. (1992). Spatial distribution of photosynthesis during drought in field-grown and acclimated and nonacclimated chamber-grown cotton. *Plant Physiol.* **100,** 26–32.

Wong, S. C. (1979). Elevated atmospheric partial pressure of CO_2 and plant growth. I. Interactions of nitrogen nutrition and photosynthetic capacity in C_3 and C_4 plants. *Oecologia (Berl.)* **44,** 68–74.

Wong, S.-C. (1990). Elevated atmospheric partial pressure of CO_2 and plant growth. *Photosynth. Res.* **23,** 171–180.

Wong, S.-C., Cowan, I. R., and Farquhar, G. D. (1979). Stomatal conductance correlates with photosynthetic capacity. *Nature* **282,** 424–426.

Younis, H. M., Boyer, J. S., and Govindjee, (1979). Conformation and activity of chloroplast coupling factor exposed to low chemical potential of water in cells. *Biochem. Biophys. Acta* **548,** 328–340.

2

Increasing Atmospheric CO_2 Concentration, Water Use, and Water Stress: Scaling Up from the Plant to the Landscape

Jeffrey S. Amthor

Water is the most nourishing food a garden can have.
—Plato (ca. 428–347 B.C.E.), The Laws

Without seasonable showers the earth cannot send up gladdening growths.
—Titus Lucretius Carus (ca. 99–55 B.C.E.), On the Nature of Things

Plants without water perish; they also die without CO_2.
—Jonathan R. Cumming, in laboratorium, 1986

I. Introduction

Definitions of stress abound. For present purposes, I take environmental stress to be an environmental limitation on ecosystem net primary production (NPP). Annual NPPs in deserts and rainforests, for example, differ significantly. This large divergence in NPP in those two ecosystem types is presumably due in large part to differences in availability of soil water and therefore degree of water stress, although other environmental differences may also be important. Indeed, one simple and commonly used model of terrestrial ecosystem NPP—the "Miami model" (see Lieth, 1975)—relates annual NPP to annual precipitation and annual mean temperature. This might be interpreted as a general model of landscape water and temperature stresses. Water stress can be reduced by an increase in precipitation or irrigation, increased access to existing soil water that would otherwise be unavailable to plants, or reduced evapotranspiration.

The amount of NPP or plant growth per unit environmental resource used is an important measurement of the relationship between plants and their environment. Although it does not correspond to its thermodynamic definition, the word *efficiency* is used to denote phytomass produced per unit resource used. Common examples are radiation-use efficiency, which is plant growth per unit solar radiation absorbed by the plants in an ecosystem (see, e.g., Monteith, 1972; Loomis and Amthor, 1996), and water use efficiency (WUE), which is the amount of growth per unit water evaporated (see, e.g., Jones, 1992), Thus, if WUE increases, an increase in productivity per unit water added to an ecosystem in precipitation and irrigation can occur. An increase in WUE may have a result similar to a decrease in water stress.

This chapter considers the potential effects of atmospheric CO_2 concentration ($[CO_2]_a$) increase on ecosystem water use, WUE, and water stress. A chief concern about increasing $[CO_2]_a$ is that global climatic changes may result (Manabe, 1983). In addition, direct effects of increasing $[CO_2]_a$ on plant processes are nearly certain (Amthor, 1995). Effects of increasing $[CO_2]_a$ on the supply of water through precipitation and the general effects of increasing-CO_2-induced global warming on "demand" for ecosystem water use are considered only briefly herein. My primary aim is to outline issues associated with the scaling up of the effects of increasing $[CO_2]_a$ on stomates and individual-plant water use to levels of ecosystems and landscapes.

II. Ecosystem Hydrology and Global Climatic Change

The release of water vapor to the atmosphere from terrestrial ecosystems is called *evapotranspiration* (ET). It is the sum of several processes: interception losses (see later discussion), evaporation of water from the soil, and transpiration. Evapotranspiration often represents a significant ecosystem water loss, but it is not the only fate of water in ecosystems. Some water may leave the ecosystem as surface flow, eventually contributing to streamflow. Water may also move laterally through soil or cracks in bedrock and enter streambeds. In many ecosystems, subsurface flow to streambeds is larger than surface flow. Deep seepage of soil water and storage in aquifers also occur in some ecosystems.

Terrestrial ecosystems gain water mainly in precipitation processes such as rain, snow, and dew formation. Some of that precipitation is intercepted by plant leaves, branches, and stems, and some of that intercepted water remains on those shoot organs from which it evaporates or sublimates rather than being transferred to the soil or absorbed into the plant. Precipitation captured by plant shoots and returned to the atmosphere as water vapor directly from the external surfaces of those shoots is termed *canopy interception loss* (e.g., Helvey and Patric, 1965). Canopy interception loss is significant in many ecosystems (Table I) and prevents water from entering soil where it might be absorbed by roots. In ecosystems with a litter layer, some precipitation is also intercepted by and retained on that litter from which it evaporates. This water is also prevented from entering the soil. Thus, in spite of the significance of transpiration to ET, interception losses can be as large as transpiration. Precipitation not lost to canopy or litter interception is transferred to the soil, some of it as stemflow. Once on or in the soil, water can evaporate, be absorbed by roots, drain by surface or subsurface processes, or remain there. Snow and ice on the surface of a soil must melt

Table I Estimates of Canopy Interception Loss[a,b]

Ecosystem	Annual canopy interception loss as a fraction of annual precipitation
Tropical rainforest	0.10–0.36
Other tropical forests	0.11–0.38
Temperate coniferous forest	0.26–0.42
Picea plantation	0.38
Picea stand (summer)	0.24–0.82[c]
Temperate deciduous forest	0.08–0.31
Quercus forest	
In leaf	0.22[d]
Bare branches	0.11[e]
Fagus stand (summer)	0.18–0.72[c]
Mediterranean forest	0.05–0.34
Boreal coniferous forest	0.06–0.30
Dwarf-shrub heath	As much as 0.50
Subtropical shrub formations	0.05–0.15
Palm groves	0.10–0.15
Grassland	0.03–0.05
Cultivated land	<0.10

[a] From reviews by Benecke, 1975; Schnock, 1981, *in* Galoux *et al.*, 1981; Calder, 1990; Larcher, 1995.

[b] Canopy interception loss is affected by factors such as the seasonal timing of canopy development, plant density (area of leaves, stems, and branches per unit ground area), intensity and duration of precipitation events, and nature of the precipitation (snow, ice, fine mist, large water drops, and so on). In general, interception loss as a fraction of incident precipitation associated with a particular precipitation event is negatively related to precipitation amount during that event.

[c] Range of canopy interception loss for individual rain events.

[d] Fraction of summer precipitation.

[e] Fraction of winter precipitation.

before that water can enter that soil, and ice within a soil must melt before that water can be absorbed by roots.

A main feature of predicted increasing-$[CO_2]_a$-induced climatic change is warming at the earth's surface. That is, present atmosphere or coupled atmosphere–ocean general circulation models (GCMs) agree that an increasing-$[CO_2]_a$-induced global warming is probable (IPCC, 1996). In addition to warming, GCMs predict an increase in global mean precipitation, although it is also predicted that precipitation will decrease in some regions due to increased $[CO_2]_a$ and related gases and aerosols (IPCC, 1996). GCMs differ greatly with respect to predictions of future precipitation in specific regions, and as a result, future precipitation patterns are largely uncertain.

In general, an increase in precipitation in a region is likely to increase the amount of water added to soils. It might also increase the amount of canopy and litter interception losses. As a first approximation, an increase in precipitation may be expected to reduce ecosystem water stress and enhance NPP (cf. Miami model), although other climatic changes associated with increasing $[CO_2]_a$ may alter such a cause-and-effect relationship.

Importantly, warming may elicit an increase in demand for ET. The GCMs predict that relative humidity will be fairly stable with increasing $[CO_2]_a$. This would increase water vapor content of the atmosphere with warming, but it would also increase the amount of additional water vapor that the atmosphere could hold and would support more ET. Many GCMs therefore predict drier surface soil in summer for the northern midlatitudes due to enhanced ET (IPCC, 1996). But when atmospheric aerosols—not just CO_2—are included in some GCMs, predicted summer soil moisture in North America and Europe increases with elevated $[CO_2]_a$ levels (IPCC, 1996). Terrestrial hydrology submodels and predictions of soil moisture, however, are generally overly simplistic in GCMs. Moreover, typical GCM grid cells are so large (e.g., several degrees latitude by several degrees longitude) that they say little about the effects of increasing $[CO_2]_a$ on the many important local processes determining weather, or about which ecosystems in a region might experience significant changes in precipitation or demand for ET. Present GCMs are, nonetheless, the best available tools for predicting future climate, and they provide numerical predictions that may be useful in considering general future climatic changes, including some amount of warming and potentially altered patterns of precipitation and ET. It remains unclear, however, whether climatic change itself will favor greater ecosystem water stress, reduced water stress, or no net change in water stress for any particular region.

Increasing $[CO_2]_a$ itself, as opposed to any resulting climatic change, can have a direct effect on transpiration through its effects on stomatal aperture (Section III). Growth responses to elevated $[CO_2]_a$ can also be important to transpiration, and they might affect surface and subsurface flow, interception loss, and soil evaporation too (Section IV). In short, plant responses to increasing $[CO_2]_a$ can affect all aspects of terrestrial ecosystem water use and, by extension, water stress.

III. Stomatal Response to CO_2 Concentration

It is nearly axiomatic that an increase in $[CO_2]$ around a leaf causes partial stomatal closure (e.g., Morison, 1987), although this is not always the case (Amthor, 1995; Field *et al.*, 1995). When stomates do close in response to elevated $[CO_2]$, the result is a decrease in gaseous diffusion through stomatal pores.

Stomata respond to substomatal cavity CO_2 concentration (C_i) rather than leaf-surface or ambient $[CO_2]$ (Mott, 1988), although C_i and $[CO_2]_a$ are related, and this often results in a strong relationship between $[CO_2]_a$ and stomatal conductance. The mechanism of action of CO_2 on stomatal opening and closing is unknown. Because factors stimulating photosynthesis tend to reduce C_i (because the CO_2 in substomatal cavities is then used more rapidly in photosynthesis), those same factors tend to drive stomatal opening, and vice versa. Apparently, C_3 and C_4 species respond similarly to a short-term change in C_i (e.g., Morison and Gifford, 1983).

Transpiration at the individual-leaf scale is the product of leaf conductance and the water vapor mole fraction (W) difference between the air in leaf intercellular spaces and the air outside the leaf boundary layer. Leaf conductance is composed of stomatal, cuticular, and boundary layer components. Thus, a reduction in stomatal conductance can lead to an approximately proportional reduction in leaf transpiration when (1) leaf boundary layer conductance is much larger than leaf surface conductance, (2) cuticular conductance is much smaller than stomatal conductance, (3) a change in transpiration does not significantly alter the humidity of the air enveloping the leaf, and (4) leaf temperature is unchanged. These conditions are usually met in a well-mixed leaf cuvette used in plant physiological research when it receives a steady supply of preconditioned air. This means that in most leaf-cuvette experiments, any stomatal response to elevated $[CO_2]_a$ results in a nearly proportional change in transpiration rate. A reduction in transpiration rate will in turn increase leaf water potential (i.e., make it less negative) and may therefore limit leaf water stress, at least for a time (Hsiao, 1973; but see Bhattacharya *et al.*, 1994, for results from a field experiment involving elevated $[CO_2]_a$). If a reduction in individual-leaf transpiration leads to a reduction in whole-plant transpiration, the soil water use rate will be slowed.

Many environmental factors can induce stomatal closure, not just elevated $[CO_2]_a$. For example, low soil water content can lead to stomatal closure (Jones, 1992). This response may be in part hormonally mediated. It is a form of negative feedback on soil water use in transpiration. That is, as soil moisture declines, stomates close and slow transpiration, which slows the rate of soil water use. In water-limited ecosystems, a reduction in stomatal conductance brought about by elevated $[CO_2]_a$ might delay the onset of water stress and the stomatal closure resulting from water stress itself. But even with elevated $[CO_2]_a$, soil moisture *in water-limited ecosystems* may still eventually limit plant growth. Nonetheless, a delay in the onset of water stress provides plants with greater opportunities for growth (Hättenschwiler *et al.*, 1997).

In general then, stomatal closure will reduce individual-leaf water use through transpiration, and in so doing, perhaps delay or prevent water stress, depending in part on whole-plant and community responses to stomatal closure (see Section V). If reduced stomatal conductance is not accompanied (or caused) by an increase in $[CO_2]_a$, then photosynthesis, and probably growth, will also be limited because the daytime uptake of CO_2 will be slowed. With elevated compared to present ambient $[CO_2]_a$, however, stomatal closure can reduce transpiration even while photosynthesis is stimulated. This is because an increase in the CO_2 mole fraction gradient from outside to inside a leaf may more than compensate for reduced leaf conductance of CO_2 resulting from stomatal closure due to an increase in $[CO_2]_a$. This is a win–win process; water stress is potentially reduced or delayed while potential growth, set by photosynthate production, is enhanced. Even if stomatal closure and the increase in the CO_2 mole fraction gradient from outside to inside the leaf balance each other with the result that photosynthesis is unchanged, water use per unit of photosynthate generation is reduced and WUE is increased. And even for C_4 species, for which photosynthesis is nearly CO_2-saturated at present ambient $[CO_2]_a$, WUE generally increases with an elevated-$[CO_2]_a$-induced closure of stomata. This is the essence of the most significant effect of increasing $[CO_2]_a$ on the relationship between water use and photosynthesis: Water is used more efficiently—with respect to CO_2 assimilation—as $[CO_2]_a$ increases, and this increased efficiency may be coupled to delay, or perhaps prevention, of the onset of water stress. The implications for improved ecosystem NPP and water yield are legion. But this begs the question: Do individual-leaf responses to elevated $[CO_2]_a$ translate into significant changes in whole-plant or ecosystem-scale processes or, more specifically, does increasing $[CO_2]_a$ slow ecosystem transpiration? Because the regulation of individual-leaf and whole-canopy or landscape water use can differ, it follows that the answer is complex.

IV. Effects of Plant Growth Responses to Elevated CO_2 Concentration on Water Use

Because elevated $[CO_2]_a$ stimulates photosynthesis in C_3 species, it can in turn enhance C_3 plant growth (e.g., Ceulemans and Mousseau, 1994; Rogers *et al.*, 1994; Amthor and Koch, 1996). One component of increased plant growth is larger, often deeper, roots (e.g., Prior *et al.*, 1994a,b; Rogers *et al.*, 1994; and see the insightful review by Norby, 1994, concerning root responses to elevated $[CO_2]_a$). It is unnecessary for root-shoot partitioning to be affected by elevated $[CO_2]_a$ in order for root growth to be stimulated. Indeed, a constant root : shoot combined with increased whole-plant growth leads directly to increased root growth, although elevated $[CO_2]_a$ does often affect root : shoot (e.g., Rogers *et al.*, 1996).

Increased root growth can enhance plant access to soil water. This has the potential to reduce water stress directly and to increase water use. In the long-term, however, soil water used by plants must be replenished, so long-term alleviation of water stress via greater root growth will operate only when water normally lost from the "root zone" (via subsurface flow to streambeds, for example) is absorbed by new, larger, or deeper roots. Thus, increased root growth can potentially increase transpiration at the expense of streamflow and deep seepage. Increased root growth due to elevated $[CO_2]_a$ might therefore reduce watershed water yield (i.e., water leaving a watershed in streamflow) if other factors are unchanged. Many other factors can be changed, however.

Another aspect of increased plant growth in elevated $[CO_2]_a$ that might affect ecosystem water use is increased leaf area index (LAI, which is leaf area per unit ground area). Longer periods of leaf display in cold-deciduous ecosystems brought about by increasing-$[CO_2]_a$-induced global warming might also alter water use. Either response could enhance transpiration because transpiration rate is related to leaf area and leaf duration. Few experimental data bear on the issue of increased leaf area or leaf duration at the ecosystem scale due to elevated $[CO_2]_a$; the most comprehensive data are for crops. In *Gossypium hirsutum* (cotton), relative LAI increases due to elevated $[CO_2]_a$ may be more prevalent with water stress than in well-watered crops (Pinter *et al.*, 1996). In *Triticum aestivum* (wheat), LAI may be slightly enhanced by elevated $[CO_2]_a$, and again the relative enhancement can be positively related to the degree of water stress (Kimball *et al.*, 1995; Pinter *et al.*, 1996; B. A. Kimball and P. J. Pinter, Jr., personal communications). Arguments for a relatively small effect of elevated $[CO_2]_a$ on LAI increase in many ecosystems are outlined by Field *et al.* (1995).

An increase in transpiration resulting from increased root and (or) leaf growth or duration might contribute to more frequent, or earlier, water stress in water-limited ecosystems. Stomatal closure in elevated $[CO_2]_a$, however, might compensate for greater access to water by larger root systems or greater potential transpiration by larger LAI or longer leaf duration. Data are scarce, but increased $[CO_2]_a$ may actually reduce soil water extraction by some plant communities, in spite of increased plant growth (e.g., Morgan *et al.*, 1994; Hungate *et al.*, 1997). This will tend to enhance subsurface flow to streambeds and deep seepage, thus potentially increasing watershed water yields. It may also delay water stress. In circumstances resulting in mild, short episodes of water stress in an ambient $[CO_2]_a$, an increase in $[CO_2]_a$ might even eliminate the occurrence of water stress all together.

Evaporation of soil water is not affected directly by $[CO_2]_a$. It is, however, affected by plant coverage of the soil, which affects micrometeorological conditions near the soil surface. Consequently, any change in canopy development caused by increased $[CO_2]_a$ may affect soil water evaporation. For example, increased early season leaf area growth (e.g., Kimball *et al.*, 1995) in elevated $[CO_2]_a$ might limit soil evaporation at that same time. Moreover, soil water evaporation might be affected by any climatic changes (e.g., warming) that are associated with, or caused by, increasing $[CO_2]_a$. In general, soil evaporation is a small component of the water balance of ecosystems with dense vegetation. Significant relative changes (if any) in soil evaporation due to increasing $[CO_2]_a$ might therefore be expected primarily in ecosystems with sparser vegetation where significant changes in ground cover might occur. These ecosystems, however, are generally dry and often already have slow soil evaporation rates.

A change in canopy structure (e.g., leaf, branch, or stem area) that occurs due to elevated $[CO_2]_a$ could alter interception losses and therefore affect water supply to soils. I am unaware of experimental field data related to this issue.

V. Considerations in Scaling Up Effects of Atmospheric CO_2 Concentration on Stomatal Conductance to Effects on Ecosystem and Regional Transpiration

The simplest assumption about scaling up a stomatal response to increasing $[CO_2]_a$ and its effect on ecosystem transpiration and water use is that ecosystem transpiration decreases in proportion to the degree of stomatal closure. For example, a long-term doubling of present $[CO_2]_a$ may cause a 20–25% reduction in daytime individual-leaf stomatal (actually leaf-surface) conductance by tree species (reviewed in Field *et al.*, 1995), and under the simplest assumption, such a decrease in leaf-surface conductance by all the leaves in a forest canopy would result in a 20–25% reduction in forest ecosystem transpiration, other factors being equal. This could conserve soil water, that is, increase its duration in the soil, and so reduce or delay water stress.

In this discussion, it is assumed that elevated-$[CO_2]_a$-induced stomatal closure occurs at the whole-plant, canopy, and landscape scales. For present purposes, any stomatal acclimation (phenotypic adjustment) or adaptation (genotypic adjustment) to long-term elevated $[CO_2]_a$ is ignored (see references in Amthor, 1995; Santrucek and Sage, 1996); the focus is on scaling up the transpirational response to a given reduction in leaf-surface conductance as a result of an increase in $[CO_2]_a$.

Ecosystem transpiration is affected by more than leaf-surface conductance, however. Factors affecting ecosystem transpiration include canopy temperature (and therefore W in leaf intercellular spaces), total leaf-surface conductance of all the leaves in a canopy per unit ground area (which is called *canopy conductance* herein), conductance of the boundary layers of all the leaves in a canopy, aerodynamic conductance within and above the canopy, and W of the air above the ecosystem. At the landscape to regional scales, W in the well-mixed planetary boundary layer (PBL) is a relevant measure of above-ecosystem water vapor. (The PBL is thin early in the morning, grows during the day to perhaps a few kilometers of thickness, and then contracts in the evening and night.) Also at the landscape scale, aerodynamic conductance from outside the boundary layer of individual leaves to the bulk, well-mixed PBL can be a significant regulator of transpiration. The structure of the underlying ecosystem(s) affects the aerodynamic conductance and the coupling of the vegetation to the PBL (see, e.g., Jarvis and McNaughton, 1986) and this can be important to scaling up changes in individual-leaf stomatal conductance to landscape-scale transpiration. For example, short, smooth vegetation such as grassland or pasture is relatively poorly coupled with the atmosphere, and a fractional reduction in canopy conductance is expected to result in a smaller fractional reduction in transpiration (other factors being equal). Leaves in tall, rough ecosystems such as forests, on the other hand, are more tightly coupled to the atmosphere, and a fractional decrease in canopy conductance can be expected to result in a nearly proportional decrease in transpiration (again, with other factors equal). Note, however, that simulations by Martin *et al.* (1989) indicate an equal or greater sensitivity of transpiration to a change in stomatal conductance in grassland compared to broad-leaved forest. Also, other factors are rarely equal and because various feedback processes can operate between transpiring vegetation and the atmosphere, scaling up by Nature—as compared with simple human-developed conceptual or numerical models—can be multifarious.

Several physical feedback processes may be important, and they do not operate in isolation of each other or in isolation of many plant physiological processes. Some of these are discussed in more detail elsewhere (e.g., De

Bruin, 1983; Brutsaert, 1986; Jarvis and McNaughton, 1986; McNaughton and Spriggs, 1986; Shuttleworth, 1988; McNaughton, 1989; Dickinson, 1991; McNaughton and Jarvis, 1991; Jacobs and De Bruin, 1992; Jacobs, 1994; Monteith, 1995b; Raupach, 1995). The crux of accounting for important feedbacks in the scaling up of transpiration and water use is consideration of one- and two-way links between decreases in stomatal conductance and larger scale (canopy to landscape to region) transpiration and atmospheric state. That is to say, canopy transpiration affects the canopy environment, and extensive areas of vegetation influence the state of the PBL. At even larger scales, ecosystem physiological processes influence the whole troposphere.

If elevated $[CO_2]_a$ causes stomatal closure and reduces transpiration, leaf and canopy temperatures will tend to rise due to reduced latent heat loss. This affect of stomatal closure and reduced transpiration itself can partly compensate for the stomatal-closure-induced reduction in transpiration. An increased canopy temperature will increase W within leaves—air inside leaves is generally saturated with water vapor and saturation vapor pressure is a positive function of temperature—and this in turn will increase the gradient in W from inside leaves to the bulk PBL. The increased gradient in W will stimulate transpiration. (Leaf temperature is generally unaffected by stomatal closure in elevated-CO_2 leaf-cuvette experiments because leaf temperature is then controlled by an exaggerated convective energy exchange.)

If transpiration is reduced at the regional scale, the PBL will tend to be relatively dry. The PBL will then also tend to be relatively warm because of an increase in sensible heat exchange at the underlying plant-canopy surface; that is, for a given radiation input, a reduction in transpiration (i.e., latent heat exchange) will increase the sensible heat flux from the underlying ecosystems into the PBL as well as warm the underlying ecosystems. A relatively dry, warm PBL will stimulate transpiration for a given canopy conductance because of a large gradient in W and will also act as a negative feedback on a stomatal-closure-induced reduction of regional transpiration. This feedback mechanism was called *PBL feedback* by Jacobs (1994). This PBL feedback is more important at the regional scale than at the individual ecosystem scale. It can partially compensate for the effects of lower stomatal conductance on large-scale transpiration rates. At least it can in mathematical models.

Warming of the ecosystems underlying the PBL due to stomatal closure results in a decrease in net radiation ($\mathbf{R}_n$, W m^{-2}) because the outgoing ecosystem long-wave radiation flux is increased. This reduces the sum of sensible and latent heat exchanges needed to balance $\mathbf{R}_n$. This might limit transpiration in some ecosystems and therefore may act as a negative feedback on the PBL feedback, reducing somewhat the quantitative significance of the PBL feedback. But, a reduction in canopy conductance and increased sensible heat flux into the atmosphere from the warmer underlying ecosystems can accelerate PBL growth and the entrainment of dry (and often warm) air from above the PBL. This tends to increase the PBL water vapor deficit and might therefore partially counteract reductions in large-scale transpiration due to stomatal closure.

Drying of the air in and above ecosystems brought about by elevated-$[CO_2]_a$-induced stomatal closure might have other effects on transpiration, including a positive feedback on stomatal closure. Because "the stomata of most plants close as the vapor pressure deficit between the leaf and the surrounding air increases" (Mansfield and Atkinson, 1990; and see Monteith, 1995a), a decrease in canopy conductance could itself lead to a further decrease in canopy conductance due to drying of the air surrounding leaves. Thus, even as drying of air can stimulate transpiration due to a larger W gradient from inside leaves to the bulk PBL, that same drying can also lead to further stomatal closure in a positive feedback. This positive feedback is likely to be small in smooth vegetation with a large canopy conductance.

The likely overall influence of feedbacks between transpiring vegetation and the atmosphere is to limit somewhat the effects of increasing $[CO_2]_a$ on landscape-to-regional scale transpiration compared with effects on canopy conductance (see also Section VII). Quantitative effects of increasing $[CO_2]_a$ on regional canopy conductance and the resulting positive and negative feedbacks on transpiration are, however, uncertain. Feedbacks are likely to be affected by vegetation type and other factors, and elevated-$[CO_2]_a$-induced stomatal closure may have different effects on water use in different ecosystems and regions. For example, a fractional decrease in canopy conductance is likely to have relatively small effects on transpiration in aerodynamically smooth canopies with large canopy conductance, irrespective of most of the potential feedbacks outlined earlier. In aerodynamically rough forests, on the other hand, a change in canopy conductance can bring with it a nearly proportional change in transpiration, and PBL feedbacks may therefore be of greater relative significance. In dry ecosystems, with low leaf-surface conductance, feedbacks on stomatal control of transpiration are likely to be small (McNaughton and Jarvis, 1991). In those ecosystems, transpiration is controlled to a large extent by water availability. It is in those ecosystems, I think, that elevated-$[CO_2]_a$-induced stomatal closure could delay the onset of water stress, but perhaps only briefly.

Landscape heterogeneity must also be considered in scaling up ecosystem water use (see, e.g., Raupach, 1995) in response to increasing $[CO_2]_a$. Suffice it to say here that many landscapes are not uniform, but are composed of a variety of vegetation types, structures, and sizes. Land surface heterogeneity across landscapes can have significant effects on the coupling of the component ecosystems to the atmosphere through effects on the structure of the lower atmosphere. A related issue is that of advection. In many analyses of scaling up of the control of transpiration by stomatal conductance, advection is ignored (see, e.g., Raupach, 1995), even though it can be important. This may be particularly significant for irrigated crops in arid regions.

VI. Effects of Atmospheric CO_2 Concentration on Plot-Scale Water Use: Experimental Results

Experimental manipulation of $[CO_2]_a$ at the scale of the ecosystem, landscape, or region is impossible with present technology and resources, but several techniques are available for controlled manipulation of $[CO_2]_a$ at the plot scale. Perhaps the most important method available for studying short-term (months to a few years) plot-scale plant responses to elevated $[CO_2]_a$ in the field is the free-air CO_2 enrichment (FACE) technique (Allen *et al.*, 1992). This is a modification of open-air exposure systems used with air pollutants (McLeod, 1993). The most comprehensive studies to date using FACE were in southern Arizona, USA, with cotton (1989, 1990, 1991) and spring wheat (1993, 1994, 1996, 1997; wheat was sown in December before the year of each experiment) (e.g., Kimball *et al.*, 1995; Pinter *et al.*, 1996). Most of the water use studies in those experiments were for well-watered plots. Mean daytime $[CO_2]_a$ of the air above FACE plots was 550–570 ppm; it was 360–370 ppm in the air above the ambient-$[CO_2]$ plots (e.g., Kimball *et al.*, 1995). In both cotton and wheat, FACE during the daytime caused stomatal closure and increased leaf temperature (Pinter *et al.*, 1996).

According to measurements made with canopy chambers covering 1 m^2 of crop within the FACE plots, midday cotton ET was unaffected by elevated $[CO_2]_a$ during the 1989 experiment, in which the crop was well watered (Hileman *et al.*, 1994). Neither midday nor daily-total cotton ET measured with canopy chambers was affected by elevated $[CO_2]_a$ in the 1990 experiment, except late in the season when ET was slightly reduced by FACE (Hileman *et al.*, 1994). It is possible that the effects of decreased stomatal conductance on individual-leaf transpiration due to FACE were counteracted by increased leaf area early in the 1990 growing season (see Hileman *et al.*, 1994; Pinter *et al.*, 1994) although feedbacks through elevated canopy temperature may also have been important. Cotton transpiration per plant (which due to uniform plant density can be expressed equally per unit ground area) was measured by sap flow gauges. It was unaffected by FACE in the 1990 and 1991 experiments (Dugas *et al.*, 1994). The measured plants were well watered and supplied with ample nutrients. In the same experiments, a soil water balance analysis indicated no effect of FACE on weekly or seasonal whole-plot ET (Hunsaker *et al.*, 1994). This was true for well-watered and water-stressed plots during both years. With a fourth method of estimating cotton crop ET (i.e., an energy balance analysis), a small *enhancement* of ET in 1991 due to FACE was calculated. That enhancement was at the limit of discernment of the method, however, and it was concluded that any affects of FACE in 1991 "on the ET of cotton were too small to be detected" (Kimball *et al.*, 1994).

Generalizing the results from four measurement methods, some used in multiple years, neither ET nor its component transpiration was significantly affected by 550-ppm-CO_{2a} FACE treatment in cotton supplied with ample soil water or subjected to soil water stress. Whole-plant growth and economic yield were stimulated by FACE for both well-watered and water-stressed cotton crops (Pinter *et al.*, 1996), indicating that WUE was enhanced even though water use was unaffected. At present, I discourage extrapolating results of cotton stomatal and transpirational responses to FACE to other species because stomates of modern cotton cultivars may be less sensitive to $[CO_2]_a$ compared with other species (e.g., cf. Radin *et al.*, 1987, with Morison, 1987.

In both cotton and wheat subjected to water stress, FACE sometimes led to less negative leaf water potential compared to ambient $[CO_2]_a$ plots (Pinter *et al.*, 1996). This implies a partial alleviation of water stress in leaves by elevated $[CO_2]_a$ that might have applicability to other water-stressed vegetation.

Midseason daily $\mathbf{R}_n$ was reduced 4% by FACE in the 1993 well-watered wheat experiment (Kimball *et al.*, 1995). The reduction in $\mathbf{R}_n$ was accompanied by an increase in sensible heat flux—the daytime canopy was warmer with FACE than without—and an 8% decrease in latent heat flux, as determined by energy balance analysis (Kimball *et al.*, 1995; see also Senock *et al.*, 1996, for sap flow measurements in those experiments, and a discussion of methodologic difficulties). A decrease in ET of 11% due to FACE was reported for well-watered plots in the 1994 experiment according to an energy balance (Kimball *et al.*, as cited by Pinter *et al.*, 1996). According to a soil water balance, however, FACE reduced seasonal ET only 4.5% (1993) and 5.8% (1994) in well-watered plots (Hunsaker *et al.*, 1996). As opposed to cotton, effects of reduced wheat stomatal conductance on crop transpiration in well-watered FACE plots were not completely compensated for by increased leaf area, with the result that transpiration was reduced by FACE (Kimball *et al.*, 1995). In water-stressed wheat, however, seasonal water use was *increased* 4.8% in 1993 and 0.9% in 1994 by FACE compared to ambient $[CO_2]_a$ plots (Hunsaker *et al.*, 1996). This was attributed to increased access to soil water due to larger root systems (Hunsaker *et al.*, 1996; Pinter *et al.*, 1996) as outlined earlier. These results emphasize the complexity of scaling up from changes in individual-leaf stomatal conductance to transpiration by canopies and communities; elevated $[CO_2]_a$ can alter transpiration by means other than that of changing stomatal conductance.

Wheat growth was stimulated by FACE in each of the experiments (Kimball *et al.*, 1995; Pinter *et al.*, 1996; P. J. Pinter, Jr., and B. A. Kimball, personal communications), which resulted in an increase in WUE expressed on a grain yield basis (Hunsaker *et al.*, 1996). Season-long water use and growth of wheat exposed to FACE in 1993 and 1994 were complicated, however, by a significant decrease in time to crop maturity in the well-watered FACE plots (6–7 d in 1993; Kimball *et al.*, 1995) but a much smaller reduction in time to maturity in the water-stressed FACE plots (1 d in 1993; Kimball *et al.*, 1995) compared with ambient $[CO_2]_a$ plots. Thus, the larger percentage grain growth stimulation due to FACE in the water-stressed plots compared with the well-watered plots (e.g., Pinter *et al.*, 1996) might have been the result of an interactive effect of FACE and water stress on crop development. It is now apparent that in those experiments the FACE apparatus per se—as opposed to elevated $[CO_2]_a$—significantly affected wheat development and senescence, resulting in earlier maturity in the well-watered FACE plots compared with the well-watered ambient $[CO_2]_a$ plots (P. J. Pinter, Jr., and B. A. Kimball, personal communications).

Present controlled manipulations of $[CO_2]_a$ are limited to the plot (or smaller) scale. Free-air CO_2 enrichment is one useful way (among others) of studying plant community responses to elevated $[CO_2]_a$—witness the important work summarized by Kimball *et al.*, (1995), Pinter *et al.* (1996), and references therein—but it does not allow for large-scale environment–plant interactions. Although a FACE experiment can allow the operation of some feedbacks on the transpirational response to elevated-$[CO_2]_a$-induced stomatal closure, other potentially important feedbacks cannot be expressed. For example, effects of stomatal closure on plot-scale energy balance and transpiration can be studied (e.g., Kimball *et al.*, 1992, 1995), but any effects of elevated $[CO_2]_a$ in FACE experiments on altered canopy micrometerorology, including leaf temperature, are only at the scale of small plots (e.g., 0.0005 km^2; see Kimball *et al.*, 1995) rather than at the scales of fields, forests, landscapes, or regions. Thus, FACE experiments are unable to address the issue of scaling up from the plot to the landscape, where additional and important feedback mechanisms may occur (Section V). That is to say, only small-scale canopy micrometeorologic factors can be studied with FACE experiments and these will be limited to the particular plant community being studied in the experiment rather than the assembly of plant communities and ecosystems that make up a landscape. Consequently, FACE experiments cannot address regional stomatal conductance changes and PBL feedbacks. Moreover, because small plots are used in

FACE experiments, plants within the areas treated with elevated $[CO_2]_a$ can be exposed to atmospheric conditions set largely by the surrounding landscape, which is existing in (and responding to) an ambient $[CO_2]_a$. In general, data concerning water use and WUE derived from any plot-scale experiment are applicable to larger scales only with consideration of the larger scale feedback processes.

As stated earlier, the effects of stomatal closure on transpiration by crops (smooth vegetation) is expected to be generally small relative to many other ecosystems. Nonetheless, canopy conductance may be an important regulator of transpiration in irrigated crops in Arizona because of the generally advective conditions, landscape heterogeneity, and low atmospheric humidity there. Also, advection of sensible heat or water vapor can complicate the scaling up of plot-scale field experiments (see, e.g., Monteith, 1995b), which may be particularly important for extrapolation from irrigated crops grown in dry regions—such as the Arizona FACE experiments with cotton and wheat—to other biomes. In any case, these experiments are an important first step; they are not the last word on the effects of elevated $[CO_2]_a$ on plot-scale or ecosystem-scale water use and water stress.

VII. Scaling Up with Global Atmospheric Models

One tool for assessing the large-scale effects of $[CO_2]_a$ on transpiration is the atmospheric GCM, *when it contains plant physiological responses to $[CO_2]_a$ and other atmospheric variables.* An example is the SiB2-GCM used by Sellers *et al.* (1996) to evaluate effects of a doubled $[CO_2]_a$ on global photosynthesis, transpiration, surface air temperature, and precipitation. The model was implemented with a very coarse spatial resolution. In the simulations, doubled $[CO_2]_a$ (700 vs. 350 ppm) increased global photosynthesis on land, as expected. It also reduced canopy conductance 24–34% on a global basis. Effects of doubled $[CO_2]_a$ on ET, however, were small. Indeed, doubled $[CO_2]_a$ *enhanced* global terrestrial ET by 0.3–3.3%, but in W m^{-2} rather than amount of water. (Herein, I consider results of only the RP and RPV experiments of Sellers *et al.* because they are the most relevant to $[CO_2]_a$ increase.) Clearly, reductions in simulated global canopy conductance due to doubled $[CO_2]_a$ did not result in comparable reductions in simulated latent heat exchange with this implementation of this GCM.

For a doubled $[CO_2]_a$, the SiB2-GCM simulations predicted 3.0–6.5% increases in precipitation over land when aggregated globally (Sellers *et al.*, 1996). This might reduce water stress depending on the seasonal timing of the additional precipitation. In the simulations, global mean land surface air temperature increased with doubled $[CO_2]_a$ by 2.7–2.8°C.

Bear in mind that GCMs are crude, with large grid cells that make it difficult to account for many aspects of spatial heterogeneity at the landscape scale. Also, different GCMs predict different effects of increasing $[CO_2]_a$ on transpiration and surface energy balance (see, e.g., Notes in Sellers *et al.*, 1996). GCMs may prove to be important in understanding CO_2–water interactions in terrestrial ecosystems, but not without supporting field-based ecological research. In particular, GCMs account for plant water stress poorly, if at all, and this topic needs better model–experiment collaborative work.

VIII. Forest Water Use and Atmospheric CO_2 Concentration Increase during the Past Several Decades: The Real Thing

Because there are no large-scale controlled manipulations of $[CO_2]_a$, there is no firsthand knowledge of large-scale water use responses to an experimental change in $[CO_2]_a$. A glimpse may be gained, however, into ecosystem-scale water use responses to $[CO_2]_a$ by examining watershed-scale water use over time as global $[CO_2]_a$ increases. That is not to say that the increase in $[CO_2]_a$ during the past 150–200 years is an ecological experiment; it is not, inadvertent or otherwise. Nonetheless, perhaps something can be learned from the past in order to better predict the future.

To estimate the effects of a global $[CO_2]_a$ increase on ecosystem water use, and in particular ET, I analyzed relationships between global $[CO_2]_a$ data and hydrologic data from small forested watersheds (Amthor, 1998). The watershed hydrology data are from the Hubbard Brook Experimental Forest (HBEF) in the White Mountain National Forest of central New Hampshire, USA (for details, see Bormann and Likens, 1979); Likens and Bormann, 1995). Precipitation (ca. 30% of which is snow) and streamflow are measured in eight watersheds (designated W1–W8) in the HBEF. These watersheds are well suited to the study of ecosystem ET. Soils are shallow, with low water storage capacity, and fill with water each spring during snowmelt. Bedrock beneath the soil is thought to be nearly watertight and the valleys are rimmed by walls of bedrock (Likens and Bormann, 1995). Because of this, precipitation is nearly completely disposed of through local streamflow and ET; deep seepage is minor. Gauging weirs used to monitor streamflow are built on bedrock exposures in streambeds. Thus, in the HBEF, the difference between annual precipitation and streamflow is a good measure of annual ecosystem ET, although transpiration, interception loss, soil evaporation, and sublimation of snow cannot be distinguished by this method. According to Likens and Bormann (1995), it is useful to define the HBEF "water-year" as beginning 1 June and ending the following 31 May, and that is how this analysis is based.

The first hydrologic measurements at the HBEF were made in January 1956 (in W1), and they continue today. Whole-watershed biomass manipulation experiments were conducted in W2, W4, and W5 during the measurement period and because of this these watersheds are excluded from my analysis. The five other watersheds have not been significantly disturbed since about 1917, when salable timber was removed from the site. The HBEF represents a typical unevenly aged and well-stocked aggrading northern hardwood forest (Bormann and Likens, 1979). Aboveground biomass accumulation in the HBEF averaged 485 g m^{-2} yr^{-1} during 1965–1977, but declined to 89 g m^{-2} yr^{-1} during 1987–1992 (Likens and Bormann, 1995).

Almost concurrently with these small-watershed-scale hydrologic measurements, continuous measurements of $[CO_2]_a$ were made—beginning in 1958—at the Mauna Loa Observatory, Hawaii (e.g., Keeling, 1960; Keeling *et al.*, 1976; Keeling and Whorf, 1994, 1997). For correspondence with HBEF water-years, I averaged Mauna Loa $[CO_2]_a$ values over the same 12-month periods used to sum the HBEF hydrologic variables into water-year totals. I used $[CO_2]_a$ values given by Keeling and Whorf (1997). Mean $[CO_2]_a$ at Mauna Loa for the HBEF water-years increased monotonically during the 1958–1996 period. Five monthly mean values of $[CO_2]_a$ at Mauna Loa, from a total of 456 used in this analysis, are missing from the record of Keeling and Whorf (1997); I interpolated values for those five months.

This analysis covers the period from June 1956 to May 1996, but only W1 has a complete record for the period. Also, Mauna Loa $[CO_2]_a$ measurements are not available for the HBEF water-years beginning 1 June 1956 and 1957. To extend the $[CO_2]_a$ data to include the first two water-years, I interpolated the Mauna Loa $[CO_2]_a$ data and measurements of CO_2 trapped in Antarctic ice during the 1940s and 1950s (Etheridge *et al.*, 1996) to arrive at values of 314.0 and 314.7 for global $[CO_2]_a$ during the water-years beginning 1 June 1956 and 1957, respectively. The 1 June 1995–31 May 1996 mean Mauna Loa $[CO_2]_a$ was 361.7 ppm, so the analysis of W1 hydrology corresponds to a $[CO_2]_a$ increase from about 314 ppm to about 362 ppm—a 15% increase or more than half the increase in $[CO_2]_a$ since the preindustrial period (see Etheridge *et al.*, 1996).

If $[CO_2]_a$ has a marked effect on forest ET, it might be reflected in the HBEF data, although at most only a limited response is seen. Within a watershed, streamflow and precipitation are insignificantly correlated with $[CO_2]_a$ (Table II). As expected, annual streamflow is well related to precipitation, whereas ET is relatively insensitive to precipitation over the range of annual precipitation in the HBEF since 1956 (Fig. 1). There is little indication from this analysis that the 15% increase in $[CO_2]_a$ during the past four decades affected precipitation or streamflow in the HBEF.

Water-year ET in W1 and W3, the two longest records in this analysis, is negatively correlated with $[CO_2]_a$, though only the W3 correlation is

Table II Correlation of Water-Year Precipitation, Streamflow, and Evapotranspiration with Atmospheric CO_2 Concentration in Six Watersheds in the Hubbard Brook Experimental Forest[a]

	Correlation[c] with atmospheric CO_2 concentration					
Watershed[b]	Precipitation	Streamflow	Evapotranspiration	Area (km^2)	Elevation (m)	Years in analysis[d]
1	+0.19 (0.24)	+0.24 (0.13)	−0.24 (0.14)	0.118	488–747	40
3	+0.21 (0.21)	+0.30 (0.07)	−0.47 (0.003)	0.424	527–732	38
6	+0.15 (0.42)	+0.19 (0.28)	−0.03 (0.85)	0.132	549–792	32 (33)[e]
7	+0.13 (0.48)	+0.13 (0.50)	+0.07 (0.71)	0.764	619–899	31
8	−0.01 (0.97)	+0.02 (0.90)	+0.11 (0.55)	0.594	610–905	27 (28)[e]

[a] Annual totals of precipitation, streamflow, and evapotranspiration (given by precipitation minus streamflow) are for water years beginning on 1 June and ending on 31 May (see Likens and Bormann, 1995). The CO_2 concentrations used in this analysis were derived mainly from measurements at Mauna Loa Observatory, Hawaii, and are the mean value for the same 12-month water years (see text).

[b] Forest co-dominants include *Fagus grandifolia* (American beech), *Betula alleghaniensis* (yellow birch), and *Acer saccharum* (sugar maple). Watersheds 1, 3, and 6 are south facing. Watersheds 7 and 8 are north facing and include significant numbers of *Picea rubens* (red spruce), *Abies balsamea* (balsam fir), *B. papyrifera* (paper birch), and *Tsuga canadensis* (Eastern hemlock).

[c] The correlation coefficient is given. The probability that the variable is uncorrelated with $[CO_2]_a$ is given in parentheses (*t* test).

[d] Each analysis ends in May 1996 (for correspondence with the water-year), so a 40-year analysis begins June 1956.

[e] The watershed 6 streamflow analysis includes 33 water-years but the precipitation analysis is for 32 water-years; the watershed 8 streamflow analysis includes 28 water-years but the precipitation analysis is for 27 years.

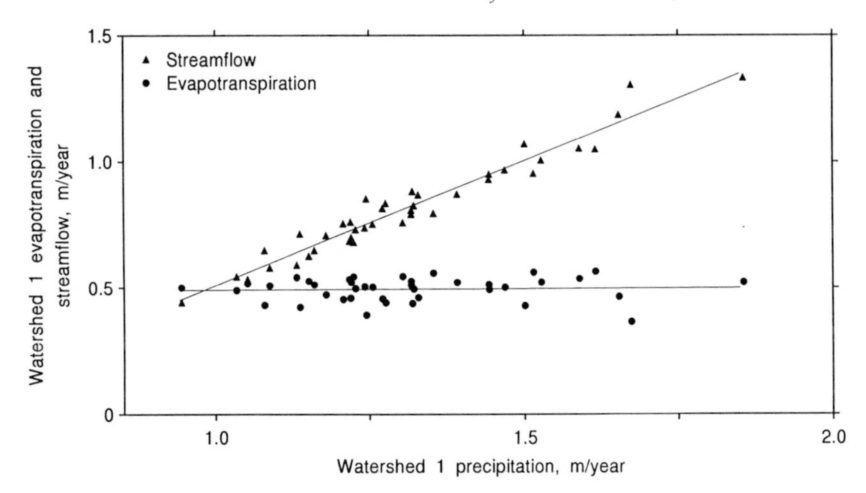

Figure 1 Water-year streamflow and evapotranspiration (given by precipitation minus streamflow) as a function of precipitation for water-years beginning 1 June in watershed 1 in the Hubbard Brook Experimental Forest. The symbols show values for 40 water-years (June 1956–May 1996). The solid lines show least-squares linear regressions. These results are similar to those of Bormann and Likens (1979) for a record about half as long as this (see their Fig. 2-6); over a nearly twofold range in water-year precipitation, water-year ET is almost constant in this forest.

statistically significant (Table II; and see regression equations in Fig. 2). Watershed 6 ET is also negatively, though insignificantly, correlated with $[CO_2]_a$ (Table II). On the other hand, ET in W7 and W8 is positively, but insignificantly, correlated with $[CO_2]_a$ (Table II). Because stomatal closure induced by increasing $[CO_2]_a$ (or by any other cause) in aerodynamically rough landscapes such as the White Mountains in New Hampshire may be expected to result in nearly proportional reductions in transpiration, it might be tempting to attribute the significant negative relationship between W3 water-year ET and $[CO_2]_a$ to elevated-$[CO_2]_a$-induced stomatal closure. But correlations between ET and $[CO_2]_a$ in the other four watersheds are insignificant. In short, trends are inconsistent among HBEF watersheds, with water-year ET in four of five watersheds apparently unaffected by a 15% increase in global $[CO_2]_a$. Moreover, annual ET by a *Picea sitchensis* (sitka spruce) plantation in western England (Stocks Reservoir; Calder, 1990) was *positively* related with a $[CO_2]_a$ increase from 1956 to 1967, but that time series is too short to be meaningful.

Any slowing of ET with an increase in $[CO_2]_a$, which may be the case for W3 in the HBEF, is not necessarily a result (or cause) of reduced water stress. This analysis for W3 itself sheds no light on inssues of water stress

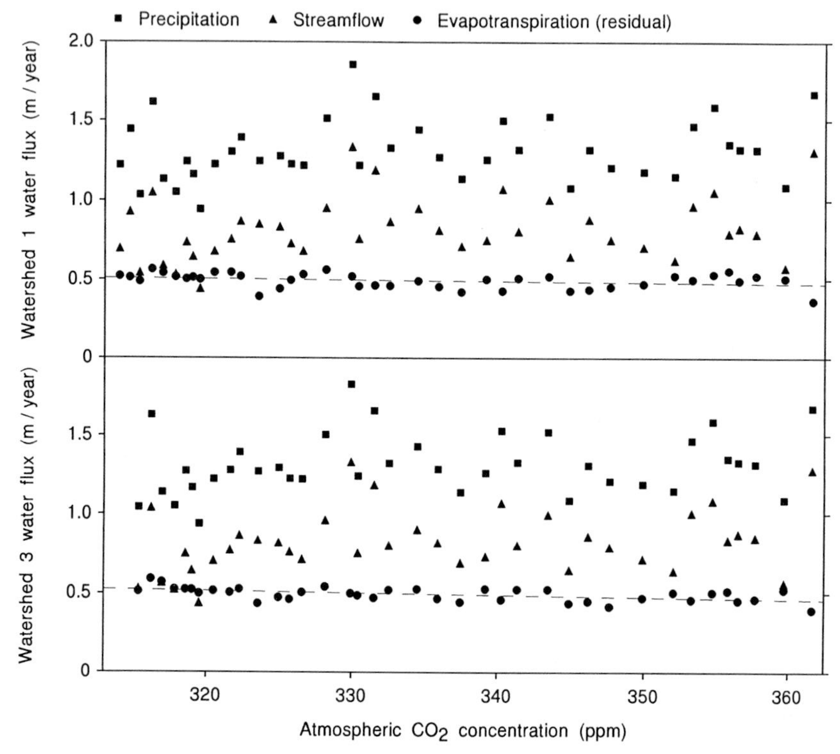

Figure 2 Water-year precipitation, streamflow, and evapotranspiration (given by precipitation minus streamflow) in watersheds 1 (W1; top panel) and 3 (W3; bottom panel) in the Hubbard Brook Experimental Forest. The dashed lines show regression equations for ET as a linear function of atmospheric CO_2 concentration, $[CO_2]_a$ (W1: ET = 0.74531 − 0.00075141 $[CO_2]_a$, $r^2 \approx 0.06$; W3: ET = 0.94659 − 0.0013463 $[CO_2]_a$, $r^2 \approx 0.22$). The probability that the slope of the linear regression line is equal to zero is ca. 0.14 for W1 and ca. 0.003 for W3 (two-tailed t tests). Additional statistics are given in Table II.

of WUE—it is related simply to watershed water use. Forest NPP must be included in the analysis for it to go further.

If increasing $[CO_2]_a$ does slow ET, it might be viewed as a water savings. As a result, drainage and streamflow might be enhanced. But an increase in drainage, in addition to increasing the supply of water downstream, has the potential to enhance watershed nutrient and soil losses (Likens and Bormann, 1995). This has potentially negative implications for long-term ecosystem productivity or sustainability that must be considered in a balanced analysis. conversely, tree growth stimulation due to increasing $[CO_2]_a$ could result in more extensive roots systems that might stabilize soils.

IX. Discussion and Summary

Increasing $[CO_2]_a$ clearly has the potential to stimulate plant growth and ecosystem NPP (Amthor, 1995). Although it has yet to be demonstrated conclusively that the 25–30% increase in $[CO_2]_a$ during the past 200 years (see Etheridge *et al.*, 1996; Keeling and Whorf, 1997) increased NPP in nature, it seems that this must have occurred to some degree (Amthor, 1995). In addition to enhancing growth, elevated $[CO_2]_a$ might alleviate or postpone water stress and increase WUE (Eamus, 1991).

Because elevated $[CO_2]_a$ can increase WUE and might delay or prevent water stress, there is much speculation about plant growth responses to elevated $[CO_2]_a$ with and without limited water supply. A common assumption is that elevated $[CO_2]_a$ will cause a relatively larger growth stimulation in stressed plants, although the literature survey of Wullschleger *et al.* (1995) indicates that the relative enhancement (fractional increase) of growth caused by elevated $[CO_2]_a$ is at least as large in well-watered trees compared with water-stressed trees grown inside various chambers. In addition, relative whole-plant (Amthor and Koch, 1996) and lint (cotton fiber) (Pinter *et al.*, 1996) growth stimulation in the Arizona cotton FACE experiments were at least as great in the well-watered plots as in the water-stressed plots. In the wheat FACE experiments, whole-plant growth response to elevated $[CO_2]_a$ was relatively larger for well-watered (compared to water-stressed) plants in one of two years (Amthor and Koch, 1996), but the relative increase in grain yield due to elevated $[CO_2]_a$ was greater with water stress in both years (Pinter *et al.*, 1996). Note that the effects of the FACE apparatus itself on wheat development may invalidate comparative results in 1993 and 1994.) In the field in a naturally CO_2-enriched environment, tree stem growth responses to elevated $[CO_2]_a$ were relatively greater in dry compared with wet years (Hättenschwiler *et al.*, 1997). In spite of this range of experimental results, it is clear that plant responses to $[CO_2]_a$ can be affected by water availability and (or) water stress (e.g., Owensby *et al.*, 1993).

With respect to ecosystem water use, as opposed to water stress, all major components of the water balance must be considered. In some ecosystems, canopy interception loss is greater than transpiration (e.g., Calder, 1990), so even if elevated $[CO_2]_a$ reduced transpiration by 10% the effect on landscape ET could be much less. For landscapes that experience a spatial and (or) temporal increase in ground cover by vegetation as a result of increasing $[CO_2]_a$ and climatic change, canopy interception loss may increase even as transpiration decreases and this might compensate for reductions in stomatal conductance on the annual whole-ecosystem ET. An increase in ground cover might also reduce soil evaporation. In sparse vegetation that remains sparse even as $[CO_2]_a$ increases, soil evaporation can make a significant contribution to ET and it will not be directly influenced by stomatal closure.

Because present experiments do not address large-scale responses to increasing $[CO_2]_a$, data bearing on the effects of increasing $[CO_2]_a$ on ecosystem-scale ET are limited. Indeed, the examination of HBEF watershed ET as a function of $[CO_2]_a$ during the 1956–1996 period carried out herein is the only directly applicable, moderately long-term analysis of which I am aware. (See Amthor, 1998, for a longer, more complete analysis.) It indicates that watershed ET may have slowed with increasing $[CO_2]_a$ in only one (W3) of five forested watersheds. Before much more can be said about general real-world watershed-scale water use responses (if any) to increasing $[CO_2]_a$, more analyses of watershed water balances are required. I cannot discern from the present analysis whether forest water stress (if any) was affected by the 15% increase in $[CO_2]_a$ that has occurred since 1956; consideration of watershed NPP will be needed for that. *If* increasing $[CO_2]_a$ is causing a reduction in ET according to the regression analysis for W3 in the HBEF (see Fig. 2, bottom panel), the recent increase in $[CO_2]_a$ from 310 to 362 ppm (corresponding to the past ca. 55 yr; see Etheridge *et al.*, 1996) resulted in a 0.070-m (or 13%) reduction in ET, which would clearly be significant to regional water balances. But only data for W3 indicate a significant response by ET to increasing $[CO_2]_a$; any effect of $[CO_2]_a$ on ET in the other four watersheds analyzed was not detected. There is, therefore, only weak evidence that increasing $[CO_2]_a$ is reducing ET in the HBEF and, by extension, in other temperate deciduous forests.

To the extent that greater soil water content results from future $[CO_2]_a$ increases (see Hungate *et al.*, 1997, for very small chamber experiments) species composition may change in terrestrial ecosystems. For instance, shrubs may be able to invade grasslands if soil water contents remain relatively high (see Williams *et al.*, 1987) in elevated $[CO_2]_a$. The relative advantage of C_3 vegetation over C_4 species *may* also be increased in elevated $[CO_2]_a$ due to a relatively greater positive photosynthetic response (see references in Amthor, 1995), but recent invasions of C_4 grasslands by C_3 shrubs is likely due mainly to human land management rather than plant physiological responses to $[CO_2]_a$ or changes in soil water content (see Archer *et al.*, 1995; Amthor, 1995). Other effects of any enhanced soil water content caused by increasing $[CO_2]_a$ are possible. For example, greater retention of water by soils may affect nutrient cycling and availability (Hungate *et al.*, 1997) and enhance soil organic matter decomposition processes (Sommers *et al.*, 1980).

Returning to the climatic implications of increasing $[CO_2]_a$, it must be noted that increasing-$[CO_2]_a$-induced global warming has the potential to counteract any water-conserving effects of increasing-$[CO_2]_a$-induced stomatal closure. This is because warming tends to stimulate ET. Also, warming may extend leaf duration in cold-deciduous ecosystems, which could increase both annual interception loss and transpiration. The presently unanswerable question at hand is this: Which will dominate changes in future ecosystem ET, stomatal closure and the associated tendency for reduced ET, or surface warming and the associated tendencies for enhanced landscape ET? The implications of globally increasing $[CO_2]_a$ on landscape-scale water use and water stress are therefore unclear. Nonetheless, a general increase in WUE can be expected in many ecosystems, and NPP is also likely to be stimulated even if ecosystem-scale water use is unchanged. The largest reductions (or seasonal delays) in water stress may occur in presently water-limited ecosystems, which cover a large fraction of earth's ice-free land. Also, any significant future global warming might increase the area of water-limited ecosystems even while increased WUE from increasing $[CO_2]_a$ could stimulate NPP in those ecosystems.

Acknowledgments

Preparation of this chapter was supported by the DOE/NSF/NASA/USDA/EPA Interagency Program on Terrestrial Ecology and Global Change (TECO) by the U.S. Department of Energy's Office of Biological and Environmental Research under contract W-7405-Eng-48 with the University of California and contract DE-AC05-96OR22464 with Lockheed Martin Energy Research Corporation. Accordingly, the U.S. government retains a nonexclusive, royalty-free license to publish or reproduce the published form of this contribution, or allow others to do so, for U.S. government purposes.

Bruce Kimball and Paul Pinter were helpful, as always, in discussing results of the DOE/USDA FACE experiments in Arizona. All data shown in Fig. 1 and some data summarized in Fig. 2 and Table II were obtained by scientists of the Hubbard Brook Ecosystem Study; this chapter was not reviewed by those scientists. The Hubbard Brook Experimental Forest (HBEF) is operated and maintained by the USDA Northeastern Forest Experiment Station, Radnor, Pennsylvania. The HBEF data were obtained from the World Wide Web. I thank Hugo Rogers, Bruce Kimball, Margaret Goodbody, Paul Hanson, Rich Norby, and Yiqi Luo for reviewing various drafts of this chapter. The opinions expressed in this chapter are my own. Publication No. 4839, Environmental Sciences Division, Oak Ridge National Laboratory.

References

Allen, L. H., Jr., Drake, B. G., Rogers, H. H., and Shinn, J. H. (1992). Field techniques for exposure of plants and ecosystems to elevated CO_2 and other trace gases. *Crit. Rev. Plant Sci.* **11,** 85–119.

Amthor, J. S. (1995). Terrestrial higher-plant response to increasing atmospheric $[CO_2]$ in relation to the global carbon cycle. *Global Change Biol.* **1,** 243–274.

Amthor, J. S. (1998). "Searching for a Relationship Between Forest Water Use and Increasing Atmospheric CO_2 Concentration with Long-Term Hydrologic Data from the Hubbard Brook Experimental Forest." ORNL/TM-13708. Oak Ridge National Laboratory, Oak Ridge, Tennessee.

Amthor, J. S., and Koch, G. W. (1996). Biota growth factor β: Stimulation of terrestrial ecosystem net primary production by elevated atmospheric CO_2. *In* "Carbon Dioxide and Terrestrial Ecosystems" (G. W. Koch and H. A. Mooney, eds.), pp. 399–414. Academic Press, San Diego.

Archer, S., Schimel, D. S., and Holland, E. A. (1995). Mechanisms of shrubland expansion: Land use, climate or CO_2. *Climatic Change* **29,** 91–99.

Benecke, P. (1975). Soil water relations and water exchange of forest ecosystems. *In* "Water and Plant Life" (O. L. Lange, L. Kappen, and E.-D. Schulze, eds.), pp. 101–131. Springer-Verlag, Berlin.

Bhattacharya, N. C., Radin, J. W., Kimball, B. A., Mauney, J. R., Hendrey, G. R., Nagy, J., Lewin, K. F., and Ponce, D. C. (1994). Leaf water relations of cotton in a free-air CO_2-enriched environment. *Agric. For. Meteorol.* **70,** 171–182.

Bormann, F. H., and Likens, G. E. (1979). "Pattern and Process in a Forested Ecosystem." Springer-Verlag, New York.

Brutsaert, W. (1986). Catchment-scale evaporation and the atmospheric boundary layer. *Water Resource Res.* **22,** 39S–45S.

Calder, I. R. (1990). "Evaporation in the Uplands." Wiley, Chichester.

Ceulemans, R., and Mousseau, M. (1994). Effects of elevated atmospheric CO_2 on woody plants. *New Phytol.* **127,** 425–446.

De Bruin, H. A. R. (1983). A model for the Priestly–Taylor parameter α. *J. Climate Appl. Meteorol.* **22,** 572–578.

Dickinson, R. E. (1991). Global change and terrestrial hydrology—a review. *Tellus* **43AB,** 176–181.

Dugas, W. A., Heuer, M. L., Hunsaker, D. Kimball, B. A., Lewin, K. F., Nagy, J., and Johnson, M. (1994). Sap flow measurements of transpiration from cotton grown under ambient and enriched CO_2 concentrations. *Agric. For. Meterorol.* **70,** 231–245.

Eamus, D. (1991). The interaction of rising CO_2 and temperatures with water use efficiency. *Plant Cell Environ.* **14,** 843–852.

Etheridge, D. M., Steele, L. P., Langenfelds, R. L., Francey, R. J., Barnola, J.-M., and Morgan, V. I. (1996). Natural and anthropogenic changes in atmospheric CO_2 over the last 1000 years from air in Antarctic ice and firn. *J. Geophys. Res.* **101,** 4115–4128.

Field, C. B., Jackson, R. B., and Mooney, H. A. (1995). Stomatal responses to increased CO_2: Implications from the plant to the global scale. *Plant Cell Environ.* **18,** 1214–1225.

Galoux, A., Benecke, P., Gietl, G., Hager, H., Kayser, C., Kiese, O., Knoerr, K. R., Murphy, C. E., Schnock, G., and Sinclair, T. R. (1981). Radiation, heat, water and carbon dioxide balances. *In* "Dynamic Properties of Forest Ecosystems" (D. E. Reichle, ed.), pp. 87–204. Cambridge Univ. Press, Cambridge.

Hättenschwiler, S., Miglietta, F., Raschi, A., and Körner, C. (1997). Thirty years of *in situ* tree growth under elevated CO_2: A model for future forest responses? *Global Change Biol.* **3,** 463–471.

Helvey, J. D., and Patric, J. H. (1965). Canopy and litter interception of rainfall by hardwoods of eastern United States. *Water Resource Res.* **1,** 193–206.

Hileman, D. R., Huluka, G., Kenjige, P. K., Sinha, N., Bhattacharya, N. C., Biswas, P. K., Lewin, K. F., Nagy, J., and Hendrey, G. R. (1994). Canopy photosynthesis and transpiration of field-grown cotton exposed to free-air CO_2 enrichment (FACE) and differential irrigation. *Agric. For. Meteorol.* **70,** 189–207.

Hsiao, T. C. (1973). Plant responses to water stress. *Annu. Rev. Plant Physiol.* **24,** 519–570.

Hungate, B. A., Chapin, F. S., III, Zhong, H., Holland, E. A., and Field, C. B. (1997). Stimulation of grassland nitrogen cycling under carbon dioxide enrichment. *Oecologia* **109,** 149–153.

Hunsaker, D. J., Hendrey, G. R., Kimball, B. A., Lewin, K. F., Mauney, J. R., and Nagy, J. (1994). Cotton evapotranspiration under field conditions with CO_2 enrichment and variable soil moisture regimes. *Agric. For. Meterorol.* **70,** 247–258.

Hunsaker, D. J., Kimball, B. A., Pinter Jr., P. J., LaMorte, R. L., and Wall, G. W. (1996). Carbon dioxide enrichment and irrigation effects on wheat evapotranspiration and water use efficiency. *Trans. Am. Soc. Agric. Engineers* **39,** 1345–1355.

IPCC (Intergovernmental Panel on Climate Change). (1996). "Climate Change 1995: The Science of Climate Change." Cambridge Univ. Press, Cambridge.

Jacobs, C. M. J. (1994). "Direct Impact of Atmospheric CO_2 Enrichment on Regional Transpiration." Doctoral thesis, Wageningen Agricultural Univ, The Netherlands.

Jacobs, C. M. J., and De Bruin, H. A. R. (1992). The sensitivity of regional transpiration to land-surface characteristics: Significance of feedback. *J. Climate* **5,** 683–698.

Jarvis, P. G., and McNaughton, K. G. (1986). Stomatal control of transpiration: Scaling up from leaf to region. *Adv. Ecol. Res.* **15,** 1–49.

Jones, H. G. (1992). "Plants and Microclimate," 2nd ed. Cambridge Univ. Press, Cambridge.

Keeling, C. D. (1960). The concentration and isotopic abundances of carbon dioxide in the atmosphere. *Tellus* **12,** 200–203.

Keeling, C. D., and Whorf, T. P. (1994). Atmospheric CO_2 records from sites in the SIO air sampling network. *In* "Trends '93: A Compendium of Data on Global Change" (T. A. Boden, D. P. Kaiser, R. J. Sepanski, and F. W. Stoss, eds.), pp. 16–26. Carbon Dioxide Information Analysis Center, Oak Ridge National Laboratory, Oak Ridge, TN.

Keeling, C. D., and Whorf, T. P. (1997). Atmospheric CO_2 concentrations (ppmv) derived from *in situ* air samples collected at Mauna Loa Observatory, Hawaii. Obtained from http://cdiac.esd.ornl.gov/ (November 1997).

Keeling, C. D., Bacastow, R. B., Bainbridge, A. E., Ekdahl Jr., C. A., Guenther, P. R., Waterman, L. S., and Chin, J. F. S. (1976). Atmospheric carbon dioxide variations at Mauna Loa Observatory, Hawaii. *Tellus* **28,** 538–551.

Kimball, B. A., Pinter Jr., P. J., and Mauney, J. R. (1992). Cotton leaf and boll temperatures in the 1989 FACE experiment. *Crit. Rev. Plant Sci.* **11,** 233–240.

Kimball, B. A., LaMorte, R. L., Seay, R. S., Pinter Jr., P. J., Rokey, R. R., Hunsaker, D. J., Dugas, W. A., Heuer, M. L., Mauney, J. R., Hendrey, G. R., Lewin, K. F. and Nagy, J. (1994). Effects of free-air CO_2 enrichment on energy balance and evapotranspiration of cotton. *Agric. For. Meteorol.* **70,** 259–278.

Kimball, B. A., Pinter Jr., P. J., Garcia, R. L., LaMorte, R. L., Wall, G. W., Hunsaker, D. J., Wechsung, G., Wechsung, F., and Kartschall, T. (1995). Productivity and water use of wheat under free-air CO_2 enrichment. *Global Change Biol.* **1,** 429–442.

Larcher, W. (1995). "Physiological Plant Ecology," 3rd ed. (J. Wieser, trans.). Springer-Verlag, Berlin.

Lieth, H. (1975). Modeling the primary productivity of the world. *In* "Primary Productivity of the Biosphere" (H. Lieth and R. H. Whittaker, eds.), pp. 237–263. Springer-Verlag, New York.

Likens, G. E., and Bormann, F. H. (1995). "Biogeochemistry of a Forested Ecosystem," 2nd ed. Springer-Verlag, New York.

Loomis, R. S., and Amthor, J. S. (1996). Limits to yield revisited. *In* "Increasing Yield Potential in Wheat: Breaking the Barriers" (M. P. Reynolds, S. Rajaram, and A. McNab, eds.), pp. 76–89. Centro Internacional de Mejoramiento de Maíz y Trigo (CIMMYT), Ciudad de México.

Manabe, S. (1983). Carbon dioxide and climatic change. *Adv. Geophys.* **25,** 39–82.

Mansfield, T. A., and Atkinson, C. J. (1990). Stomatal behaviour in water stressed plants. *In* "Stress Responses in Plants: Adaptation and Acclimation Mechanisms" (R. G. Alscher and J. R. Cumming, eds.), pp. 241–264. Wiley-Liss, New York.

Martin, P., Rosenberg, N. J., and McKenney, M. S. (1989). Sensitivity of evapotranspiration in a wheat field, a forest, and a grassland to changes in climate and direct effects of carbon dioxide. *Climatic Change* **14,** 117–151.

McLeod, A. R. (1993). Open-air exposure systems for air pollutant studies—their potential and limitations. *In* "Design and Execution of Experiments on CO_2 Enrichment" (E. D. Schulze and H. A. Mooney, eds.), pp. 353–365. Commission of the European communities, Luxembourg.

McNaughton, K. G. (1989). Regional interactions between canopies and the atmosphere. *In* "Plant Canopies: Their Growth, Form and Function" (G. Russell, B. Marshall, and P. G. Jarvis, eds.), pp. 63–81. Cambridge Univ. Press, Cambridge.

McNaughton, K. G., and Jarvis, P. G. (1991). Effects of spatial scale on stomatal control of transpiration. *Agric. For. Meteorol.* **54,** 279–301.

McNaughton, K. G., and Spriggs, T. W. (1986). A mixed layer model for regional evaporation. *Boundary-Layer Meteorol.* **34,** 243–262.

Monteith, J. L. (1972). Solar radiation and productivity in tropical ecosystems. *J. Appl. Ecol.* **9,** 747–766.

Monteith, J. L. (1995a). A reinterpretation of stomatal responses to humidity. *Plant Cell Environ.* **18,** 357–364.

Monteith, J. L. (1995b). Accommodation between transpiring vegetation and the convective boundary layer. *J. Hydrol.* **166,** 251–263.

Morgan, J. A., Knight, W. G., Dudley, L. M., and Hunt, H. W. (1994). Enhanced root system C-sink activity, water relations and aspects of nutrient acquisition in mycotrophic *Bouteloua gracilis* subjected to CO_2 enrichment. *Plant Soil* **165,** 139–146.

Morison, J. I. L. (1987). Intercellular CO_2 concentration and stomatal response to CO_2. *In* "Stomatal Function" (E. Zeiger, C. D. Farquhar, and I. R. Cowan, eds.), pp. 229–252. Standford Univ. Press, Standford, CA.

Morison, J. I. L., and Gifford, R. M. (1983). Stomatal sensitivity to carbon dioxide and humidity. A comparison of two C_3 and two C_4 grass species. *Plant Physiol.* **71,** 789–796.

Mott, K. A. (1988). Do stomata respond to CO_2 concentrations other than intercellular? *Plant Physiol.* **86,** 200–203.

Norby, R. J. (1994). Issues and perspectives for investigating root responses to elevated atmospheric carbon dioxide. *Plant Soil* **165,** 9–20.

Owensby, C. E., Coyne, P. I., Ham, J. M., Auen, L. M., and Knapp, A. K. (1993). Biomass production in a tallgrass prairie ecosystem exposed to ambient and elevated CO_2. *Ecol. Appl.* **3,** 644–653.

Pinter Jr., P. J., Kimball, B. A., Mauney, J. R., Hendrey, G. R., Lewin, K. F., and Nagy, J. (1994). Effects of free-air carbon dioxide enrichment on PAR absorption and conversion efficiency by cotton. *Agric. For. Meteorol.* **70,** 209–230.

Pinter Jr., P. J. Kimball, B. A., Garcia, R. L., Wall, G. W., Hunsaker, D. J., and LaMorte, R. L. (1996). Free-air CO_2 enrichment: Responses of cotton and wheat crops. *In* "Carbon Dioxide and Terrestrial Ecosystems" (G. W. Koch and H. A. Mooney, eds.), pp. 215–249. Academic Press, San Diego.

Prior, S. A., Rogers, H. H., Runion, G. B., and Hendrey, G. R. (1994a) Free-air CO_2 enrichment of cotton: Vertical and lateral root distribution patterns. *Plant Soil* **165,** 33–44.

Prior, S. A., Rogers, H. H., Runion, G. B., and Mauney, J. R. (1994b). Effects of free-air CO_2 enrichment on cotton root growth. *Agric. For. Meteorol.* **70,** 69–86.

Radin, J. W., Kimball, B. A., Hendrix, D. L., and Mauney, J. R. (1987). Photosynthesis of cotton plants exposed to elevated levels of carbon dioxide in the field. *Photosynth. Res.* **12,** 191–203.

Raupach, M. R. (1995). Vegetation–atmosphere interaction and surface conductance at leaf, canopy and regional scales. *Agric. For. Meteorol.* **73,** 151–179.

Rogers, H. H., Runion, G. B., and Krupa, S. V. (1994). Plant responses to atmospheric CO_2 enrichment with emphasis on roots and the rhizosphere. *Environ. Pollut.* **83,** 155–189.

Rogers, H. H., Prior, S. A., Runion, G. B., and Mitchell, R. J. (1996). Root to shoot ratio of crops as influenced by CO_2. *Plant Soil* **187,** 229–248.

Santrucek, J., and Sage, R. F. (1996). Acclimation of stomatal conductance to a CO_2-enriched atmosphere and elevated temperature in *Chenopodium album. Aust. J. Plant Physiol.* **23,** 467–478.

Sellers, P. J., Bounoua, L., Collatz, G. J., Randall, D. A., Dazlich, D. A., Los, S. O., Berry, J. A., Fung, I., Tucker, C. J., Field, C. B., and Jensen, T. G. (1996). Comparison of radiative and physiological effects of doubled atmospheric CO_2 on climate. *Science* **271,** 1402–1406.

Senock, R. S., Ham, J. M., Loughin, T. M., Kimball, B. A., Hunsaker, D. J., Pinter, P. J., Wall, G. W., Garcia, R. L., and LaMorte, R. L. (1996). Sap flow in wheat under free-air CO_2 enrichment. *Plant Cell Environ.* **19,** 147–158.

Shuttleworth, W. J. (1988). Macrohydrology—the new challenge for process hydrology. *J. Hydrol.* **100,** 31–56.

Sommers, L. E., Gilmour, C. M., Wildung, R. E., and Beck, S. M. (1980). The effect of water potential on decomposition processes in soils. *In* "Water Potential Relations in Soil Microbiology" (J. F. Parr, W. R. Gardner, and L. F. Elliot, eds.), pp. 97–117. Soil Sci. Soc. Am., Madison, WI.

Williams, K., Hobbs, R. J., and Mooney, H. A. (1987). Invasion of an annual grassland in northern California by *Baccharis pilularis* ssp. *consanguinea. Oecologia* **72,** 461–465.

Wullschleger, S. D., Post, W. M., and King, A. W. (1995). On the potential for a CO_2 fertilization effect in forests: Estimates of the biotic growth factor based on 58 controlled-exposure studies. *In* "Biotic Feedbacks in the Global Climatic System" (G. M. Woodwell and F. T. Mackenzie, eds.), pp. 85–107. Oxford Univ. Press, New York.

3

Temperature: Cellular to Whole-Plant and Population Responses

R. M. M. Crawford and D. W. Wolfe

I. Introduction

The temperature regime of habitats occupied by plants ranges from just over freezing during the active growth period in arctic regions to daytime temperatures exceeding 50°C during the growth cycle of some equatorial desert species. Temperature plays a major role in determining plant productivity and species distribution, and it can be examined on several scales from processes occurring at the molecular (e.g., enzyme kinetics) to population and biogeographical changes (e.g., migration). The overwhelming evidence that atmospheric CO_2 is increasing exponentially and will likely double within the next century (Schimel *et al.*, 1995) demands that any attempt to predict the effect of climate change on terrestrial vegetation must take into account the possible interactive effects between increasing atmospheric CO_2 and temperature on plants. In addition to being the most important of the greenhouse gases, accounting for about 60% of the global warming potential (Schimel *et al.*, 1995), elevated CO_2 has profound direct effects on plant physiology that could alter some plant responses to temperature. Conversely, temperature can affect plant responses to CO_2.

Ecologically, the ultimate plant response to change in temperature will be on species distribution and the geographical limits to crop production. It is therefore necessary to go beyond the immediate effects of temperature and CO_2 on metabolic processes and consider the possible impact of temperature change at the plant population and species level. In predicting long-term consequences of climatic warming, both experimentation and modeling have their limitations. Perturbation experiments can give opposing results depending on the length of time the observations are carried out (Chapin *et al.*, 1995). Similarly, recent advances in modeling future plant distributions demonstrate that species migration is highly sensitive to temperature seasonality, and similar patterns of annual change can even cause species in different parts of their ranges to migrate in opposite directions (Jeffree and Jeffree, 1996). There is therefore a need to relate modeling studies closely to the actual physiological and ecological responses of plants in the field.

Long-term evidence for plant responses to temperature change can be sought in terms of past and present distributions. Historical biogeography therefore has a unique authority in providing evidence from previous periods of climatic change. However, the responses of plants during periods of climatic warming (or cooling) are not determined solely by temperature. Other factors, both physical and biological, have to be brought into consideration, because ecological factors may differ between past and present periods of climatic warming. Projections on the future distribution of plants with regard to temperature that fail to take into account the diversity of response that can exist within plant populations neglect one of the most important features of plant adaptation to changing conditions, namely, the capacity of polymorphic species to change their morph frequencies (Crawford *et al.*, 1993).

It is not possible in any one single review to provide an adequate treatment of every biome and its probable responses to rising temperatures and elevated levels of atmospheric carbon dioxide. For this reason the present discussion of climate change in relation to ecology and plant populations is limited to the low-temperature environments of the arctic and boreal zone where there are already discernible changes both in climate and its impact on the vegetation.

II. Fundamental Temperature Effects on Plants and Interactions with Elevated CO_2

At the most basic level, sensing of temperature by plants can usually be traced to a combination of metabolic (e.g., enzyme and membrane function) and physical (e.g., chemical potential gradients, water viscosity) changes. In this section we examine some of the fundamental temperature responses of plants, and how, in some cases, increases in ambient CO_2 may alter these responses.

A. Enzymes

1. Temperature Effects on Enzyme Kinetics Plant biochemistry is driven by enzymatic reactions. The temperature dependence of enzyme rate constants (k) can often be characterized by an exponential equation of the form:

$$k_2 = k_1 Q_{10}((T_2 - T_1)/10),$$

where T is temperature (°C), and Q_{10} is an empirically derived value for the magnitude of change in rate constant for every 10°C change in temperature (Farrar and Williams, 1991). The biophysical basis of this temperature effect on rate constants is related to (absolute) temperature effects on kinetic energy of enzymes and substrates, and energies of activation (E_a) of specific chemical reactions (*see* Berry and Raison, 1981 for a more thorough discussion).

Because different enzymes have distinct Q_{10} values, a temperature shift can shift the balance between alternative biochemical pathways. This has been well documented for several key carbon metabolism enzymes. For example, elevated temperatures increase sucrose phosphate–synthase (SPS) activity of source leaves in potato (Lafta and Lornenzen, 1995) and soybean (Rufty *et al.*, 1958), while decreasing activity of other carbon metabolism enzymes, such as ADPG-phosphorylase in potato tuber sink organs (Lafta and Lorenzen, 1995; Kraus and Marschner, 1984).

Increasing temperatures increase respiration rates in a Q_{10} fashion. Respiration costs per unit of new growth may be unaffected by temperature, but growth rate and activity of many enzymes affiliated with growth respiration and synthetic processes increase with nonstress warming (Farrar and Williams, 1991). Maintenance respiration rates also often increase as temperatures warm due to accelerated rates of catalysis and synthesis (accelerated cycling) of enzymes and other metabolites (Amthor, 1989). At low temperatures, concentrations of glycolytic products that can accumulate and damage membranes and membrane function (Berry and Raison, 1981).

*2. **Enzyme Synthesis and Stability at Temperature Extremes*** Enzymes, being proteins, disassociate or denature and become inactive at the temperature extremes. Different enzymes have different temperature thresholds. For example, the oligomeric phosphofructokinase of potato tubers disassociates at low temperature (Rees *et al.*, 1988), while the cytosolic fructose 1,6-bisphosphate of spinach and pea is cold labile (Weeden and Buchanan, 1983). The activation state of the key photosynthetic enzyme, Rubisco, becomes unstable at elevated temperatures, possibly in response to limitations in thylakoid membrane processes (Weis and Berry, 1988).

Inactivation of enzymes involved in synthesis of hormones can have long-term effects on growth and development, or affect response to temperature and other stresses. For example, low temperatures inhibit the synthesis of ABA (Vernieri *et al.*, 1991), and in some species this may lead to temporary loss of stomatal control during chilling, leaf dehydration, and irreversible chilling damage (Eamus *et al.*, 1983).

At low temperatures, light absorbed by PSII is sometimes in excess of that utilized by the Calvin cycle (Ortiz-Lopez *et al.*, 1990; Van Hasselt and Van Berlo, 1980). This can lead to production of superoxide radicals and photo-oxidative damage to proteins, nucleic acids, and the chloroplast membrane. Species vary in their sensitivity to this type of chilling injury. There is some limited evidence to suggest that chilling-induced photodamage might be partially mitigated or exacerbated at elevated compared to ambient CO_2 levels depending on Calvin cycle activity during the chilling event (Rowley and Taylor, 1974; Boese *et al.*, 1997). This potential temperature × CO_2 interaction has not been well investigated.

3. Temperature × CO_2 Interactive Effects on Rubisco and Photosynthesis Temperature affects photosynthesis by affecting the capacity of thylakoid membranes for electron transport and by affecting Rubisco kinetics and the efficiency of carbon assimilation (Berry and Raison, 1981). In C_3 plants, Rubisco is the primary enzyme involved in carbon assimilation, but oxygen (O_2) competes with CO_2 for sites on the enzyme. Oxygenation promotes the glycolate biosynthetic pathway and photorespiratory carbon loss. As temperatures increase, the oxygenation to carboxylation ratio increases. This is because an increasing temperature differentially affects the Michaelis–Menten constants of Rubisco for O_2 vs. CO_2 such that the affinity of Rubisco for O_2 increases relative to CO_2 (Badger and Collatz, 1977; Jordan and Ogren, 1984). Another factor is that increasing temperature increases the O_2-to-CO_2 solubility ratio within the chloroplast (Ku and Edwards, 1977).

One of the most important and most frequently observed CO_2 × temperature interactions is that the stimulation of C_3 photosynthesis by elevated CO_2 increases as temperatures rise within the approximate temperature range of 20–35°C (Berry and Raison, 1981; Sage and Reid, 1994). One reason for this is that at warm temperatures, photorespiration of C_3 plants is inhibited proportionally more by increasing CO_2 than at cool temperatures (Jordan and Ogren, 1984). The second reason that the magnitude of beneficial effects of elevated CO_2 on photosynthesis is greater at warm as compared to cool temperatures is that metabolism and sink demand for photosynthates are usually greater at warm temperatures (Farrar and Williams, 1991). Consequently, there is less accumulation of immediate photosynthetic products, triose and hexose phosphates, and sucrose and starch in the leaves at warm as compared to cool temperatures. Accumulation of triose and hexose phosphate intermediates can have a negative feedback effect on photosynthetic capacity because there is insufficient cycling of inorganic phosphate (P_i) for ATP synthesis required for photosynthesis

(Sage and Sharkey, 1987). Accumulation of sugar hexoses in the leaves can also inhibit photosynthetic capacity by repressing genes transcribing for Rubisco and other photosynthetic enzymes (Sheen, 1994). At the temperature extremes, beneficial effects on photosynthesis from elevated CO_2 can become minimal because of direct damage to the photosynthetic apparatus, as well as due to slowed sink utilization of carbon at extreme high (Kraus and Marschner, 1984) or low (Paul et al., 1991) temperatures.

The $CO_2 \times$ temperature interaction is illustrated in the simulations of Fig. 1, based on the photosynthesis model of Farquhar *et al.* (1980) and modified by Long (1991). The benefit from CO_2 enrichment becomes very small at low temperatures, even assuming no negative feedback effects on photosynthetic capacity (Rubisco activity) due to accumulation of leaf carbohydrates at elevated CO_2. In simulations incorporating the assumption that (downward) acclimation to high levels of CO_2 reduces Rubisco activity by 20 and 40%, the photosynthetic rate of a CO_2-enriched plant becomes lower than the control at temperatures of 12.5 and 22.5°C, respectively. Note also that the temperature optimum for photosynthesis tends to increase at increased CO_2 concentrations. This effect may be beneficial in a future high-CO_2 world, where air temperatures may be higher and leaf temperatures may increase because of the partial stomatal closure response to increased CO_2 (Mott, 1990).

The nature of the $CO_2 \times$ temperature interactive effect on photosynthesis can vary with species, duration of acclimation to the environment, and experimental methods. In a recent study by Greer *et al.* (1995), when cool

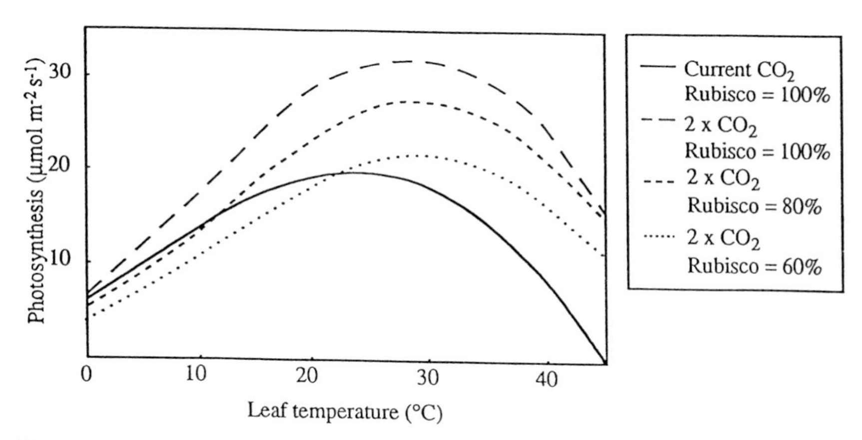

Figure 1 Computer simulations of net photosynthesis per unit leaf area in relation to leaf temperature, CO_2 concentration, and activity level of the key photosynthetic enzyme, Rubisco. Photosynthetic acclimation to elevated CO_2 sometimes involves reductions in Rubisco content or activity (see text). (From Long, 1991, as adapted by Wolfe, 1994.)

temperature-adapted pasture species were grown for an extended period at 700 μmol mol^{-1} CO_2, the temperature maximum for photosynthesis shifted upward [as predicted by the Farquhar *et al.* (1980) model] in six of the species, but the shift was downward in the other five species examined. In this same study, photosynthetic stimulation by elevated CO_2 was surprisingly high, even in a 12°C temperature treatment (48% increase), but relative growth rate was stimulated very little (less than 10% increase) at cool temperatures. Contrary to the notion that photosynthetic stimulation by CO_2 enrichment always increases at warmer temperatures, in a long-term study with the cool-season C_3 grass, *Pascopyrum smithii,* Morgan *et al.* (1994) found that photosynthesis was less stimulated by CO_2 doubling at a warm temperature (4°C above ambient) compared to the ambient cool temperature regime. Acclimation of this cool season-adapted species to warm temperatures for several seasons had led to reduced photosynthetic capacity of the leaves and thereby constrained CO_2 response.

B. Membranes

Membranes are highly selective permeability barriers separating cells from their external environment and creating organelle boundaries within individual cells. They are comprised primarily of proteins imbedded in a fluid, sheetlike lipid structure. The proteins of membranes serve many functions essential to life. Some membrane proteins are enzymes, such as the ATPases. Others make up solute channels, act as proteinaceous carriers, or are involved in energy conversion processes such as photosynthesis and oxidative phosphorylation. At low temperatures, the degree of saturation of membrane lipids and lipid species ratios shifts, causing a transition of the membrane from a fluid to a gel-like state (DeKok and Kuiper, 1977). This affects protein–lipid interactions, ion permeability, and diffusion properties of the membrane. At temperature extremes, normal membrane function is disrupted.

Membrane transport activity may respond to temperature in a Q_{10} fashion similar to enzyme rate constants, although values for specific membrane functions have not been well characterized. It has been documented that phloem loading, an important membrane transport activity, is faster at warm as compared to cool temperatures (Giaquinta, 1980; Ntsika and Delrot, 1986). Johnson and Thornley (1985) estimated the Q_{10} of sucrose diffusion to and from phloem to be approximately 2. Viscosity of phloem sap decreases with a Q_{10} of about 1.3 (Farrar, 1988) as temperature increases, contributing to increased flux at warmer temperatures. Plant acclimation to changes in temperature are often associated with changes in plant membranes. Within limits, the optimum temperature for photosynthesis can shift up to 1°C for each 3°C shift in growth temperature; this is associated with changes in fatty acid composition of chloroplast membrane lipids (Berry and Raison, 1981).

C. Plant–Water Relations

1. Biophysical Temperature Effects on Water Relations Transpiration is driven by the leaf-to-air vapor pressure gradient (vpg). Warmer air temperatures often increase leaf temperatures, which increases the saturated water vapor pressure within the leaf and the vpg. The degree of coupling between leaf and air temperature depends, in part, on the amount of evaporative cooling (i.e., transpiration rate), which in turn depends on stomatal conductance, air humidity, leaf temperature, leaf morphology, and wind speed. The stomates of some plant species respond directly to vpg, closing as vpg rises (Schulze, 1986). This leads to a feedback effect in which stomatal closure and slowed transpiration reduce evaporative cooling of the leaf, leading to an increase in leaf temperature and a counteracting increase in the vpg.

Low temperatures affect plant water relations in a number of ways. A decline in air temperature below the dew point leads to a decline in air humidity that affects the vpg and transpiration. Plant water stress and visible wilting of the leaves are early symptoms of chilling stress in many plant species, including bean (Wolfe, 1991; Eamus *et al.,* 1983; McWilliam *et al.,* 1982), soybean (Markhart, 1984), and cotton (McWilliam *et al.,* 1982). This has been associated with reduced hydraulic conductance of root membranes (Markhart, 1984; McWilliam *et al.,* 1982; Kuiper, 1964) and/or a temporary stomatal dysfunction leading to continued water loss from leaves despite declining leaf water potentials (Pardossi *et al.,* 1992; Eamus *et al.,* 1983). This chilling-induced water stress is exacerbated by the increased viscosity of water at low temperature.

Temperature effects on root and stomatal function may be mediated by temperature effects on abscisic acid (ABA). Low temperatures inhibit synthesis of ABA (Vernieri *et al.,* 1991), diminish stomatal sensitivity to ABA (Pardossi *et al.,* 1992), and impair movement of ABA to guard cells by slowing the transpiration stream to leaves (Radin and Ackerson, 1982). Additional evidence for involvement of ABA in temperature effects on water relations is based on studies in which exogenous applications of ABA increase chilling tolerance by improving root hydraulic conductance (Markhart, 1984), or by causing partial closure of the stomates (Semeniuk *et al.,* 1986; Eamus and Wilson, 1984).

Temperature affects the solubility of osmotically active compounds within plant tissues and can thereby affect potential gradients driving inter- and intracellular water and ion movement within the plant, and cell turgor. This is most dramatically illustrated in events leading to freezing injury. Ice formation usually begins in the extracellular solution and increases the extracellular solute concentration because solutes are excluded from the forming ice crystals. Because of the semipermeable nature of the plasma membrane, the cell behaves as an osmometer and dehydrates in response to the lower chemical potential of the extracellular solution (Steponkus, 1984). Equilibrium is established by continued water flux from the cell or by intracellular ice formation. Various hypotheses have been suggested regarding the specific mechanisms of freezing injury. Expansion-induced lysis of the plasma membrane during freeze–thaw cycles may be the most common mechanism of freezing injury, although loss of osmotic responsiveness and direct damage to cellular membranes by intracellular ice formation are other mechanisms that have been reported.

2. Temperature × CO_2 Interactive Effects on Plant–Water Relations In absolute terms, the greatest benefit from CO_2 enrichment occurs when water is not limiting, but, on a relative basis, the percent growth increase with CO_2 enrichment is often as large or larger under mild to moderate water stress as it is under nonstress conditions. This is presumably due to the partial stomatal closure response to increases in CO_2 (Mott, 1990) and improved water use efficiency. Studies with cotton (Wong, 1993; Radin *et al.*, 1987) showed that the stomates of CO_2-enriched plants were more sensitive to water deficits, closing more quickly or completely than those of plants grown at ambient CO_2. Two other plant responses to elevated CO_2 that are sometimes observed, and that may be beneficial under water-limiting conditions, are a decrease in stomatal density (Oberbauer *et al.*, 1986) and an increase in the root-to-shoot ratio (Tognoni *et al.*, 1967). In a future high-CO_2 world, the interactions between temperature and CO_2 effects on plant–water relations will be important in some environments, but these interactions have not been well investigated.

An example of a potentially important interaction is the recent evidence that elevated CO_2 can mitigate chilling-induced water stress in some species (Boese *et al.*, 1997). The rate of water loss from bean and cucumber leaves during chilling (24 h at 5°C) was less at elevated compared to ambient CO_2, and the decline in leaf water potential Table I) and relative water content were significantly less at elevated CO_2. Leaf osmotic potentials were more negative at elevated CO_2 (presumably associated with higher photosynthetic rates and leaf carbohydrate levels at elevated CO_2), leading to higher leaf turgor for any given leaf water potential as leaf water potential declined during chilling. After 24 h of chilling, photosynthetic reductions and leaf damage were also significantly less at elevated compared to ambient CO_2 (Table I). These experiments were conducted in growth chambers. The implications for long-term growth and reproductive success in managed and natural ecosystems will require evaluation under field conditions.

D. Plant Development

1. Temperature and CO_2 Effects on Developmental Rate Within the nonstress temperature range, plant developmental processes frequently respond to

Table I Effect of Elevated CO_2 on the Response of 4- to 6-Week-Old Plants to a 24-h Chilling (5°C) Treatment[a]

Crop	Growth CO_2	ΨL(MPa) 24-h chill	Percent P_n reduction		Leaf damage (%)
			24-h chill	24-h recovery	
Bean	350	−1.72	83	62	21
	700	−1.23	53	0	10
Cucumber	350	−1.34	89	46	32
	700	−1.15	40	45	24
Maize	350	−0.63	24	25	0
	700	−0.60	4	0	0

[a] From Boese *et al.* 1997. Data shown are leaf water potential (ΨL) after chilling; percent reduction in photosynthetic rate (P_n) of leaves of chilled plants compared to nonchilled controls; and percent of total plant leaf area with irreversible chilling damage.

accumulated temperature such that specific phenological stages (e.g., leaf number, first flower, first ripe fruit) are reached sooner at warmer temperatures (Johnson and Thornley, 1985). The mechanisms for this are not well understood, but temperature effects on enzyme kinetics (see Section II,A,1) are undoubtedly a major factor. The "heat unit" or "day-degree" concepts (Johnson and Thornley, 1985) involve totaling mean temperature (above a threshold) over days and using this to predict the rate of plant development and crop maturity. This method has proved reliable in some regions for predicting harvest date of some crop species (Perry *et al.*, 1986; Boswell, 1935; Magoon and Culpepper, 1932).

Most crop models used in climate change research incorporate temperature effects on development in some manner. This often leads to a prediction of reduced yields with increasing temperatures for determinate crop species, because plants reach maturity much sooner and the total growing season length is shortened. Simulations by Rosenzweig and Parry (1993), for example, predicted lower yields for grain crops with increasing temperature scenarios of various global climate models. Only when an assumption was made that farmers could adapt by selection for longer growing season varieties was a benefit from global warming predicted for some regions.

Several CO_2 studies have reported accelerated ontogeny of CO_2-enriched plants (Conroy *et al.*, 1986; Radoglou and Jarvis, 1990), but this is not always observed, and mechanisms (e.g., CO_2 effects on hormone physiology) have not been identified. In one recent study with carrot (Wheeler *et al.*, 1994), effects of both temperature and CO_2 on development were evaluated. These researchers found that the increase in biomass at warm temperature was due almost entirely to temperature effects on rate of crop development, whereas increased biomass associated with elevated CO_2 was due partially to larger shoot and root size, as well as faster development. When plants were compared at the same leaf stage, there was almost no temperature effect, but still a 25% benefit in terms of biomass from elevated CO_2. More research on the CO_2 direct effects on phenology and possible $CO_2 \times$ temperature interactive effects is needed to improve plant growth models used in climate change research.

2. Temperature and CO_2 Effects on Partitioning and Morphology Both temperature and CO_2 have been reported to affect leaf morphology, the root-to-shoot (R:S) ratio, and partitioning of C and N among plant organs (Wolfe *et al.*, 1998). In some cases increasing temperature and increasing CO_2 have similar effects, while in others the effects are in the opposite direction. Very few studies have focused on the interaction between these two environmental variables with regard to partitioning and morphology.

An increase in CO_2 tends to have the opposite effect on leaf development as an increase in temperature. Plants grown at warm compared to cool temperatures often have thinner leaves, with higher specific leaf area (SLA, leaf area per unit leaf dry wt.) (Wolfe, 1991; Wolfe and Kelly, 1992; Dale, 1965) and fewer cell layers (Boese and Huner, 1990). In contrast, elevated CO_2 has been reported to decrease SLA, lead to extra palisade layer development (Mousseau and Enoch, 1989), increase mesophyll cell size (Thomas and Harvey, 1983; Conroy *et al.*, 1986), and increase internal surface area for CO_2 absorption (Radoglou and Jarvis, 1990).

Although the leaves of plants grown at elevated CO_2 are usually thicker with more leaf mass per unit leaf area, leaf N per unit leaf area often declines because of a decrease in leaf N per unit leaf mass (Conroy, 1992; Luo *et al.*, 1994). The reduction in leaf N concentration at elevated CO_2 is part of a photosynthetic down-regulation acclimation response. Genes encoding for some N-containing photosynthetic enzymes are repressed when leaf carbohydrates accumulate (Sheen, 1994) due to a source–sink imbalance at elevated CO_2. With regard to leaf N per unit leaf area response to environment, increasing CO_2 and increasing temperature may have similar effects. In a study with bean, Wolfe and Kelly (1992) found that plants grown at warm compared to cool temperatures had less leaf N per unit leaf area.

Davidson (1969) suggested R:S is minimum at the temperature range optimum for root function (usually between 20 and 30°C), and R:S increases at growth temperatures below or above this optimum. This shift in allocation may be adaptive and help plants maintain a balance between root function (e.g., water and nutrient uptake) and shoot function (e.g., carbon assimilation) (Thornley, 1977). The effect of elevated CO_2 on R:S varies with species and experimental methods. An increase in partitioning to the roots at elevated CO_2 is frequently observed in herbaceous species, such as tomato (Tognoni *et al.*, 1967) and soybean (Patterson and Flint, 1980), and is most pronounced in root crops, such as sugar beet (Wyse, 1980) and carrot (Wheeler *et al.*, 1994). This effect may be explained by the strong root sink strength for carbon in these crops, and less downward acclimation of photosynthesis. Relatively less R:S response has been observed in cereal crops (Cure and Acock, 1986) and temperate tree species (Eamus and Jarvis, 1989).

E. Temperature Effects on Whole-Plant Growth Response to CO_2

The effects of temperature on growth response to elevated CO_2 involves an integration of many of the basic plant responses discussed earlier, and often reflects the photosynthetic $CO_2 \times$ temperature interaction (Section II,A; Fig. 1). Most studies have found that, within the nonstress temperature range for a particular species, beneficial effects of CO_2 enrichment on growth increase with increasing temperature. There is little growth or yield increase from CO_2 enrichment at extremely low temperatures or high temperatures. The minimum temperature required for beneficial effects from CO_2 enrichment varies with species, period of acclimation, and experimental methods.

Using temperature gradient tunnels in the field, Rawson (1995) observed that wheat grain yields increased only 7% with a CO_2 doubling in winter plantings where temperatures averaged about 12.5°C. A temperature increase of 2°C above ambient in these winter trials had little impact on CO_2 response. In contrast, in summer plantings when temperatures averaged about 20°C, yields increased by 34%. Another study with wheat (Krezner and Moss, 1975) found no significant increase in wheat yield from CO_2 enrichment at the cool temperature regime of 13/7°C (day/night) temperature. Idso and Kimball (1989) reported no CO_2 response in carrot and radish at temperatures below 12.5°C. Rawson (1992) compiled data from several sources and found the minimum temperature for CO_2-beneficial effects to be about 10°C. Several CO_2 growth response curves relative to growth temperature reported in the literature are shown in Fig. 2.

Less information is available regarding temperature effects on growth response to CO_2 for natural ecosystems. Little benefit from long-term CO_2 enrichment was observed in natural tundra vegetation at a high-latitude site with cool temperatures (Grulke *et al.*, 1990), whereas enrichment of warm wetland vegetation at a low-latitude site resulted in significant increases in productivity (Drake and Leadley, 1991). Other factors, such as limited nutrient availability at the high-latitude site, may also have been involved (see Section III,B,2).

High-temperature stress has a severe negative effect on reproductive development in some species, affecting growth and yield response to CO_2 enrichment. Recent greenhouse experiments with beans (Jifon and Wolfe,

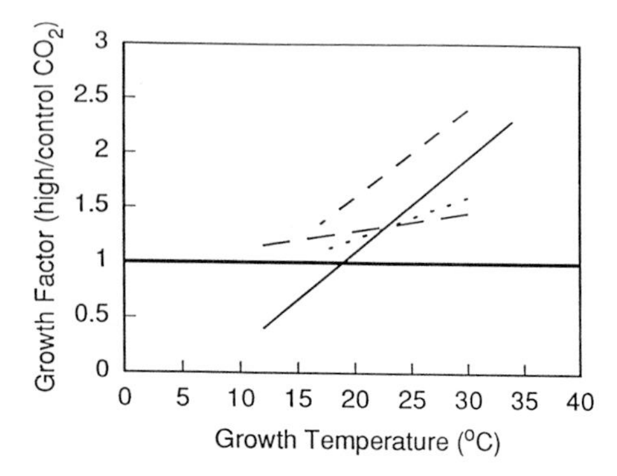

Figure 2 Growth response of plants to a doubling of atmospheric CO_2 concentration in relation to average air temperature. A growth factor of 1.0 indicates no response to elevated CO_2; values below or above 1.0 indicate negative or positive responses to CO_2, respectively. Results are based on combined data for five plant species (solid line) from Idso *et al.* (1987); data for carrot (medium dashes) and radish (dotted line) from Idso and Kimball (1989); and combined data for six plant species (long dashes) from Rawson (1992).

unpublished) have found no beneficial effect from a CO_2 doubling on pod yield at high temperatures (35/21°C, day/night) because few pods are produced at high temperature regardless of CO_2 environment (Table II). Total biomass production also is not stimulated by elevated CO_2 at high compared to optimum temperature. Despite lower photosynthetic rates at high temperature, accumulation of carbohydrates in the leaves of CO_2-enriched plants is similar to that observed at optimum temperature (Table II). Downward acclimation of photosynthesis is usually greater at high compared to optimum temperature, perhaps associated with reduced sink demand for photosynthates at high temperature. In a similar study, Ahmed *et al.* (1993) found greater leaf carbohydrate accumulation, more pronounced downward photosynthetic acclimation, and little growth and yield benefit from elevated CO_2 imposed at high temperatures in a heat-sensitive bean line compared to a tolerant line, which had less impairment of reproductive development in the high-temperature treatments.

III. Population Responses in Arctic and Alpine Habitats

A. Sources of Evidence

Any discussion of plant responses to increased temperature and carbon dioxide levels would be incomplete if it considered only the immediate experimental response that can be produced by laboratory simulations. To

Table II Temperature and CO_2 Effects on Photosynthesis, Leaf Carbohydrate Accumulation, Total Biomass, and Pod Yield of Bean[a]

Growth temperature (°C)	Growth CO_2 (μmol mol^{-1})	Photosynthesis (μmol m^{-2}s^{-1}) measured at		Leaf carbohydrates		Total biomass (g plt^{-1})	Pod yield (g plt^{-1})
		350	700	Soluble sugars (μg g^{-1})[b]	Starch (μg g^{-1})		
25/15	350	29.6 ± 6.7	53.6 ± 5.0	12.3 ± 2.0	91.5 ± 7.9	98.6 ± 5.8	34.9 ± 3.6
	700	18.7 ± 1.3	34.5 ± 2.3	22.4 ± 4.7	151.2 ± 21.3	122.2 ± 9.7	54.2 ± 4.5
35/21	350	7.2 ± 0.9	8.6 ± 1.9	12.1 ± 2.6	89.0 ± 24.9	112.8 ± 13.5	22.9 ± 6.5
	700	2.9 ± 0.3	11.8 ± 2.6	26.7 ± 8.3	130.4 ± 28.4	83.1 ± 18.7	23.4 ± 3.0

[a] CV. Redkloud, 54 days after planting; Jifon and Wolfe unpublished.
[b] Glucose + sucrose as glucose equivalents.

answer some of the wide-ranging and important questions that are raised by climatic change, such as impact on species distribution and consequences for natural productivity of ecosystems and crops, it is necessary to go beyond the particular focus of the preceding section (Section II) and consider the effects of temperature and seasonality on processes operating at the plant population and species level. A fundamental question with regard to the response of whole plants to climatic change is whether or not the natural distribution of species will be altered. Answering such a global question in a precise scientific manner presents many practical difficulties. Consequently, experiments on the ecological consequences of climatic change are usually directed at subsidiary questions, namely, the impact of rising temperatures and CO_2 availability on plant growth and development. Results obtained on these latter problems are then used to predict possible answers to the first basic question, on the expected impact of climatic change on species distribution. Although such experiments may describe the initial effects of climatic warming, there are three possible basic reasons which suggest that caution should be used when accepting their long-term validity.

1. Investigating effects of increased temperature or CO_2 on sample plots observes only phenotypic responses to short-term perturbations.
2. Past vegetation migrations and, presumably, also those of the future are demographic movements that will involve population and genotypic alterations.
3. The capacity to migrate and survive in new habitats as a result of alterations in climate may not be directly related to positive growth responses to increased temperatures or higher atmospheric CO_2 levels as observed experimentally.

There is, therefore, a need to examine the contrary argument, that facility for migration is not necessarily a property of populations that show rapid growth responses to temperature change, but may instead take place more readily in those populations where completion of the life cycle and population survival is possible over a broad range of temperatures.

Current investigations can be considered under two categories, modeling and experimentation. On one hand, climatic models with their emphasis on physical parameters and assessment of productivity potential of growing seasons (Prentice *et al.*, 1992) tend to produce predictions in which climatic warming is considered likely to bring about rapid ecological change, especially at high latitudes. Model-based predictions that large areas of tundra in Alaska, Canada, and northern Eurasia will be rapidly colonized by subarctic boreal trees (Huntley, 1997) are not yet supported by any marked changes in tree distribution in these regions (see later discussion). On the other hand, controlled experiments on plant responses to temperature and CO_2 levels are at first sight scientifically appealing, and possibly more realistic, because they examine direct responses of plants to strictly defined changes in temperature. Such experiments are usually carried out either under controlled environments or by some artificial perturbation of the natural environment from soil warming cables or the emplacement of shelters around patches of vegetation in order to provide "passive warming." This latter method has many imperfections in the manner in which the environment is altered (Kennedy, 1995). Even in temperature-controlled chambers, uncertainty arises through the conflicting responses that can be obtained depending on the relative degree of soil or air warming. Further problems arise from the limited length of time that these perturbations are usually maintained and the frequent inability to imitate changes in seasonality. Initial responses to warming under tents can disappear after a few years as growth becomes limited by soil nutrients (Chapin *et al.*, 1995; Wookey and Robinson, 1997). In all of these growth perturbation analyses, irrespective of whether they are carried out in the laboratory or in the field, the results are based on a few individual plants from a particular geographical location (Grime, 1996) which may not necessarily be representative of the species as a whole. These few examples serve to highlight the limitations of experiments and the need for observing and analyzing wherever possible long-term natural responses of plants to past and present climatic alterations.

B. Species Distribution and Climate Change

1. Effects of Temperature on Northern Plant Migrations The ease with which temperature records can now be collated and compared with past and present plant distribution maps can create an impression that cartographic representation combined with mathematical modeling is all that is needed to forecast the probable effects of climatic warming on future plant distribution. An expectation of large migrational activity with temperature change is perhaps understandable, given the detailed information that is available on early Holocene plant migrations. The rapid advance of boreal trees at rates of 1 to 2 km yr^{-1} in both Europe and Canada around 9 ka B.P. caused a substantial reduction in the area of the Tundra biome (Huntley and Birks, 1983; Huntley and Webb, 1988; Ritchie, 1987). However, plant migration in a modern world differs from the recolonization that took place after the Pleistocene *tabula rasa.* Resistance to poleward migration can be expected now due to the presence of existing plant communities and their specific soil types (Billings, 1997) as well as anthropogenic disturbance. Bog formation is a striking example of a climax community that has increased its extent, particularly in arctic and subarctic locations, during the past 6000 years and which will have marked inhibiting effect on the northward migration of boreal forest (see later discussion).

The biogeographical history of the Holocene provides possibly more information on the effects of climatic cooling than on climatic warming.

The rapid rise in temperature that took place as the ice retreated up to the period known as the hypsithermal (approx. 10–7 ka B.P. depending on location) (Edwards and Barker, 1994; Feng and Epstein, 1994), was followed by a general tendency for a decline to lower temperatures (Pielou, 1991). The consequences of this progressive temperature decline during the last 5000–7000 years have varied depending on the region. In some areas, small reductions in temperature have been followed by marked vegetational changes due to a parallel onset of more oceanic conditions. Thus, in northern Sweden there was a regression from a boreal birch and pine forest to subarctic birch woodland tundra ca. 3400 B.P. that has existed to the present day due, not so much to a change in temperature, as to the onset of a more oceanic climate with greater soil leaching and erosion and the establishment of heath vegetation in place of forest (Holmgren and Tjus, 1996). A similar contemporaneous change took place in Shetland (U.K.) and has been considered as part of an autogenic succession under oceanic conditions rather than a direct response to decrease in temperature (Bennett *et al.*, 1992).

The present global warming trend will be likely to have its eventual impact through reversal of the changes brought about by the Little Ice Age. In Finland, it was already realized in the early 1950s that this northern region of Scandinavia had just undergone a significant climatic warming during the first half of the twentieth century. A north–south orientation, on mainly level ground, with a low level of rural disturbance, makes Finland ideally suited to mapping the effects of climatic changes on natural limits to plant distribution. Comparision of past distribution records shows that some of the principal changes taking place at this northern latitude are related to changes in the moisture regime that have allowed a number of aquatic plant species to migrate inland as a response to increased oceanicity rather than north as might have been expected (Erkamo, 1956). Figure 3 shows the example of *Typha latifolia* which since 1900 has expanded eastwards rather than north. Such a migration would have been unlikely to be anticipated in computer simulations such as BIOME (Prentice *et al.*, 1992) because this model's predictions are based principally on growth potentials and not on dispersal and colonization of previously unavailable habitats. The effects of winter water table levels and water movement, which have had such marked effects on Finland would also have been ignored by most contemporary model predictions on vegetation change.

Predicting plant migration in terms of temperature change demands correlative observations from a range of variables. The classical studies of Iversen (1944) pioneered the matching of plant distribution to summer and winter isotherms. The choice of isotherm, however, is often arbitrary and the coincidence of a species boundary and an isotherm is not necessarily causal (Crawford, 1989; Hengeveld, 1990) and may over-estimate the geo-

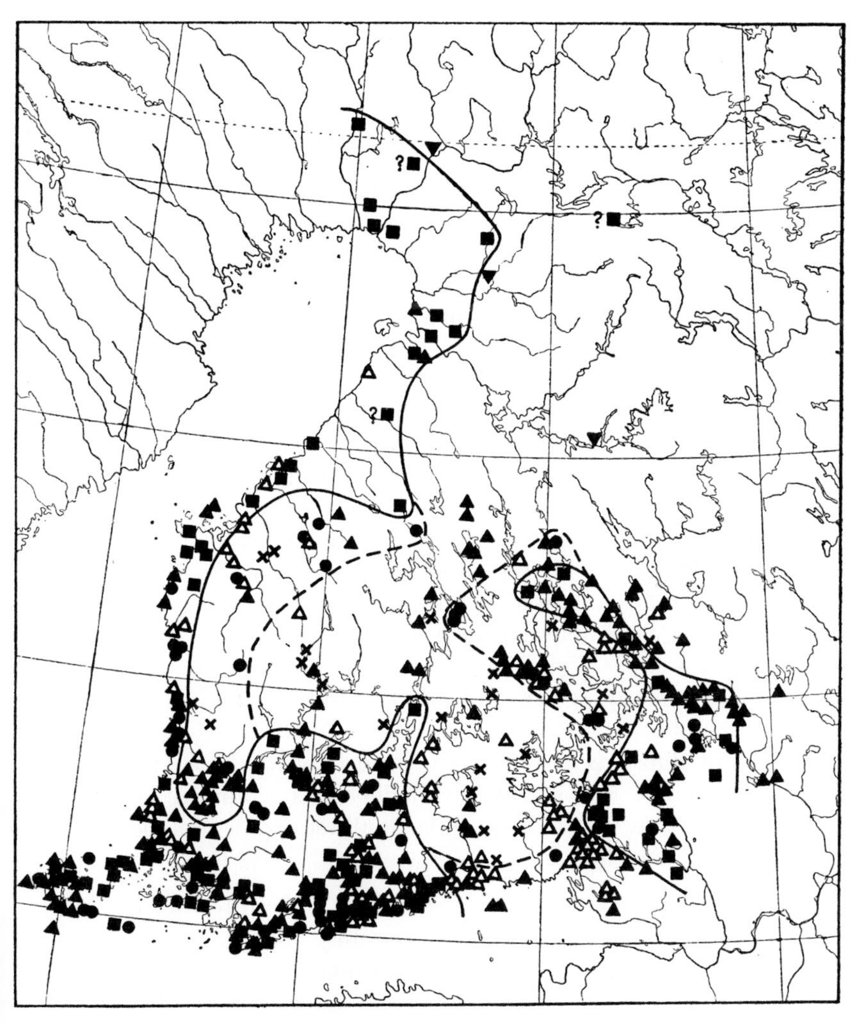

Figure 3 Spread of *Typha latifolia* in Finland 1900–1952; ■, at or before 1900; ●, new records 1901–1925; ▲, new records 1926–1950, x, special survey carried out in 1949. Open symbols represent Erkamo's observations; solid symbols other observations. (Reproduced with permission from Erkamo, 1956.)

graphical area occupied by species. Isotherms are equivalent to straight lines in *x, y* Cartesian space. Therefore, sets of four isotherms (upper and lower *x* and *y* limits) map out rectangles of temperature space. Species, however, occupy scatters which have the form of inclined ellipses. Consequently, the fit of an isotherm to an ellipse is good only locally and any isotherm-box overestimates species distributions in environment space by adding unoccupied corners (Jeffree and Jeffree, 1994).

A more objective manner of comparing species distribution to the interaction of winter and summer temperatures has recently been described and examined for a number of European species (Jeffree and Jeffree, 1994, 1996), in which climatic temperature preferences of species are described in terms of the temperatures of the coldest (t_x) and warmest (t_y) months of the year recorded at locations within a species' geographical distribution. Plotted on Cartesian coordinates for t_x and t_y, the temperature data for locations occupied by a species form a bell-shaped probability density distribution described about the bivariate mean t_x bar, t_y, bar. Elliptical contours on this surface, representing defined limits in the continuously diminishing scale of probability that a species may occur at increasing distance from the bivariate mean, may be calculated using the equations given by Jeffree and Jeffree (1994) for any desired level of probability. Consequently, ellipses, calculated from temperatures within species distributions may be used to define and map those parts of the world's surface in which temperatures are potentially suitable for the species at a specified level of probability. Thus, a temperature ellipsoid calculated for a species distribution under current climatic conditions may be used as a template for identifying those geographical areas that might fall within the temperature range suitable for the species in any scenario of past or future climate change (Jeffree and Jeffree, 1996).

The use of ellipsoids derived from t_x and t_y for mistletoe (*Viscum album*, Figs. 4A–D; Jeffree and Jeffree, 1996) illustrates a number of important features concerning the direction and magnitude of potential species migration under climate change.

1. Species migration is sensitive to existing temperature *seasonality* and the superimposed *seasonality* of temperature change.
2. Migration cannot be predicted from annual mean temperature alone.
3. For the same patterns of change, in different parts of their ranges, species could migrate in *opposite* directions.
4. Seasonality gradients may present barriers to migration notwithstanding overall warming.

According to these authors, mean annual temperature alone proves to have little value as a basis for predicting species response to climate change, falsely indicating the likely direction of change relative to the region of t_x, t_y temperature space, which represents species temperature preferences. The potential ranges of species with southerly British distributions and subcontinental distributions in Europe are most likely to extend into the British Isles if summer warming exceeds that in winter, inducing a more continental climate type than at present. However, GCM scenarios of cli-

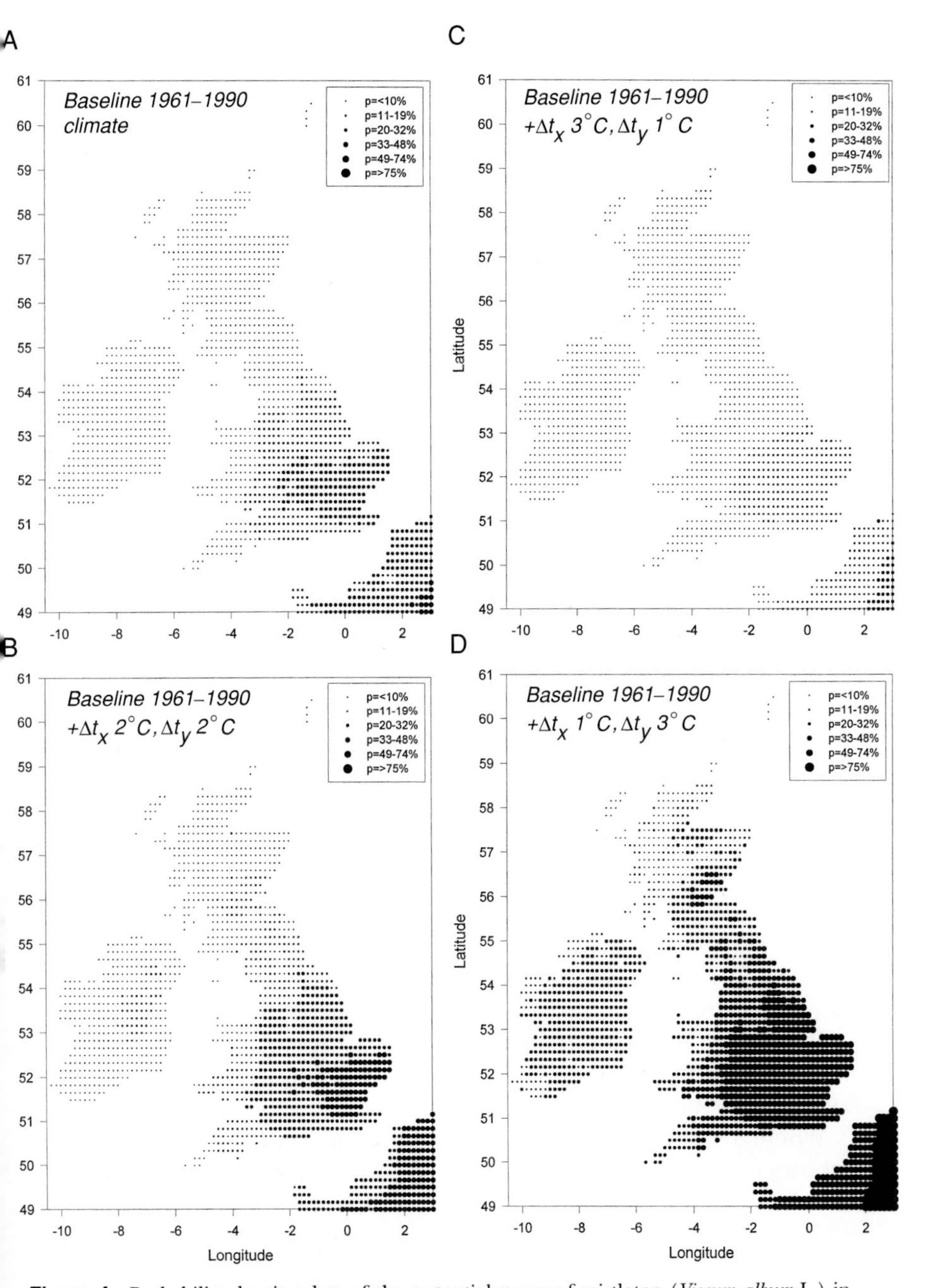

Figure 4 Probability density plots of the potential range of mistletoe (*Viscum album* L.) in the British Isles, showing the consequences of an increase in mean annual temperature of 2°C obtained by distributing the relative amounts of summer and winter warming in different ways to obtain three contrasting amounts of additional seasonality. (A) The modeled probability distribution of *Viscum* in the British Isles under current (1961–1990) climate. (B) The temperature increase is applied equally, 2°C each to summer and winter temperatures. (C) The winter temperature is increased by 3°C and summer temperature by only 1°C. (D) The summer temperature is increased by 3°C and the winter temperature by 1°C.

mate change indicate that the likely outcome of greenhouse warming will be greater in winter than in summer in much of Northern Europe. Thus in Fig. 4 (Jeffree, unpublished) an extension of the range of mistletoe in the British Isles will take place only if summer warming is equal or greater than winter warming. The more oceanic climate with warmer winters (the current meteorological expectation) is predicted as leading to a reduction of the species area in the U.K. and adjoining areas of the Netherlands, Belgium, and France (Figs. 4B and D).

2. Effects of Increased Atmospheric CO_2 on Vegetation in Low-Temperature Habitats Despite their low growth rates and small stature, plants from montane and high-latitude habitats have been consistently observed as having higher rates of photosynthesis and respiration as compared with lowland or more southern species (Körner and Larcher, 1988; Crawford, 1989). In montane species from the European Alps the photosynthetic capacities of pairs of species from low- and high-altitude sites have been compared for their response to light, temperature, and increased atmospheric CO_2. The plants from high altitudes were found to have a higher rate of CO_2 uptake as determined on a leaf area basis (Körner and Diemer, 1987). Further studies have shown that alpine plants may be able to obtain, at least initially, greater carbon gains in a CO_2-enriched atmosphere than comparable lowland plants (Körner and Diemer, 1994). As with studies on plant respiration in warm and cold sites at high latitudes (Crawford *et al.*, 1993, Crawford and Abbott, 1994) this increase in metabolic activity can be regarded as enhanced *capacity adaptation* (Hochachka and Somero, 1973), which allows the plants to compensate metabolically for both low temperatures and the brevity of the growing season. This capacity enhancement will be due in part to differences in morphology of lowland and upland leaves (Körner *et al.*, 1989) and in part to increased specific activity of Rubisco (Crawford, 1989). Thicker leaves, with a greater development of palisade tissues and a higher internal volume, will probably account for the greater assimilation of CO_2 at high altitudes in montane species (Fig. 5) (Körner *et al.*, 1989) and can be compared to the greater *vital capacity* (air expired after inspiration) of Quechua Indians born and raised at high altitudes in the Andes (Heath and Williams, 1977).

Whether or not changes in Rubisco activity will lead automatically to increased productivity is open to question. In tobacco plants in which Rubisco activity has been reduced by being transformed with antisense *rbcS* (Stitt and Schulze, 1994), it can be demonstrated that reduced Rubisco levels do not necessarily lead to decreased photosynthesis or growth. Decreased starch accumulation in leaves may actually allow a more efficient use of fixed carbon as lower amounts of Rubisco can also make available nitrogen that could otherwise be limiting growth due to sequestration in Rubisco protein.

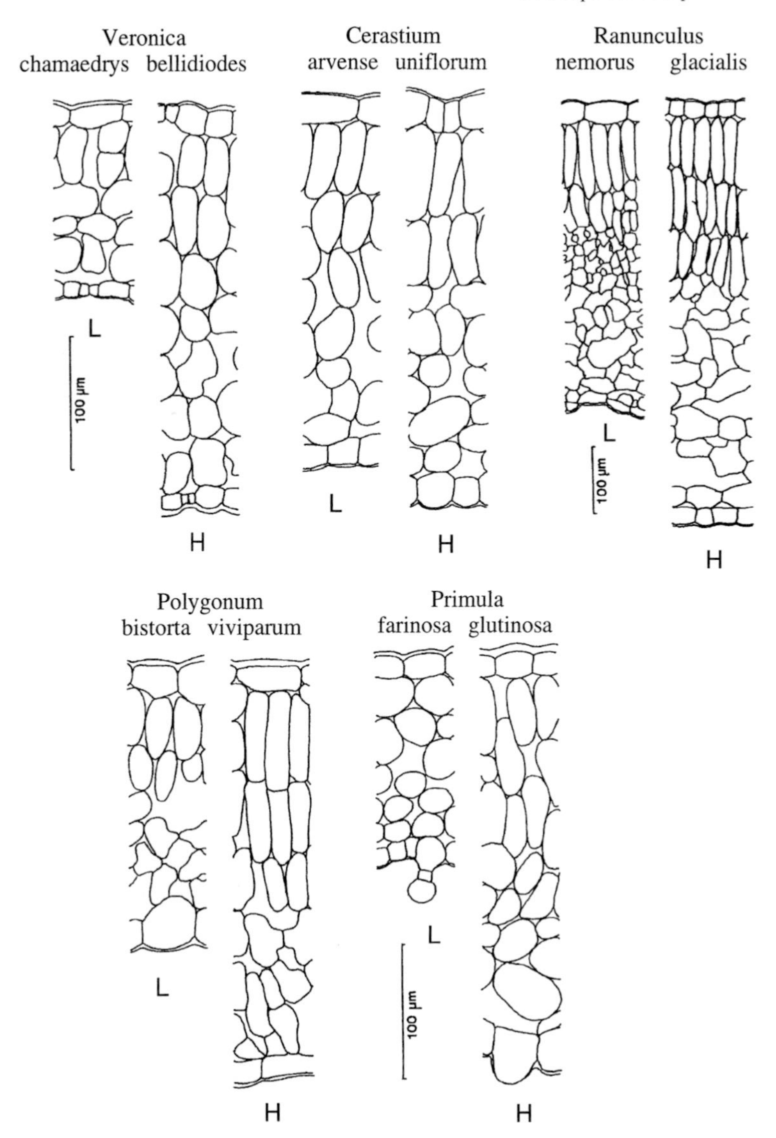

Figure 5 Comparison of leaf thickness and internal development lowland and montane pairs of species. (Reproduced with permission from Körner *et al.*, 1989).

In perennial plants from arctic and montane habitats, growth is not directly related to photosynthesis and many plants retain a small structure despite active photosynthetic activity. In alpine plants it has been clearly demonstrated that even although elevated CO_2 levels may increase photosynthetic activity it is not necessarily translated into higher biomass production (Schäppi and Körner, 1996). If lower levels of Rubisco can increase growth it therefore follows that higher levels of Rubisco could cause a reduction. In many habitats a reduction in growth and an increase in stored carbohydrate from increased photosynthetic activity could have a significant effect on survival. In uncertain environments starch accumulation without growth is important in aiding plants to overcome long nonproductive periods. Some plants that live in arctic snow patches can survive 2–3 yr without emerging from the snow bank (Pielou, 1994). Plants in high-latitude or high-altitude habitats can be metabolically active without necessarily producing extensive growth. In the context of this present discussion, it would appear that an enhanced *capacity adaptation* (whether of enzymatic or morphological origin) would allow montane species to profit from the current rise in atmospheric CO_2 in terms of carbon balance even if it did not enhance growth. In a number of ways these findings are counterintuitive because they suggest that positive *biomass* responses to CO_2 enrichment are unlikely in current dominant alpine species (Schäppi and Körner, 1996).

Notice also has to be taken of the morphology of prostrate and cushion plants in high Arctic habitats in relation to the manner of access to sources of CO_2. In the high Arctic the bulk of terrestrial photosynthesis is carried out by algae and nitrogen-fixing cyanobacteria (Getsen *et al.*, 1997). The carbon that is fixed by the soil microflora may contribute significantly to the CO_2 resources of prostrate and cushion forms of higher plants through soil respiration and make them less likely to be influenced by increases in global atmospheric levels of CO_2. Estimation of isotope discrimination ratios in *Saxifraga oppositifolia* growing at 79°N in Spitsbergen gave a $\partial^{13}C$ value of $-19.5‰ \pm 0.32$ ($n = 8$) and suggests that soil CO_2 is a significant contributor to the carbon content of this species (Crawford *et al.*, 1995). A case, therefore, emerges which implies that both high-latitude and high-altitude plants are already either preadapted to profit from current increases in atmospheric CO_2 or else are indifferent to atmospheric CO_2 (as the soil is their main source) and can therefore be expected to maintain their present ecological position in a CO_2-enriched world.

At a global level, the marked seasonality of photosynthesis in the far north causes atmospheric CO_2 levels to fall in summer and rise in winter to a greater degree than at lower latitudes. The current global range of this cycle is between 15 and 20 ppm by volume. A recent examination of changes during the last 30 yr shows that the annual amplitude of these seasonal cycles has increased by 20% in Hawaii and by 40% in the Arctic (Keeling *et al.*, 1996). It has also been noted that the annual amplitudes show a sensitivity to global warming that peaked in 1981 and 1990 and are also accompanied by phase advances of about 7 d, suggesting a lengthening of the growing season. However, it has to be recognized that climatic warming at high latitudes is most marked in its effect on winter temperatures. Winter CO_2 emissions from soil and forests could therefore account for 40% of the seasonal amplitude in atmospheric CO_2 north of 60°N. Thus, an increase in plant respiration in winter could account for the greater amplitude of atmospheric CO_2 without any change in summer primary production (Chapin *et al.*, 1996). Here again, the importance of examining winter-warming emerges as possibly the dominant factor in assessing the future ecological consequences for plant growth and distribution.

3. Effect of Climate Change on Regional Crop Potential As with natural vegetation it is possible with crop plants, by using models based on present limits to distribution, to make a calculation that might reflect the potential poleward extension of cultivation of crops such as wheat for a given climatic warming scenario. On this basis it is possible to suggest from studies on the growth of specific varieties of spring wheat (e.g., *Triticum aestivum* cv "Kadett") that per 1°C of warming the cultivation of this crop could shift northwards in eastern Finland by 290 km and in western Finland by 110 km. However, due to the phase shift in the ripening period there would be a reduction in grain yield unless new slower maturing and higher yielding cultivars became available (Carter and Saariko, 1996). Similar prognostications can be made for other crops such as rice for which a higher temperature, although it could increase crop biomass, may nevertheless result in a reduction in yield due to an increase in spikelet sterility (Matthews *et al.*, 1997). These are but two examples from many others that could have been cited which illustrate the complexity of factors of crop production which will certainly involve extensive breeding programs before profitable arable farming can be extended to new areas at high latitudes. When economic factors are added to problems of productivity and the more specialized concept of yield, any prediction about arable crops has to be considered with extreme caution.

Despite the qualifications given for arable farming, a more optimistic prognostication is possible for pastoral farming in northern maritime regions. In many northern areas the overwintering of livestock has been limited in the recent past (the Little Ice Age) by the length of the winter coupled by the low productivity of adjoining upland areas in summer. This has been particularly the case in oceanic environments due to the high lapse rate (e.g., in Scotland 0.6–0.8°C 100^{-1} m of altitude), which also limits the winter pasturage to ground below 150 m. The probability of higher temperatures will alleviate this disadvantage. Similarly, the probability of increased drought frequencies in more southerly regions of Europe and North America will create a premium for the summer productivity of montane pastures, particularly in oceanic areas that are free of drought.

C. Montane Vegetation and Temperature Change

Concern is being expressed about the impending fate of montane vegetation should there be a sustained and marked climatic warming of up to 2°C (Grabherr *et al.*, 1995). The specific sources of danger fall into two categories: (1) an upward migration of forest that will eliminate the communities from mountain summits (Grabherr *et al.*, 1994) and (2) the disappearance of snow patches and their associated species in the subnival zone (Guisan *et al.*, 1995). In Austria, mean annual temperatures have already increased by 0.7°C in the last 70–90 yr and, based on a lapse rate of 0.5°C 100 m^{-1}, should be, in theory, sufficient to move the vegetation zone upwards by 8–10 m per decade. However, careful comparison of present species locations with previous precise records shows maximum values of only 4 m per decade with most species moving upward at only 1 m per decade (Grabherr *et al.*, 1994).

Due to the height of the Alps, and most mountains in northern Scandinavia, the extent of the nival zone is sufficient to accommodate much of this upward migration without imminent danger of species extinctions. Within the alpine zone of most mountains there is also sufficient variation in microclimate to accommodate the changes that may arise from global warming (Körner, 1995). There are at present few examples of species that are likely to become extinct in the higher mountain ranges within the next century (Körner, 1995). On the contrary, instead of species disappearing off the tops of mountains there has been in the Alps an increase in species richness on mountain summits during this century. On Mt. Linard (3411 m), there was one summit species in 1835, four in 1895, and 10 in 1992 (Grabherr *et al.*, 1995). It is in some of our lower and less-studied mountains, however, that examples can be found of species that are on the verge of extinction. In the French Massif Central the highest mountain is the Mezenc (1754 m). Just under the summit of the mountain on its north-facing slope (Fig. 6) grows the last French colony of the rare montane *Senecio leucophyllus*. The few remaining plants do not exhibit any loss of vigor or reproductive capacity as a result of climatic warming, but nevertheless appear unlikely to survive due to the invasion of their preferred habitat by grasses (Fig. 7). Fortunately, the species still exists in the eastern Pyrenees and its likely disappearance from its last French locality will not lead to its extinction.

Paradoxically, the greatest danger to montane floras from rising temperatures is probably not to be found in continental mountains with marked temperature rises, as in the central Swiss Alps, but in mountains with oceanic climates. It might have been expected that oceanic mountains would be buffered against climatic change due to a reduction in annual temperature range. However, the cause of the relatively species-poor mountain flora of

Figure 6 The isolated summit of the Mezenc (1754 m) in the French Massif Central. Just under the summit on the right-hand (north) side in a last patch of open scree is the last French colony of the rare montane *Senecio leucophyllus* DC.

the Scottish Highlands and southwest Norway is considered to be due, at least in part, to the mild periods of winter weather that encourage premature spring growth when there is still a risk of exposure to spring frosts. In Norway, those montane species that are absent from more oceanic mountains have been described as *southwest coast avoiders* (Dahl, 1951, 1990).

In Scotland a negative influence of warm winters has been investigated physiologically in both moorland and coastal vegetation. In heath vegetation (Stewart and Bannister, 1973) the disappearance of some heath species from lowland areas was related to the difficulty of maintaining carbohydrate levels at warmer temperatures. In a study of four umbelliferous coastal species, two of northern and two of southern distribution, it was found that the overwintering roots of the northern species are much more responsive to temperature increase than the southern species (Fig. 8). Consequently, should the northern species be exposed to warmer temperatures this will also lead to increased winter carbohydrate consumption, spring-starvation, and a subsequent northward retreat (Table III) (Crawford, 1996).

Figure 7 Close up view of *Senecio leucophyllus* DC, growing at 1744 m on the north slope of the Mezenc, French Massif Central.

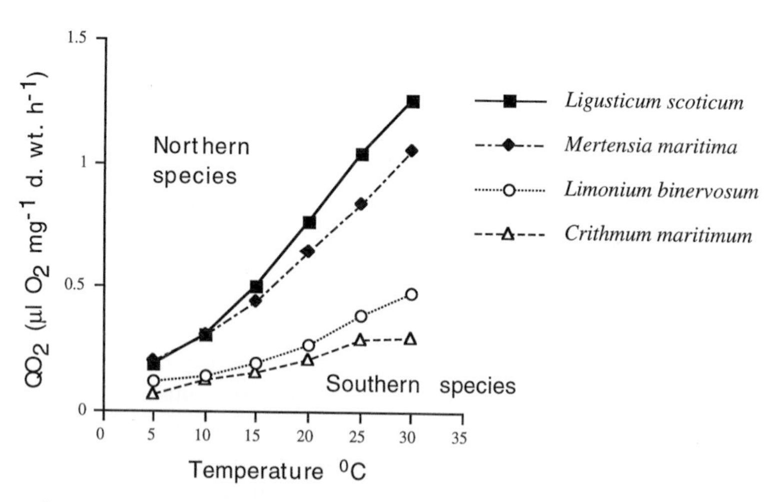

Figure 8 Respiration (oxygen uptake) by roots of four coastal Umbelliferae species, two with northern distributions reaching their southern limit in Scotland, *Ligusticum scoticum* and *Mertensia maritima,* and two with southern distributions reaching their northern limits of distribution in England, *Limonium binervosum* and *Crithmum maritimum.* (Data from Crawford and Palin, 1981.)

Table III Calculated Loss of Carbohydrate (as Glucose)[a]

	1 day		1 week		Time to lose 50% dry weight in days	
	5°C	25°C	5°C	25°C	5°C	25°C
Ligusticum scoticum	0.7	3.0	5.0	20.9	70	17
Mertensia maritima	1.0	3.5	7.0	24.3	50	14
Crithmum maritimum	0.4	1.3	2.7	9.4	131	37
Limonium binervosum	0.5	1.5	3.5	10.5	100	33

[a] Determined from respiration data and expressed as percent of original (total) root dry weight. The time for loss of 50% of dry weight was calculated directly from these values (Crawford and Palin, 1981).

C. Treelines and Climate Change

Attention is currently being given to the detection of northward or upward movement of arctic and alpine treelines as a possible source of confirmatory evidence of the reality of climatic warming. Throughout the northern world, from Mongolia (Jacoby *et al.*, 1996) to Siberia (Briffa *et al.*, 1995) and through central Europe (Spiecker *et al.*, 1996) and North America (Lavoie and Payette, 1994), there is extensive evidence of improved boreal tree growth in recent years. Although this improved growth is usually coincident with increased temperatures, there are also cases with a demonstrable influence of increased precipitation (Graumlich, 1987; Whitlock, 1993). Despite the improvement in growth, even at marginal sites, observations of either a marked upward or northward extension of the treelines remain minimal.

For montane treelines, provided there is an accurate estimate of the lapse rate, the application of Köppen's rule that the treeline approximates to the 10°C isotherm for the warmest month in the year (Köppen, 1931) provides a basis for predicting the effect of temperature change. In continental areas with a lapse rate of 0.6°C per 100 m, a 1°C rise in temperature would be expected to elevate the treeline by ca. 170 m. For a higher oceanic-type lapse rate of 0.8°C 100 m^{-1} a 1°C rise in temperature would move the treeline upward by only 125 m. Current predictions of a climatic warming of 0.3°C per decade (Houghton *et al.*, 1992) should result in a global upslope movement of trees by 30–50 m per decade. Such rapid movement has not so far been recorded. In the Swedish subarctic a ca. 1.5°C rise in mean annual air temperature between 1910 and 1940 gives a calculated projection of more than 200 m in the upper level of birch colonization. However, this has not materialized, because there has been only a small rise of 20–50 m (Holmgren and Tjus, 1996). Similarly, responses to recent climatic warming of *Pinus sylvestris* and *Pinus cembra* within their montane transition zone in the Swiss Alps showed no demographic trends in seedling establishment between 1902 and 1991 despite an 0.8°C rise during the preceding 30 yr. The montane pine ecocline appears to be stabilized by species interactions and not directly responsive to moderate climatic change (Hättenschwiller and Körner, 1955).

Movement of the treeline boundary at the taiga–tundra interface is difficult to assess. In northwestern Canada there has been increased establishment of white spruce (*Picea glauca*) resulting in increasing density, but no substantial increase in the altitude of the treeline during the past 150 yr (Szeicz and Macdonald, 1995). Similar situations have been reported in central Sweden (Kullman, 1987) and northern Québec (Payette and Filion, 1985). The northern limit of the boreal forest (the continental arctic treeline), is usually described as a distinct phytogeographical boundary related to the median July position of the polar front and is approximately equivalent to the 10°C July isotherm (Bryson, 1966) (Fig. 9). This northern limit of the boreal forest is usually considered to be dependent on the summer position of the polar front. However, it appears that the preferred position of the polar front at the edge of the boreal forest is in turn influenced by the nature of the vegetation. The lower albedo of the forest contributes to greater heating as contrasted with the weaker heating and higher albedo over the adjacent tundra. Thus, in contrast to previous assumptions in which the polar front has been presumed to determine the northern limit of the boreal forest, it has been suggested (Pielke and Vidale, 1995) that it is the boreal forest itself that influences the preferred position of the polar front. Hence, the resistance to migration of the treeline may be due to a mutual stabilization of the polar front with the boreal forest-tundra interface. As with the stability of the treeline in the Alps discussed earlier, plant communities may provide a powerful buffer against climatic change (Hättenschwiller and Körner, 1995).

There has been a lengthy debate among Russian ecologists as to whether climatic warming will cause an advance or retreat of the treeline (Kriuchkov, 1971). Those who believe that climatic warming causes an advance of the treeline northwards subscribe to the eventual overriding influence of climatic factors. There is, however, in the north Siberian Lowlands a powerful argument that climatic warming can cause a retreat of the treeline among those who believe that the presence or absence of trees is due to edaphic factors influenced by the proximity of the Arctic Ocean and the long-term persistence of permafrost. The rise in sea level at the end of the Pleistocene resulted in the presence of a huge mass of open cold water in summer in the coastal areas of northern Siberia. This source of cold air with high humidity is a limiting factor affecting the condition of the present lowland Siberian tundra with its excessively wet and treeless terrain supporting only an impoverished flora and fauna (Sher, 1996). In such an environ-

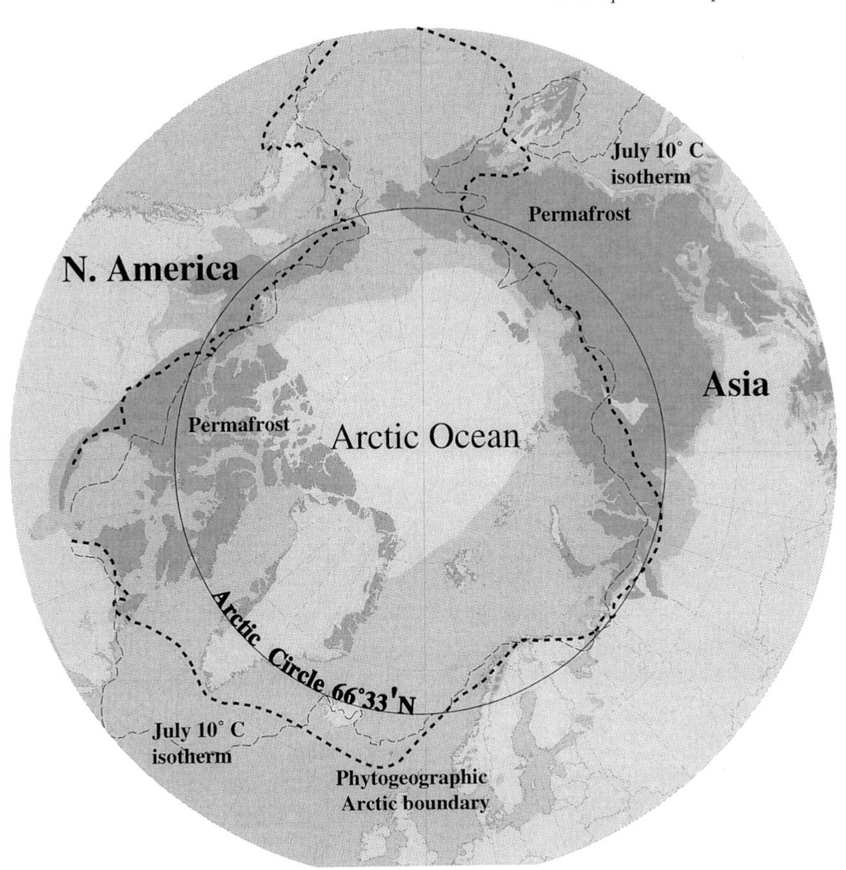

Figure 9 The Arctic region as defined by the Arctic Circle, and the July 10°C isotherm corresponding to the northern limit to tree growth. Note the greater extension of permafrost in Siberia and the proximity of the Arctic Ocean as compared with North America (see text for implication for the position of the treeline). (Map adapted from data compiled by the Norwegian Directorate for Nature management.)

ment any advance of tree vegetation is limited to better drained sites along river banks (Vilchek, 1994). The shade that these trees provide in such areas, where extensive permafrost remains not far below the soil surface, can bring the frozen layer nearer to the surface with the consequent destabilization and death of the trees, followed by reestablishment of tundra vegetation, leading under these humid cool conditions to bog growth (*paludification*).

The relation between climate and permafrost is not direct, but is mediated by a buffer layer, consisting of the snow, vegetation, and organic cover. Increase in vegetation and in the organic layer results in less thaw and greater permafrost aggradation (Brown, 1997). A reversal of succession from forest to bog is a common feature of boreal forests both in North America (Klinger, 1996) and northern Siberia. The Siberian forests, however, may be particularly vulnerable due to proximity of the Arctic Ocean. Boreal forest could exist under present climatic conditions within much of the southern zone of the subarctic tundra. In the Eurasian zone there is a region that attains a north–south treeless expanse of more than 200 km where colonization by native trees (*Larix sibirica, L. gmelinii*) is not limited by warmth (Vilchek, personal communication). These species are able to survive with a shallow active layer above the permafrost and their absence may therefore be due to the counterargument that climatic warming leads to a retreat of the forests due to bog and swamp formation. This form of reversal in tree succession appears to be particularly prevalent in Siberia, possibly due to the greater extent of permafrost as compared with North America (Fig. 9). Notwithstanding the more positive predictions of numerous climate and vegetation models (Huntley, 1997) of marked climatic warming in this region, the establishment of fully developed boreal forest on the shores of the Arctic Ocean may be delayed, possibly for centuries.

It is impossible to make global generalizations, even within the Arctic, as to the probable extent of treeline movement at present, or in the foreseeable future, because the climatic history of the different regions in relation to trees is highly varied. In Alaska, the treeline is currently at its most northerly Holocene extent, while in northwestern Canada there has been a retreat since the mid-Holocene (Edwards and Barker, 1994). Where recent advances in treeline have been observed it is frequently in areas where the species is already present in the Krummholz form and climatic warming has resulted in the emergence of vertical trees. This has been noticed both in subarctic Québec (Lavoie and Payette, 1994) and on the western shore of Hudson Bay. In this latter site, the *Picea mariana* treeline runs north–south, parallel to the shore, and has advanced eastwards by 12 km since the late 1800s due to the development of vertical trees within already established Krummholz (Lescop-Sinclair and Payette, 1996).

There appears therefore to be considerable resistance to both expansion and retraction of boreal forest in response to climatic warming, both at present and in the past. During cooling periods treeline-altitude decrease generally lags changes in solar activity levels by 400–500 yr in the Sierra Nevada (Scuderi, 1994) and for up to 650 yr in the mixed forests of southern Ontario (Campbell and McAndrews, 1993). Similarly, responses to temperature increase where recorded (see earlier discussion and Holmgren and

Tjus, 1996) are less than would be predicted from meteorological data. Treelines, it would appear, represent ecological boundaries of great complexity that develop over long periods of plant–environment interaction. Consequently, it seems doubtful that forest migration will be directly proportional to short-term climatic change.

E. Plant Phenology and Climate Change

Thermal time, as expressed either in day-degrees, or phenologically in growing season length, provides a quantification of the degree of warmth needed for certain species to complete their life cycle or maintain their presence in natural communities. In natural ecosystems it is also possible to relate the survival of species of thermal time and make similar optimistic projections as to the future spread of many species. In Scotland the growth of *Betula pendula* requires 1100 day-degrees above 5.6°C while *B. pubescens* can survive with as little as 600–670 day-degrees (Forbes and Kenworthy, 1973). An expansion of *B. pendula* populations might be expected were it not for the fact that climatic warming in Scotland is associated with increased oceanicity (Harrison, 1997), with greater winter rainfall and increased flooding, which will be inimical to the spread of *B. pendula,* which requires well-drained sites.

As discussed earlier, plant responses to climatic change will depend on a number of factors which are outside the range of meteorological data normally used to describe the growing season. It is therefore salutary to examine past phenological records for some actual, and not merely projected, responses to climatic fluctuation. Britain is fortunate in having not only the oldest homogeneous climatic data in the world, but also phenological records that have been collected in the same area for more than 200 yr. The Markham family, living in central England, recorded a wide range of phenological phenomena from 1736 to 1947 (Sparks and Carey, 1995). A more detailed but shorter record (1954–1989) has been collected and analyzed by the Fitter family (Fitter *et al.,* (1995). A striking feature that emerges from analysis of these data is that periods of warming can have both positive and negative effects in terms of advancing or retarding plant development. Table IV records phenomena that have been both advanced and retarded by periods of warm weather. Paradoxically, it is the more southern species that suffer the greatest delay in their development as a result of warm weather.

This paradox is well illustrated in the comparative day-degrees needed for bud burst in hawthorn (*Crataegus monogyna*) compared with beech (*Fagus sylvatica*) where warm weather advances hawthorn but retards beech (Murray *et al.,* 1989). The explanation would appear to be that beech, which requires winter chilling, delays bud opening after mild winters. This behavior may be the result of selection against premature bud opening after

Table IV Species Showing the Greatest Effects of a Simulated 1 C° Warming in All Months[a]

Delayed species			Advanced species		
Species	Days delayed	Mean FFD	Species	Days advanced	Mean FFD
Petasites hybridus	42	31 March	*Geranium robertianum*	37	1 May
Corydalis lutea	36	12 April	*Byronia cretica*	26	29 April
Helleborus viridis	33	25 February	*Carlina vulgaris*	24	30 July
Prunus spinosa	32	31 March	*Alnus glutinosa*	23	26 Feb.
Polygonum hydropiper	27	1 August	*Raphanus raphanistrum*	22	27 May
Adoxa moschatellina	22	30 March	*Trifolium repens*	19	20 May
Euphorbia amygdaloides	21	14 April	*Alopecurus geniculatus*	19	4 June
Tussilago farfara	20	21 February	*Sorbus aria*	19	25 May
Veronica hederifolia	19	18 March	*Agrimonia eupatoria*	19	26 June
Viburnum lantana	18	30 April	*Lycopus europaeus*	18	24 July

[a] Mean FFD is the actual mean first flowering date from the data set (from Fitter *et al.*, 1995).

mild winters in northern-adapted populations. The phenological records mentioned make it clear that predicting plant responses to climate change purely on the basis of climatic (GCM) models, will not take into account past evolutionary histories where selection for nonresponsiveness may be firmly embedded, not only in the current populations, but also in the genetic memory of populations as held in seed banks (McGraw, 1993). The survival value of nonresponsiveness was well formulated many years ago in respect to selection in North American cereals as the *Montgomery effect* (Montgomery, 1912), where *ecological advantage is conferred by low growth rates in areas of low environmental potential.*

F. Arctic Plant Population Adaptations to Climate Change

The Arctic provides a model situation for observing the effects of climatic change on existing plant populations. In high Arctic areas such as Spitsbergen (77–80°N) there exist in proximity to each other warm and cold habitats with different populations of the same species living in sites with distinct differences in temperature as well as growing season length. Differentiation in the extent of summer warming in these juxtaposed habitats has lead to distinct phenological divisions between plants that can inhabit warmer areas with longer growing seasons on drought-prone slopes, moraines, and beach ridges as compared with those that live in wet valley bottoms and low shores, where late snow cover delays the onset of the growing season. One much studied case in relation to phenology is *Dryas octopetala* where at high latitudes differences in lateness of snow cover have resulted in the evolution of two mutually exclusive ecotypes, namely, fellfield and snow-bed forms (McGraw and Antonovics, 1983), which, although interfertile and separated by only 2 weeks in flowering time, remain distinct due to the extremely short growing seasons of the high Arctic (Fig. 10a).

An example similar to *Dryas octopetala* in having fellfield and snow-bed ecotypes is found in high Arctic populations of *Saxifraga oppositifolia,* one of the hardiest plants of polar regions. On dry, exposed ridges in Spitsbergen with warmer temperatures and a longer growing season due to earlier snow melt than the adjacent low shore, the purple saxifrage has a semierect form that starts growing before the populations on the cold-low shore. The population on the shore are predominantly made up of trailing prostrate plants (Fig. 10b) and do not give an impression of being particularly robust. However, appearances are misleading and this frail-looking prostrate saxifrage has a facility for *metabolic rate compensation* (Hochachka and Somero, 1973), (see Section III,B). This property of metabolic rate compensation allows the cold-wet shore ecotype to outperform the more robust type from the beach ridge in gross photosynthetic capacity, respiration, and shoot growth (Fig. 11) (Crawford *et al.,* 1993, 1995). Physiologically, the semierect form, on the beach ridge, with its well-developed tap root, is much more

Figure 10 Snow patch (left) and exposed ridge (right) forms of (a) *Dryas octopetala* and (b) *Saxifraga oppositifolia*. (The *D. octopetala* figure is reproduced with permission from McGraw and Antonovics, 1983, and the drawing by Dagny Tagne Lid of *S. oppositifolia* is from Lid and Lid, 1994.)

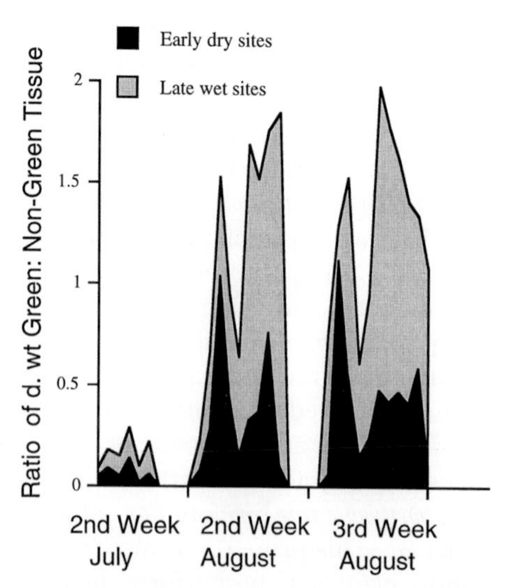

Figure 11 Ratio of d.wt of green to non-green tissues in plants of the semierect and prostrate forms of *Saxifraga oppositifolia* inhabiting, respectively, dry warm ridge sites and cold, wet shore sites at Kongsfjord, Spitsbergen. Note the greater proportion of green tissue in the prostrate form that developed during the peak of the growing season, which will compensate in part for the shortness of the growing season on the wet shore site.

drought tolerant, and conserves carbohydrate for periods of stress. The plants of the shore habitat are, by contrast, less able to conserve carbon resources and use a much greater proportion of their energy gains for immediate rapid growth. The two forms have developed opposing survival strategies, one which saves carbohydrate resources (the ridge populations), the other which spends them (the low-shore populations). For each population its own particular survival strategy aids survival in its particular microhabitat. However, what is possibly more important, for the persistence of the species in the long term, is the contemporary existence of populations adapted to short-cold and warm-dry habitats, which provide the ecotypic diversity that allows the species to adapt rapidly to climatic change. Each population derives long-term advantages from the presence of the other as a refuge for genetic diversity. In cold, unfavorable years the plants in the low-lying shore habitat are unable to set seed, but nevertheless produce pollen which fertilizes some of the adjacent plants on the beach ridge and thus ensures the continuation of its adaptations in a genetically mixed rain of seeds that washes down to the shore where habitat selection results in the preferential survival of the creeping ecotype (Teeri, 1973).

1. Immediate versus Long-Term Fitness in Subspecies Populations The differentiation of populations in relation to those two thermally contrasting habitats (as discussed for *Saxifraga oppositifolia* and *Dryas octopetala*) is also clearly seen in other arctic species. The genus *Draba* in the Nordic area alone (including Spitsbergen) has at least 16 recognizable species, 3 diploid and 13 polyploid (Brochmann and Elven, 1992). There is much interbreeding particularly among the polyploid species and within this subgroup there can be found marked differentiation in preferences for late and early sites. In all of these ecotype-rich species or closely related interfertile species, polymorphism increases the habitat range available, and therefore confers *immediate fitness*. It also has an important additional long-term significance. By maintaining separate populations with opposing thermal strategies that are adapted, respectively, for survival in ultra-short and longer growing seasons, these polymorphic species preserve the capacity to adjust rapidly to extensive variations in growing season length. Consequently, this degree of ecotype variation has a long-term survival value as it *preadapts* the species as a whole to withstand oscillating climatic conditions and can be considered as *long-term fitness* (Crawford, 1997; Crawford and Abbott, 1994). The long-term relationship of these arctic ecotypes to each other is not therefore the generally supposed situation of direct competition. They are in a sense an example of mutualism because each benefits in the long term from the presence of the other.

G. Oceanic Climates and Alternating Temperatures

Oceanicity is a complicating factor in many studies on climatic warming. In the past, some of the major influences of warm temperatures observed on plant distribution have been due to oceanic influences as manifested through increased precipitation and milder winters. Already, examples are available where milder conditions at high latitudes are producing more variable winters. One serious consequence of climatic warming has been recent severe cases of ice-encasement. Instead of a continuous covering of snow, thaw periods with rain falling on frozen ground have resulted in vegetation ice-encasement, which can completely destroy large areas of improved pasture with disastrous results for agriculture in the North. The winter of 1994 resulted in the destruction of much of the improved pasture land in northern Norway and the absence of grass in the ensuing growing season necessitated extensive shipments of fodder from southern parts of the country. Reindeer (caribou) herds in eastern Siberia have recently been badly affected by vegetation ice-encasement as have wild populations in Spitsbergen. In the high Arctic, however, although the reindeer suffer through the inaccessibility of winter fodder, many of the plants are highly resistant to ice-encasement and the consequent imposition of long periods of oxygen deprivation (anoxia). The ability of even the shoots and leaves of some of these arctic populations to withstand total anoxia is a distinct feature of the high Arctic, as populations of the same species from further south are not anoxia tolerant (Crawford *et al.*, 1994).

Mild winters can also create physiological problems for some overwintering trees. Experimentation on overwintering sitka spruce shows that milder winter conditions, combined with flooding can result in a significant drop in root carbohydrate levels which is accompanied by a fall in root vitality (Fig. 12). Flood-induced oxygen deprivation, especially to overwintering roots that are not fully dormant due to warmer winter conditions, can lead to extensive spring die-back of the root system. Flooding increases the risk of starvation-injury through accelerated consumption of carbohydrate reserves from anaerobic metabolism (Smith, 1994). In addition, there is the damage that is caused by postanoxic injury when winter water tables subside (Crawford, 1996). Death of the whole tree may not be immediate, but successive die-back of roots over a number of winters produces a tree that is not sufficiently well anchored to resist the strong winds of oceanic regions and so the trees become unstable and prone to wind-throw.

Physiologically, reduced survival in warm, prolonged winters may be brought about by (1) overconsumption of overwintering carbohydrate reserves, (2) flooding injury, (3) ice-encasement, (4) postanoxic injury on renewed soil aeration in spring, and (5) frost or exposure injury from precocious shoot growth in spring. Ecologically adverse factors include (1) competition from flood-tolerant species, (2) bog growth, (3) wind-throw, and (4) soil leaching. Accordingly, the apparent benign conditions of oceanic environments have a number of hidden stresses for plant survival

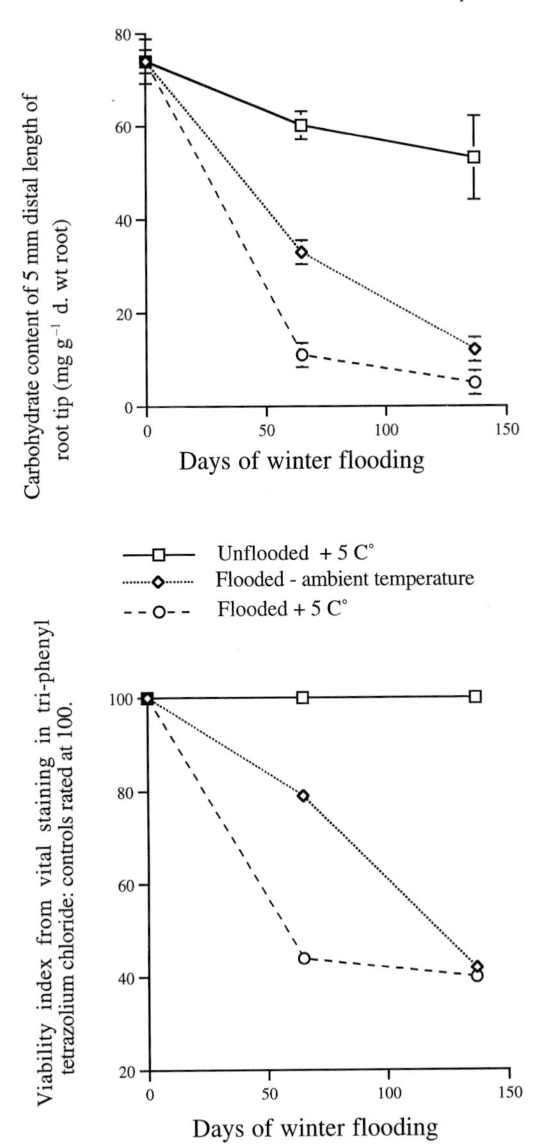

Figure 12 Comparison of the carbohydrate content of 5-mm distal length of root tips of overwintering 3-yr-old trees of *Picea sitchensis* with viability of the same portion of roots as tested by vital staining with triphenyl tetrazolium chloride when flooded at ambient temperature and ambient temperature +5°C. (Data from Smith, 1994.)

that are likely to be increased should there be any significant increase in temperature.

IV. Conclusions

At the molecular and cellular levels, temperature affects plants primarily through the function of enzymes and membranes. Plant response to elevated CO_2 can be affected by temperature, and conversely, atmospheric CO_2 concentration can have an impact on plant response to temperature. A fundamental and well-understood temperature $\times$ CO_2 interaction involves the key photosynthetic enzyme Rubisco. Temperature differentially affects the carboxylation and oxygenation kinetics of Rubisco such that the magnitude of the stimulation of photosynthesis and growth by CO_2 enrichment in C_3 plants tends to increase with increasing temperature. This temperature effect on CO_2 response is also due, in part, to increased utilization of photosynthates by sink organs at warm temperatures, which results in less accumulation of leaf carbohydrates and less downward acclimation of photosynthesis under high-CO_2 conditions. At very low and high temperature extremes the benefits from elevated CO_2 on photosynthesis and growth become negligible. High-temperature stress impairment of reproductive development can exacerbate downward acclimation of photosynthesis in CO_2-enriched plants and further limit growth response to CO_2. Temperature and CO_2 interactive effects on plant–water relations, and on plant development and morphology, may have significant impacts on productivity, survival, or reproductive fitness in a future high-CO_2 world.

Case studies from the arctic and boreal regions illustrate that at the species or population level there are many facets of physiological, morphological, and phenological adaptation which buffer plant populations against changes in temperature. Models that seek to predict the likely effects of increasing temperature and atmospheric CO_2 on ecosystems need to include some assessment of the impact of potentially adverse factors that come from prolonged winters, particularly at high latitudes and in oceanic environments. Warmer winters are a major component of current climatic change, and there is evidence to suggest that if winter warming is greater than summer warming, some species may retreat rather than advance.

A high degree of polymorphism, gene migration between adjacent populations in contrasting microsites, a genetic memory in the seed bank and in hybrid populations, together with longevity, are just some of the features that are particularly noticeable in alpine and arctic regions and contribute to the capacity of plant populations to withstand the impact of climatic change. *Preadaptation* to temperature change also exists in arctic and alpine plant populations because the same properties that confer *immediate fitness* for surviving in regions with marked habitat and climatic variation also serve to confer *long-term fitness* to climatic change. Thus, the ecological sensitivity to climatic warming when examined at a species or population level suggests that there is a high degree of homeostasis in native plant populations that will retard any immediate major changes in vegetation distribution.

References

Ahmed, F. E., Hall, A. E., and Madore, M. A. (1993). Interactive effects of high temperature and elevated carbon dioxide concentration on cowpea (*Vigna unguiculata,* (L.) Walp). *Plant Cell Environ.* **16,** 835–842.

Amthor, J. S. (1989). "Respiration and Crop Productivity." Springer-Verlag. Berlin.

ap Rees, T., Burrell, M. M., Entwhistle, T. G., Hammond, J. B., Kirk, D., and Kruger, N. J. (1988). Effect of low temperature on the respiratory metabolism of carbohydrate by plants. *In* "Plants and Temperature," Vol. 32 (S. P. Long and F. I. Woodward, eds.), pp. 3–394. Society for Experimental Biology, Cambridge.

Badger, M. R., and Collatz, G. J. (1977). Studies on the kinetic mechanism of ribulose-1,5-bisphosphate carboxylase and oxygenase reactions, with particular reference to the effect of temperature on kinetic parameters. *Carnegie Institution of Washington Yearbook* **76,** 355–361.

Bennett, K. D., Boreham, S., Sharp, M. J., and Switsur, V. R. (1992). Holocene history of environment, vegetation and human settlement on Catta Ness, Lunnasting, Shetland. *J. Ecol.* **80,** 241–273.

Berry, J. A., and Raison, J. K. (1981). Responses of macrophytes to temperature. *In* "Encyclopedia of Plant Physiol." Vol. 12A (O. L. Lange, P. S. Nobel, C. B. Osmond, and H. Zeigler, eds.), pp. 278–338.

Billings, W. D. (1997). Arctic phytogeography. *In* "Disturbance and Recovery in Arctic Lands: An Ecological Perspective" (R. M. M. Crawford, ed.), pp. 25–45. Kluwer, Dordrecht.

Boese, S. R., and Huner, N. P. A. (1990). Effects of growth temperature and temperature shifts on spinach leaf morphology and photosynthesis. *Plant Physiol.* **94,** 1830–1836.

Boese, S. R., Wolfe, D. W., and Melkonian, J. J. (1997). Elevated CO_2 mitigates chilling-induced water stress and photosynthetic reduction during chilling. *Plant Cell Environ.* **20,** 625–632.

Boswell, V. R. (1935). A study of temperature, day length and development interrelationships of spinach varieties in the field. *Proc. Am. Soc. Hort. Sci.* **32,** 549–557.

Briffa, K. R., Jones, P. D., Schweingruber, F. H., Shiyatov, S. G. and Cook, E. R. (1995). Unusual 20th-century summer warmth in a 1,000-year temperature record from Siberia. *Nature* **376,** 156–159.

Brochmann, C., and Elven, R. (1992). Ecological and genetic consequences of polyploidy in arctic *Draba* (Brassiceae). *Evolutionary Trends Plants* **6,** 111–124.

Brown, J. (1997) Disturbance and recovery of permafrost terrain. *In* "Disturbance and Recovery in Arctic Lands: An Ecological Perspective" (R. M. M. Crawford, ed.), pp. 167–178 Kluwer, Dordrecht.

Bryson, R. A. (1966). Air masses, streamlines, and the boreal forest. *Geograph. Bull.* **8,** 228–2269.

Campbell, I. D., and McAndrews, J. H. (1993). Forest disequilibrium caused by rapid Little Ice Age cooling. *Nature* **366,** 336–338.

Carter, T. R., and Saarikko, R. A. (1996). Estimating regional crop potential in Finland under a changing climate. *Agric. For. Meteorol.* **79,** 301–313.

Chapin, III, F. S., Shaver, G. R., Giblin, A., Nadelhoffer, K. G., and Laundre, J. A. (1995). Response of Arctic tundra to experimental and observed changes in climate. *Ecology* **76,** 694–711.

Chapin, III, F. S., Zomov, S. A., Shaver, G. R., and Hobbie, S. E. (1996). CO_2 fluctuation at high latitudes. *Nature* **383,** 585–586.

Conroy, J. P. (1992). Influence of elevated atmospheric CO_2 concentration on plant nutrition. *Aust. J. Bot.* **40,** 445–456.

Conroy, J. P., Smillie, R. M., Küppers, M., Bevege, D. I., and Barlow, E. W. R. (1986). Chlorophyll *a* fluorescence and photosynthetic and growth responses of *Pinus radiata* to phosphorous deficiency, drought stress and high CO_2. *Plant Physiol.* **81,** 423–429.

Crawford, R. M. M. (1989). "Studies in Plant Survival." Blackwell Scientific Publications, Oxford.

Crawford, R. M. M. (1996). Whole plant adaptations to fluctuating water tables (first published *Folia Geobotanica & Phytoaxonomica* (1996) **31,** 7–24). *In* "Adaptation Strategies in Wetland Plants: Links between Ecology and Physiology" (R. Brändle, H. Cizkova, and J. Pokorny, eds.), pp. 13–30. Opulus Press, Uppsala.

Crawford, R. M. M. (1997). Habitat fragility as an aid to long term survival in arctic vegetation. *In* "Ecology of Arctic Environment" (S. J. Woodin and M. Marquiss, eds.), Special Publication No. 13 of the British Ecological Society, pp. 113–136. Blackwell Scientific Ltd., Oxford.

Crawford, R. M. M., and Abbott, R. J. (1994). Pre-adaptation of Arctic plants to climate change. *Botanica Acta* **107,** 271–278.

Crawford, R. M. M., and Palin, M. A. (1981). Root respiration and temperature limits to the north south distribution of four perennial maritime species. *Flora* **171,** 338–354.

Crawford, R. M. M., Chapman, H. M., Abbott, R. J., and Balfour, J. (1993). Potential impact of climatic warming on Arctic vegetation. *Flora* **43,** 367–381.

Crawford, R. M. M., Chapman, H. M., and Hodge, H. (1994). Anoxia tolerance in high Arctic vegetation. *Arctic Alpine Res.* **26,** 308–312.

Crawford, R. M. M., Chapman, H. M., and Smith, L. C. (1995). Adaptation to variation in growing season length in Arctic populations of *Saxifraga oppositifolia* L. *Bot. J. Scotland* **41,** 177–192.

Cure, J. D., and Acock, B. (1986). Crop responses to carbon dioxide doubling: A literature survey. *Agric. For. Meteor.* **38,** 127–145.

Dahl, E. (1951). On the relation between summer temperatures and the distribution of alpine plants in the lowlands of Fennoscandinavia. *Oikos,* **3,** 22–52.

Dahl, E. (1990). Probable effects of climatic change due to the greenhouse effect on plant productivity and survival in North Europe. *In* "Effects of Climate Change on Terrestrial Ecosystems," Vol. 4, pp. 7–18. Norsk Institut For Naturforskning, Trondeim.

Dale, J. E. (1965). Leaf growth in *Phaseolus vulgaris.* 2. Temperature effects and the light factor. *Annals Bot.* **29,** 293–308.

Davidson, R. L. (1969). Effect of root/leaf temperature differentials on root/shoot ratios in some pasture grasses and clover. *Annals Bot.* **33,** 561–569.

DeKok, J. L., and Kuiper, P. J. C. (1977). Glycolipid degradation in leaves of the thermophilic *Cucumis sativus* as affected by light and low temperature treatment. *Physiologia Plantarum* **39,** 123–128.

Drake, B. G., and Leadley, P. W. (1991). Canopy photosynthesis of crops and native plant communities exposed to long-term elevated CO_2 treatment. *Plant Cell Environ.* **14,** 853–860.

Eamus, D., and Jarvis, P. G. (1989). The direct effects of increase in the global atmospheric CO_2 concentration on natural and commercial temperate trees and forests. *Adv. Ecol. Res.* **19,** 2–55.

Eamus, D., and Wilson, J. M. (1984). A model for the interaction of low temperature, ABA, IAA and CO_2 in the control of stomatal behavior. *J. Exp. Bot.* **35,** 91–98.

Eamus, D., Fenton, R., and Wilson, J. M. (1983). Stomatal behavior and water relations of chilled *Phaseolus vulgaris* L. and *Pisum sativuum* L. *J. Exp. Bot.* **34,** 434–441.

Edwards, M. E., and Barker, E. D. (1994). Climate and vegetation in Northeastern Alaska 18,000 yr BP–present. *Palaeogeog. Palaeoclimatol. Palaeoecol.* **109,** 127–135.

Erkamo, V. (1956). Untersuchungen über die Pflanzenbiologischen und einige andere Folgeerscheinungen der neuzeitlichen Klimaschwankung in Finnland. *Annales Botanici Societas-Zoologicae Botanicae Fennicae 'Vanamo'* **28,** 1–283.

Farquhar, G. D., Von Caemmerer, S., and Berry, J. A. (1980). A biochemical model of photosynthetic CO_2 assimilation in leaves of C_3 species. *Planta* **149,** 78–90.

Farrar, J. F. (1988). Temperature and the partitioning and translocation of carbon. *In* "Plants and Temperature," Vol. 42 (S. P. Long and F. I. Woodward, eds.), pp. 203–235. Society for Experimental Biology, Cambridge.

Farrar, J. F., and Williams, M. L. (1991). The effects of increased atmospheric carbon dioxide and temperature on carbon partitioning, source–sink relations and respiration. *Plant Cell Environ.* **14,** 819–830.

Feng, X. H. and Epstein, S. (1994). Climatic implications of an 8000-year hydrogen isotope time series from Bristlecone-pine trees. *Science* **265,** 1079–1081.

Fitter, A. H., Fitter, R. S. R., Harris, I. T. B., and Williamson, M. H. (1995). Relationships between first flowering date and temperature in the flora of a locality in central England. *Functional Ecol.* **9,** 55–60.

Forbes, J. C., and Kenworthy, J. B. (1973). Distribution of two species of birch forming stands in Deeside, Aberdeenshire. *Trans. Bot. Soc. Einburgh* **42,** 101–110.

Getsen, M. V., Kostyaev, V. J., and Patova, E. N. (1997). Role of nitrogen-fixing cryptogammic plants in the tundra. *In* "Disturbance and Recovery in Arctic Lands: An Ecological Perspective" (R. M. M. Crawford, ed.), pp. 135–150. Kluwer, Dordrecht.

Giaquinta, R. (1980). Mechanism and control of phloem loading of sucrose. *Berichte Deutsche Botanische Gessellschaft* **93,** 187–201.

Grabherr, G., Gottfried, M., and Pauli, H. (1994). Climate effects on mountain plants. *Nature* **369,** 448.

Grabherr, G., Gottfried, M., Gruber, A., and Pauli, H. (1995). Patterns and current changes in alpine plant diversity. *In* "Arctic and Alpine Biodiversity" (F. S. Chapin III and C. Körner, eds.), pp. 167–181. Springer-Verlag, Berlin.

Graumlich, L. J. (1987). Precipitation variation in the Pacific Northwest (1675–1975) as reconstructed from tree rings. *Annals Assoc. Am. Geographers* **77,** 19–29.

Greer, D. H., Laing, W. A., and Campbell, B. D. (1995). Photosynthetic responses of thirteen pasture species to elevated CO_2 and temperature. *Aust. J. Plant Physiol.* **22,** 713–722.

Grime, J. P. (1996) Mechanisms of vegetation response to climate change. *In* "Past and Future Rapid Environmental Changes" (B. Huntely, W. Cramer, A. V. Morgan, H. C. Prentice, and J. R. M. Allen, eds.), pp. 195–204. Springer-Verlag, Berlin.

Grulke, N. E., Riechers, G. H., Oechel, W. C., Hjelm, U., and Jaeger, C. (1990). Carbon balance in tussock tundra under ambient and elevated atmospheric CO_2. *Oecologia* **83,** 485–494.

Guisan, A., Holten, J. L., Spichiger, R., and Tessier, L. Eds., (1995). "Potential Ecological Impacts of Climate Change in the Alps and Fennoscandian Mountains," p. 194. Ville de Généve, Geneva.

Harrison, S. J. (1997) Changes in the Scottish climate. *Bot. J. Scotland* **49,** 287–300.

Hättenschwiller, S., and Körner, C. (1995). Responses to recent climatic warming of *Pinus sylvestris* and *Pinus cembra* within their montane transition zone in the Swiss Alps. *J. Vegetation Sci.* **3,** 357–368.

Heath, D. and Williams, D. R. (1977). "Man at high altitude." Churchill Livingstone, Edinburgh.

Hengeveld, R. (1990). "Dynamic Biogeography." Cambridge University Press, Cambridge.

Hochachka, P. W., and Somero, G. N. (1973). "Strategies of Biochemical Adaptation." W. B. Saunders Company, Philadelphia.

Holmgren, B., and Tjus, M. (1996). Summer air temperatures and tree line dynamics at Abisko. *In* "Plant Ecology in the Subarctic Swedish Lapland" (P. S. Karlsson and T. V. Callaghan, eds.), Ecological Bulletins No. 45, pp. 159–169. Munksgaard, Copenhagen.

Houghton, J. Y., Callander, B. A., and Varney, K. (1992). "Climate Change 1992. The Supplementary Report to the IPCC Scientific Assessment." Cambridge University Press, Cambridge.

Huntley, B. (1997). Arctic ecosystems and environmental change. *In* "Disturbance and Recovery in Arctic Lands: An Ecological Perspective" (R. M. M. Crawford, ed.), pp. 1–24. Kluwer, Dordrecht.

Huntley, B., and Birks, H. J. B. (1983). "An Atlas of Past and Present Pollen Maps for Europe: 0–13000 Years Ago." Cambridge Univ. Press, Cambridge.

Huntley, B., and Webb III, T. (1988). "Vegetation History" (L. H. ed.). Kluwer, Dordrecht.

Idso, S. B., and Kimball, B. A. (1989). Growth response of carrot and radish to atmospheric CO_2 enrichment. *Environ. Exp. Bot.* **29,** 135–141.

Idso, S. B., Kimball, B. A., Anderson, M. G., and Mauney, J. R. (1987). Effects of atmospheric CO_2 enrichment on plant growth: The interactive role of air temperature. *Agric. Ecosys. Environ.* **20,** 1–10.

Iversen, J. (1944). Viscum, Hedera and Ilex as climatic indicators. *Geol. Foren. Stockh. Forh.* **66,** 463.

Jacoby, G. C., Darrigo, R. D., and Davaajamts, T. (1996). Mongolian tree rings and 20th-century warming. *Science* **273,** 771–773.

Jeffree, C. E., and Jeffree, E. P. (1996). Redistribution of the potential geographical ranges of mistletoe and Colorado beetle in Europe in response to the temperature component of climate change. *Functional Ecol.* **10,** 562–577.

Jeffree, E. P., and Jeffree, C. E. (1994). Temperature and the biogeographical distribution of species. *Functional Ecol.* **8,** 640–650.

Johnson, I. R., and Thornley, J. H. M. (1985). Temperature dependence of plant and crop processes. *Annals Bot.* **55,** 1024.

Jordan, D. B., and Ogren, W. L. (1984). The CO_2/O_2 specificity of ribulose 1,5-bisphosphate carboxylase/oxygenase. Dependence on ribulose-bisphosphate concentration, pH and temperature. *Planta* **161,** 308–313.

Keeling, C. D., Chin, J. F. S., and Whorf, T. P. (1996). Increased activity of northern vegetation inferred from atmospheric CO_2 measurements. *Nature* **382,** 146–149.

Kennedy, A. A. (1995). Simulated climate change: Are passive greenhouses a valid microcosm for testing biological effects of environmental perturbations? *Global Change Biol.* **1,** 29–42.

Klinger, L. F. (1996). Coupling of soils and vegetation in peatland succession. *Arctic Alpine Res.* **28,** 380–387.

Köppen, W. (1931). "Grundriss der Klimakunde," Verbesserte Auflagender Klimate der Der Erde. Walter Gruyter, Berlin.

Körner, C. (1995). Alpine plant diversity: A global survey and functional interpretations. *In* "Arctic and Alpine Biodiversity: Patterns, Causes, and Ecosystem Consequences" (F. S. Chapin III and C. Körner, eds.), Ecological Studies No. 113, pp. 45–62. Springer-Verlag, Berlin.

Körner, C., and Diemer, M. (1987). *In situ* photosynthetic responses to light, temperature and carbon dioxide in herbaceous plants from low and high altitude. *Functional Ecol.* **1,** 179–194.

Körner, C., and Diemer, M. (1994). Evidence that plants from high altitudes retain their greater photosynthetic efficiency under elevated CO_2. *Functional Ecol.* **8,** 58–68.

Körner, C., and Larcher, W. (1988). Plant life in cold climates. *In* "Plants and Temperature" (S. P. Long and F. I. Woodward, eds.). *Symp. Soc. Exptl. Biol.* **42,** 22–57.

Körner, C., Neumayer, M., Pelaez Menendez-Riedl, S., and Smeets-Scheel, A. (1989). Functional morphology of mountain plants. *Flora* **182,** 353–383.

Kraus, A., and Marschner, H. (1984). Growth rate and carbohydrate metabolism of potato tubers exposed to high temperature. *Potato Res.* **27,** 297–303.

Krezner, E. G., and Moss, D. N. (1975). Carbon dioxide enrichment effects upon yield and yield components in wheat. *Crop Sci.* **15,** 71–74.

Kriuchkov, V. V. (1971). Woodland communities in tundra, possibilities for its establishment and dynamics. *Ekologiya* (*Soviet J. Ecol.*) **6,** 9–19 (in Russian).

Ku, S.-B., and Edwards, G. E. (1977). Oxygen inhibition of photosynthesis. II. Kinetic characteristics as affected by temperature. *Plant Physiol.* **59,** 986–990.

Kuiper, P. J. C. (1964). Water uptake of higher plants as affected by root temperature. *Meded Landbouwhogesch. Wageningen* **63,** 1–11.

Kullman, L. (1987). Little Ice-Age decline of a cold marginal *Pinus sylvestris* forest in the Swedish Scandes. *New Phytologist* **106,** 567–584.

Lafta, A. M., and Lorenzen, J. H. (1995). Effect of high temperature on plant growth and carbohydrate metabolism in potato. *Plant Physiol.* **109,** 637–642.

Lavoie, C. and Payette, S. (1994) Recent fluctuations of the lichen-spruce forest limit in subarctic Québec. *J. Ecol.* **82,** 725–734.

Lescop-Sinclair, K., and Payette, S. (1996). Recent advance of the arctic treeline along the eastern coast of Hudson-Bay. *J. Ecol.* **83,** 929–936.

Lid, J., and Lid, D. T. (1994) "Norsk Flora," (R. Elven, ed.), 6th ed. Edn. Det Norske Samlaget, Oslo.

Long, S. P. (1991). Modification of the response of photosynthetic productivity to rising temperature by atmospheric CO_2 concentrations: Has its importance been underestimated? *Plant Cell Environ.* **14,** 729–739.

Luo, Y., Field, C. B., and Mooney, H. A. (1994). Predicting responses of photosynthesis and root fraction to elevated CO_2: Interaction among carbon, nitrogen, and growth. *Plant Cell Environ.* **17,** 1195–1204.

Magoon, C. A., and Culpepper, C. W. (1932). Response of sweet corn to varying temperature from time of planting to canning maturity. *USDA Tech. Bull.* **312.**

Markhart III, A. H. (1984). Amelioration of chilling-induced water stress by abscisic acid-induced changes in root hydraulic conductance. *Plant Physiol.* **74,** 81–83.

Matthews, R. B., Kropff, M. J., Horie, T., and Bachele, M. D. (1997). Simulating the impact of climate change on rice production in Asia and evaluating options for adaptation. *Agric. Sys.* **54,** 399–425.

McGraw, B. (1993). Ecological genetic variation in seed banks. IV. Differentiation of extant and seed bank-derived populations of *Eriophorum vaginatum. Arctic Alpine Res.* **25,** 45–49.

McGraw, J. B., and Antonovics, J. (1983). Experimental ecology of *Dryas octopetela* ecotypes. I. Ecotypic differentiation and life cycle stages of selection. *J. Ecol.* **71,** 879–897.

McWilliam, J. R., Kramer, P. J., and Musser, R. L. (1982). Temperature-induced water stress in chilling-sensitive plants. *Aust. J. Plant Physiol.* **9,** 343–352.

Montgomery, E. G. (1912). Competition in cereals. *Bull. Nebraska Argric. Station* **26,** 1–12.

Morgan, J. A., Hunt, H. W., Monz, C. A., and Lecain, D. R. (1994). Consequences of growth at two carbon dioxide concentrations and two temperatures for leaf gas exchange of Pascopyrum smithii (C_3) and *Bouteloua gracillis* (C_4). *Plant Cell Environ.* **17,** 1023–1033.

Morrison, J. L. (1990). Intercellular CO_2 concentration and stomatal response to CO_2. *In* "Stomatal Function" (E. Zeiger, G. D. Farquhar, and I. R. Cowan, eds.), pp. 229–252. Stanford Univ. Press, Stanford, CA.

Mott, K. A. (1990). Sensing of atmospheric CO_2 by plants. *Plant Cell Environ.* **13,** 731–737.

Mousseau, M., Enoch, H. Z. (1989). Carbon dioxide enrichment reduced shoot growth in sweet chestnut (*Castanea sativa* Mill.). *Vegetatio* **104,** 413–419.

Murray, M. B., Cannell, M. G. R., and Smith, R. I. (1989). Date of budburst of fifteen tree species in Britain following climatic warming. *J. Appl. Ecol.* **26,** 693–700.

Ntsika, G., and Delrot, S. (1986). Changes in apoplastic and intracellular leaf sugars induced by the blocking of export in Vicia faba. *Physiologia Plantarum* **68,** 145–153.

Oberbauer, W., Sionit, N., Hastings, S. J., and Oechel, W. C. (1986). Effects of CO_2 enrichment and nutrition on growth, photosynthesis, and nutrient concentration of Alaskan tundra plant species. *Can. J. Bot.* **64,** 2993–2998.

Ortiz-Lopez, A., Nie, G.-Y., Ort, D. R., and Baker, N. R. (1990). The involvement of the photoinhibition of photosystem II and impaired membrane energization in the reduced quantum yield of carbon assimilation in chilled maize. *Planta* **181,** 78–84.

Pardossi, A., Vernieri, P., and Tognoni, F. (1992). Involvement of abscisic acid in regulating water status in *Phaseolus vulgaris* L. during chilling. *Plant Physiol.* **100,** 1243–1250.

Patterson, D. T., and Flint, E. P. (1980). Potential effects of global atmospheric CO_2 enrichment on the growth and competitiveness of C3 and C4 weed and crop plants. *Weed Sci.* **28,** 71–75.

Paul, M. J., Driscoll, S. D., and Lawlor, D. W. (1991). The effects of cooling on photosynthesis, amounts of carbohydrate and assimilate export in sunflower. *J. Exp. Bot.* **42,** 845–852.

Payette, S., and Filion, L. (1985). White spruce expansion at the treeline and recent climatic change. *Can. J. For. Res.* **15,** 241–251.

Perry, K. B., Wehner, T. C., and Johnson, G. L. (1986). Comparison of 14 methods to determine heat unit requirements for cucumber harvest. *Hort. Sci.* **21,** 419–423.

Pielke, R. A. and Vidale, P. L. (1995) The boreal forest and the polar front. *J. Geophys. Res. Atmos.* **100,** 25755–25758.

Pielou, E. C. (1991). "After the Ice Age." Univ. of Chicago Press, Chicago.

Pielou, E. C. (1994). "A Naturalist's Guide to the Arctic." Univ. of Chicago Press, Chicago.

Prentice, I. C., Cramer, W., Harrison, S. P., Leemans, R., Monserud, R. A., and Solomon, A. M. (1992). A global biome model based on plant physiology and dominance, soil properties and climate. *J. Biogeog.* **19,** 117–134.

Radin, J. W., and Ackerson, R. C. (1982). Does abscisic acid control stomatal closure during water stress? *What's New Plant Physiol.* **13,** 9–12.

Radin, J. W., Kimball, B. A., Hendrix, D. L., and Mauney, J. R. (1987). Photosynthesis of cotton plants exposed to elevated levels of CO_2 in the field. *Photosynth. Res.* **12,** 191–203.

Radoglou, K. M., and Jarvis, P. G. (1990). Effects of CO_2 enrichment on four poplar clones. I. Growth and leaf anatomy. *Annals Bot.* **65,** 617–626.

Rawson, H. M. (1992). Plant responses to temperature under conditions of elevated CO_2. *Aust. J. Bot.* **40,** 473–490.

Rawson, H. M. (1995). Yield responses of two wheat genotypes to carbon dioxide and temperature in field studies using temperature gradient tunnels. *Aust. J. Plant Physiol.* **22,** 23–32.

Ritchie, J. C. (1987). "Postglacial Vegetation of Canada." Cambridge Univ. Press, Cambridge.

Rosenzweig, C., and Parry, M. L. (1993). Potential impacts of climate change on world food supply: A summary of a recent international study. *In* "Agricultural Dimensions of Global Climate Change" (H. M. Kaiser and T. E. Drennen, eds.), pp. 87–116. St. Lucie Press, Delray Beach, FL.

Rowley, J. A., and Taylor, A. O. (1972). Plants under climatic stress. IV. Effects of CO_2 and O_2 on photosynthesis under high-light, low-temperature stress. *New Phytol.* **71,** 477–481.

Rufty, T. W., Huber, S. C., and Kerr, P. S. (1985). Association between sucrose-phosphate synthase activity in leaves and plant growth rate in response to altered aerial temperature. *Plant Sci.* **39,** 7–12.

Sage, R. F., and Reid, C. D. (1994). Photosynthetic response mechanisms to environmental change in C3 plants. *In* "Plant–Environment Interactions" (R. E. Wilkinson, ed.), pp. 413–499. Marcel Dekker, New York.

Sage, R. F., and Sharkey, T. D. (1987). The effect of temperature on the occurrence of O_2 and CO_2 insensitive photosynthesis in field grown plants. *Plant Physiol.* **84,** 658–664.

Schäppi, B., and Körner, C. (1996). Growth responses of an alpine grassland to evelated CO_2. *Oecologia* **105,** 43–52.
Schimel, D., Alves, D., Entig, I., *et al.* (1995). Radiative forcing of climate. *In* "Climate Change 1995, Second Assessment Report of the IPCC". (J. T. Houghton, L. G. Meira Filho, B. A. Callander *et al.*, eds.), pp. 69–131. Cambridge Univ. Press, Cambridge.
Schulze, E.-D. (1986). Carbon dioxide and water vapor exchange in response to drought in the atmosphere and the soil. *Annu. Rev. Plant Physiol.* **37,** 247–274.
Scuderi, L. A. (1994) Solar influence in holocene treeline altitude variability in the 'Sierra-Nevada. *Physical Geog.* **15,** 146–165.
Semeniuk P., Moline, H. E., and Abbott, J. A. (1986). A comparison of the effects of ABA and an antitranspirant on chilling injury of coleus, cucumbers, and diffenbachia. *J. Am. Soc. Hort. Sci.* **111,** 866–868.
Sheen, J. (1994). Feedback control of gene expression. *Photosynth. Res.* **39,** 427–438.
Sher, A. V. (1996) Late-quaternary extinction of large mammals in northern Eurasia: A new look at the Siberian contribution. *In* "Past and Future Rapid Environmental Changes" (B. Huntely, W. Cramer, A. V. Morgan, H. C. Prentice, and J. R. M. Allen, eds.), pp. 319–339. Springer-Verlag, Berlin.
Smith, L. C. (1994). "The Root in Winter." Ph. D. thesis, Univ. of St. Andrews.
Sparks, T. H., and Carey, P. D. (1995). The responses of species over two centuries: An analysis of the Marsham phenological record. *J. Ecol.* **83,** 321–329.
Spiecker, H., Mielikäinen, K., Köhl, M., and Skovsgaard, J., eds. (1996). "Growth Trends in European Forests," pp. 372. Springer-Verlag, Berlin.
Steponkus, P. L. (1984). Role of the plasma membrane in freezing injury and cold acclimation. *Ann. Rev. Plant Physiol.* **35,** 543–584.
Stewart, W. S., and Bannister, P. (1973). Seasonal changes in carbohydrate content of three *Vaccinium* species with particular reference to *V. uliginosum* and its distribution in the British Isles. *Flora* **162,** 134–155.
Stitt, M., and Schulze, D. (1994). Does Rubisco control the rate of photosynthesis and plant growth? An exercise in molecular ecophysiology. *Plant Cell Environ.* **17,** 465–487.
Szeicz, J. M., and Macdonald, G. M. (1995). Recent white spruce dynamics at the subarctic alpine treeline of north-western Canada. *J. Ecol.* **83,** 873–885.
Teeri, J. A. (1973). Polar desert adaptations of a high Arctic plant species. *Science* **179,** 496–497.
Thomas, R. B., and Harvey, C. N. (1983). Leaf anatomy of four species grown under continuous CO_2 enrichment. *Bot. Gaz.* **144,** 303–309.
Thornley, J. H. M. (1977). Root:shoot interactions. *In* "Integration of Activity in the Higher Plant" (D. H. Jennings, ed.), Vol. 31, pp. 367–390. Society for Experimental Biology, Cambridge.
Tognoni, F., Halevy, A. H., and Winter, S. H. (1967). Growth of bean and tomato plants as affected by root absorbed growth substances and atmospheric carbon dioxide. *Planta* **72,** 43–52.
Van Hasselt, P. R., and Van Berlo, H. A. C. (1980). Photooxidative damage to the photosynthetic apparatus during chilling. *Physiol. Plantarum* **50,** 52–56.
Vernieri, P., Pardossi, A., and Tognoni, F. (1991). Influence of chilling and drought on water relations and abscisic acid accumulation in bean. *Aust. J. Plant Physiol.* **18,** 25–35.
Vilchek, G. E. (1994) Taiga/tundra ecotone structure and dynamics: the case study of Western Siberia. *In* "Intecol CongressVI 21–26th August, 1994 Manchester" (J. H. Tallis, H. J. Norman, and R. A. Benton, eds.).
Weeden, N. F., and Buchanan, B. B. (1983). Leaf cytosolic fructose 1,6-bisphosphatase: A potential target size in low temperature stress. *Plant Physiol.* **72,** 259–261.
Weis, E., and Berry, J. A. (1988). Plants and high temperature stress. *In* "Plants and Temperature" (S. P. Long and F. I. Woodward, eds.), pp. 329–346. The Company of Biologists, Ltd., Cambridge.

Wheeler, T. R., Morison, J. I. L., Ellis, R. G., and Hadley, P. (1994). The effects of CO_2, temperature and their interaction on the growth and yield of carrot (*Daucus carota* L.). *Plant Cell Environ.* **17,** 1275–1284.

Whitlock, C. (1993) Postglacial vegetation and climate of Grandteton and Southern Yellowstone National Parks. *Ecol. Monogr.* **63,** 173–198.

Wolfe, D. W. (1991). Low temperature effects on early vegetative growth, leaf gas exchange and water potential of chilling-sensitive and chilling-tolerant crop species. *Ann. Bot.* **67,** 205–212.

Wolfe, D. W. (1994). Physiological and growth responses to atmospheric carbon dioxide concentration. *In* "Handbook of Plant and Crop Physiology" (M. Pessarakli, ed.). pp. 223–242. Marcel Dekker, New York.

Wolfe, D. W., and Kelly, M. O. (1992). Photosynthesis of *Phaseolus vulgaris* L. in relation to leaf nitrogen and chlorophyll accumulation at low growth temperature. *Photosynthetica* **26,** 475–478.

Wolfe, D. W., Gifford, R. M., Hilbert, D., and Luo, Y. (1998). Integration of photosynthetic acclimation to elevated CO_2 at the whole-plant level. *Global Change Biol.* **4,** 879–893.

Wong, S. C. (1993). Interaction between elevated atmospheric concentration of CO2 and humidity on plant growth: Comparison between cotton and radish. *Vegetation* **104,** 211–221.

Wookey, P. A., and Robinson, C. H. (1997). Interpreting environmental manipulation experiments in arctic ecosystems: Are 'disturbance' perspectives properly accounted for? In "Disturbance and Recovery in Arctic Lands: An Ecological Perspective" (R. M. M. Crawford, ed.), pp. 115–134. Kluwer, Dordrecht.

Wyse, R. (1980). Growth of sugarbeet seedlings in various atmospheres of oxygen and carbon dioxide. *Crop Sci.* **20,** 456–458.

4

Effects of Elevated CO_2 and Temperature Stress on Ecosystem Processes

Stanley D. Smith, Dean N. Jordan, and Erik P. Hamerlynck

I. Introduction

In this chapter, we examine the effects of temperature stress and elevated CO_2 on ecosystem processes, including primary production, nutrient cycling, and landscape water and energy balance. Although we examine the interplay between changing CO_2 and rising temperature on ecosystem processes, an important focus of this book is on the effects of environmental stress in concert with elevated CO_2 on biological systems. Environmental stress has been defined in many ways, and often varies depending on whether it is applied to a cultivated production system or to an ecosystem. In its broadest sense, environmental stress is any environmental factor that lowers primary production below its optimum (Osmond *et al.*, 1987). In this context, any nonoptimum temperature can be considered stressful. However, a more narrow view of environmental stress would require that an environmental condition (e.g., freezing/chilling temperatures, heat stress) cause some type of damage to the plant in question. Severe damage to plants, causing either mortality or an irreversible loss in productivity, can be seen to have clear impacts on ecosystem processes. However, there are very few elevated-CO_2 studies that have examined this important type of temperature response in plants. Also of interest from an ecosystem perspective is recurring environmental stress, which can result in either physiological hardening in plants (which may be irreversible within a given growing season) or genetic responses of populations over time, such that the ecosystem equilibrates to nonoptimum conditions through a loss of productivity and/or a shift to more stress-tolerant genotypes and functional types (Chapin, 1991).

One important current aspect of the global change debate is the potential rapidity with which global climate is predicted to change and what effects that may have on ecological systems. The challenge for ecologists is to ascertain what the biological and ecological responses to potentially rapid global change might be. Paleoecologists have been reluctant to investigate rapid change in ecosystems (Gajewski, 1993) because of the assumption that vegetation responds to climate change very slowly and with long time lags (Davis, 1989). Factors that could be important in determining how rapidly vegetation may respond to climate changes include (1) processes, such as migration and mortality, that influence the rate of change of range boundaries of species, (2) disturbance frequency, most notably from fire, and (3) ecosystem feedback processes, such as nutrient cycling and site water balance. The prevailing view is that ecological responses to global climate change will first be observed at certain types of "sensitive" sites (Brubaker, 1986) and, if global change first affects migration fronts or is manifested primarily through increased disturbance, then ecological responses to climate change may initially be detected at ecotones (Neilson, 1993). However, ecosystem properties such as primary production may best be studied away from ecotones, where change is slower and trajectories may be more responsive to long-term changes in the environment (Gajewski, 1993). We should also point out that rapid century-scale and even interdecadal oscillations in climate are intrinsic to the natural global climate system (Dansgaard *et al.*, 1993; Taylor *et al.*, 1993; Mann *et al.*, 1995). Therefore, future increases in global temperatures as a consequence of greenhouse gases may not be unprecedented and thus outside the "evolutionary memory" of today's biological and ecological systems. The historical record shows that past climate change has continuously forced major rearrangements in plant and animal assemblages (Webb and Bartlein, 1992), and a disequilibrium condition for many ecosystem types has apparently been the norm rather than the exception (Campbell and McAndrews, 1993). Future climate change may be anticipated potentially to exacerbate such a pattern, rather than introduce an anomalous disturbance regime.

We also pay particular attention to fire in this chapter, because elevated CO_2 and global warming are each predicted to increase the incidence of fire; together they may dramatically increase its importance worldwide. We then combine the information from these process studies and discuss the potential effects of global change on four broad ecosystem types: forests, grasslands, arctic-alpine tundra, and deserts. Each of these biomes has been predicted to respond differently to elevated CO_2 (Strain and Bazzaz, 1983; Melillo *et al.*, 1993). This is because global warming and its associated climatic effects will not be of the same magnitude at the latitudinal ranges that these four broad biome types occupy (Emanuel *et al.*, 1985), and each of these ecosystems has intrinsically different abiotic and biotic stresses that influence the structure and function of the ecosystem (Osmond *et al.*, 1987).

II. Ecosystem Processes

A. Primary Production and Ecosystem Feedbacks on Production

Growth in C_3 herbaceous crop plants at elevated CO_2 increases linearly as growth temperature increases, with growth enhancement above about 18°C and a doubling of growth above about 32°C (Fig. 1). There are a variety of reasons for this response. Elevation of CO_2 to double ambient concentration substantially reduces photorespiration in C_3 plants, so that their photosynthetic temperature optimum often approaches that of C_4 plants (Long, 1991; Bowes, 1993; Chapter 3, this volume). $CO_2 \times$ temperature effects on respiration rates are more difficult to predict, because respiration may increase or decrease at elevated CO_2 (Amthor, 1991; Wullschleger *et al.*, 1994). Results to date indicate that respiration rates may

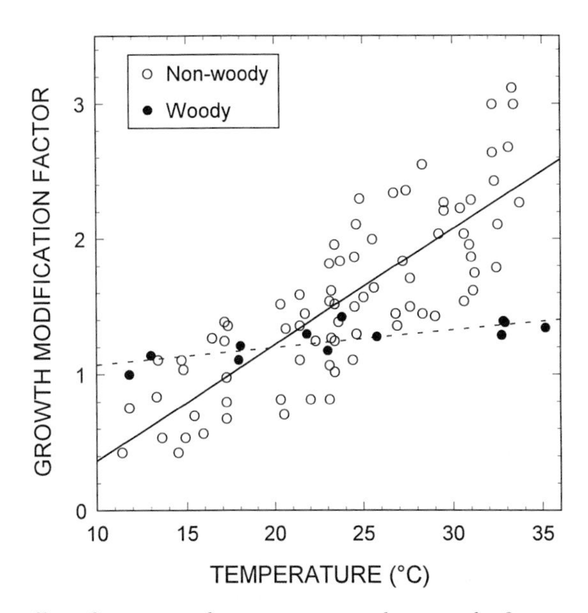

Figure 1 The effect of mean growth temperature on plant growth of nonwoody crop plants (open circles, solid line) and a woody plant (sour orange tree; circles, dashed line) at double ambient CO_2 concentration. Relative growth rate is shown as the "growth modification factor", which is defined as the ratio of total plant growth at elevated CO_2 at each indicated temperature to growth at ambient CO_2 at 25°C. (Data are redrawn from Kimball *et al.*, 1993, for nonwoody plants and from Idso *et al.* (1987) for the woody plant.)

decline at elevated CO_2 when plants are exposed to low temperatures, but respiration increases or remains unchanged at higher temperatures (Ziska and Bunce, 1994). Further complicating the issue is the fact that maintenance and growth respiration may respond differently to elevated CO_2 and temperature (Eamus, 1996). Of interest in an elevated $CO_2 \times$ temperature analysis is that the enhancement of growth at elevated CO_2 has a much flatter temperature response curve for sour orange trees, a woody cultivated species, than for herbaceous crop plants (Fig. 1). This could be because of the greater allocation demands of a perennial woody species, and perhaps greater structural respiration costs at warmer temperatures. This has potentially important implications and makes extrapolation of growth results from herbaceous species to intact ecosystems dominated by woody perennials a questionable practice. Nevertheless, based on growth experiment data (Fig. 1), we may expect that elevated CO_2 enhancements of productivity may be more pronounced in warmer ecosystems (i.e., at tropical and subtropical latitudes) than in cold ecosystems. Indeed, this seems to be borne out by the limited ecosystem-level experiments that have been conducted to date. For example, a coastal salt marsh dominated by herbaceous plants and with a warm growing season and abundant nutrients exhibited a strong growth response to elevated CO_2 (Drake, 1992a), whereas Arctic tussock tundra, with a cold growing season, exhibited little or no productivity response to elevated CO_2 (Tissue and Oechel, 1987; Grulke *et al.*, 1990).

The occurrence of ecosystem stress will have important implications when determining how ecosystems will respond to elevated CO_2 and warmer temperatures. Although global warming may alleviate cold temperature stress in some ecosystems, it may increase high temperature stress in others. Elevated CO_2 appears not to alter heat shock responses in plants (Coleman *et al.*, 1991), but it may ameliorate high-temperature inhibition of photosynthesis (Hogan *et al.*, 1991). If it indeed reduces respiration at high temperatures (Sage, 1996a), it may also mitigate chilling-induced water stress and photosynthetic reduction during chilling due to its effects on stomatal conductance and carbohydrate accumulation (Potvin, 1985; Boese *et al.*, 1997). However, decreased stomatal conductance at elevated CO_2 in C_3 plants would reduce latent cooling at high air temperatures, which could result in greater heat stress. In such situations, genotypes with high stomatal conductance and/or smaller leaf size may be favored if high temperature becomes an important factor in selection (Lu *et al.*, 1994).

It is well accepted that there are fundamental trade-offs between photosynthetic capacity and stress tolerance in native plants (Chapin, 1991). Results showing that plants exhibit a negative correlation between stress (drought or temperature) resistance and maximum growth rate (MacGillivray and Grime, 1995) suggest that an increased incidence of environmental stress over time, due to either higher temperatures or temperature-induced increases in drought, may favor selection of slow-growth, stress-tolerant species. These species would not be expected to show the levels of growth enhancement at elevated CO_2 that fast-growth species from less stressful habitats do. Therefore, the observation that enhancement of primary production at elevated CO_2 may be even greater in stressful environments than in managed agricultural systems (Idso and Idso, 1994) should be viewed with caution; the opposite may indeed be the case.

If elevated CO_2 does indeed result in increased photosynthesis and primary production in a given ecosystem, increased quantities of litter may have important impacts on nutrient cycling and the storage of soil carbon, nitrogen, and other resources. Elevated CO_2 may also result in greater availability of nonstructural carbohydrates, which could be utilized to support greater activity by N-fixing mutualists or mycorrhizae. However, elevated CO_2 may also result in altered C : N : P ratios of litter (in favor of C), which can decrease decomposition rates and therefore N and P availability in soils (Enriquez *et al.*, 1993).

Global warming will also have potentially important impacts on nutrient cycling processes. Whereas elevated CO_2 may affect nutrient cycling primarily through changes in primary production and litter, potentially resulting in greater carbon sequestration in many ecosystems, global warming should have important effects on microbial-based decomposition rates, which would act to increase carbon flux to the atmosphere (Dixon *et al.*, 1994). For example, Peterjohn *et al.* (1993) found that experimentally heating a forest soil to 5°C above ambient increased soil CO_2 emission by a factor of 1.6 and decreased soil carbon concentration by as much as 36%; similar heating of intact cores of arctic tundra resulted in a 50% reduction in net carbon storage (Billings *et al.*, 1982). Even more pronounced, Jones *et al.* (1995) found that methanogenesis from floodplain soils in Arizona was 10- to 15-fold higher at 32°C than it was at 24°C, and this temperature effect was most pronounced in soils with high organic matter (Fig. 2). By increasing decomposition rates, global warming may also act to increase N and P availability in soils that are currently nutrient poor, particularly cold soils such as those from tundra and boreal forest ecosystems.

An important but potentially unpredictable indirect effect of increasing temperatures and CO_2 when considering production processes is herbivory. Increasing temperatures may affect patterns of herbivory, but the mechanisms are not clear. For example, Ayers (1993) predicted that global warming may lead to outbreaks of many insect herbivores because insects tend to be more sensitive to temperature increases than are their host plants. However, herbivore–host interactions tend to be complex and highly responsive to stress conditions (Waring and Cobb, 1992) and to other biotic interactions, such as predators and parasitoids of the herbivores, or to mutualists of the host plants such as mycorrhizae (Gehring and Whitham,

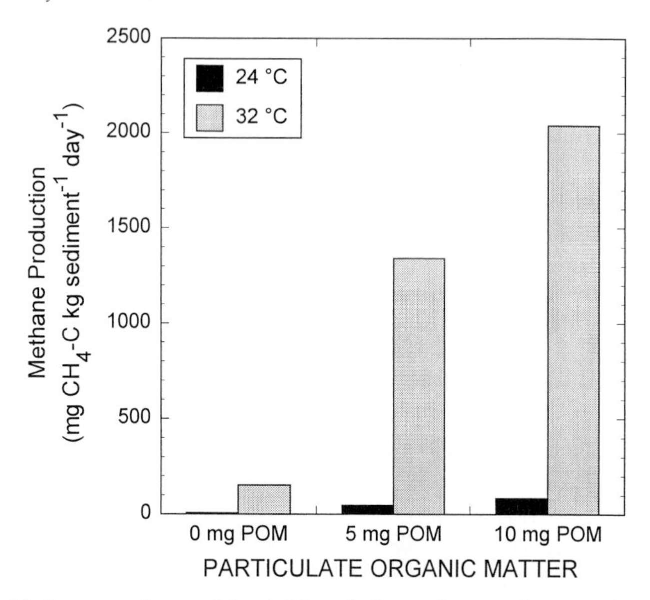

Figure 2 Methanogenesis rate of floodplain soils from a Sonoran Desert wash ecosystem as a function of particulate organic matter at two temperatures, 24°C (solid bars) and 32°C (shaded bars). (Data are redrawn from Jones *et al.*, 1995.)

1995). At present, an insufficient number of studies have been done to form a basis for a predictive capability of how plant-herbivore systems will respond to global warming, and whether or not general rules can be applied to these systems. As for CO_2, if elevated CO_2 results in increased primary production, then herbivory would also be expected to increase. However, if elevated CO_2 results in lower quality, higher C:N ratio foliage, this could result in either higher or lower herbivore consumption, depending on whether the excess carbon is invested in carbon-based structural compounds or in secondary chemicals that deter herbivory. A number of studies have shown an increased C:N ratio of high CO_2-grown plants to result in higher consumption by herbivores, particularly insect larvae (Lincoln *et al.*, 1993; Lindroth *et al.*, 1993). Higher herbivore consumption of live vegetation and litter would act to accelerate biomass and litter turnover, which would result in more rapid litter decomposition. However, if elevated CO_2 resulted in increased secondary chemical production, then herbivores may consume a smaller percentage of primary production, thus increasing litter input into the system.

B. Water Balance

With greater availability of CO_2 in the atmosphere, stomatal and canopy conductances are predicted to decrease, leading to reduced transpiration rates per unit leaf surface area (Bazzaz, 1990; Kimball *et al.*, 1993; Field *et al.*, 1995). However, other potential regulators of evapotranspiration (ET) include leaf area index (LAI), canopy temperature, dynamics of the boundary layer, and moisture-holding capacity of the soil (Field *et al.*, 1992), all of which may be indirectly affected by elevated CO_2. The dynamics of the boundary layer and LAI can both be expected to be influenced by changes in atmospheric CO_2. Higher LAI may counteract the effects of reduced stomatal conductance at the canopy level, an interaction term that has been ignored in many simulation studies that seem to correlate directly decreased canopy transpiration with reductions in stomatal conductance. This may be because CO_2-induced changes in LAI make scaling leaf conductance responses to the canopy level a difficult predictive process (Oechel *et al.*, 1995). LAI plays a key role in the water balance of ecosystems, particularly semiarid ecosystems without continuous canopy cover (Rambal and Debussche, 1995). In addition, potential increases in allocation of biomass to roots at elevated CO_2 may result in higher soil organic matter and greater soil moisture holding capacity over the long term (Post *et al.*, 1992; Oechel *et al.*, 1995; Harte *et al.*, 1996), which could in turn result in greater water availability. All of these factors together (lower stomatal conductance at the leaf level, higher LAI, and perhaps higher soil moisture holding capacity) lead to the conservative prediction that ET rates may not be responsive to increased CO_2 concentration alone, a prediction that appears to be supported by empirical ET data from free-air CO_2 enrichment (FACE) experiments (Grant *et al.*, 1995).

Global warming will also strongly affect ET, and will tend to counteract any potential CO_2 effects on stomata by increasing transpirational water loss. Increases in leaf temperature may act to nullify changes in water loss rates from decreased stomatal conductances (Bazzaz, 1990; Oechel *et al.*, 1995), or leaf conductances may adjust upward to maintain leaf temperature set points in a warmer environment. Indeed, recent studies have shown that genotypes of cotton that have been bred for greater performance at high temperatures have higher stomatal conductances and smaller leaves than do ancestral varieties that fare poorly at high temperature (Lu *et al.*, 1994; Lu and Zeiger, 1994). Smaller leaves would increase the amount of convective heat exchange relative to latent heat loss, which may be particularly important for plants adapting to drought periods.

Global warming may have important effects on landscape-level water balance in both natural and agroecosystems. Global warming alone (without CO_2 effects) should act to increase ET in the growing season and also to stimulate important hydrological processes, such as snowmelt in higher altitude or latitude regions, to occur earlier in the year. This has been predicted to result in reductions in summer soil moisture in continental temperate climates (Manabe and Wetherald, 1986), which in turn could increase both heat and water stress conditions for plants. What is more problematic is predicting the role elevated CO_2 will play in growing season soil moisture content and runoff from watersheds. Idso and Brazel (1985) predicted that streamflow would increase in central Arizona under high CO_2 conditions due to the higher water use efficiency (WUE) of the vegetation. However, this analysis ignored important feedback effects such as potential increases in LAI of the vegetation, scaling stomatal conductance from individual leaves to canopies, and direct responses of global warming on soil moisture and streamflow. Although hydrologic models using global change model (GCM) climatic inputs are difficult to calibrate and impossible to validate, they do tend to point toward shorter winter seasons, larger winter–spring floods, and drier summers with lower runoff (Fig. 3; Loaiciga *et al.*, 1996). Elevated CO_2 may have little direct effect on these landscape-level water balance responses (Skiles and Hanson, 1994). Therefore, plant functional types that are capable of exploiting cool-season moisture conditions, but then are able to avoid or tolerate stressful summer conditions, may be best suited to these altered climatic regimes.

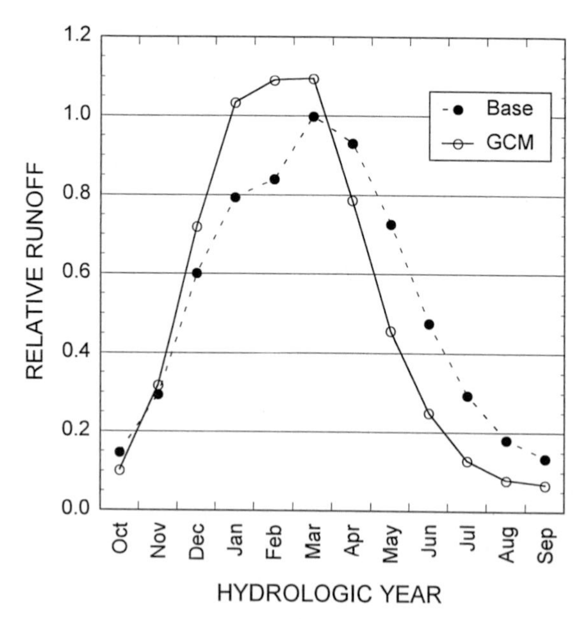

Figure 3 Relative average monthly runoff (versus peak base flow under ambient climatic conditions) for a Sierra Nevada watershed under base flow conditions (solid circles, dashed line) and under simulated global warming (open circles, solid line; mean of three GCM temperature simulations, assuming no change in precipitation). Note that the data are plotted for a standard hydrologic year (Oct.–Sept.). (Redrawn from Gleick, 1987.)

C. Fire

Given that most GCMs predict warmer and drier global conditions (King and Neilson, 1992; Neilson, 1993), increased fire frequency and intensity are expected in many ecosystems (Bergeron and Flannigan, 1995; Strain and Thomas, 1995; Sage, 1996b). The most direct effects of fire are on species composition and canopy architecture, often shifting a community's structure toward species having fire-resistant life history patterns or physical characteristics and/or physiological attributes that allow exploitation of gaps produced by fire (Reich *et al.*, 1990; Abrams, 1992; Bell, 1994). Very often, increased fire frequency reduces system species diversity to levels observed in earlier seral stages (Abrams, 1992; D'Antonio and Vitousek, 1992). In the Great Basin of the western United States, *Bromus tectorum* (cheatgrass), an invasive annual C_3 grass, has accelerated burn cycles to the point where it has completely replaced native shrubs and grasses in some areas (Young and Evans, 1978). *Bromus tectorum* may respond more strongly than native grasses to elevated CO_2 (Smith *et al.*, 1987), and thus may further exacerbate fire cycles in the Great Basin and encroach more rapidly into these regions. Fire also often maintains ecotones (Abrams, 1992; Knight *et al.*, 1994). Under elevated CO_2 and warmer, drier conditions, accelerated fire cycles may cause changes in ecotones, often toward species better able to cope with more open successional habitats, or possibly moving them to new positions in the landscape (Neilson, 1993; Bergeron and Flannigan, 1995; Hamerlynck and Knapp, 1996).

Warmer temperatures, moister soils, elevated CO_2, and fire may strongly affect nutrient dynamics, especially nitrogen (Rice *et al.*, 1994; Oechel *et al.*, 1995). Net nitrogen mineralization often increases after fires, stimulating productivity (Knapp and Seastedt, 1986; Van Cleve *et al.*, 1996), though continued frequent burning often leads to long-term nitrogen limitation (Seastedt *et al.*, 1991; Dewar and McMurtie, 1996). Volatilization losses would be expected to increase with accelerated fire frequency and stronger fire intensity, especially if significant amounts of nitrogen are stored in aboveground biomass (Dewar and McMurtie, 1996). Both fire and elevated CO_2 alter soil microbial populations, which strongly control nutrient cycling (Seastedt *et al.*, 1991; Rice *et al.*, 1994). Tissue C : N ratios, fire, and temperature could act either synergistically or antagonistically to regulate soil processes mediating nutrient cycling and nutrient availability in the future.

Increased fire severity, especially under dry, warm conditions, could have especially profound effects on woody ecosystems. Besides the potential for increased volatilization losses (Dewar and McMurtie, 1996), very severe burns in forest systems often result in complete removal of all detrital layers and leave little organic residue (Schimmel and Granstrom, 1996). Such losses often outstrip any immediate gains in soil nitrogen mineralization following fires, and contribute to dramatically higher erosion rates, especially after dry periods (Meyer *et al.*, 1992). For some species, such burns exceed the tolerance limits of seeds that normally germinate after fires (Bell, 1994). Given that woody plants show marked gains in biomass production under elevated CO_2, and that respiration rates (a major carbon loss for woody species) are often reduced in woody species (Wullschleger *et al.*, 1994), such fires may become serious disturbances in the future, even in woody systems well adapted to fire (Strain and Thomas, 1995).

III. Case Studies

A. Forests

Responses of forested ecosystems to increased CO_2 and predicted changes in temperature and precipitation will probably be dependent on several factors; in particular, the base temperatures at which studies of forest responses are conducted may have profound effects on the results. Certainly temperature changes alone can cause changes in forest growth, particularly in colder climates or in ecotonal regions. However, the interaction between increased atmospheric CO_2 and low temperatures is poorly understood (Eamus, 1996). Few studies have addressed the joint effects on plant growth of unusually high and unusually low temperatures in conjunction with elevated CO_2 (Bazzaz, 1990). This type of study may be important for accurate estimates of the effects of climate change on future plant communities. However, the impracticality of applying temperature and CO_2 treatments to areas greater than 0.1 ha may require individual plant responses to be extrapolated to the ecosystem level for modeling purposes.

Modeled responses of temperate forest ecosystems predict that coniferous forests may respond most strongly to global warming, while broadleaf hardwood forests will respond most strongly to elevated CO_2 and decreased precipitation (Aber *et al.*, 1995). In response to all three factors, broadleaf forests were predicted to respond most strongly in production, whereas all forest types showed a decline in water yield (Fig. 4). Tropical rainforests may not be affected directly by CO_2 because of existing light and nutrient limitations, but may be directly influenced by secondary climate effects or by changes in canopy architecture (Bazzaz, 1990; Körner and Arnone, 1992). In studies of temperate conifer forests, changes in canopy architecture have been shown in response to temperature (Callaway *et al.*, 1994) and CO_2 (Berninger *et al.*, 1995).

Researchers have predicted poleward shifts in forests and other ecosystems that would follow temperature regimes now found at lower latitudes (Dale and Rauscher, 1994; Aber *et al.*, 1995; Hättenschwiler and Körner, 1995). However, others have suggested that a predicted increase in opti-

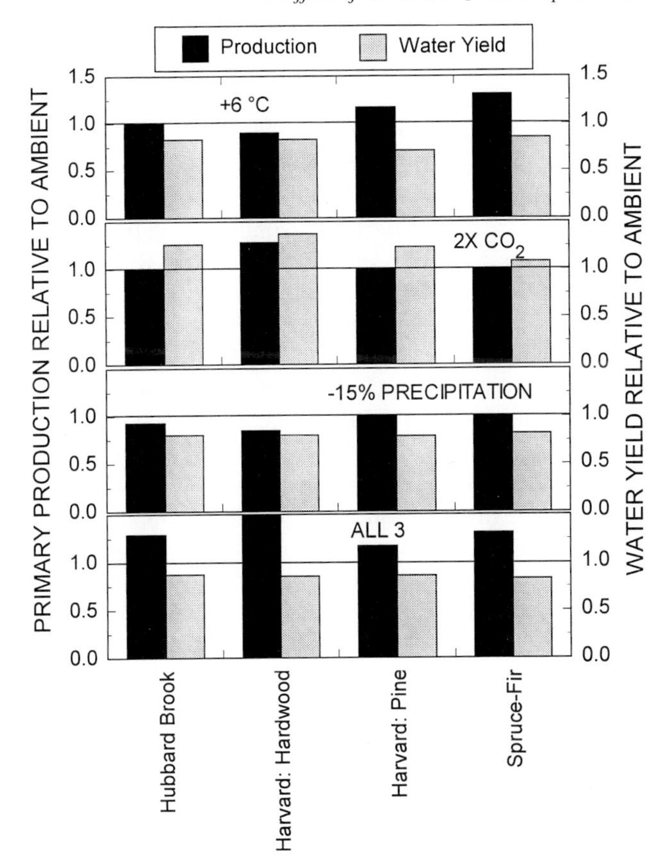

Figure 4 Relative response of primary production (solid bars) and water yield (shaded bars) for four experimental forests representing different forest types (Hubbard Brook deciduous forest, Harvard deciduous forest, Harvard pine forest, and spruce–fir forest) to an increase in annual temperature of 6°C (top), double ambient CO_2 concentration (top-middle), a decrease in annual precipitation of 15% (bottom-middle), and the imposition of all three changes (bottom). (Redrawn from Aber *et al.*, 1995.)

mum photosynthetic temperatures of individual plants or shifts in phenotypes within species will offset increases in ambient temperature, maintaining ecosystems in their approximate current ranges or at least curtailing rates of migration (Crawford *et al.*, 1993; Nowak *et al.*, 1994). It has also been suggested that higher net assimilation rates and optimum temperatures for C_3 photosynthesis may lead to competitive enhancement of forest species relative to those in C_4 grasslands (Drake, 1992b; Idso and Idso, 1994; Kirschbaum, 1994), although these potential advantages of C_3 plants at elevated

CO_2 and warmer temperatures may be offset by increasing fire cycles within herbaceous systems.

B. Grasslands

Grasslands occur in areas with seasonal precipitation having a high degree of interannual variation, so most perennial grasses are well adapted to endure prolonged seasonal dry periods, usually during the warmest portion of the growing season, as well as episodic droughts (Risser, 1985). Disturbances such as fire can maintain grass dominance, depending on growing season precipitation and the evolutionary history of the system (Hulbert, 1988). Grasslands have highly diverse community structure patterns that are strongly influenced by system-level driving variables such as fire frequency, grazing, and rainfall (Monson *et al.*, 1983; Knapp, 1985; Seastedt *et al.*, 1991; Briggs and Knapp, 1995). Historically, these factors have made grassland form, and subsequent function, particularly sensitive to rapid (5–25 yr) climate changes (Albertson *et al.*, 1957; Parton *et al.*, 1994).

Several models suggest grasslands will be fairly sensitive to elevated CO_2, and that semiarid grassland systems should respond most strongly to increased CO_2 if global warming occurs (Strain and Bazzaz, 1983; Melillo *et al.*, 1993). However, the same assumptions that make such predictions tenuous for desert systems (see later discussion) apply to grassland systems as well, because it has not been firmly established that increased WUE is directly tied to increased productivity (but see Polley *et al.*, 1993). For example, there is no evidence that shortgrass prairie, a semiarid grassland, is a more water use efficient system than tallgrass prairie (Briggs and Knapp, 1995), and thus might be expected to show proportionally greater stimulation of biomass accumulation at elevated CO_2. Indeed, detailed models including warmer growing season temperatures predict that greater CO_2 will not offset losses in production incurred by higher temperatures in semiarid grassland systems (Baker *et al.*, 1993; Parton *et al.*, 1994). This difference in response could be due to the physical structure of grasslands as it relates to control of primary productivity. For instance, ANPP in shortgrass prairie is tightly coupled to annual precipitation, while ANPP in more mesic tallgrass prairie is less correlated with precipitation (Briggs and Knapp, 1995). Detrital biomass accumulation in tallgrass prairie leads to cooler and moister soils, which eventually reduces productivity by increasing light and nutrient limitations, effectively decoupling productivity from total annual precipitation in some areas (Knapp and Seastedt, 1986; Briggs and Knapp, 1995). Increased biomass accumulation and reduced stomatal conductances in tallgrass prairie canopy dominants may exacerbate such limitations, but could also buffer tallgrass prairie from higher temperatures by increasing soil moisture reserves (Knapp *et al.*, 1993, 1996). Shortgrass prairie, on the other hand, has far less canopy development and is more water and temperature limited than tallgrass prairie (Monson *et al.*, 1983), and might be less buffered from higher temperatures in a warmer future climate.

There is little evidence that elevated CO_2 directly affects belowground processes in grassland systems. Elevated CO_2 indirectly increases microbial activity, mainly through increasing soil moisture (Rice *et al.*, 1994), but few direct effects on nutrient dynamics have been found (Owensby *et al.*, 1993a), though there is some evidence that C_3 grass species allocate more nitrogen to belowground biomass under elevated CO_2 (Read and Morgan, 1996). This suggests, as most models assert, that belowground processes will respond more directly to alterations in temperature and water balance in grassland systems (Parton *et al.*, 1994; Chen *et al.*, 1996). Parton *et al.* (1994) ran CENTURY model simulations of soil C dynamics for most of the world's grassland systems. In all cases, even if productivity increased under higher CO_2 and warmer temperatures, grasslands showed net losses in soil carbon stores. They suggested that this would mobilize nutrient stores, stimulating short-term productivity but eventually leading to longer term nutrient limitations that might impact the ability of primary producers to endure increased heat stress (Long, 1991; Drake, 1992b).

Several community-level interactions could be affected by higher temperature and increasing atmospheric CO_2, which might alter ecosystem processes in grasslands. Grasslands are unique among terrestrial ecosystems in that they are often dominated by C_4 species during warmer periods, but by C_3 grasses during cooler periods of the growing season (Risser, 1985). Thus, a significant body of work has focused on the impacts of elevated CO_2 and temperature on C_3 and C_4 species. In shortgrass prairie species, high temperature limitations to photosynthesis have been found to be offset more in a C_4 grass than in a C_3, most likely due to nonstomatal photosynthetic limitations brought on by a down-regulation of the photosynthetic capacity of the C_3 species under elevated CO_2 (Morgan *et al.*, 1994; Hunt *et al.*, 1996). Elevated temperatures also markedly decreased the effect of elevated CO_2 on biomass accumulation in the C_3 grass (Read and Morgan, 1996). Additionally, imposition of water stress with higher temperature under elevated CO_2 resulted in better plant performance in C_4 than in C_3 grasses (Hunt *et al.*, 1996). This is important, because seasonal water stress in grassland systems generally accompanies heat stress (Knapp, 1985). These findings suggest that C_4 grasses may persist, and even be favored, in grasslands subjected to increased temperature and elevated CO_2. As valuable as these studies have been, we should note that they have relied on controlled growth chamber studies, and until similar field studies have been undertaken, caution should be exerted when applying these results to complex natural grassland systems. Indeed, field studies of intact tallgrass prairie have shown that C_4 grasses receive consistent benefits from increased carbon dioxide throughout the growing season, especially in drier years (Knapp *et al.*, 1993, 1996; Owensby *et al.*, 1993b).

Grasslands are not composed of perennial grasses alone; in fact most of the diversity of these systems is in other growth forms, such as annuals, herbaceous forbs, and woody shrubs, each with unique physiological, morphological, and phenological adaptations allowing them to persist in a "sea of grass." In tallgrass prairie, total productivity increases under elevated CO_2; in wet years, this is due to increases in forb and woody plant biomass (Owensby *et al.*, 1993b), and to stimulated C_4 photosynthesis improving dominant grass biomass accumulation during dry years (Knapp *et al.*, 1993). In a study of stomatal conductance and water relations in 12 tallgrass prairie species, Knapp *et al.* (1996) found that stomatal conductances were lower in elevated CO_2 for all species during the wettest portion of the season, but increased over conductances seen in ambient CO_2 conditions in two shrub species in dry conditions, suggesting that woody species in tallgrass prairie maintained favorable water status longer through the season than the dominant C_4 grasses. These findings suggest that the indirect effects of elevated CO_2, via increased soil moisture reserves, had more effect on plant performance than direct effects on photosynthesis. Soil moisture is strongly determined by evaporative demand, and the warmer conditions that are predicted to accompany higher CO_2 in this region (Parton *et al.*, 1994) could strongly affect woody plant performance. Broad-leaved woody species growing in grasslands could be most susceptible to heat stress, especially if stomatal conductances are reduced under elevated CO_2 and higher temperatures. Indeed, there is evidence that high temperatures play an important role in the distribution of woody species at ecotones of the tallgrass prairie (Hamerlynck and Knapp, 1996). Thus, not only could broad-leaved grassland species be sensitive to the interaction of elevated carbon dioxide and warmer temperatures, the ecotones at the edges of grasslands could be strongly affected as well. Warmer climatic conditions would not only directly affect physiological function of ecotonal species and potentially ecotonal structure (Hamerlynck and Knapp, 1996), but would also be conducive to increasing fire and drought frequency, causing potentially wide shifts in ecotonal zones (Neilson, 1993; Noble, 1993).

Taken together, these findings suggest that increasing CO_2 and temperature may have marked effects on grassland productivity and structure. Historically, grassland structure has been found to change rapidly, on the order of 5–25 yr, with changes in climatic patterns (Albertson *et al.*, 1957). These changes were far more rapid than the increases in CO_2 and temperature currently being tracked. As a result, climate will most likely set up the interactions that determine future community structure in grasslands well before elevated CO_2 becomes a direct mediator of grassland system function.

C. Arctic-Alpine

Climatic warming predicted to occur with increased CO_2 may be greatest (6–12°C) during the winter months in Arctic ecosystems (Schlesinger and Mitchell, 1987; Bengtsson, 1994). Smaller increases are predicted during Arctic summers and at lower latitudes. Note that high-latitude and high-altitude ecosystems are relatively insensitive to winter temperatures except for indirect effects on snowmelt timing and volume.

Paleoecological evidence for changes at the boreal treeline during an episode of climate warming that occurred 5000 yr B.P. showed that a transformation from tundra to forest-tundra took about 150 yr, similar to the time period modeled by GCMs (MacDonald *et al.*, 1993). The warming episode was not correlated with a change in atmospheric CO_2 such as is expected in the next 100–200 yr. Vegetation models based on temperature alone indicate that northern forests could eventually reach a new equilibrium between 500 and 1000 km north of current ranges (Overpeck *et al.*, 1991; but see Chapter 3, this volume). The potential rate of vegetation migration is much more uncertain, because seedlings are much more sensitive to environmental stresses than are mature plants.

Arctic and alpine ecosystems are inherently dependent on the hardiness of overwintering tissues and vegetative reproduction. Forecast changes in temperature, particularly with the mitigating effects of increased CO_2, may not approach damaging thresholds for established populations for many years, even those on the warm boundary of their current geographical range (Hättenschwiler and Körner, 1995). Thus, established systems of perennials may not be under great abiotic migratory pressure. More realistically, seedling establishment of existing species under changing temperature regimes may become less successful near one ecotone, and mortality of mature plants may eventually cause the ecosystem boundary to shift. However, no evidence of any differential species success was found in a recent montane ecotonal study (Hättenschwiler and Körner, 1995). Phenotypes with least plasticity to adapt to higher temperatures and modified seasonality of precipitation may exhibit the earliest mortality, blurring the ecosystem boundary and making niches available for the seedlings of better adapted individuals. Therefore, abiotic pressures are likely to influence phenotypic shifts of warm-adapted individuals in the existing species complex, or to expand species ranges as temperatures and high CO_2 interact with germination and establishment rates. Conversely, if temperature variability increases with increasing mean temperatures, as suggested by Katz and Brown (1992), extreme cold events may favor existing species distributions, limiting the success of warm-adapted exotics, or even permitting expansion of cold-limited systems to lower latitudes or elevations. Existing experimental evidence indicates that the view of upward elevational and latitudinal migration of ecosystems with warming may be premature (see Chapter 3, this volume), because it is not currently possible to predict the fate of individual alpine species as CO_2 and global temperatures continue to increase (Woodward, 1993). The fact that primary production has been predicted to remain fairly constant under elevated CO_2 and global warming in Arctic and alpine ecosystems while mean global primary production is expected to increase (Melillo *et al.*, 1993) supports the view that change will be slow in Arctic-alpine ecosystems.

The smaller predicted temperature increases at lower latitudes will present a contrast of climate change effects between temperate alpine tundra (increased CO_2 with little temperature change) and Arctic tundra (increased CO_2 and increased temperature). Harte and Shaw (1995) performed the other permutation of this combination by experimentally warming plots in a subalpine meadow at current CO_2 concentrations. They found that the dominance of woody shrubs was dramatically enhanced, results similar to the boreal tree expansion at low CO_2 (5000 yr B.P.) reported by MacDonald *et al.* (1993). In montane wet meadows, the shrub increase was at the expense of forbs, while in dry meadows, both forbs and graminoids showed reductions with heating (Fig. 5). In a study of the dominant shrub

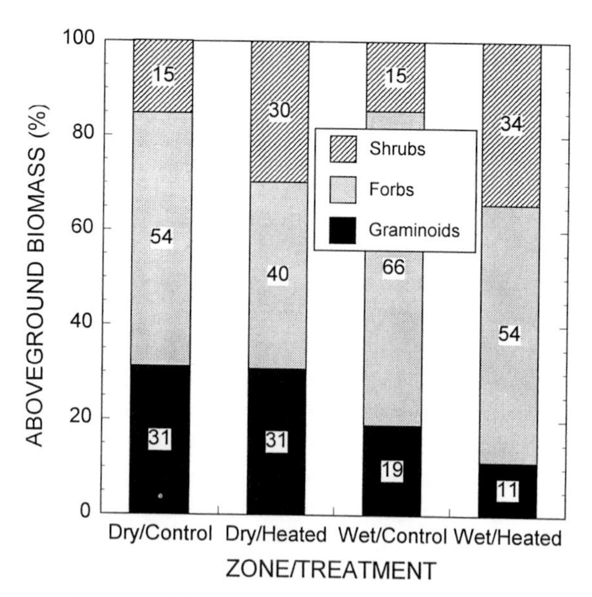

Figure 5 Percent aboveground biomass in a montane meadow community in Colorado made up of shrubs (top), forbs (middle), and graminoids (bottom) in four experimental treatments combining two experimental regimes: a moisture gradient from dry to wet meadow types; and a heating treatment made up of either control plots (nonheated) or plots heated 5°C above ambient temperature. Data were obtained after two full years of treatment. (Redrawn from Harte and Shaw, 1995.)

(*Artemisia tridentata*) and forb (*Potentilla gracilis*) species, Loik and Harte (1996) found that there were no differences in leaf-level physiological tolerance to elevated temperatures, but the shrub tended to operate at lower leaf temperatures than did the forb in midsummer on both experimentally warmed and control plots. Microhabitat and morphological factors may thus have been more important than physiology in affecting the differential response of these two growth forms to a warming treatment.

With ecosystem warming, decomposition rates of the extensive peat and soil organic matter stores in the Arctic are expected to increase (Shaver *et al.*, 1992; Billings *et al.*, 1994). With early signs of regional warming, the Arctic tundra and taiga have recently switched from a carbon sink to a carbon source (Oechel *et al.*, 1993). The net loss of soil carbon by tundra ecosystems should result in short-term increases in the availability of nitrogen and phosphorus from decomposition. Warming in the tundra appears likely to increase the length of the growing season, increase the depth of annual thaw, and permit greater water loss from the upper soil (Rouse *et al.*, 1992). By increasing effective soil depth in the growing season and increasing the rate of organic matter decomposition, this could provide increased nutrient availability and a more favorable rooting environment for vascular plants, which could significantly increase their importance relative to mosses in Arctic ecosystems.

D. Deserts

A number of studies have predicted that desert ecosystems may be among the most responsive to elevated CO_2 and associated climatic change. Strain and Bazzaz (1983) predicted that desert ecosystems may be the most responsive single biome type. In their conceptual model, they plotted biome types along two axes, one of increasing responsiveness to elevated CO_2 as annual aridity increases (due to known increases of plant WUE at elevated CO_2), and one of increasing responsiveness to soil nitrogen availability (due to increased plant responsiveness to elevated CO_2 at high nutrient availability). They predicted that alluvial desert ecosystems, with presumed higher nutrient contents, would be very responsive to elevated CO_2, whereas older piedmont surfaces, with low nutrient contents, would show only intermediate responsiveness relative to other ecosystem types. Recently, Strain and Thomas (1995) reiterated the hypothesis that semiarid Mediterranean ecosystems would be expected to exhibit "large and significant responses to CO_2 increase and other factors of global change" because "[stomatal] closure [at] high CO_2 will greatly increase WUE and improve plant and ecosystem water status." Using similar reasoning of enhanced WUE at elevated CO_2, Melillo *et al.* (1993) predicted that desert ecosystems would exhibit the greatest relative increase in annual net primary production of any biome type, even when future GCM climate scenarios were factored into the analysis (Table I). However, these models are based on a very important assumption—that water use efficiency of plants is the most important determinant of plant productivity in water-limited systems. To date, there are no studies that have definitively shown that plant production correlates with plant WUE in aridland ecosystems; in fact, the opposite may be the case (Smith *et al.*, 1997) because plant WUE is highest when plant water stress is most pronounced (Ehleringer and Cooper, 1988).

Bearing these caveats in mind, it is still reasonable to assume that if atmospheric CO_2 were to increase in aridland systems without any change in climate, enhanced WUE would be expected to increase plant production, although the magnitude of such a direct CO_2 effect may be much less than has been predicted (Table I). What clearly complicates the issue is determining the effects of concomitant increases in annual temperature in already water-limited systems and also potential changes in the seasonality of rainfall. These climatic effects have the potential to swamp elevated CO_2 effects on desert plants and ecosystems, particularly in regions where biological activities are strongly controlled by the episodic occurrence of water and nutrient availability in the soil. In this context, it is useful to discuss what the salient features of desert ecosystems are and, in particular, what are potentially unique attributes of these systems. Smith *et al.* (1997) recently proposed three "unifying assumptions" that act to structure desert ecosystems: (1) Desert ecosystem function is largely dictated by episodic events; (2) the processes of competition and facilitation are critically important; and (3) desert communities are "surface limited", that is, their primary productivity is most limited by low leaf area (and probably low root

Table I Predicted Percent Increases in Annual Net Primary Production of the World's Major Biome Types under Elevated CO_2 (Double Ambient Concentration) and Global Climate Change[a]

Biome type	Climate	
	Contemporary	GCMs[b]
Deserts	50	70
Tropical forest	22	35
Temperate grassland	9	11
Temperate deciduous forest	5	17
Wet tundra	0	25
Boreal forest	0	0

[a] From Melillo *et al.* (1993).

[b] Mean global temperature increases between 2.8 and 4.2°C; mean global precipitation increases between 7.8 and 11%; and mean global cloudiness decreases between 0.4 and 3.4%.

area), not low photosynthetic capacity of the vegetation. These assumptions make extrapolations from steady-state responses of plants from controlled environments of limited utility in predicting how plants and ecosystems will respond to natural variation in arid climates. For example, the notion that enhanced WUE under elevated CO_2 will dramatically enhance ecosystem water balance and thus primary production implies that desert ecosystems may indeed reach new set points of primary productivity and standing biomass based on increasing CO_2 alone. However, several potential problems are associated with this assumption. First, if elevated CO_2 did indeed substantially increase plant production in desert ecosystems, this would also result in significantly higher leaf area during the growing season, which would tend to offset lower transpiration rates on a leaf area basis. Second, such an analysis ignores the importance of episodic drought on long-term biomass dynamics in desert ecosystems; under severe climatic drought conditions, desert perennials tend to shed significant quantities of aboveground and belowground biomass (Orshan, 1954), and it is doubtful that elevated CO_2 will have a pronounced effect on plants under such extreme stress conditions. It is also under conditions of water stress that soil nutrients are generally unavailable to plants, and so the plants would not be expected to be as responsive to elevated CO_2 even though plant WUE may respond strongly. It should thus be clear that elevated CO_2 may indeed result in increased growth rates of desert plants during times of the year when resources are readily available, but the extreme stress conditions that characterize episodic droughts should act to limit long-term ecosystem gains in stored water, soil nutrients, and standing biomass. If this logic is correct, then climatic changes that act to minimize the frequency and magnitude of drought conditions would have far greater impact on long-term plant production and biomass in desert ecosystems than does elevated CO_2 per se. Without these climatic changes, the increases in productivity and biomass of desert ecosystems in response to elevated CO_2 may be smaller than have been predicted to date.

The preceding discussion has focused primarily on the effects of elevated CO_2 on desert plants because that is what most predictions to date have been based on. However, increases in environmental temperatures with global warming may also have important effects on desert ecosystems. Although there have been no experimental warming experiments in desert ecosystems, it can be predicted with some confidence that significant global warming may have positive effects on ecosystem processes in high-latitude cold deserts but would tend to have negative effects in most hot deserts of the world. Subtropical deserts may be particularly vulnerable to increasing temperature, and it is in these ecosystems that true high temperature stress may be most pronounced as global warming takes effect. In an analysis of mean annual temperature as a function of latitude, Clark (1992) found that desert ecosystems tend to have mean annual temperatures that are 5–8°C warmer than rainforests at similar latitudes (Fig. 6). This implies that feedback processes between the ecosystem and atmosphere are substantially different in the two systems, with deserts much less buffered from temperature extremes than are tropical and subtropical forests. Because open desert landscapes exhibit enhanced cooling at night and therefore lower minima, one would expect the temperature difference between deserts and forests to be significantly greater than 5–8°C when daily maxima are considered. Furthermore, studies have shown that high temperature tolerance in many desert plants may be only slightly greater than the extreme maxima that they encounter in their natural environments (Downton *et al.*, 1984; Smith *et al.*, 1984). It is thus reasonable to assume that heat stress may become an increasingly important abiotic factor in driving species distributions and vegetation seasonality in open desert ecosystems, and the importance of high temperature stress may be more pronounced in deserts than in other ecosystem types as global warming proceeds.

Water balance in desert ecosystems may also be affected by global warming. Simulations have indicated that runoff from semiarid watersheds may decrease in a warmer climate (Gleick, 1987; Nash and Gleick, 1991), and in Mediterranean climates the summer dry season may start earlier in the year. If this were to happen, it may favor plants that are opportunistically

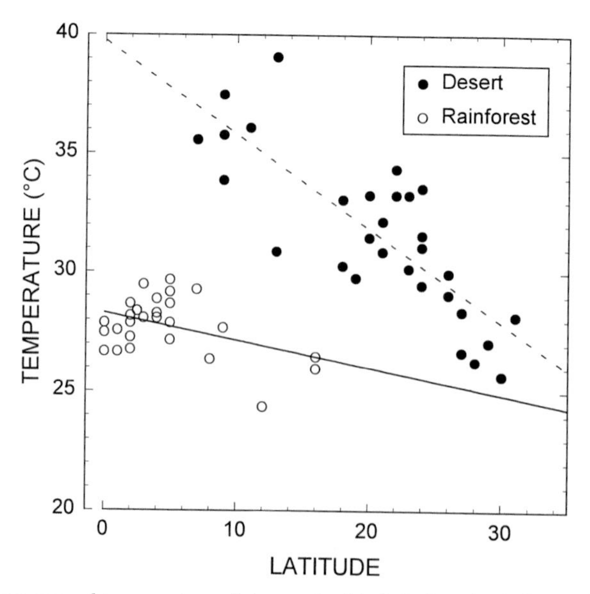

Figure 6 Mean annual temperature of deserts (solid circles) and rainforests (open circles) as a function of latitude within the tropics of America. (Redrawn from Clark, 1992.)

capable of exploiting early-season moisture sources, whereas perennials that do not come out of dormancy until later in the year may be at a disadvantage. One group of taxa in western North America that are well known to be capable of rapid early-season growth and soil moisture depletion are the annual brome grasses (*Bromus tectorum* and *B. rubens*). These exotic species are also potentially highly responsive to elevated CO_2, which has led several groups of investigators to speculate that elevated CO_2 and climatic warming may differentially favor these exotic grasses (Mayeux *et al.*, 1994; Smith *et al.*, 1997), which would in turn promote a fire cycle that could result in a physiognomic shift from shrublands to annual grassland in many regions of the West. Changing water balance due to global warming has important implications for desert ecosystems, particularly considering that the prevailing opinion holds that elevated CO_2 will result in enhanced water balance in water-limited systems. In a simulation analysis of warming effects in a subtropical arid climate, Abderrahman *et al.* (1991) found that a 5°C increase in mean annual temperature increased potential evapotranspiration by 8–16%. By increasing crop irrigation demands, which in turn would exacerbate soil salinity problems, these authors concluded that warming would accelerate the desertification process in aridland regions with inadequate water supplies to keep pace with increasing atmospheric evaporative demand. These studies show that the interactive effects of elevated CO_2 and climatic warming on water balance in desert ecosystems will be hard to predict, and may be site specific. What is clear, though, is that if global change also results in increased precipitation in arid regions, we may see significant increases in primary production, but if no precipitation increases occur then the direct enhancement effect of elevated CO_2 in water-limited environments may be overwhelmed by the effects of global warming and therefore overstated.

IV. Summary and Conclusions

Few studies, either experimental or simulation based, have addressed the combined effects of elevated CO_2 and temperature stress on ecosystem processes. We have focused in this review on elevated CO_2 and global warming, which may have opposite effects on ecosystem primary production, nutrient cycling, and water balance. Although much of the literature suggests that plant growth will respond positively to elevated CO_2 and increased temperatures, there are crucial ecosystem-level feedbacks that must be considered in such an analysis. Elevated CO_2 is predicted to stimulate primary production most in water-limited ecosystems due to a substantial enhancement of plant WUE and has been predicted to result in lower stand-level transpiration rates due to lower stomatal conductances, though this may be offset at the landscape level by higher leaf area production. Increased C : N ratios of plant tissues may negatively affect nutrient cycling, since lower quality litter has slower turnover rates, which in turn may feed back on primary production. In contrast to elevated CO_2, higher temperatures could have markedly negative effects on plant production in water-limited ecosystems due to either higher ET rates at warmer temperatures or more frequent temperature stress. In cold ecosystems, warming might stimulate ecosystem production by directly increasing metabolic rates as well as increasing nutrient availability by stimulating microbially mediated decomposition. Therefore, a warmer climate may be depicted by accelerated nutrient cycling, but also increased ET and reduced soil moisture content during the growing season. Decreased soil moisture will, in turn, have important feedback effects on production and nutrient cycling processes. Because elevated CO_2 and global warming seem to have opposite effects on these three ecosystem processes, and because these processes are interdependent in space and time, regional predictions are difficult with any level of precision because overall ecosystem responses will probably vary widely across different ecosystem types (i.e., climates).

To examine ecosystem responses to global change in a holistic framework, we have adapted a conceptual model of the effects of elevated CO_2 on primary production (Long and Hutchin, 1991), and expanded it to include the potential effects of elevated CO_2 *and* temperature on primary production and ecosystem water balance (Fig. 7). What can be seen from this analysis is that the direct effects of elevated CO_2 should act to increase primary production (through increased photosynthesis) in most environments, whereas the direct effects of elevated temperatures may act to either increase or decrease production, depending on climate. For ecosystem water balance, elevated CO_2 has been predicted to result in water savings, mediated through leaf-level reductions in transpiration rate, whereas higher temperature should cause increased plant transpiration and therefore a reduction (more negative) ecosystem water balance in all climate types (Fig. 7).

From a regional perspective, we hypothesize that temperature alone would act to increase production in cool or moist-temperate environments, where water balance is of less relative importance, but could act to decrease production in dry environments where water balance more strongly feeds back on production processes. In moist environments, elevated CO_2 may be expected to have a minimal effect on production (except in successional gap environments) because leaf area already tends to be maximal, photosynthesis is limited by light and nutrients, and water is usually not limiting. However, in dry environments elevated CO_2 may stimulate production due to its strong effect on plant WUE, but this may be counteracted by the effects of higher temperatures on plant water stress and therefore the

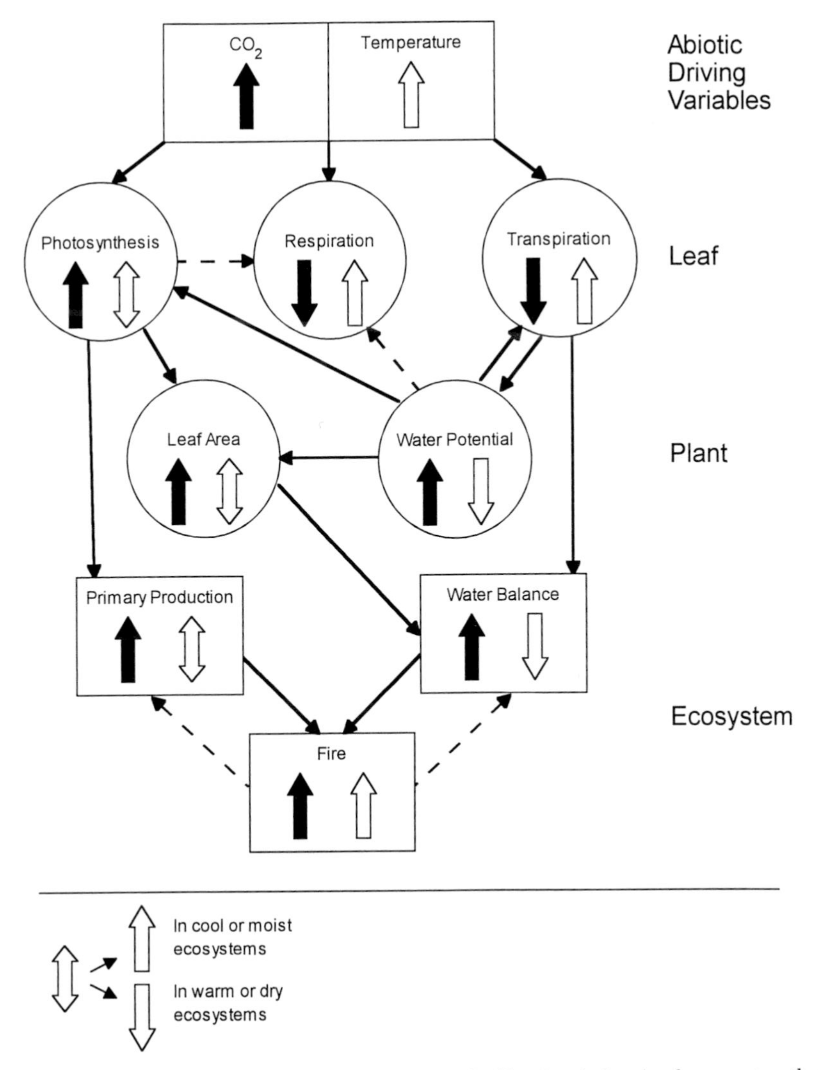

Figure 7 A conceptual model showing how various leaf-level and plant-level parameters that have a direct influence on ecosystem-level primary production, water balance, and fire may be influenced by elevated CO_2 (solid arrows) and increasing global temperatures (open arrows). Upward arrows indicate an increase in the parameter, downward arrows a decrease; arrows pointing in both directions indicate that the parameter is predicted to increase in cool or in moist-temperate ecosystems, but will decrease in warm and/or dry ecosystems. The conceptual model is adapted from Long and Hutchin's (1991) analysis of elevated CO_2 effects on primary production, and has been expanded to include the effects of elevated temperature and to include effects on ecosystem water balance and fire frequency. An increase in water balance results in an increase in soil moisture storage over time; a decrease means less water availability over time.

maintenance of leaf area (Fig. 7). Finally, the one ecosystem process most likely to be stimulated by both elevated CO_2 and higher temperatures is fire, which could increase in frequency and severity in most of the world's ecosystem types (Fig. 7), potentially resulting in dramatic changes in the structure, function, and species composition of those affected ecosystems.

When considering temperature *stress,* it is usually episodic, extreme events that are most important in determining ecosystem boundaries, and so global change predictions need to incorporate the frequency and severity of stress events rather than just increases in mean annual temperature. For example, Markham (1996) identified stress-related thresholds for various ecosystem types, and then coupled these thresholds with ecological tolerances in relation to global change. In Table II, important biome-specific threshold responses that potentially effect major shifts in species composition and/or physiognomy of the vegetation are listed. To those, we add predictions as to how elevated CO_2 and global warming may influence each type of threshold response (and therefore impact that ecosystem type). When utilizing such an approach, we find that key thresholds may often be exceeded or relaxed with climate change, potentially resulting in significant changes in species composition or perhaps even physiognomy of the vegetation.

Of the primary world biome types, the drier biomes such as deserts and grasslands have been predicted to respond most strongly to elevated CO_2 because it has been assumed that higher WUE at elevated CO_2 will have its most positive impact on production processes where water is most limiting (Strain and Bazzaz, 1983; Melillo *et al.,* 1993). In contrast, elevated CO_2 should have a limited stimulatory effect in moist forests, where light is most limiting to production, or in tundra/taiga ecosystems where lack of nutrients primarily limits production. It is in cold ecosystems that warming may have its greatest effect, primarily by increasing soil temperatures and depth to permafrost, thereby alleviating waterlogging effects and stimulating nutrient cycling. Therefore, dryland ecosystems may be most stimulated by elevated CO_2, cold ecosystems by increasing temperatures, and warm, moist forests by neither. It may be semiarid, cool regions such as shrub-steppe ecosystems that respond most strongly to both elevated CO_2 and global warming.

Table II Examples of Ecological Tolerances and Thresholds in Relation to Global Climate Change for Important Biome Types of the World[a]

Biome type	Threshold	Effect(s)	Potential response to: Elevated CO_2	Potential response to: Global warming
Tropical forest	Hurricane frequency	Shift to early succession	No direct effect	Increase frequency
Temperate forest	Episodic severe drought	Increase canopy gaps	Decrease severity	Increase severity
	Fire return frequency	Rapid decline of certain stands	Increase	Increase
Temperate grassland	Episodic severe drought	Differential mortality	Decrease severity	Increase severity
	Fire frequency/intensity	Reduce diversity	Increase	Increase
Warm desert	Episodic severe drought	Differential mortality; loss of biomass	Decrease severity	Increase severity
	Episodic hard freeze	Northward extension of subtropical taxa	No effect	Decrease incidence
Alpine	January isotherm of −10°C	Threatens group of taxa requiring cold temperatures	No effect	Increase
Arctic	Mean temperature of 10°C for the warmest month and 30 days above 10°C	Poleward extension of treeline	No effect	Increase

[a] Thresholds and effects (how that threshold may affect species distributions) are from Markham (1996).

Acknowledgments

We are grateful to Rowan Sage and David Wolfe for critical review of an earlier draft of the manuscript. The National Science Foundation provided support for this project through the Nevada EPSCoR Program and the NSF/DOE/NASA/USDA-TECO Program (NSF grant IBN-9524036).

References

Abderrahman, W., Bader, T., Kahn, A., and Ajward, M. (1991). Weather modification on reference evapotranspiration, soil salinity and desertification in arid regions: A case study. *J. Arid Environ.* **20,** 277–286.

Aber, J., Ollinger, S., Federer, C., Reich, P., Goulden, M., Kicklighter, D., Melillo, J., and Lathrop, R. (1995). Predicting the effects of climate change on water yield and forest production in the northeastern United States. *Climate Res.* **5,** 207–222.

Abrams, M. (1992). Fire and the development of oak forests. *BioScience* **42,** 346–353.

Albertson, F., Tomanek, G., and Riegel, A. (1957). Ecology of drought cycles and grazing intensity on grasslands of the central Great Plains. *Ecol. Monogr.* **27,** 27–44.

Amthor, J. (1991). Respiration in a future, higher-CO_2 world. *Plant Cell Environ.* **14,** 13–20.

Ayers, M. (1993). Plant defense, herbivory, and climate change. *In* "Biotic Interactions and Global Change" (P. Kareiva, J. Kingsolver, and R. Huey, eds.), pp. 75–94. Sinauer Associates, Sunderland, MA.

Baker, B., Hanson, J., Bourdon, R., and Eckert, J. (1993). The potential effects of climate change on ecosystem processes and cattle production on U.S. rangelands. *Climatic Change* **25,** 97–117.

Bazzaz, F. (1990). The response of natural ecosystems to the rising global CO_2 levels. *Annu. Rev. Ecol. Syst.* **21,** 167–196.

Bell, D. (1994). Interaction of fire, temperature and light in the germination response of 16 species from the *Eucalyptus marginata* forest of south-western Australia. *Aust. J. Bot.* **42,** 501–509.

Bengtsson, L. (1994). Climate change: Climate of the 21st century. *Agric. For. Meteorol.* **72,** 3–29.

Bergeron, Y., and Flannigan, M. (1995). Predicting the effects of climate change on fire frequency in the southeastern Canadian boreal forest. *Water Air Soil Poll.* **82,** 437–444.

Berninger, F., Mencuccini, M., Nikinmaa, E., Grace, J., and Hari, P. (1995). Evaporative demand determines branchiness in Scots pine. *Oecologia* **102,** 164–168.

Billings, W., Luken, J., Mortensen, D., and Peterson, K. (1982). Arctic tundra: A source or sink for atmospheric carbon dioxide in a changing environment? *Oecologia* **53,** 7–11.

Billings, W., Peterson, K., Luken, J., and Mortensen, D. (1994). Interaction of increasing atmospheric carbon dioxide and soil nitrogen on the carbon balance of tundra microcosms. *Oecologia* **65,** 26–29.

Boese, S., Wolfe, D., and Melkonian, J. (1997). Elevated CO_2 mitigates chilling-induced water stress and photosynthetic reduction during chilling. *Plant Cell Environ.* **20,** 625–632.

Bowes, G. (1993). Facing the inevitable: Plants and increasing atmospheric CO_2. *Annu. Rev. Plant Physiol. Plant Mol. Biol.* **44,** 309–332.

Briggs, J., and Knapp, A. (1995). Interannual variability in primary production in tallgrass prairie: Climate, soil moisture, topographic position, and fire as determinants of above-ground biomass. *Am. J. Bot.* **82,** 1024–1030.

Brubaker, L. (1986). Responses of tree populations to climatic change. *Vegetatio* **67,** 119–130.

Callaway, R., DeLucia, E., and Schlesinger, W. (1994). Biomass allocation of montane and desert ponderosa pine: An analog for response to climatic change. *Ecology* **75,** 1474–1481.

Campbell, I., and McAndrews, J. (1993). Forest disequilibrium caused by rapid Little Ice Age cooling. *Nature* **366,** 336–338.

Chapin III, F. S. (1991). Integrated responses of plants to stress. *BioScience* **41,** 29–36.

Chen, D., Hunt, H., and Morgan, J. (1996). Responses of C_3 and C_4 perennial grasses to CO_2 enrichment and climate change: Comparison between model predictions and experimental data. *Ecol. Model.* **87,** 11–27.

Clark C. (1992). Empirical evidence for the effect of tropical deforestation on climatic change. *Environ. Conserv.* **19,** 39–47.

Coleman, J., Rochefort, L., Bazzaz, F., and Woodward, F. (1991). Atmospheric CO_2, plant nitrogen status and the susceptibility of plants to an acute increase in temperature. *Plant Cell Environ.* **14,** 667–674.

Crawford, R., Chapman, H., Abbott, R., and Balfour, J. (1993). Potential impact of climatic warming on Arctic vegetation. *Flora* **188,** 367–381.

Dale, V., and Rauscher, H. (1994). Assessing impacts of climate change on forests: The state of biological modeling. *Climatic Change* **28,** 65–90.

Dansgaard, W., Johnsen, S., Clausen, H., Dahl-Jensen, D., Gundestrup, N., Hammer, C., Hvidberg, C., Steffensen, J., Sveinbjornsdottir, A., Jouzel, J., and Bond, G. (1993). Evidence for general instability of past climate from a 250-kyr ice-core record. *Nature* **364,** 218–220.

D'Antonio, C., and Vitousek, P. (1992). Biological invasions by exotic grasses, the grass/fire cycle, and global change. *Annu. Rev. Ecol. Syst.* **23,** 63–87.

Davis, M. (1989). Lags in vegetation response to greenhouse warming. *Climatic Change* **15,** 75–82.

Dewar, R., and McMurtie, R. (1996). Sustainable stemwood yield in relation to the nitrogen balance of forest plantations—a model analysis. *Tree Physiol.* **16,** 173–182.

Dixon, R., Brown, S., Houghton, R., Solomon, A., Trexler, M., and Wisniewski, J. (1994). Carbon pools and flux of global forest ecosystems. *Science* **263,** 185–190.

Downton, W., Berry, J., and Seemann, J. (1984). Tolerance of photosynthesis to high temperature in desert plants. *Plant Physiol.* **74,** 786–790.

Drake, B. (1992a). A field study of the effects of elevated CO_2 on ecosystem processes in a Chesapeake Bay wetland. *Aust. J. Bot.* **40,** 579–595.

Drake, B. (1992b). The impact of rising CO_2 on ecosystem production. *Water Air Soil Poll.* **64,** 25–44.

Eamus, D. (1996). Tree responses to CO_2 enrichment: CO_2 and temperature interactions, biomass allocation and stand-scale modeling. *Tree Physiol.* **16,** 43–47.

Ehleringer, J., and Cooper, T. (1988). Correlations between carbon isotope ratio and microhabitat in desert plants. *Oecologia* **76,** 562–566.

Emanuel, W., Shugart, H., and Stevenson, M. (1985). Climate change and the broad-scale distribution of terrestrial ecosystem complexes. *Climatic Change* **7,** 29–43.

Enriquez, S., Duarte, C., and Sand-Jensen, K. (1993). Patterns in decomposition rates among photosynthetic organisms: The importance of detritus C:N:P content. *Oecologia* **94,** 457–471.

Field, C., Chapin, F., Matson, P., and Mooney, H. (1992). Responses of terrestrial ecosystems to the changing atmosphere: A resource-based approach. *Annu. Rev. Ecol. Syst.* **23,** 201–235.

Field, C., Jackson, R., and Mooney, H. (1995). Stomatal responses to increased CO_2: Implications from the plant to the global scale. *Plant Cell Environ.* **18,** 1214–1225.

Gajewski, K. (1993). The role of paleoecology in the study of global climatic change. *Rev. Paleobot. Palynol.* **79,** 141–151.

Gehring, C., and Whitham, T. (1995). Environmental stress influences aboveground pest attack and mycorrhizal mutualism in pinon-juniper woodlands: Implications for management in the event of global warming. *In* "Desired Future Conditions for Pinon-Juniper Ecosystems" (D. Shaw, E. Aldon, and C. LoSapio, eds.), pp. 30–37. USDA Forest Service, Rocky Mtn. For. Range Exp. Stn., Ft. Collins, CO.

Gleick, P. (1987). Regional hydrological consequences of increases in atmospheric CO_2 and other trace gases. *Climatic Change* **10,** 137–161.

Grant, R., Kimball, B., Pinter, P., Wall, G., Garcia, R., LaMorte, R., and Hunsaker, D. (1995). Carbon dioxide effects on crop energy balance: Testing *ecosys* with a free-air CO_2 enrichment (FACE) experiment. *Agron. J.* **87,** 446–457.

Grulke, N., Riechers, G., Oechel, W., Hjelm, U., and Jaeger, C. (1990). Carbon balance in tussock tundra under ambient and elevated atmospheric CO_2. *Oecologia* **83,** 485–494.

Hamerlynck, E., and Knapp, A. (1996). Photosynthetic and stomatal responses to high temperature and light in two oaks at the western limit of their range. *Tree Physiol.* **16,** 557–565.

Harte, J., and Shaw, R. (1995). Shifting dominance within a montane vegetation community: Results of a climate-warming experiment. *Science* **267,** 876–880.

Harte, J., Rawa, A., and Price, V. (1996). Effects of manipulated soil microclimate on mesofaunal biomass and diversity. *Soil Biol. Biochem.* **28,** 313–322.

Hättenschwiler, S., and Körner, C. (1995). Responses to recent climate warming of *Pinus sylvestris* and *Pinus cembra* within their montane transition zone in the Swiss Alps. *J. Veg. Sci.* **6,** 357–368.

Hogan, K., Smith, A., and Ziska, L. (1991). Potential effects of elevated CO_2 and changes in temperature on tropical plants. *Plant Cell Environ.* **14,** 763–778.

Hulbert, L. (1988). Causes of fire effects in tallgrass prairie. *Ecology* **69,** 46–58.

Hunt, H., Elliott, E., Detling, J., Morgan, J., and Chen, D.-X. (1996). Responses of a C_3 and a C_4 perennial grass to elevated CO_2 and temperature under different water regimes. *Global Change Biol.* **2,** 35–47.

Idso, K., and Idso, S. (1994). Plant responses to atmospheric CO_2 enrichment in the face of environmental constraints: A review of the past 10 years' research. *Agric. For. Meteorol.* **69,** 153–203.

Idso, S., and Brazel, A. (1985). Rising atmospheric carbon dioxide concentration may increase streamflow. *Nature* **312,** 51–53.

Idso, S., Kimball, B., Anderson, M., and Mauney, J. (1987). Effects of atmospheric CO_2 enrichment on plant growth: The interactive role of air temperature. *Agric. Ecosyst. Environ.* **20,** 1–10.

Jones, J., Holmes, R., Fisher, S., Grimm, N., and Greene, D. (1995). Methanogenesis in Arizona, USA dryland streams. *Biogeochemistry* **31,** 155–173.

Katz, R., and Brown, B. (1992). Extreme events in a changing climate: Variability is more important than averages. *Climatic Change* **21,** 289–302.

Kimball, B., Mauney, J., Nakayama, F., and Idso, S. (1993). Effects of increasing atmospheric CO_2 on vegetation. *Vegetatio* **104/105,** 65–75.

King, G., and Neilson, R. (1992). The transient response of vegetation to climate change: A potential source of CO_2 to the atmosphere. *Water Air Soil Poll.* **64,** 365–383.

Kirschbaum, M. (1994). The sensitivity of C_3 photosynthesis to increasing CO_2 concentration: A theoretical analysis of its dependence on temperature and background CO_2 concentration. *Plant Cell Environ.* **17,** 747–754.

Knapp, A. (1985). Effect of fire and drought on the ecophysiology of *Andropogon gerardii* and *Panicum virgatum* in a tallgrass prairie. *Ecology* **66,** 1309–1320.

Knapp, A., and Seastedt, T. (1986). Detritus accumulation limits the productivity of tallgrass prairie. *BioScience* **36,** 662–668.

Knapp, A., Hamerlynck, E., and Owensby, C. (1993). Photosynthetic and water relations responses to elevated CO_2 in the C_4 grass *Andropogon gerardii. Int. J. Plant Sci.* **154,** 459–466.

Knapp, A., Hamerlynck, E., Ham, J., and Owensby, C. (1996). Responses in stomatal conductance to elevated CO_2 in 12 grassland species that differ in growth form. *Vegetatio* **125,** 31–41.

Knight, C., Briggs, J., and Nellis, M. (1994). Expansion of gallery forest on Konza Prairie Research Natural Area, Kansas, U.S.A. *Landscape Ecol.* **9,** 117–125.

Körner C., and Arnone, J. (1992). Responses to elevated carbon dioxide in artificial tropical ecosystems. *Science* **257,** 1672–1675.

Lincoln, D., Fajer, E., and Johnson, R. (1993). Plant-insect herbivore interactions at elevated CO_2 environments. *Trends Ecol. Evol.* **8,** 64–68.

Lindroth, R., Kinney, K., and Platz, C. (1993). Responses of deciduous trees to elevated atmospheric CO_2: Productivity, phytochemistry and insect performance. *Ecology,* **74,** 763–777.

Loaiciga, H., Valdes, J., Vogel, R., Garvey, J., and Schwarz, H. (1996). Global warming and the hydrologic cycle. *J. Hydrol.* **174,** 83–127.

Loik, M., and Harte, J. (1996). High-temperature tolerance of *Artemisia tridentata* and *Potentilla gracilis* under a climate change manipulation. *Oecologia* **108,** 224–231.

Long, S. (1991). Modification of the response of photosynthetic productivity to rising temperature by atmospheric CO_2 concentrations: Has its importance been underestimated? *Plant Cell Environ.* **14,** 729–739.

Long, S., and Hutchin, P. (1991). Primary production in grasslands and coniferous forests with climate change: An overview. *Ecol. Applic.* **1,** 139–156.

Lu, Z., and Zeiger, E. (1994). Selection for higher yields and heat resistance in Pima cotton has caused genetically determined changes in stomatal conductances. *Physiol. Plant* **92,** 273–278.

Lu, Z., Radin, J., Turcotte, E., Percy, R., and Zeiger, E. (1994). High yields in advanced lines of Pima cotton are associated with higher stomatal conductance, reduced leaf area and lower leaf temperature. *Physiol. Plant* **92,** 266–272.

MacDonald, G., Edwards, T., Moser, K., Pienitz, R., and Smol, J. (1993). Rapid response of treeline vegetation and lakes to past climate warming. *Nature* **361,** 243–246.

MacGillivray, C., and Grime, J. (1995). Testing predictions of the resistance and resilience of vegetation subjected to extreme events. *Funct. Ecol.* **9,** 640–649.

Manabe, S., and Wetherald, R. (1986). Reduction in summer soil wetness induced by an increase in atmospheric carbon dioxide. *Science* **232,** 626–628.

Mann, M., Park, J., and Bradley, R. (1995). Global interdecadal and century-scale climate oscillations during the past five centuries. *Nature* **378,** 266–270.

Markham, A. (1996). Potential impacts of climate change on ecosystems: A review of implications for policymakers and conservation biologist. *Climate Res.* **6,** 179–191.

Mayeux, H., Johnson, H., and Polley, H. (1994). Potential interactions between global change and Intermountain annual grasslands. *In:* "Ecology and Management of Annual Grasslands" (S. Monsen and S. Kitchen, eds.), pp. 95–110. Intermountain Research Station, Ogden, UT.

Melillo, J., McGuire, A., Kicklighter, D., Moore, B., Vorosmarty, C., and Schloss, A. (1993). Global climate change and terrestrial net primary production. *Nature* **363,** 234–240.

Meyer, G., Wells, S., Balling, R., and Jull, A. (1992). Response of alluvial systems to fire and climate change in Yellowstone National Park. *Nature* **357,** 147–150.

Monson, R., Littlejohn, R., and Williams, G. (1983). Photosynthetic adaptation to temperature in four species from the Colorado shortgrass steppe: A physiological model for coexistence. *Oecologia* **58,** 43–51.

Morgan, J., Hunt, H., Monz, C., and LeCain, D. (1994). Consequences of growth at two carbon dioxide concentrations and two temperatures for leaf gas exchange in *Pascopyrum smithii* (C_3) and *Bouteloua gracilis* (C_4). *Plant Cell Environ.* **17,** 1023–1033.

Nash, L., and Gleick, P. (1991). Sensitivity of streamflow in the Colorado Basin to climatic changes. *J. Hydrol.* **125,** 221–241.

Neilson, R. (1993). Transient ecotone response to climatic change: Some conceptual and modelling approaches. *Ecol. Applic.* **3,** 385–395.

Noble, I. (1993). A model of the responses of ecotones to climate change. *Ecol. Applic.* **3,** 396–403.

Nowak, C., Nowak, R., Tausch, R., and Wigand, P. (1994). Tree and shrub dynamics in north western Great Basin woodland and shrub steppe during the Late-Pleistocene and Holocene. *Am. J. Bot.* **81,** 265–277.

Oechel, W., Hastings, S., Vourlitis, G., Jenkins, M., Riechers, G., and Grulke, N. (1993). Recent change of Arctic tundra ecosystems from a net carbon dioxide sink to a source. *Nature* **361,** 520–523.

Oechel, W., Hastings, S., Vourlitis, G., Jenkins, M., and Hinkson, C. (1995). Direct effects of elevated CO_2 in chaparral and Mediterranean-type ecosystems. *In* "Global Change and Mediterranean-Type Ecosystems" (J. Moreno and W. Oechel, eds.), pp. 58–75. Springer-Verlag, Berlin.

Orshan, G. (1954). Surface reduction and its significance as a hydroecological factor. *J. Ecol.* **42,** 442–444.

Osmond, C., Austin, M., Berry, J., Billings, W., Boyer, J., Dacey, J., Nobel, P., Smith, S., and Winner, W. (1987). Stress physiology and the distribution of plants. *BioScience* **37,** 38–48.

Overpeck, J., Bartlein, P., and Webb, T. (1991). Potential magnitude of future vegetation change in eastern North America: Comparisons with the past. *Science* **254,** 692–695.

Owensby, C., Coyne, P., and Auen, L. (1993a). Nitrogen and phosphorus dynamics of a tallgrass prairie ecosystem exposed to elevated carbon dioxide. *Plant Cell Environ.* **16,** 843–850.

Owensby, C., Coyne, P., Ham, J., Auen, L., and Knapp, A. (1993b). Biomass production in a tallgrass prairie ecosystem exposed to ambient and elevated CO_2. *Ecol. Applic.* **3,** 644–653.

Parton, W., Ojima, D., and Schimel, D. (1994). Environmental change in grasslands: Assessment using models. *Climatic Change* **28,** 111–141.

Peterjohn, W., Melillo, J., Bowles, F., and Steudler, P. (1993). Soil warming and trace gas fluxes: Experimental design and preliminary flux results. *Oecologia* **93,** 18–24.

Polley, H., Johnson, H., Marino, B., and Mayeux, H. (1993). Increase in C_3 plant water-use efficiency and biomass over glacial to present CO_2 concentrations. *Nature* **361,** 61–64.

Post, W., Pastor, J., King, A., and Emanuel, W. (1992). Aspects of the interaction between vegetation and soil under global change. *Water Air Soil Poll.* **64,** 345–363.

Potvin, C. (1985). Amelioration of chilling effects by CO_2 enrichment. *Physiol. Veg.* **23,** 345–352.

Rambal, S., and Debussche, G. (1995). Water balance of Mediterranean ecosystems under a changing climate. *In* "Global Change and Mediterranean-Type Ecosystems" (J. Moreno and W. Oechel, eds.), pp. 386–407. Springer-Verlag, Berlin.

Read, J., and Morgan, J. (1996). Growth and partitioning in *Pascopyrum smithii* (C_3) and *Bouteloua gracilis* (C_4) as influenced by carbon dioxide and temperature. *Ann. Bot.* **77,** 487–496.

Reich, P., Abrams, M., Ellsworth, D., Kruger, E., and Tabone, T. (1990). Fire effects ecophysiology and community dynamics of central Wisconsin oak forest regeneration. *Ecology* **71,** 2179–2190.

Rice, C., Garcia, F., Hampton, C., and Owensby, C. (1994). Soil microbial response in tallgrass prairie to elevated CO_2. *Plant Soil* **165,** 67–74.

Risser, P. (1985). Grasslands. *In* "Physiological Ecology of North American Plant Communities" (B. Chabot and H. Mooney, eds.), pp. 232–256. Chapman and Hall, New York.

Rouse, W., Carlson, D., and Weick, E. (1992). Impacts of summer warming on the energy and water balance of wetland tundra. *Climatic Change* **22,** 305–326.

Sage, R. (1996a). Atmospheric modification and vegetation responses to environmental stress. *Global Change Biol.* **2,** 79–83.

Sage, R. (1996b). Modification of fire disturbance by elevated CO_2. *In* "Carbon Dioxide, Populations, and Communities" (C. Körner and F. Bazzaz, eds.), pp. 231–249. Academic Press, San Diego.

Schimmel, J., and Granstrom, A. (1996). Fire severity and vegetation response in the boreal Swedish forest. *Ecology* **77,** 1436–1450.

Schlesinger, M., and Mitchell, J. (1987). Climate model simulations of the equilibrium climatic response to increased carbon dioxide. *Rev. Geophys.* **25,** 760–798.

Seastedt, T., Briggs, J., and Gibson, D. (1991). Controls of nitrogen limitation in tallgrass prairie. *Oecologia* **87,** 72–79.

Shaver, G., Billings, W., Chapin, F. S. III, Giblin, A., Nadelhoffer, K., Oechel, W., and Rastetter, E. (1992). Global change and the carbon balance of arctic ecosystems. *BioScience* **42,** 433–441.

Skiles, J., and Hanson, J. (1994). Responses of arid and semiarid watersheds to increasing carbon dioxide and climate change as shown by simulation studies. *Climatic Change* **26,** 377–397.

Smith, S., Didden-Zopfy, B., and Nobel, P. (1984). High-temperature responses of North American cacti. *Ecology* **65,** 643–651.

Smith, S., Strain, B., and Sharkey, T. (1987). Effects of CO_2 enrichment on four Great Basin grasses. *Funct. Ecol.* **1,** 139–143.

Smith, S., Monson, R., and Anderson, J. (1997). "Physiological Ecology of North American Desert Plants." Springer-Verlag, Berlin.

Strain, B., and Bazzaz, F. (1983). Terrestrial plant communities. *In* "CO_2 and Plants: The Response of Plants to Rising Levels of Carbon Dioxide" (E. Lemon, ed.), pp. 177–222. Am. Assoc. Adv. Sci., Washington, DC.

Strain, B., and Thomas, R. (1995). Anticipated effects of elevated CO_2 and climate change on plants from Mediterranean-type ecosystems utilizing results of studies in other ecosystems. *In* "Global Change and Mediterranean-Type Ecosystems" (J. Moreno and W. Oechel, eds.), pp. 121–139. Springer-Verlag, Berlin.

Taylor, K., Lamorey, G., Doyle, G., Alley, R., Grootes, P., Mayewski, P., White, J., and Barlow, L. (1993). The 'flickering switch' of late Pleistocene climate change. *Nature* **361,** 432–436.

Tissue, D. T., and Oechel, W. C. (1987). Response of *Eriophorum vaginatum* to elevated CO_2 and temperature in the Alaskan tussock tundra. *Ecology* **68,** 401–410.

Van Cleve, K., Viereck, L., and Dyrness, C. (1996). State factor control of soils and forest succession along the Tanana River in interior Alaska. *Arctic Alpine Res.* **28,** 388–400.

Waring, G., and Cobb, N. (1992). The impact of plant stress on herbivore population dynamics. *In* "Plant-Insect Interactions" (E. Bernays, ed.), pp. 167–226. CRC Press, Boca Raton, FL.

Webb, T., and Bartlein, P. (1992). Global changes during the last 3 million years: Climatic controls and biotic responses. *Annu. Rev. Ecol. Syst.* **23,** 141–173.

Woodward, F. (1993). The lowland-to-upland transition—modelling plant responses to environmental change. *Ecol. Applic.* **3,** 404–408.

Wullschleger, S., Ziska, L., and Bunce, J. (1994). Respiration responses of higher plants to atmospheric CO_2. *Physiol. Plant.* **90,** 221–229.

Young, J., and Evans, R. (1978). Population dynamics after wildfires in sagebrush grasslands. *J. Range Manage.* **31,** 283–289.

Ziska, L., and Bunce, J. (1994). Direct and indirect inhibition of single leaf respiration by elevated CO_2 concentrations: Interaction with temperature. *Physiol. Plant.* **90,** 130–138.

5

Interactions between Rising CO_2, Soil Salinity, and Plant Growth

Rana Munns, Grant R. Cramer, and Marilyn C. Ball

I. Introduction

Much of the saline land in the world has been caused by human activities: by clearing, overgrazing, or the installation of irrigation schemes. The result of these activities is euphemistically termed "secondary salinization." The causes are well known, and revegetation of saline land is taking place on a small scale in some countries, but generally the pressure of increasing populations and increasing demand for food means that land is still being cleared and irrigated, and salinity is still increasing. Small increases in food production can be made by bringing new land into agricultural production, but these increases are generally unsustainable. The long-term productivity of the global agricultural system as a whole is declining, as is ecosystem stability.

Rising CO_2 levels could increase the productivity of cultivated species and natural vegetation on saline soils, but it may also increase soil salinity. Elevated CO_2 can increase the growth rate of crops, pastures, and trees, but because it also increases their water use efficiency, it may alter the soil–plant water balance in a way that has adverse consequences for land with saline groundwater, where any factor that causes the water table to rise will increase the rate of salinization of the topsoil. This applies to large areas of land on most continents.

In this chapter we describe the global impact of salinity and the likely consequences of elevated CO_2 on plant growth and soil salinity. To aid in predictions and models of the effects of elevated CO_2 on plant growth in saline soil, the cellular and whole plant mechanisms by which salinity affects growth are described, with emphasis on the mechanisms by which elevated CO_2 might interact with salinity. We then summarize the few data that have been published on the effects of elevated CO_2 on plant growth and water use in saline soil, and on the response of natural communities.

II. Global Extent of Salt-Affected Land

A significant amount of the earth's land is saline, sodic, or alkaline. The total area of this salt-affected land has been estimated at up to 950 Mha, or 7% of the world land area (Szabolcs, 1994; Flowers and Yeo, 1995). Much of this is caused by clearing of perennial vegetation, and the problem has increased with development of irrigation systems. The problem therefore does not apply to timber production, but is a threat to the world's food production, because it is affecting the most productive agricultural systems in the world.

Human-induced salinity arises in arid and semiarid regions, where groundwaters are often rich in soluble salts from ancient marine or alluvial sediments (Szabolcs, 1994). When land is vegetated with trees or deep-rooted perennial grasses, the water table is kept well below the surface, 10–30 m below it. When land is cleared of perennial vegetation, or irrigation schemes are installed, the water table rises because more water then enters the ground (with rain or irrigation) than leaves it (by surface evaporation or plant transpiration). Salts move up with the water, and the soil becomes saline, sodic, or alkaline. Salinity can also rise in low-rainfall pastoral zones. Overgrazing can cause soil erosion that exposes a subsoil that is saline or sodic. The result is termed *scald.*

Saline soil is classified as having a high concentration of soluble salts, with an EC_e of 4 dS m^{-1} or more (EC_e is the electrical conductivity of the solution extracted from a soil sample after being mixed with sufficient water to produce a saturated paste). The dominant soluble salts in saline soil are chlorides, sulfates, and bicarbonates of sodium, calcium, and magnesium. Of all the salts, NaCl is the most common. The value of 4 dS m^{-1} (equivalent to about 40 mM NaCl) is generally used worldwide as a definition of salinity because this level would affect the yield of most crops. However, the Soil Science Society of America has recommended the limit be reduced to 2 dS m^{-1} because many crops can be damaged in the range 2–4 dS m^{-1} (Ghassemi *et al.*, 1995). Sodic soils have a low EC_e, but a high percent of Na^+ on the soil complex; they have an ESP of 15 or more (ESP is the percentage of exchangeable sodium ions to the total exchangeable cations). Alkaline soils are a type of sodic soil with a high pH due to carbonate salts, and are defined as having an ESP of 15 or more and a pH of 8.5–10. A high ESP produces a poor soil structure that does not drain well, and waterlogging is a major problem with sodic soils. Infiltration too may be a problem, and irrigation water or rainwater tends to puddle, causing anaerobiosis and poor water and nutrient uptake by roots (Ghassemi *et al.*, 1995).

Continued cultivation of salt-affected soils is sustained only by leaching (if saline) or by gypsum application (if sodic), coupled with drainage. Because drainage is costly, decisions may be made to abandon salt-affected areas for agricultural production, and attempt revegation with salt-tolerant perennials. This may be a wise decision in the global context, but could have adverse effects on the local economy, because much of the land now salinized was once highly productive.

The picture is particularly bleak for sustained productivity of irrigated agriculture. There are large areas of irrigated land not yet classified as saline only because they are managed by leaching and drainage, a practice that is often not sustainable in regional terms because disposal of drainage water may cause harm to another region. Irrigated lands of the world in 1987 totaled 227 Mha (Table I). This is only 15% of total cultivated land, but because irrigated land has at least twice the productivity of rainfed land, it may produce one-third of the world's food. According to estimates of FAO and UNESCO, as many as half of the existing irrigation systems of the world are under the influence of secondary salinity, sodicity, or waterlogging (Szabolcs, 1994).

And what of dryland salinity? The figure quoted earlier of 7% for the world's salt-affected land includes land that is uncultivated (e.g., land degraded by ancient civilizations, or naturally barren), and there are no data to specify the extent of arable land that is currently salt affected. However, there are data for the area of land that has become saline due to recent agricultural practices, which are an indication of future trends. In a UNEP

Table I Cultivated and Irrigated Land of the World, Taken from FAO Data, 1989[a]

Continent	Cultivated land (Mha)	Irrigated land (Mha)
Asia	451	142
Africa	185	11
South America	142	9
North America	274	26
Europe	140	17
Australasia	49	2
Former USSR	233	20
Total	1474	227

[a] The total land area of the world is 13,077 Mha. Data from Ghassemi *et al.*, 1995.

study on the global assessment of soil degradation, soil scientists were asked to categorize soils degraded during the past 45 yr as a result of human intervention. The total area of human-induced degraded soils was assessed at 1964 Mha, an alarmingly high figure (Ghassemi *et al.,* 1995). Most of the degradation was due to wind and water erosion. Chemical degradation accounted for 239 Mha, of which loss of nutrients was the major factor (135 Mha) and salinization was the next most important at 77 Mha (Table II). Of this, only 35 Mha was classed as still manageable by farmers using appropriate cropping systems, including drainage or perennial forage plants. The remainder needed reclamation with salt-tolerant trees.

Dryland salinity can be managed by land use planning at the local and catchment level. It involves planting trees in both the recharge and discharge areas to control the rise of water tables. (The recharge area is the highest point of the catchment and the origin of the rainfall that enters the water table; the discharge area is the point where the water table rises to the surface.) Planting deep-rooted perennial pasture species that continue to use water all year round can also control subsoil water movement and provide some income for pastoralists. Water tables in discharge areas can be lowered by pumps (e.g., windmills) or by subsurface drainage, but in many regions the saline discharge is at the lowest point of the landscape and there is no obvious place to dump the solution that is removed from a saline field. Pipes to the ocean have been suggested, but the cost is prohibitive.

This area of man-made salinity over the last 45 yr, of 77 Mha, seems relatively small at a global level, but as a percentage of the cultivated land of the world, it is quite significant—5%. This trend is likely to continue

Table II Global Extent of Human-Induced Salinization, Derived from Maps Prepared by the International Soil Reference and Information Centre[a]

Extent of land degradation	Light (Mha)	Moderate (Mha)	Strong (Mha)	Extreme (Mha)	Total (Mha)
Asia	26.8	8.5	17.0	0.4	52.7
Africa	4.7	7.7	2.4	—	14.8
South America	1.8	0.3	—	—	2.1
North America	0.3	1.5	0.5	—	2.3
Europe	1.0	2.3	0.5	—	3.8
Australasia	—	0.5	—	0.4	0.9
Total	34.6	20.8	20.4	0.8	76.6

[a] Lightly degraded soils have reduced productivity, but are still manageable by local farming systems. Moderately degraded soils have greatly reduced productivity, and can only be improved by major and costly restoration. Strongly degraded soils are not reclaimable at the farm level; extremely degraded soils are beyond reclamation by any means. Data from Ghassemi *et al.,* 1995.

during the next 45 yr, and another 77 Mha will become salt affected, because forests are still being cleared and irrigation schemes are still being developed.

III. Effect of Salinity on Plant Production

A. Food Plants

The three most important crops in the world are wheat, rice, and maize. Asia is the most productive continent (FAO, 1995). Asia produces 41% of the world's wheat, 91% of the world's rice, 47% of the world's pulses, 60% of the world's vegetables, and 36% of the world's fruits. It has one-third of the world's arable land, but also one-third of the world's saline soils and two-thirds of the world's population. Of the irrigated land in Asia (Table I), most is from India and China: 44 and 48 Mha, respectively. The sustainability of food production in these countries depends largely on how salinization is controlled.

Differences in the growth response of various species are shown in Fig. 1. Wheat is one of the more salt-tolerant crop species, and many cultivars

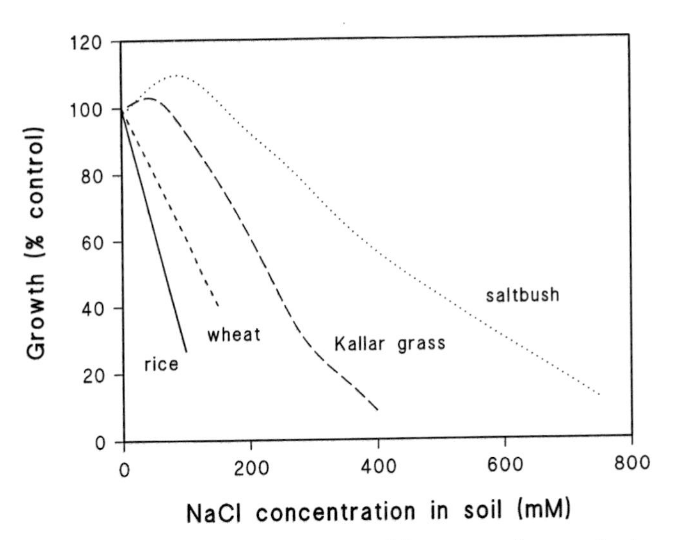

Figure 1 Biomass production of four diverse and important plant species in a range of salinities. Two crop species are considered: wheat, one of the more salt-tolerant crops (Munns *et al.*, 1995), and rice, one of the more salt-sensitive crops (Aslam *et al.*, 1993); and two halophytes: a saltbush species *Atriplex amnicola* (Aslam *et al.*, 1986) and a grass *Diplachne* (syn. *Leptochloa*) *fusca* or Kallar grass (Myers and West, 1990). Both halophytes show outstanding salt tolerance with high growth rates and are being used in Australia and Asia to support grazing on saline land.

that have been selected for yield in water-limited conditions do not suffer a 50% reduction in biomass until salinities reach 15 dS m^{-1} (approximately 150 m*M* NaCl). Rice is more salt sensitive, and many cultivars suffer a 50% reduction in growth at half this concentration of salts (Fig. 1). Maize falls between these two species in terms of salt sensitivity (e.g., Cramer *et al.*, 1994).

A large part of the Asian diet is rice, which is one of the more salt-sensitive crop species. Much of the paddy rice is produced in low-level, coastal regions and some regions will be inundated if there is a rise in sea levels due to global warming. The demand for food due to population pressure in Asia is still increasing, yet food production has reached a threshold and the sustainability of its current production is in doubt. Globally, the area under rice cultivation is not expected to decrease, because global warming will shift the area suitable for rice (Bachelet and Kropff, 1995), but there is clearly a threat to some parts of Asia. Is elevated CO_2 likely to compensate for the presence of salinity in coastal or low-lying areas?

B. Halophytes, and Reclamation Plants

When land becomes salinized it is abandoned for cultivation. During the last decade, efforts have been made to reclaim salinized land for some sort of economic return. Halophytic plants have been investigated for this purpose. Halophytes are the native flora of saline soils and can complete their life cycle at salinities above 250 m*M* NaCl. Most halophytes need at least 1 m*M* NaCl to grow well, many need 10–50 m*M* NaCl to reach maximum growth, and a few grow best at 200–300 m*M* NaCl (Flowers *et al.*, 1986). Many halophytes are very slow growing even at their optimal salinity, but a few are suitable for reclamation. For instance, Kallar grass (*Diplachne fusca*) is widespread in many continents and now planted as a fodder species in salt-affected soil (Fig. 1). Another species being planted for fodder is the dicotyledenous halophyte, *Atriplex amnicola,* a saltbush species native to Western Australia (Fig. 1). Some halophytes are more tolerant than this, but *A. amnicola* shows the growth maximum at low salinity and the extended growth at very high salinities that is typical of many dicotyledenous halophytes.

Recently, interest has been strong in the salt tolerance of trees that are not halophytes but have good salt tolerance and high transpiration rates that can be planted in saline soils to lower water tables. Such species need also to be tolerant of waterlogging because secondary salinity occurs with rising water tables, so the ability to withstand periods of waterlogging and to continue high transpiration rates is essential. In Australia, river red gum (*Eucalyptus camaldulensis*) has been the most widely used. However, there are a number of other species that are better able to tolerate saline and waterlogged soils. Many *Acacia, Casuarina,* and *Melaleuca* species are tolerant of high salinity; for example, *A. stenophylla* did not suffer a 50% growth reduction until an EC_e of 20 dS m^{-1}, equivalent to about 200 m*M* NaCl (Marcar *et al.*, 1995). Such species are recommended for reclamation of land that has become saline because of irrigation, because they provide some income to farmers as fodder and fuel wood.

IV. Mechanisms of Salt Tolerance That Relate to Effects of Elevated CO_2

Salts in the soil water may inhibit plant growth for two reasons. First, the presence of salt in the soil solution reduces the ability of the plant to take up water, and this leads to reductions in the growth rate. This is referred to as the *osmotic* or *water-deficit effect* of salinity. Second, if excessive amounts of salt enter the plant in the transpiration stream, injury will occur to cells in the transpiring leaves and this may cause further reductions in growth. This is called the *salt-specific* or *ion-excess effect* of salinity (Greenway and Munns, 1980; Munns, 1993). The definition of salt tolerance is usually the percent biomass production in saline soil relative to plants in nonsaline soil, after growth for an extended period of time. For slow-growing, long-lived, or uncultivated species it is often difficult to assess the reduction in biomass production, so percent survival is often used. To grow in saline conditions, plants must maintain a high water status in the face of soil water deficits and potential ion toxicity. A plant can only grow or survive in a saline soil if it can both continue to take up water *and* exclude a large proportion of the salt in the soil solution.

A. The Importance of Salt Exclusion

Plants take up much more water from the soil than they retain in their shoot tissues. If, for example, they take up 50 times more water than they retain, any solute dissolved in the soil water that is not excluded by the roots as the plants take up water will be deposited in the shoot as the water evaporates and end up at a concentration 50 times that in the soil solution.

Roots, therefore, must exclude most of the Na^+ and Cl^- present in solution in a saline soil or the salt will gradually build up with time in the leaves and become so high that it kills the leaf. Roots themselves are the least vulnerable part of the plant: The Na^+ and Cl^- concentration in roots is rarely higher than in the external solution, and often is lower. In some species, notably in halophytes adapted to highly saline regions, salt glands or bladders excrete NaCl and so help to avoid excessive accumulation of salt in the mesophyll cells of leaves, but the role of salt-excreting mechanisms is minor in comparison to the salt exclusion performed by roots that, as explained later, must exclude close to 98% of the salt from the soil solution.

The percentage of transpired water that is retained in the shoot can be calculated from the product of the water use efficiency (WUE; mg dry weight of shoot produced per g H_2O transpired) and the water content (WC; g H_2O per g dry weight):

$$\% \text{ water retained} = \text{WUE} \times \text{WC} \times 100. \tag{1}$$

The term *water use efficiency* in this case refers to the net increase of total shoot dry weight per unit of water used by the plant over a substantial period of time (not the gas exchange ratio over a short period). It thus takes into account allocation of carbon within the shoot, and respiratory losses. The water content in this case is the shoot average. We assume for now that there is no partitioning of water or salt within the shoot; this level of complexity is discussed later.

Water use efficiencies of plants growing at moderate evaporate demand are usually in the range of 3–6 mg/g, whether they are for wheat (Farquhar and Richards, 1984) or mangroves (Ball, 1988). Although variation can exist within a species (Farquhar and Richards, 1984), it is relatively small, and values higher or lower than 3–6 mg/g are more likely due to extremes of evaporative demand, rather than a peculiarity of the species. For example, the WUE for wheat increased from 3 to 9 as the pan evaporation rate over a season fell from 9 to 1 mm/d (Richards, 1991). For a WUE of 4 mg/g (i.e., for a C_3 plant in a fairly dry environment) and an H_2O/DW ratio of 5 : 1 (i.e., for a grass or young herbaceous species), 20 mg of water is retained in the shoot for every g of water transpired [Eq. (1)]. That is, the shoot transpires about 50 times more water than it retains. The range of WUE of 3–9 mg/g gives a range of 1.5–4.5% for the percentage of water retained in shoots, which translates to shoots transpiring 22–100 times more than they retain.

The concentration at which NaCl accumulates in the shoot depends simply on the salt concentration in the soil solution, the percentage of salt taken up (not excluded) by roots, and the percentage of water retained:

$$[\text{NaCl}]_{\text{shoot}} = [\text{NaCl}]_{\text{soil}} \times \frac{\% \text{ salt taken up}}{\% \text{ water retained}}. \tag{2}$$

To maintain a salt concentration in the shoot that is no greater than that in the soil, the percent of the salt taken up must be the same as the percent of transpired water retained. If 2% of the water is retained in the shoot, as in the preceding example, then only 2% of the salt should be allowed into the shoot; that is, 98% should be excluded. Taking the example of a plant growing in a saline soil of 100 m*M* NaCl, that excludes 98% of the salt in the soil solution, then the concentration of salt in the shoot will be 100 m*M* [Eq. (2)]. If a plant excludes only 96% of the salt in the soil solution, the concentration in the shoot will be 200 m*M* [Eq. (2)].

The degree of exclusion that plants achieve can be gained from measuring or calculating the concentration of salt in the xylem sap flowing from roots to shoots in relation to that in the external solution. Two sets of data are illustrated in Fig. 2 for two diverse species: barley and a mangrove. The degree of exclusion is remarkably similar. For barley, the concentration of Cl^- in the xylem increased from 5 to 7 m*M* while the external NaCl increased from 25 to 200 m*M* (Munns, 1985), so that at 200 m*M* the exclusion was about 97%. Na^+ showed a similar degree of exclusion. For the mangrove *Avicennia marina,* the Cl^- in the xylem increased from 3 to 9 m*M* while the external NaCl concentration increased from 50 to 500 m*M* (Ball, 1988), and so at 200 m*M* the exclusion was again about 97%. In contrast, the degree of exclusion in the salt-sensitive lupine species *Lupinus albus* growing in 100 m*M* NaCl was only 90%, and the plants survived only a few days at that salinity (Munns, 1988).

For most plants, a concentration of 100 m*M* NaCl on a whole shoot basis is about as high as desirable because it will include some old leaves with much higher salt concentrations, as well as younger leaves or other tissues with lower concentrations. Salt in some species is retained in the upper roots, the stem, or the leaf sheaths or petioles, which reduces the rate at which salts build up in the photosynthetic tissue (Greenway and Munns, 1980). This is particularly important for trees (e.g., citrus species; Storey, 1995). Nevertheless, salt concentrations eventually increase with time in fully expanded leaves until the concentration of salt becomes toxic and

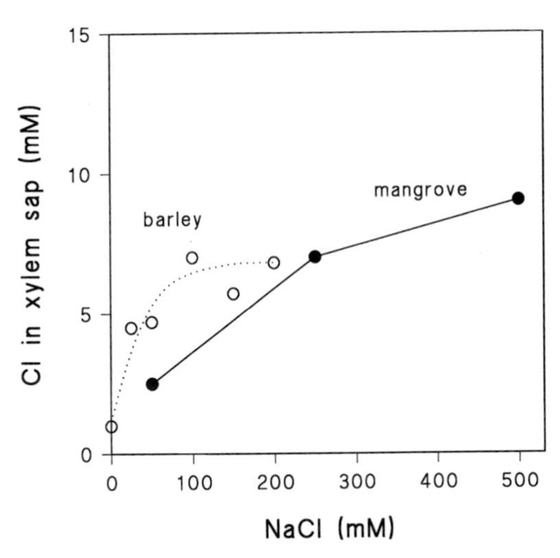

Figure 2 Cl^- concentration in the xylem sap of the mangrove *Avicennia maria* (Ball, 1988) compared with barley (Munns, 1985) grown at a range of salinities.

the leaves die (Greenway and Munns, 1980; Flowers and Yeo, 1986). The concentrations of salt that constitutes a toxicity, and the cellular mechanisms that exist to avoid toxicity, are discussed later.

Some halophytes excrete Na^+ and Cl^- at considerable rates through salt glands (Rozema, 1975) or bladders (Aslam *et al.*, 1986), which significantly reduces the rate at which salt accumulates in the mesophyll cells. In some species, for example, the mangrove *Avicennia marina*, the rate of excretion through glands can equal the rate at which the salt arrives in the leaves (Ball, 1988), so preventing any increase in salt concentration in the mesophyll cells. However, these are additional, not alternative, mechanisms for regulating the salt accumulation in leaves; prevention of salt entering the transpiration stream is the most important mechanism of regulation.

B. Relationship between Transpiration and Salt Uptake

If the degree of exclusion was constant, a very simple relationship would exist between elevated CO_2 and salt buildup in leaves, via its effect on WUE, which would allow a prediction of the effect of elevated CO_2 on salt tolerance. Excessive salt buildup causes death of old leaves, so elevated CO_2 would increase salt tolerance by increasing the WUE and reducing the subsequent buildup of salt in leaves. However, the rate of salt uptake by the xylem is influenced by the rate of transpiration to some extent.

The fundamental processes governing the relationship between water and ion flow through roots are complex and not well understood. NaCl does not move passively with the transpiration stream, neither is its movement entirely independent of it, at least in some species or over certain ranges of transpiration. Figure 3 shows the relationship between water and salt flow in the xylem of barley plants (Munns, 1985). As water flow increased from a very low to a moderate rate, an increase in Cl^- flux was seen, showing that the movement of the ion through the root was enhanced as water flow started to increase. However, when the water flow increased from moderate to high rates, there was little or no further increase in Cl^- flux, showing that the movement of the ion was independent of further increases in water flow. This relationship was also found for Na^+ and K^+ (Munns, 1985). A similar relationship was found between transpiration and Cl^- uptake in two mangrove species, *Avicennia marina* and *Aegiceras corniculatum* (Ball, 1988). In this case, the xylem sap was not collected, but the concentration of Cl^- in the sap and the Cl^- flux were calculated from the salt and water balance of the plant. The Cl^- flux increased as the transpiration increased from 0.5 to 1.5 mm m^{-2} s^{-1}, and then changed little with further increases in transpiration (Ball, 1988).

Measurements of ion concentrations in leaves of plants grown at different humidities are consistent with this pattern—that salt transport to leaves is substantially affected only if transpiration is greatly affected. For example,

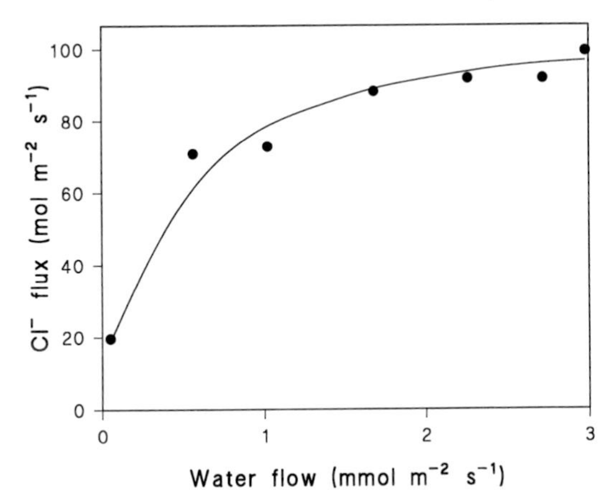

Figure 3 Dependence of Cl^- flux on water flow from roots to shoots for barley grown at 50 m*M* NaCl. Plants were grown in pots that would fit inside a pressure chamber. To collect xylem sap, a pot was placed in a pressure bomb with its shoot protruding, and water flow was varied by applying pressure to the root system. The lowest flow rate is without applied pressure, generated by root pressure. Cl^- flux was calculated from the Cl^- concentration in the sap at a given water flow rate; the water flow being expressed as the rate per unit leaf area of the plant before decapitation (Munns, 1985).

in citrus the concentration of Na^+ and Cl^- in leaves was reduced 2.5-fold by an increase in humidity, which caused transpiration rates to decrease by 4-fold (Storey, 1995), whereas in a study with five different species there was a much smaller effect on the concentration of Na^+ and Cl^- in leaves, with the transpiration rates decreased by only 2-fold or less (Salim, 1989). There were large species differences, some species showing no effect, others showing a 2-fold effect particularly of Cl^- (Salim, 1989). These results indicate that elevated CO_2 may not significantly reduce the rate at which salt arrives in leaves, unless there is a 2-fold (100%) decrease in transpiration. An effect might be seen more with species that are very poor excluders, such as lupine as mentioned earlier, and rice, which appears to carry much more salt in an apoplastic or transpirational "bypass" pathway than other species (Garcia *et al.*, 1997). In rice, the percentage of water moving through a bypass pathway from roots to shoots was estimated as 5.5% of the total water transpired, and could account for all the Na^+ transported to the shoots, whereas in wheat only 0.4% of the water moved along a bypass pathway and could not account for most of the Na^+ transported (Garcia *et al.*, 1997).

This prediction that elevated CO_2 will not substantially change the rate of salt transport to leaves is supported by studies that have examined the effect of elevated CO_2 on salt buildup in leaves, and have found little or no effect. Elevated CO_2 did not affect the accumulation of either Na^+, Cl^-, or K^+ in leaves of the C_4 grass *Andropogon glomeratus* grown at 100 m*M* NaCl, even though WUE was increased by 35% (Bowman and Strain, 1987). Neither did elevated CO_2 affect the Na^+ concentration in leaves of the mangrove species, *Rhizophora apiculata,* and *R. stylosa* (Ball and Munns, 1992), and *R. mangle* (Farnsworth *et al.,* 1996). These experiments were done in saline solution. Elevated CO_2 caused a small decrease in the concentration of Na^+ in leaves of wheat grown in saline soil that was not statistically significant for the more salt-tolerant cultivar, with a high degree of exclusion, but was significant for the salt-sensitive cultivar, a relatively poor excluder (Nicolas *et al.,* 1993).

It is likely that different effects of transpiration on salt uptake may be found in soil than in solution culture, because high rates of transpiration can cause a buildup of salts at the root–soil interface in a saline or highly fertilized soil (Stirzaker and Passioura, 1996). In that case, a lower mass flow of water through the soil due to the enhanced WUE caused by elevated CO_2 may prevent a high concentration of salt from building up locally at the root surface, and so reduce the uptake of salt by the root. In a comparison of citrus species growing in sand and solution culture, concentrations of Na^+ and Cl^- in leaves were much less when the plants were grown in sand than in solution (Storey, 1995).

C. Whole Plant Aspects: What Causes the Growth Reduction?

The effects of a saline soil are twofold: There are effects of the salt outside the roots, and there are effects of the salt inside.

The effects of the salt outside (the *osmotic stress*) reduce leaf growth and to a lesser extent root growth, and decrease stomatal conductance and thereby photosynthesis (Munns, 1993). The rate at which new leaves are produced depends largely on the water potential of the soil solution, in the same way as a drought-stressed plant. The low soil water potential leads to internal water deficits and changes in the hormonal balance of the plant that affect growth and the production of new leaves (Davies and Zhang, 1991; Munns and Sharp, 1993; Munns and Cramer, 1996). Salts themselves do not build up in the growing tissues at concentrations that inhibit growth: Meristematic tissues are fed largely in the phloem from which salt is effectively excluded, and rapidly elongating cells can dilute the salt that arrives in the xylem with their expanding vacuoles. So, the salt taken up by the plant does not directly inhibit the growth of new leaves (Munns, 1993). This occurrence also applies to halophytes at most salinity levels, with the exception of those halophytes that show enhanced growth at low salinities (Fig. 1), a phenomenon that has not been satisfactorily explained.

The salt inside causes an increased senescence of old leaves. Continued transport of salt into transpiring leaves over a long period of time eventually results in very high Na^+ and Cl^- concentrations, and they die (Flowers and Yeo, 1986; Munns and Termaat, 1986). The rate at which they die becomes the crucial issue determining the survival of the plant (Munns, 1993). If new leaves are continually produced at a rate greater than that at which old leaves die, then there are enough photosynthesizing leaves for the plant to produce flowers and seeds, although in reduced numbers. However, if the rate of leaf death exceeds the rate at which new leaves are produced, then the proportion of leaves that are injured starts to increase. There is then a race against time for the plant to initiate flowers and form seeds, or to enter some sort of dormancy condition, while there is still an adequate number of green leaves left to supply the necessary photosynthate.

The rate at which old leaves die depends on the rate at which salts accumulate to toxic levels. Thus, control of the rate at which salt arrives in leaves is essential, as are mechanisms that reduce the toxicity of the salt. The mechanism that governs the rate at which salts arrive in leaves, salt exclusion by roots, was covered earlier. The mechanism that reduces the toxic effect of salt is cellular compartmentation, whereby salts are sequestered in vacuoles of leaf cells (Flowers *et al.*, 1977, 1986; Flowers and Yeo, 1986). For species lacking the ability to compartmentalize salts in the vacuoles to high concentrations, continued transport of salt to the leaves will eventually result in either excessive buildup of salts in the cell walls of the leaf cells, causing death through dehydration, or excessive concentrations in the cytoplasm causing death through poisoning of metabolic systems such as photosynthesis or respiration (Flowers and Yeo, 1986). Salt-tolerant species have transport systems on the tonoplast that can sequester Na^+ and Cl^- at high concentrations within the vacuoles, while maintaining much lower concentrations in the cytoplasmic compartments. Estimates of the cytoplasmic ionic strength of salt-stressed plants range from below 100 m*M* to above 200 m*M* (Flowers *et al.*, 1986; Cramer, 1997). *In vitro* studies of enzyme activity in the presence of salts have indicated that most enzymes are inhibited once the NaCl concentration increases above 100 m*M*, and that enzymes from halophytes are just as salt sensitive as enzymes from other species (reviewed by Munns *et al.*, 1983). Thus, a leaf concentration of 500 m*M* Na^+ and Cl^- in a leaf that is alive and healthy must mean that cytoplasmic ion concentrations are considerably lower than vacuolar concentrations. The osmotic pressure of the cytoplasm is balanced with organic solutes that are compatible with enzyme activity (Flowers *et al.*, 1977; Greenway and Munns, 1980).

D. Summary of Mechanisms of Salt Tolerance

Roots do most of the work in protecting the plant from excessive uptake of salts; they filter out most of the salt in the soil while taking up water. Even so, there are mechanisms for coping with the continuous delivery of relatively small amounts of salt that arrive in the leaves, the most important being the cellular compartmentalization of salts in the vacuoles of the mesophyll cells. This strategy allows plants to minimize or delay the toxic effects of high concentrations of ions on important and sensitive cytoplasmic processes. The rate at which leaves die is the rate at which salts accumulate to toxic levels, so genotypes that have poor control of the rate at which salt arrives in leaves, or a poor ability to sequester that salt in cell vacuoles, have a greater rate of leaf death.

V. Effects of Elevated CO_2 on Salt Tolerance and Soil Salinity

A. Effects on Salt Tolerance

Although there have been a few studies on the effects of elevated CO_2 on plants in saline soil, it is clear that elevated CO_2 can increase the growth of many species.

1. Cultivated or Nonhalophytic Species Elevated CO_2 increases both dry weight and leaf area in salt-treated plants. For example, it increased the shoot dry weight of wheat cultivars by 76% (Fig. 4A), and the leaf area by 55% (Fig. 4B). It increased the dry matter and leaf area of the control (unsalinized) plants also, but not to the same extent (Figs. 4A and B). The leaf area was increased more by increasing the number of lateral primordia (tillers) than by the area of any one leaf, and so directly overcame the effect of salinity on leaf area which was to reduce the number of tillers rather than the area of any one leaf (Nicolas *et al.*, 1993). A similar responsiveness of tillering to elevated CO_2 was found in *Andropogon glomeratus* (Bowman and Strain, 1987). Growth of bean and xanthium also responded to high CO_2, the response of the shoots being much greater than that of the roots (Schwarz and Gale, 1984). Maize did not respond significantly to high CO_2 (Schwarz and Gale, 1984).

2. Halophytic Species As noted in a recent review (Idso and Idso, 1994), many studies have predicted that elevated CO_2 would enhance growth of plants subject to salinity stress. However, elevated CO_2 has been found variously to enhance, to suppress, and to have no effect on growth of halophytes, responses that depend on both the species and growth conditions (Schwarz and Gale, 1984; Jansen *et al.*, 1986; Lenssen and Rozema, 1990; Rozema *et al.*, 1990, 1991; Drake, 1992; Lenssen *et al.*, 1993; Farnsworth *et al.*, 1996; Ball *et al.*, 1997). Whether or not elevated CO_2 will stimulate plant growth in saline environments depends on the way in which salinity affects growth. Results of growth analyses show interspecific differences in the extent to which changes in leaf area ratio and/or net assimila-

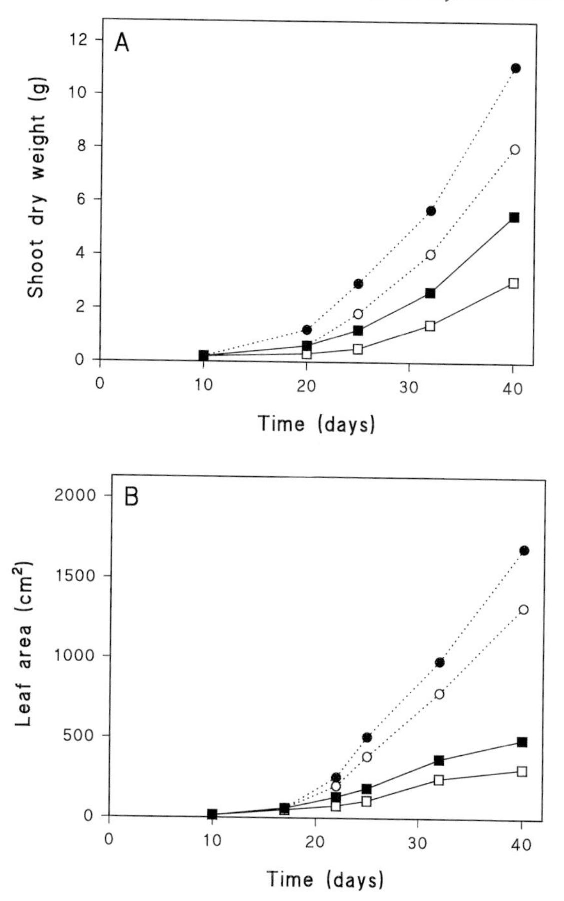

Figure 4 Effect of salinity and elevated CO_2 on growth of wheat (cv. Matong). (A) Shoot dry weight. (B) Leaf area. Dotted lines (and circular symbols) denote plants grown without NaCl; solid lines (and square symbols) denote plants grown in 150 m*M* NaCl. Open symbols represent ambient CO_2; solid symbols represent elevated CO_2 (Nicolas *et al.*, 1993).

tion rate contribute to changes in relative growth rates with increasing salinity. However, increases in the net assimilation rate accounted for most of the growth enhancement under elevated CO_2 in both field (Drake, 1992) and laboratory conditions (Jansen *et al.*, 1986; Rozema *et al.*, 1991; Farnsworth *et al.*, 1996; Ball *et al.*, 1997). In these studies, growth enhancement occurred under moderate salinity stress and was associated with increased rates of photosynthesis. Presumably, elevated CO_2 alleviated stomatal limitations to CO_2 diffusion into the leaves. However, under more stressful salinities, elevated CO_2 had few or no beneficial effects on growth

(Rozema *et al.*, 1991; Ball *et al.*, 1997). Presumably, growth under these highly saline conditions was limited by effects such as ion toxicity or induction of deficiencies in nutrients such as K^+, factors which are unlikely to be affected by elevated CO_2 (Ball and Munns, 1992). It appears that elevated CO_2 stimulates growth of plants under mild salinity stress but does not overcome limitations imposed on growth by severe salinity stress.

Interspecific differences in inherent growth rates also affect the extent to which a species can take advantage of increased supply of a resource such as CO_2 (Hunt *et al.*, 1991, 1993; Poorter, 1993). Recent work on mangroves has led to the suggestion that maximal growth rates are lower the greater the salt tolerance of the species (Ball, 1996). It follows from observations of Hunt *et al.* (1991, 1993) and Poorter (1993) that the more rapidly growing and less salt-tolerant species might be more responsive to elevated CO_2 than the slower growing and more salt-tolerant species, provided that the conditions for growth are favorable. However, because there is little evidence that elevated CO_2 would expand the range of salinities in which a species can grow (Ball *et al.*, 1997), the species that are likely to be the most responsive to elevated CO_2 at any point along a salinity gradient are those species that have the greatest potential for growth.

B. Mechanisms of Response

Hypotheses on the reasons why elevated CO_2 increases growth of salt-affected plants revolve aorund three main issues: that elevated CO_2 decreases salt accumulation in leaves, increases turgor, or increases the carbon supply to the growing tissues. These hypotheses presume that growth of salt-affected plants is limited by (1) high salt concentration, (2) low turgor, or (3) inadequate supply of carbohydrate Controversy still exists on all of these presumptions, particularly the latter. High CO_2 treatments are sometimes used as a way to test these hypotheses.

1. Salt Accumulation For halophytes, or for nonhalophytes growing at low to moderate salinity, it is unlikely that elevated CO_2 increases growth by decreasing the rate at which salts accumulate in leaves and so delaying the time at which they reach toxic levels. There are two reasons for this. One is that the elevated CO_2 may not reduce the uptake of salt significantly, as described earlier. The other is that in plants that are good excluders, the levels of salt buildup in leaves may not be hazardous. For example, in the case of the wheat cultivar shown in Fig. 4 whose growth responded dramatically to elevated CO_2, the concentration of Na^+ in the shoot in plants at 150 m*M* NaCl was far too low to be considered limiting; it was less than 0.1 mmol g^{-1} DW, or 20 m*M* in the tissue water (Nicolas *et al.*, 1993).

2. Turgor We believe it unlikely that high CO_2 increases growth of salt-affected plants by increasing the turgor of leaves, that is, by overcoming a growth limitation due to low turgor. That low turgor was limiting growth of salt-affected plants was directly tested by Termaat *et al.* (1985), who grew wheat or barley plants in pots that would fit inside a pressure chamber and applied a pneumatic pressure equal to the osmotic pressure of the salt in the soil. Growth of these pressurized plants was no greater than that for plants in saline soil that was not pressurized. Similar results were found with a number of dicotyledenous species ranging from lupines (salt-sensitive) to Egyptian clover (salt-tolerant) (Munns and Termaat, 1986). Consistent with this result is the observation that many salt-treated plants do not suffer turgor reductions, and when salt-tolerant and salt-sensitive cultivars are compared it is often found that turgor is actually higher in the salt-sensitive species because their exclusion of salt is poorer and the osmotic pressure of their leaf sap is higher (Munns, 1993).

3. Carbon Supply The third hypothesis, which is the one we consider most likely, is that elevated CO_2 increases the carbon supply of the plant in a way that stimulates the production of more tillers and leaves. The way in which elevated CO_2 does this is not clear. It is not simply that it causes an increase in the concentration of available carbohydrates (sugars, fructans, and starch) because salt-affected plants have a higher concentration of these compounds than their nonsalinized controls, in both mature organs and within growing regions (Munns and Termaat, 1986). The reason these carbon compounds accumulate in salt-affected plants is presumably because their utilization for cell wall and protein synthesis in growing regions is inhibited, and excess carbon is partitioned into reserve pools. It is unlikely that their primary function is in osmotic adjustment, because the carbon is present mainly as the complex carbohydrates, fructans and starch, rather than in low molecular weight forms (e.g., Munns *et al.*, 1982). Perhaps the growing regions respond to an increase in the *flux* of carbon compounds. (The flux of carbon to the growing regions of salt-affected plants is always lower because their leaf area is always reduced, and their photosynthesis rate per unit area is often lower as well).

A related issue is the "cost" of salt tolerance. The possibility of a limitation on growth by additional respiration required to exclude or compartmentalize NaCl is often raised. In fact, the costs of salt exclusion are small. Even the costs of taking up salt and compartmenting it at the cellular level are relatively low. For example, the cost of NaCl accumulation by the halophyte *Suaeda maritima* at its growth optimum of 340 m*M* NaCl would use little more than 10% of the respiratory ATP produced by leaves (Yeo, 1983). These costs are minimal in comparison to the costs of accumulating organic compounds for osmotic adjustment (Yeo, 1983; Raven, 1985). Calculations of the energy requirement for osmotic adjustment using inorganic or organic molecules show that about 2 ATP molecules are required to regulate the uptake and transport of Na^+ or Cl^- from roots to leaves, whereas 20 times this is required to synthesize an organic compound. Small carbon compounds such as glucose or mannitol require about 30 ATP for synthesis, whereas N-containing compounds such as proline or betaine require 40–50 ATP (Raven, 1985). Another substantial cost may be that of supporting a greater root system with a lower leaf area, as the root: shoot ratio of many species is higher in saline soil (Cheeseman, 1988).

Although we do not understand the controls of carbohydrate utilization in plants growing slowly under osmotic stress, an enhanced supply of carbon compounds to growing regions offers the most plausible explanation to date for the ameliorative effect of high CO_2 on growth in saline soil. Thus, elevated CO_2 acts by increasing the production rate of new leaves. If it reduces the rate of salt uptake, the longevity of old leaves may increase. The outcome is that the overall capacity of the plant to produce assimilate is increased.

C. Effects on Soil Salinity

While elevated CO_2 usually leads to a reduction in water use per unit leaf area, whether or not elevated CO_2 will also reduce water loss by a plant community is controversial. In one long-term study of natural salt marsh communities, Arp *et al.* (1993) reported that water loss by stands of *Scirpus olneyi* and *Spartina patens* was 17–29% lower under elevated than ambient CO_2. The impact of changes in water loss on coastal wetlands is unknown. In rainfed agricultural systems, where the soil water balance is typically unstable, elevated CO_2 may have a complex effect on both plant water use and the soil salinity.

Elevated CO_2 typically reduces stomatal conductance and transpiration per unit leaf area. Whether or not it reduces the amount of water transpired by the plant depends on the degree to which elevated CO_2 increases the leaf area of the plant. In a study on wheat grown at 150 m*M* NaCl, the reduction in transpiration rate was similar to the increase in leaf area, so there was no change in water use per plant (Nicolas *et al.*, 1993). This was found for both a salt-tolerant and a salt-sensitive wheat cultivar. The data for the salt-tolerant cultivar are given in Table III. The same outcome was observed from a parallel study with the same cultivar in drying soil (Table III). However, in another study with a different wheat cultivar carried out in different ambient conditions (lower temperature and lower light), a very different result was found. There was little increase in growth due to elevated CO_2 in either wet or drying soil, and an increase in WUE resulted in a reduction in water use per plant (Table III).

If a reduction in water use also occurs in agricultural or natural systems, it might have profound implications for depth of water tables and the stability of secondary salinity. It is difficult to extrapolate from laboratory

Table III Effect of Elevated CO_2 (700 ppm) on Growth of Spring Wheat in Saline or Dry Soil[a]

Treatment	Percent increase in dry weight (per plant)	Percent increase in water use efficiency	Percent increase in water use (per plant)
Control[b]	+41	+49	No change
Saline	+73	+64	No change
Control[c]	+62	+65	No change
Drying	+73	+68	No change
Control[d]	+6	+33	−20
Drying	+4	+30	−20

[a] Values given are the percent change due to elevated CO_2.
[b] Cv. Matong (Nicolas *et al.*, 1993).
[c] Cv. Matong (Samarakoon *et al.*, 1995).
[d] Cv. Tonic (van Vuuren *et al.*, 1997).

experiments to the field, but the responses measured in laboratory conditions with spaced plants should apply to a crop at its earlier stages of development. However, when the canopy closes, the effect of stomatal closure on WUE may weaken because of canopy heating. Furthermore, canopy closure may occur earlier under elevated CO_2. Thus, there may not be a significant change in water use over the life span of the crop. This means that any decrease in water use and increase in fixed carbon may not be realized, and the anticipated rise in global temperature may itself have adverse effects on harvest index and yield. In a current review on the interaction between salinity and climate change, particularly with respect to crop production in irrigated agriculture, Yeo (1998) concludes that elevated CO_2 appears likely to exacerbate rather than moderate the problems of secondary salinity.

A different scenario applies to dryland salinity (nonirrigated agriculture). Crops are not grown in saline soils that are rainfed, because the rise in salinity is uncontrolled. Instead, salt-tolerant pasture grasses and shrubs are established on the less saline areas (see Table II), and trees or shrubs with prolific water use are planted on the more saline site to control salinization and to revegetate saline land. To control dryland salinity, shrubs or trees (usually *Eucalyptus* species) are planted in both recharge and discharge areas. In the recharge areas, which are the regions in a catchment from which water enters the water table, the soil water is not saline, so species are chosen basically for their ability to extract water at depth, and maintain a high rate of water use throughout the year. Any reduction in water use by elevated CO_2 would be counterproductive, and species whose stomatal conductance is less sensitive to CO_2 would be preferred over those with responsive stomates. In the discharge areas, where saline groundwater is near or at the surface, trees are chosen for their salinity and waterlogging tolerance, as well as for their ability to extract water. They continue to extract water until the salt in the soil concentrates to such a high osmotic pressure that further water uptake is not possible. (For modeling purposes, it is assumed that the trees do not extract any of the salt from the soil, just the water.) Thus, the salt in the soil will concentrate until a quasi-steady state is attained, in which the trees have limited their own transpiration rate. A somewhat similar phenomenon was described for mangroves in tidal mudflats (Passioura *et al.,* 1992). Effects of elevated CO_2 would thus be small at a seasonal timescale, but may prolong the period over which the tree transpires. Elevated CO_2 might increase the amount of water extracted because as mentioned earlier it might lower the buildup of salts at the soil–root interface at periods of high evaporative demand via its effects on transpiration. These ideas have not been tested.

VI. Effects of Elevated CO_2 on Natural Communities in Saline Soil

Coastal wetlands are of enormous ecological and socioeconomic importance worldwide, with mangroves dominating the tropical and subtropical coastlines and salt marshes dominating coastal wetlands from mid to high latitudes. Changes in sea level, climate, and atmospheric CO_2 concentration will affect these resources. While there are difficulties in prediction of change in climate and sea level associated with the "Greenhouse Effect," a doubling of the concentration of atmospheric CO_2 because of its effect on both carbon gain and water use characteristics, may be sufficient to change the structure and function of coastal wetlands, as shown in elegant field studies by Drake and colleagues (Drake, 1992).

A. Responses of Halophytic Species in a Mixed Community

Prediction of community level responses to elevated CO_2 along salinity gradients suffers from a lack of experimental data and is complicated by the presence of other environmental gradients that may also influence responses. To date, the only field studies of responses of halophytic vegetation to elevated CO_2 are those conducted in open top chambers at one location in a temperate salt marsh system (Drake, 1992). Drake and colleagues studied three communities for several years: monospecific stands of the C_3 sedge, *Scirpus olneyi,* and the C_4 grass, *Spartina patens,* and a mixed community of these two species together with another C_4 grass, *Distichlis spicata.* As summarized in Drake (1992), elevation of CO_2 for 4 yr reduced water loss, increased water potential, and delayed senescence in all three species, but increased growth only in the C_3 species. Elevated CO_2 increased the CO_2 exchange rate of the C_3 species (*S. olneyi*) by 50%, but had little effect on the C_4 species (Fig. 5). Sustained increase in growth of this C_3 species over 4 yr was associated with increased quantum yield (Long and Drake, 1991), photosynthetic capacity (Ziska *et al.*, 1990; Arp and Drake, 1991), reduced dark respiration, increased numbers of shoots, roots, and rhizomes (Curtis *et al.*, 1989a), lower concentrations of nitrogen in all tissues (Curtis *et al.*, 1989b), increased nitrogen fixation, and increased ecosystem carbon accumulation (Curtis *et al.*, 1990). In the mixed community, biomass of the C_3 species increased more than 100% while biomass of the C_4 species declined (Arp *et al.*, 1993). These results were obtained in

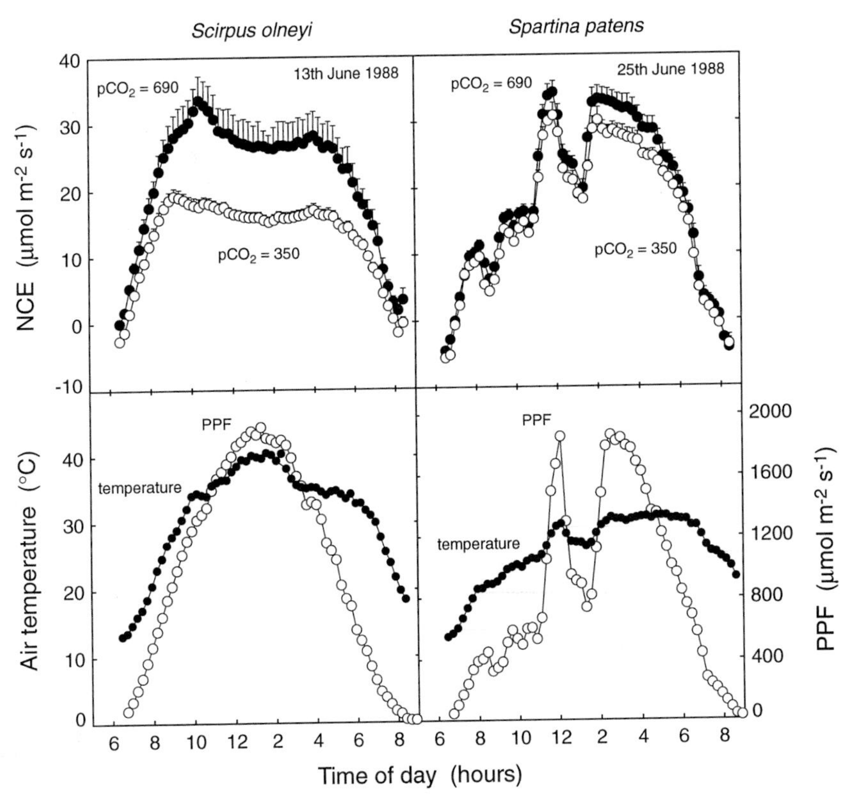

Figure 5 Net ecosystem CO_2 exchange per unit ground area (NCE) in monospecific stands of the C_3 sedge, *Scirpus olneyi,* and the C_4 grass, *Spartina patens.* Values are means ± standard errors for five chambers in each community determined at 15-min intervals throughout the day. Open circles denote plants grown and measured in normal ambient CO_2; solid circles denote plants grown and measured in elevated CO_2. (Reproduced with permission from Drake, 1992.)

a salt marsh where edaphic conditions were conducive to vigorous growth. Extrapolating from these results to how a more complex salt marsh system encompassing a broader range of edaphic conditions might respond to elevated CO_2 remains a challenge.

B. Interactions with Other Environmental Factors

In coastal wetlands, tidally maintained gradients of salinity are superimposed on gradients in waterlogging and nutrients. Waterlogging can adversely affect growth and functioning of natural wetland vegetation. These adverse effects undoubtedly result from a multitude of processes, some of which might be alleviated by elevated CO_2. For example, rates of water uptake and carbon gain are lower in waterlogged mangroves than in those grown under well-drained conditions (Naidoo, 1985). If such reductions in carbon gain are due to stomatal closure, then elevated CO_2 might overcome stomatal limitations to CO_2 diffusion into the leaves. Indeed, Rozema *et al.* (1991) found that elevated CO_2 enhanced photosynthesis and growth of the C_3 salt marsh species *Scirpus maritimus* and *Puccinella maritima* grown in anaerobic culture solutions. However, elevated CO_2 did not fully alleviate the stresses associated with waterlogging because the CO_2 enhancement of photosynthesis and growth was less than when plants were grown in aerated culture solutions.

Tidal flooding regimes also influence the form and concentration of nutrients. One field study of the effects of elevated atmospheric CO_2 on plant growth and nutrient dynamics was conducted in a natural salt marsh where edaphic conditions were conducive to vigorous growth (Curtis *et al.*, 1989a,b, 1990). Under the low to moderate salinities in which salt marshes are most productive, such as the one studied by Curtis and colleagues, nitrogen is generally considered to be the limiting nutrient. Under elevated CO_2, productivity increased while the concentration of N in shoot tissue decreased. Tissue with such a high C:N ratio was subject to slower rates of decomposition than the tissue produced under ambient CO_2, leading to accumulation of organic material under elevated CO_2. This would be expected to reduce subsequent levels of productivity, but instead enhanced levels of productivity were sustained for years under elevated CO_2 because of increased rates of rhizosphere nitrogen fixation. Presumably, the activities of nitrogen-fixing microbes were enhanced by an increase in exudation of carbohydrates from the roots of plants growing under elevated CO_2 (Curtis *et al.*, 1989b). Such plant–microbe interactions may be critical for sustaining enhanced levels of productivity in nutrient-deficient environments, but the effects are most likely to substantial where salinities are favorable for plant growth.

At larger geographic scales, structure and function of coastal wetlands vary along climatic gradients such as humidity deficits, which affect both the salinity at the roots and the evaporative demand at the leaves. Plants growing in coastal wetlands of arid or seasonally dry areas may benefit from rising CO_2. With a doubling of the atmospheric CO_2 concentration, assimilation rate in C_3 species may operate near saturation for CO_2, and hence be less affected by a decrease in stomatal conductance in response to large gradients in water vapor between leaves and air due to low atmospheric humidity (Morison and Gifford, 1983) as occurs under arid conditions. In one study, elevated CO_2 enhanced growth of the mangroves *Rhizophora apiculata* and *R. stylosa* much more when carbon gain was limited by high evaporative demand at the leaves than when it was limited by high salinity at the roots (Ball *et al.,* 1997). Such responses could affect competitive hierarchies of species whose distributions along salinity gradients are affected by aridity.

C. Effects on Community Structure

Elevated CO_2 is likely to alter species composition and biodiversity of both salt marsh and mangrove systems along salinity gradients. It is difficult to predict how it might affect competitive rankings of C_3 species, but these species are expected to benefit more from elevated CO_2 than C_4 species. Indeed, studies contrasting responses to elevated CO_2 in C_3 and C_4 salt marsh species grown under field (Drake, 1992) and laboratory (Rozema *et al.,* 1991) conditions have shown no substantial growth enhancement in the C_4 species despite increase in WUE and plant water status, whereas C_3 species typically respond with growth to elevated CO_2. Such differences can lead to changes in community structure. Under elevated CO_2, the population, leaf area, and biomass of a C_3 sedge, *Scirpus olneyi,* increased at the expense of the C_4 grasses, *Spartina patens* and *Distichlis spicata,* in a natural salt marsh system (Arp *et al.,* 1993).

Elevated CO_2 could also affect the biodiversity of functional types, particularly those that exploit understorey environments. For example, understorey vegetation in mangrove forests generally occurs in areas of low salinity (Chapman, 1975; Lugo, 1986). The basis for this distribution is unknown, but physiological attributes associated with increasing salt to tolerance (Ball, 1996) might preclude the occurrence of understorey species in highly saline environments. With increasing salinity, carbon allocation to roots can increase at the expense of leaf area (Ball, 1988; Soto, 1988), and specific leaf area can decrease (Camilleri and Ribi, 1983; Ball *et al.,* 1988). Such changes in carbon allocation, together with the reduced rate of photosynthesis associated with increasing salinity (Ball and Farquhar, 1984; Clough and Sim, 1989; Lin and Sternberg, 1992) would lower the capacity for further assimilation. All of these factors would increase the light compensation point for growth, and hence limit survival in understorey shade (Givnish, 1988). However, under elevated CO_2, inhibition of photorespiration in C_3 species would increase quantum efficiency of CO_2 fixation under light-limiting conditions (Drake *et al.*, 1997). Indeed, the quantum efficiency of *S. olneyi* grown and measured under 36 Pa CO_2, consistent with theoretical predictions (Long and Drake, 1991). As Drake *et al.* (1997) pointed out, an increase in quantum efficiency, together with a decrease in the light compensation point and increase in net photosynthetic rates under limiting irradiances, would be expected to enhance growth in understorey environments. This could increase shade tolerance at high salinities and promote the occurrence of shade-tolerant species along natural salinity gradients. However, stimulation of growth in the understorey may depend on whether or not elevated CO_2 leads to greater canopy development in the overstorey, thereby reducing light penetration to the forest floor. Studies to date, which do not include forest communities, have shown that canopy leaf area neither increased nor decreased in any of the long-term field studies of effects of elevated CO_2 on crops and native vegetation (Drake *et al.*, 1997).

VII. Summary and Future Directions

Most laboratory experiments indicate that elevated CO_2 will enhance growth of plants in saline soil. The extent of the increase will depend on many factors, such as the degree of salinity, whether there are stomatal limitations to growth, and whether growth is also affected by other edaphic factors such as waterlogging and nutrient deficiencies.

Responses of halophytes in saline communities to elevated CO_2 may be sufficient to induce substantial change in coastal wetland vegetation along natural salinity gradients. Species with greater inherent ability to use resources may show greater growth stimulation with elevated CO_2, provided that there is an adequate potential for growth. Most effect can be expected under relatively low to moderate salinity regimes in which the species already grow well. In contrast to growth enhancement under favorable salinity regimes, there is no evidence that elevated CO_2 will increase the range of salinities in which a species can grow, and halophytes are unlikely to expand into areas where salinities are too extreme to support growth. Thus, preferential enhancement of growth under conditions already supporting the most productive coastal welands may further increase disparities in community structure and productivity along natural salinity gradients.

The general optimism about the effects of rising CO_2 on plant productivity in agricultural systems should be considered carefully in relation to its effect on plant water use and thereby on the progression of soil salinity. In a review on crop responses to CO_2 enrichment (Rogers and Dahlman, 1993), an adverse side of reduced plant water use is not mentioned. The bulk of the world's food production comes from irrigated systems, many of which suffer from rising saline water tables. If water was properly managed in irrigated systems (i.e., applied sparingly), a reduced plant water use due to elevated CO_2 could result in reduced water application, with enormous benefits to the community in water conservation and soil sustainability. Unfortunately, such precise water management practices are rare, but research into soil or plant waer status monitoring devices, and increasing use of "regulated deficit" irrigation scheduling for horticulture are promoting this practice. Thus the sustainability of irrigation systems could improve under increasing CO_2, as long as they are managed in a way that reduces accessions to groundwater and keeps water tables below the level of the roots of crops.

Reductions in plant water use due to elevated CO_2 are also of concern regarding dryland salinity, because reclamation of saline land by establishment of trees and perennial grasses may be hampered by increased CO_2. Reclamation is brought about by planting perennials on the recharge areas to reduce input of water to the groundwater, and on the discharge areas to lower the water table. Tree and shrub species with prolific water use are being selected for this purpose. A reduction in water use by vegetation on the recharge areas will be counterproductive to the control of salinization, and species will need to be screened with this in mind.

Future research directions should target both the hydrology and physiology of the effects of elevated CO_2 on plant growth in saline soils. The most important issue for sustained productivity in saline soils is whether elevated CO_2 will affect plant water uptake, and how this would affect the movement of both salt and water in the soil at the catchment level. The most fascinating physiological issue is how high CO_2 affects plant growth in saline soil. Elevated CO_2 appears to operate by enhancing the carbon supply rather than by increasing turgor or lowering internal salt concentrations, but how this occurs when salt-affected plants already have apparently ample reserves of carbohydrates remains an enigma.

Acknowledgments

We thank Drs. Miko Kirschbaum, Nico Marcar, Richard Stirzaker, John Read, and Tony Yeo for critical reviews of the manuscript.

References

Arp, W. J., and Drake, B. G. (1991). Increased photosynthetic capacity of *Scirpus olneyi* after 4 years of exposure to elevated CO_2. *Plant Cell Environ.* **13,** 1003–1006.

Arp, W. J., Drake, B. G., Pockman, W. T., Curtis, P. S., and Whigham, D. F. (1993). Interactions between C_3 and C_4 salt marsh plant species during four years of exposure to elevated atmospheric CO_2. *Vegetatio* **104/105,** 133–143.

Aslam, Z., Jeschke, W. D., Barrett-Lennard, E. G., Setter, T. L., Watkin, E., and Greenway, H. (1986). Effects on external NaCl on the growth of *Atriplex amnicola* and the ion relations and carbohydrate status of the leaves. *Plant Cell Environ.* **9,** 571–580.

Aslam, Z., Qureshi, R. H., and Ahmed, N. (1993). A rapid screening technique for salt tolerance in rice (*Oryza sativa* L.). *Plant Soil* **150,** 99–107.

Bachelet, D., and Kropff, M. J. (1995). The impact of climatic change on agro-climatic zones in Asia. *In* "Modeling the Impact of Climate Change on Rice Production in Asia" (R. B. Matthews, M. J. Kropff, D. Bachelet, and H. H. van Laar, eds.), pp. 85–94.

Ball, M. C. (1988). Salinity tolerance in the mangroves, *Aegiceras corniculatum* and *Avicennia marina.* I. Water use in relation to growth, carbon partitioning and salt balance. *Aust. J. Plant Physiol.* **15,** 447–464.

Ball, M. C. (1996). Comparative ecophysiology of mangrove forest and tropical lowland moist rainforest. *In* "Tropical Forest Plant Ecophysiology" (S. S. Mulkey, R. L. Chazdon, and A. P. Smith, eds.), pp. 461–496. Chapman and Hall, New York.

Ball, M. C., and Farquhar, G. D. (1984). Photosynthetic and stomatal responses of two mangrove species, *Aegiceras corniculatum* and *Avicennia marina,* to long salinity and humidity conditions. *Plant Physiol.* **74,** 1–6.

Ball, M. C., and Munns, R. (1992). Plant responses to salinity under elevated atmospheric concentrations of CO_2. *Aust. J. Bot.* **40,** 515–525.

Ball, M. C., Cochrane, M. J., and Rawson, H. M. (1997). Growth and water use of the mangroves, *Rhizophora apiculata* and *R. stylosa,* in response to salinity and humidity under ambient and elevated concentrations of atmospheric CO_2. *Plant Cell Environ.* **20,** 1158–1166.

Ball, M. C., Cowan, I. R., and Farquhar, G. D. (1988). Maintenance of leaf temperature and the optimisation of carbon gain in relation to water loss in a tropical mangrove forest. *Aust. J. Plant Physiol.* **14,** 263–276.

Bowman, W. D., and Strain, B. R. (1987). Interaction between CO_2 enrichment and salinity stress in the C_4 non-halophyte *Andropogon glomeratus* (Walter) BSP. *Plant Cell Environ.* **10,** 267–270.

Camilleri, J. C., and Ribi, G. (1983). Leaf thickness of mangroves (*Rhizophora mangle*) growing in different salinities. *Biotropica* **15,** 139–141.

Chapman, V. J. (1975). "Mangrove Vegetation." J. Cramer, Germany.

Cheeseman, J. M. (1988). Mechanisms of salinity tolerance in plants. *Plant Physiol.* **87,** 547–550.

Clough, B. F., and Sim, R. G. (1989). Changes in gas exchange characteristics and water use efficiency of mangroves in response to salinity and vapour pressure deficit. *Oecologia* **79,** 38–44.

Cramer, G. R. (1997). Uptake and role of ions in salt tolerance. *In* "Strategies for Improving Salt Tolerance in Higher Plants" (P. K. Jaiwal, R. P. Singh, and A. Gulati, eds.), pp. 55–86. Oxford and IBH Publishing, New Delhi.

Cramer, G. R., Alberico, G. L., and Schmidt, C. (1994). Salt tolerance is not associated with the sodium accumulation of two maize hybrids. *Aust. J. Plant Physiol.* **21,** 675–692.

Curtis, P. S., Drake, B. G., Leadly, P. W., Arp, W. J., and Whigham, D. F. (1989a). Growth and senescence in plant communities exposed to elevated CO_2 concentrations on an estuarine marsh. *Oecologia* **78,** 20–26.

Curtis, P. S., Drake, B. G., and Whigham, D. F. (1989b). Nitrogen and carbon dynamics in C_3 and C_4 estuarine marsh plants grown under elevated CO_2 *in situ. Oecologia* **78,** 297–301.

Curtis, P. S., Balduman, L. M., Drake, B. G., and Whigham, D. F. (1990). Elevated atmospheric CO_2 effects on belowground processes in C_3 and C_4 estuarine marsh communities. *Ecology* **71,** 2001–2006.

Davies, W. J., and Zhang, J. (1991). Root signals and the regulation of growth and development of plants in drying soil. *Annu. Rev. Plant Physiol.* **42,** 55–76.

Drake, B. G. (1992). A field study of the effects of elevated CO_2 on ecosystem processes in a Chesapeake Bay wetland. *Aust. J. Bot.* **40,** 579–595.

Drake, B. G., Gonzalez-Meler, M. A., and Long, S. P. (1997). More efficient plants: A consequence of rising atmospheric CO_2? *Annu. Rev. Plant Physiol. Mol. Biol.* **48,** 609–639.

FAO (1995). Production Yearbook, Vol 48, FAO Statistics Series, No. 125, Rome.

Farnsworth, E. J., Ellison, A. M., and Gong, W. K. (1996). Elevated CO_2 alters anatomy, physiology, growth, and reproduction of red mangrove (*Rhizophora mangle* L.). *Oecologia* **108,** 599–609.

Farquhar, G. D., and Richards, R. A. (1984). Isotopic composition of plant carbon correlates with water-use efficiency of wheat genotypes. *Aust. J. Plant Physiol.* **11,** 539–552.

Flowers, T. J., and Yeo, A. R. (1986). Ion relations of plants under drought and salinity. *Aust. J. Plant Physiol.* **13,** 75–91.

Flowers, T. J., and Yeo, A. R. (1995). Breeding for salinity resistance in crop plants: Where next? *Aust. J. Plant Physiol.* **22,** 875–884.

Flowers, T. J., Troke, P. F., and Yeo, A. R. (1977). The mechanism of salt tolerance in halophytes. *Annu. Rev. Plant Physiol.* **28,** 89–121.

Flowers, T. J., Hajibagheri, M. A., and Clipson, N. J. W. (1986). Halophytes. *Quart. Rev. Biol.* **61,** 313–337.

Garcia, A., Rizzo, C. A., Ud-Din, J., Bartos, S. L., Senadhira, D., Flowers, T. J., and Yeo, A. R. (1997). Sodium and potassium transport to the xylem are inherited independently in rice, and the mechanism of sodium:potassium selectivity differs between rice and wheat. *Plant Cell Environ.* **20,** 1167–1174.

Ghassemi, F., Jakeman, A. J., and Nix, H. A. (1995). "Salinisation of Land and Water Resources: Human Causes, Extent, Management and Case Studies." UNSW Press, Sydney, Australia, and CAB International, Wallingford, UK.

Givnish, T. J. (1988). Adaptation to sun and shade: A whole plant perspective. *Aust. J. Plant Physiol.* **15,** 63–92.

Greenway, H., and Munns, R. (1980). Mechanisms of salt tolerance in nonhalophytes. *Annu. Rev. Plant Physiol.* **31,** 149–190.

Hunt, R., Hand, D. W., Hannah, M. A., and Neal, A. M. (1991). Response to CO_2 enrichment in 27 herbaceous species. *Funct. Ecolo.* **5,** 410–421.

Hunt, R., Hand, D. W., Hannah, M. A., and Neal, A. M. (1993). Further responses to CO_2 enrichment in British herbaceous species. *Funct. Ecol.* **7,** 661–668.

Idso, K. E., and Idso, S. B. (1994). Plant responses to atmospheric CO_2 enrichment in the face of environmental constraints: A review of the past 10 years' research. *Agric. For. Meteor.* **69,** 153–203.

Jansen, C. M., Pot, S., and Lambers, H. (1986). The influence of CO_2 enrichment of the atmosphere and NaCl on growth and metabolism of *Urtica dioica* L. *In* "Biological Control of Photosynthesis" (R. Marcelle, H. Clijsters, and M. van Pouke, eds.), pp. 143–146. Martinus Nijhoff Publishers, Dordrecht.

Lenssen, G. M., and Rozema, J. (1990). The effect of atmospheric CO_2 enrichment and salinity on growth, photosynthesis and water relations of salt marsh species. *In* The Greenhouse Effect and Primary Productivity in European Agro-ecosystems" (J. Goudriaan, H. van Keulen, and H. H. van Laar, eds.), pp. 64–67. Pudoc. Wageningen.

Lenssen, G. M., Lamers, J., Stroetenga, M., and Rozema, J. (1993). Interactive effects of atmospheric CO_2 enrichment, salinity and flooding on growth of C_3 (*Elymus athericus*) and C_4 (*Spartina anglica*) salt marsh species. *Vegetatio* **104/105,** 379–388.

Lin, G., and Sternberg, L. da S. L. (1992). Effects of growth form, salinity, nutrient, and sulphide on photosynthesis, carbon isotope discrimination, and growth of red mangrove (*Rhizophora mangle* L.). *Aust. J. Plant Physiol.* **19,** 509–517.

Long, S. P., and Drake, B. G. (1991). Effect of the long-term elevation of CO_2 concentration in the field on the quantum yeild of photosynthesis of the C_3 sedge, *Scirpus olneyi*. *Plant Physiol.* **96,** 221–226.

Lugo, A. E. (1986). Mangrove understorey: An expensive luxury? *J. Trop. Ecol.* **2,** 287–288.

Marcar, N., Crawford, D., Leppert, P., Jovanovic, T., Floyd, R., and Farrow, R. (1995). "Trees for Saltland." CSIRO Press.

Morison, J. I. L., and Gifford, R. M. (1983). Stomatal sensitivity to carbon dioxide and humidity. A comparison of two C_3 and two C_4 grass species. *Plant Physiol.* **71,** 789–796.

Munns, R. (1985). Na^+, K^+ and Cl^- in xylem sap flowing to shoots of NaCl-treated barley. *J. Exp. Bot.* **36,** 1032–1042.

Munns, R. (1988). Effect of high external NaCl concentrations on ion transport within the shoot of *Lupinus albus*. I Ions in xylem sap. *Plant Cell Environ.* **11,** 283–289.

Munns, R. (1993). Physiological processes limiting plant growth in saline soil: Some dogmas and hypotheses. *Plant Cell Environ.* **16,** 15–24.

Munns, R., and Cramer, G. R. (1996). Is coordination of leaf and root growth mediated by abscisic acid? Opinion. *Plant Soil* **185,** 33–49.

Munns, R., and Sharp, R. E. (1993). Involvement of abscisic acid in controlling plant growth in soils of low water potential. *Aust. J. Plant Physiol.* **20,** 425–437.

Munns, R., and Termaat, A. (1986). Whole-plant responses to salinity. *Aust. J. Plant Physiol.* **13,** 143–160.

Munns, R., Greenway, H., Delane, R., and Gibbs, J. (1982). Ion concentration and carbohydrate status of the elongating leaf tissue of *Hordeum vulgare* growing at high external NaCl. II. Cause of the growth reduction. *J. Exp. Bot.* **33,** 574–583.

Munns, R., Greenway, H., and Kirst, G. O. (1983). Halotolerant eukaryotes. *In* "Physiological Plant Ecology. III. Responses to the Chemical and Biological Environment" (O. L. Lange, C. B. Osmond, P. S. Nobel, and H. Zeigler, eds.), Encyclopedia of Plant Physiology, New Series, Vol. 12C, pp. 59–135. Springer Verlag, Berlin.

Munns, R., Schachtman, D. P., and Condon, A. G. (1995). The significance of a two-phase growth response to salinity in wheat and barley. *Aust. J. Plant Physiol.* **22,** 561–569.

Myers, B. A., and West, D. W. (1990). Revegetation of saline land. *In* "Proc. Workshop at Institute for Sustainable Irrigated Agriculture," Tatura, Vic.

Naidoo, G. (1985). Effects of waterlogging and salinity on plant water relations and on the accumulation of solutes in three mangrove species. *Aquatic Bot.* **22,** 133–143.

Nicolas, M. E., Munns, R., Samarakoon, A. B., and Gifford, R. M. (1993). Elevated CO_2 improves the growth of wheat under salinity. *Aust. J. Plant Physiol.* **20,** 349–360.

Passioura, J. B., Ball, M. C., and Knight, J. H. (1992). Mangroves may salinize the soil and in doing so limit their transpiration rate. *Funct. Ecol.* **6,** 476–481.

Poorter, H. (1993). Interspecific variation in the growth response of plants to an elevated ambient CO_2 concentration. *Vegetatio* **104/105,** 77–97.

Raven, J. A. (1985). Regulation of pH and generation of osmolarity in vascular plants: A cost–benefit analysis in relation to efficiency of use of energy, nitrogen and water *New Phytol* **101,** 25–77.

Richards, R. A. (1991). Crop improvement for temperate Australia: Future opportunities. *Field Crops Res.* **26,** 141–169.

Rogers, H. H., and Dahlman, R. C. (1993). Crop responses to CO_2 enrichment. *Vegetatio* **104,** 117–131.

Rozema, J. (1975). An ecophysiological investigation into the salt tolerance of *Glaux maritima*. *Acta Bot. Neerl.* **24,** 407–416.

Rozema, J., Lenseen, G. M., Broekman, R. A., and Arp, W. P. (1990). Effects of atmospheric carbon dioxide enrichment on salt-marsh plants. *In* "Expected Effects of Climatic Change on Marine Coastal Ecosystems" (J. J. Beukema), pp. 49–54. Kluwer Publishers, Amsterdam.

Rozema, J., Dorel, F., Janissen, R., Lenssen, G., Broekman, R., Arp, W., and Drake, B. G. (1991). Effect of elevated atmospheric CO_2 on growth, photosynthesis and water relations of salt marsh grass species. *Aquatic Bot.* **39,** 45–55.

Salim, M. (1989). Effects of salinity and relative humidity on growth and ionic relations of plants. *New Phytol.* **113,** 13–20.

Samarakoon, A. B., Muller, W. J., and Gifford, R. M. (1995). Transpiration and leaf area under elevated CO_2 : Effects of soil water status and genotype in wheat. *Aust. J. Plant Physiol.* **22,** 33–44.

Schwarz, M., and Gale, J. (1984). Growth response to salinity at high levels of carbon dioxide. *J. Exp. Bot.* **35,** 193–196.

Soto, R. (1988). Geometry, biomass allocation, and leaf life-span of *Avicennia germinans* (L.). (Avicennianceae) along a salinity gradient in Salinas, Puntarenas, Costa Rica. *Rev. Biol. Trop.* **36,** 309–323.

Stirzaker, R. J., and Passioura, J. B. (1996). The water relations of the root-soil interface. *Plant Cell Environ.* **19,** 201–208.

Storey, R. (1995). Salt tolerance, ion relations and the effect of root medium on the response of citrus to salinity. *Aust. J. Plant Physiol.* **22,** 101–114.

Szabolcs, I. (1994). Soils and Salinisation. *In* "Handbook of Plant and Crop Stress" (M. Pessaraki, ed.), pp. 3–11. Marcel Dekker, New York.

Termaat, A., Passioura, J. B., and Munns, R. (1985). Shoot turgor does not limit shoot growth of NaCl-affected wheat and barley. *Plant Physiol.* **77,** 869–872.

van Vuuren, M. M. I., Robinson, D., Fitter, A. H., Chasalow, S. D., Williamson, L., and Raven, J. A. (1997). Effects of elevated atmospheric CO_2 and soil water availability on root biomass, root length, and N, P and K^+ uptake by wheat. *New Phytol.* **135,** 455–465.

Yeo, A. R. (1983). Salinity resistance: Physiologies and prices. *Physiol. Plant.* **58,** 214–222.

Yeo, A. R. (1998). Predicting the interaction between the effects of salinity and climate change. *Sci. Hort.* **78,** 159–174.

Ziska, L. H., Drake, B. G., and Chamberlain, S. (1990). Long-term photosynthetic response in single leaves of a C_3 and C_4 salt marsh species grown at elevated atmospheric CO_2 *in situ*. *Oecologia* **83,** 469–472.

6

Atmospheric CO_2 Enrichment and Enhanced Solar Ultraviolet-B Radiation: Gene to Ecosystem Responses

Jelte Rozema, Alan Teramura, and Martyn Caldwell

I. Introduction

Current atmospheric CO_2 enrichment is primarily the result of emissions from fossil fuel combustion and, to a lesser extent, from large-scale deforestation. In the global carbon cycle the amount of carbon in the atmospheric pool is small compared to other pools such as in the ocean, soils, and terrestrial biota (Schlesinger, 1993). Atmospheric CO_2 enrichment affects terrestrial plant life in numerous ways. First with elevated CO_2, carboxylation is often increased and stomata usually partially close (Strain and Cure, 1985; Bazzaz, 1990; Rozema *et al.*, 1993; Rozema, 1995; Koch and Mooney, 1996). Atmospheric CO_2 is an essential resource to plants and many effects of enhanced CO_2 are regarded beneficial, although long-term effects may differ from short-term CO_2 enrichment effects (Arp and Drake, 1991; Rogers and Dahlman, 1993). Photosynthetic acclimation may reduce the positive response of C_3 plants to elevated CO_2 in long-term studies (Arp, 1991; Jacob *et al.*, 1995).

Solar ultraviolet-B is generally regarded as an environmental stress factor. Enhanced solar UV-B at the earth's surface may reduce growth of some plants (Caldwell *et al.*, 1989; Tevini and Teramura, 1989, Stapleton, 1992; Teramura and Sullivan, 1994; Björn, 1996).

The long-term history of atmospheric CO_2 is linked with the development of CO_2 fixation and O_2 release by terrestrial vegetation. The increase of biogenic oxygen in the atmosphere was also necessary for the development of the stratospheric ozone layer and its protective filtering of much of the deleterious short-wavelength UV-C and UV-B radiation (Caldwell, 1979).

Elevated atmospheric CO_2 and enhanced solar UV-B are components of global climatic change. Although increasing atmospheric CO_2 and stratospheric ozone depletion are both the result of anthropogenic emissions, the simultaneous occurrence of these global change problems is coincidence. In this chapter, CO_2 and UV-B radiation are considered together to see if and how responses to elevated CO_2 are affected by enhanced solar UV-B radiation.

The combined effects of elevated CO_2 and enhanced UV-B will be reviewed here from cellular to ecosystem responses. For more extensive recent reviews of the effects of atmospheric CO_2 enrichment or enhanced solar UV-B on terrestrial plants and ecosystems the reader is referred to Bothwell *et al.*, 1994; Mooney *et al.*, 1991; Rozema *et al.*, 1993, 1997c,d; Melillo *et al.*, 1993; Caldwell and Flint, 1994; Teramura and Sullivan, 1994; Caldwell *et al.*, 1995; Rozema, 1995; Bolker *et al.*, 1995; Björn, 1996; Körner and Bazzaz, 1996; Koch and Mooney, 1996; Sullivan, 1997.

In this chapter physiological and ecological studies on combined effects of elevated CO_2 and UV-B are surveyed. Are plant and ecosystem responses to elevated atmospheric CO_2 altered under enhanced UV-B?

Atmospheric CO_2 and UV-B may interact at various levels. First, there may be some *direct* effects of enhanced solar UV-B on photosynthesis and plant primary production, which might interact with the effects of elevated atmospheric CO_2 on photosynthesis. Second, by influencing plant secondary chemistry (the formation of a wide variety of secondary metabolites, such as lignin, cellulose, and various phenolic compounds), enhanced solar UV-B has the potential to affect *indirectly* many processes in ecosystems (Teramura and Sullivan, 1994; Beggs and Wellman, 1994; Rozema *et al.*, 1997d). Among these processes are the decomposition of plant litter, and various plant–microorganism and plant–animal interactions. In this way elevated solar UV-B may (indirectly) affect the net release of atmospheric CO_2, carbon storage, and other features of ecosystems.

In this chapter we distinguish three major pathways by which enhanced UV-B might affect ecosystem responses to elevated CO_2.

1. Direct effects of elevated solar UV-B on plant growth and, in some cases, on primary production. Such direct effects may counteract, or at least modify, the stimulating influence of elevated CO_2 on photosynthesis on plant growth.
2. Indirect effect of elevated UV-B on biomass allocation and morphogenesis of plants. In this case plants may be more responsive or less responsive to elevated CO_2 under enhanced UV-B.
3. Enhanced solar UV-B may alter the secondary metabolism in plants, which may affect processes such as litter decomposition plant fungi,

plant bacteria, and herbivory. This would indirectly interact with effects of elevated CO_2 on ecosystems.

II. Evolution of Atmospheric Oxygen and Carbon Dioxide, Terrestrial Plant Life, and Solar Ultraviolet-B Radiation

In the reducing primeval atmosphere, oxygen was absent. The atmospheric carbon dioxide content at the time was 100–1000 times higher than at present. With the development of oxygenic photosynthesis, 2.7×10^9 yr B.P. (before present), oxygen in the atmosphere increased. Development of stratospheric ozone may have been more or less in parallel with the rise in oxygen content. Atmospheric O_2 increase may have been gradual. Alternatively, it may have increased rather abruptly going from 0.001% (2.0×10^9 yr B.P.) to the present level of 21% by 350×10^6 yr B.P. (Berner, 1989, 1993).

The buildup of stratospheric ozone, as a result of the raise of oxygen in the atmosphere, is thought to have influenced the timing of the evolution of land plants. Without the ozone layer, shortwave UV-C and UV-B likely prevented much evolution of terrestrial life (Rozema *et al.*, 1997d). With development of a stratospheric zone shield, much of the damaging solar UV was removed and this is thought to have allowed terrestrial plant life to evolve. However, plants were also developing internal UV filters largely through secondary metabolism and production of phenolics that also contributed to their ability to cope with solar UV. The timing of these events and how they contributed to terrestrial plant evolution is still in question.

There is no consensus on this question, since the reconstruction of historical levels of atmospheric oxygen remains uncertain. Information on historical atmospheric O_2 levels appears to be only indirect. For example, atmospheric O_2 at the time of Fe_2O_3 deposition in Red Beds, 2.0×10^9 yr B.P., is linked with an oxidizing terrestrial environment, and atmospheric O_2 is estimated to have been 0.001% of the present level (Berner, 1989). As mentioned before, this course of O_2 increase in the atmosphere is also uncertain. Despite this uncertainty, there is evidence that during the evolution of land plants the thickness of the stratospheric ozone layer was less than at present (Margulis *et al.*, 1976; Caldwell, 1979; Rozema *et al.*, 1997d). The origin of terrestrial plant life has been dated as about 470×10^6 B.P. (Lowry *et al.*, 1980. Thus, even with a previously mentioned scenario of abrupt O_2 increase to present-day levels by 250×10^6 B.P., this is substantially later than the evolution of the first land plants. One implication of this is that the recent and future increases of solar UV-B may still not result in levels of UV-B that were as high as those during the early evolution of land plants. It may be hypothesized that early land plants have adapted to such high UV-B fluxes.

III. Photosynthesis, Plant Growth, and Primary Production

In contrast to numerous single-factor studies of atmospheric CO_2 enrichment and enhanced solar UV-B effects on plants (see Strain and Cure, 1985; Rozema *et al.*, 1993; Koch and Mooney, 1996; Caldwell *et al.*, 1989; Teramura and Sullivan, 1994), studies of the combined effects of CO_2 and UV-B are not abundant (Caldwell and Flint, 1994; Rozema *et al.*, 1997d; Sullivan, 1997; see Rozema *et al.*, 1997c,e for reviews). Increased plant growth under elevated CO_2 relates to increased CO_2 fixation and to reduced oxygenase activity of the RuBP-carboxylase-oxygenase enzyme (Jordan *et al.*, 1992; Sullivan, 1997). Indirectly, partial stomatal closure may lead to increased water use efficiency (WUE) in response to elevated CO_2 (Lenssen, 1993; Rozema, 1995).

In laboratory studies, photosynthetic damage by enhanced solar UV-B has often been attributed to disturbance of Photosystem II (PS II) (turnover of D1 polypeptide and reduction of plastoquinone) and/or to inhibition of transcription of mRNA due to reduced activity of Rubisco (Jordan *et al.*, 1992).

UV-B damage to PS II is expected to decrease the responsiveness of plants to CO_2 enrichment (Sullivan, 1997). Teramura *et al.*, (1990) and Ziska and Teramura (1992) report an increase of the apparent quantum efficiency to elevated CO_2 only at ambient UV-B. Alternatively, it has been suggested that increased photosynthesis at elevated CO_2 may protect plants from UV-B damage or may compensate for UV-B damage. For example, reduced oxygenase activity of Rubisco may compensate for reduced overall Rubisco carboxylation activity (Jordan *et al.*, 1992). However, most of the research probing effects of enhanced UV-B on photosynthesis has been in laboratory or glasshouse conditions with artificial sources of UV-B and a different spectral balance of radiation than is found in nature.

Several field studies have shown that, even for plant species that were considered UV-B sensitive in growth chamber and glasshouse experiments, depression of photosynthesis of field grown plants is more an exception than rule (Kim *et al.*, 1996; Fiscus and Booker, 1995; Day *et al.*, 1996). In other words, plants grown in the field and exposed to natural solar UV-B do not generally experience reduced photosynthesis due to supplementary UV-B emitted from lamp systems. Although this does not rule out interactions between effects of enhanced UV-B and elevated CO_2 under field conditions, some caution in extending results of laboratory studies to field situations is warranted.

UV-B absorbing flavonoids and other complex polyphenolics in epidermal cells form an effective screen for plants to UV-B (Braun and Tevini, 1993; Day, 1993; van de Staaij *et al.*, 1995). Therefore, under field conditions this UV-B screen reduces UV-B penetration, so that UV-B radiation levels in the chloroplasts of the mesophyll cells are low (Braun and Tevini, 1993). Studies with mutants lacking these compounds show them to be extremely sensitive to normal ambient solar UV (Li *et al.*, 1993; Lois and Buchanan, 1994) (Fig. 1).

Moreover, high PAR/UV-B radiation ratios that occur under field conditions, unlike in many laboratory or glasshouse studies, appear to enhance the UV-B tolerance of plants, in part perhaps because of more effective photorepair of UV-B DNA damage by the photolyase enzymes (Caldwell and Flint, 1994; Britt, 1995; Björn, 1996; Rozema *et al.*, 1997c).

Although atmospheric CO_2 enrichment and enhanced solar UV-B may both effect photosynthesis, their primary photosynthetic target differs (Fig. 1), and interactive effects of CO_2 and UV-B on plant biomass increments have only rarely been found (Teramura *et al.*, 1990; Rozema *et al.* 1990, 1997c, Ziska and Teramura 1992; Rozema 1993; van de Staaij *et al.*, 1993;

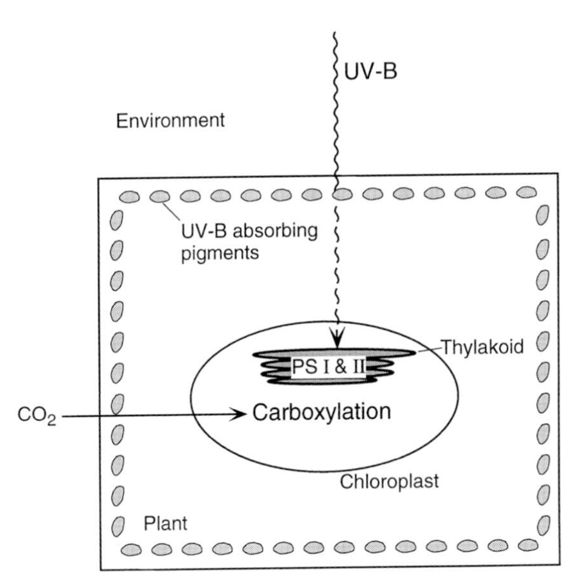

Figure 1 Diagrammatic representation of penetration of UV-B radiation from the environment into plant tissue, and diffusion of CO_2 into chloroplasts. UV-B radiation is filtered through a screen of UV-B absorbing pigments often located in (vacuoles of) epidermal cells, before reaching the chloroplasts. CO_2 is carboxylated in the stroma of chloroplasts, by the enzyme RuBP-carboxylase-oxygenase. The primary target of UV-B is possibly PS II. The greatest influence of CO_2 is on the carboxylation process.

Visser *et al.* 1997a and b; Sullivan, 1997). Combined effects of elevated CO_2 and UV-B may regarded as generally additive (Rozema *et al.,* 1997c; Sullivan, 1997). This conclusion is mainly based on studies in growth chambers and greenhouse and should not necessarily hold for field and ecosystem studies.

IV. Plant Biomass Allocation Pattern

In addition to increased photosynthesis, growth, and WUE, elevated CO_2 may also effect partitioning of biomass. The nature of these allocation changes varies among species. For example, the root: shoot ratio has been reported both to increase and to decrease under elevated CO_2.

Shifts in biomass allocation have also occasionally been reported under enhanced UV-B, and the direction of these changes also varies among species. In addition, enhanced UV-B may exert pronounced plant morphogenetic effects, such as decreased plant height, leaf length, and leaf area and increased tillering and branching, as discussed next in Section V. Sullivan (1997) and Sullivan and Teramura (1994) reported that root : shoot ratios were decreased by UV-B at ambient CO_2, but increased by UV-B at elevated CO_2. This shift in biomass allocation in loblolly pine was linked with decreased needle length under elevated UV-B. Of course, changes in biomass partitioning between aboveground and belowground plant parts may have important ecological consequences, for example, for competitive relationships, root and soil respiration, and decomposition processes.

V. Plant Morphogenesis

Enhanced UV-B may often affect biomass partitioning and plant morphogenesis than biomass increases per se. Several mechanisms for these changes have been proposed. The UV-B photons may be absorbed by growth regulators or by specific UV-B photoreceptors (Ensminger and Schäfer, 1992). In many studies UV-B reduced plant height, shoot height, and leaf length (Rozema *et al.,* 1997d,e). Cell elongation is influenced by the phytohormones auxin and indole-3-acetic acid (IAA). These growth-regulating hormones absorb UV-B radiation and may be photodegraded by high levels of UV-B radiation. Inhibition of elongation growth of sunflower seedlings and hypocotyl segments possibly relates to reduced levels of IAA and formation of growth inhibiting IAA photoproducts. Increased levels of peroxidases induced by enhanced UV-B radiation may affect growth-regulating hormones and growth in sunflower (Ros and Tevini, 1995).

Ultraviolet-B radiation appeared to inhibit hypocotyl elongation in de-etiolating tomato seedlings, and a UV-B absorbing flavin chromophore was thought to be involved (Ballaré *et al.*, 1995). Similarly, elongation of sunflower seedlings is influenced by UV-B radiation. The plant hormone ethylene, which promotes radial growth and reduces cell elongation, was increased in shoots of pear exposed to UV-B (Predieri *et al.*, 1993). By affecting apical dominance, branching of plants and tillering of grasses may increase under enchanced UV-B (e.g., Barnes *et al.*, 1988, 1995). This may, as in the case of elongation growth, relate to UV-B absorbing plant hormones (e.g., auxin and IAA) or a UV-B receptor. Other photosystems, such as cryptochrome or phytochrome, which absorb in other parts of the solar spectrum, also effect many of these characteristics such as elongation, leaf inclination, and canopy structure, which will alter the penetration of the radiation into the canopy.

In a greenhouse study, Tosserams *et al.*, (1997) found that the dune grassland species *Bromus hordeaceus* exposed to a low daily dose of UV-B had more planophilous leaves and plants not exposed at all to UV-B had more erectophilous leaves. Planophilous leaves will intercept more photosynthetically active radiation than erectophilous leaves. Because *Bromus hordeaceus,* like other dune grassland plants (Tosserams *et al.*, 1997), is not sensitive to enhanced UV-B the photomorphogenic effect of UV-B on leaf inclination helps to explain a positive growth response to elevated UV-B (see Section VII).

Such studies indicate that different levels of solar UV-B radiation may affect leaf length growth, internode elongation, and other morphological properties. These, in turn, may affect canopy structure and competitive balance of plants (see Fig. 8 in a later section).

It is far less clear whether atmospheric CO_2 enrichment leads to changes in plant morphogenesis. For the future of global change ecosystem research, studies of CO_2 by UV-B interactions on plant morphogenetic characteristics should be carefully considered and examined.

VI. Secondary Chemistry, Litter Decomposition, and Carbon Cycling

Solar UV-B radiation is known to stimulate the enzymes PAL and CHS and other branch-point enzymes of the phenylpropanoid pathway (Beggs and Wellman, 1994). PAL catalyzes the transformation of phenylalanine into transcinnamic acid, which may lead to the formation of complex phenolic compounds such as flavonoids, tannins, and lignin (Fig. 2).

UV absorption in the UV-B wavelengths by cinnamic acid (maximum absorption at 308 nm) exceeds that of phenylalanine (maximum absorption

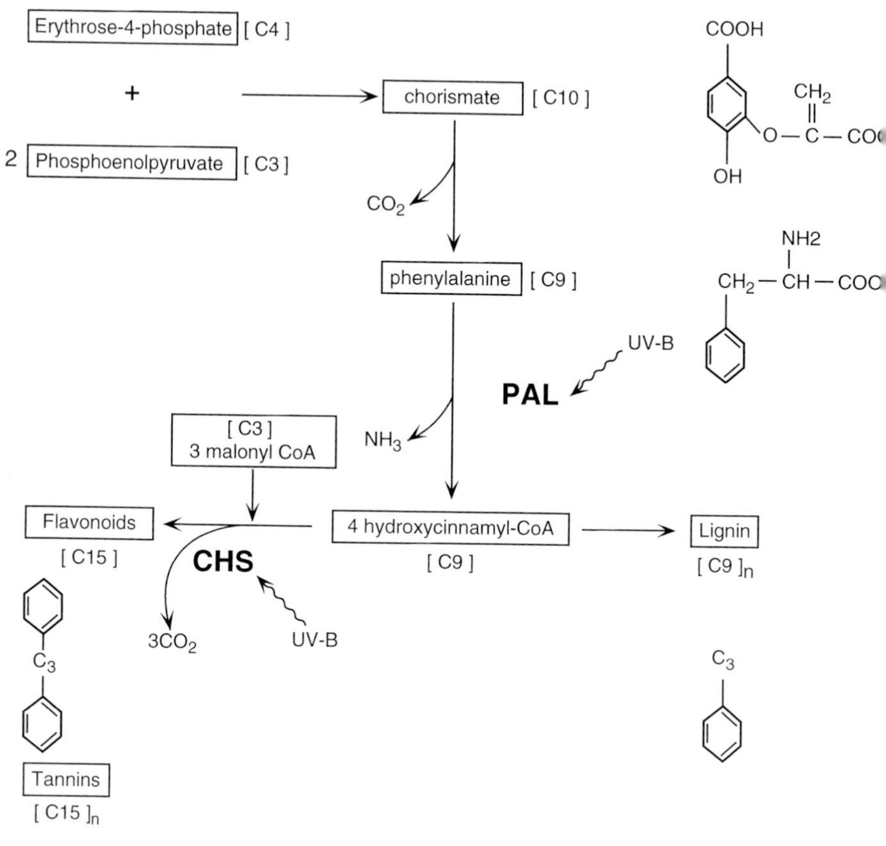

Figure 2 Diagram of major steps in the phenylpropanoid pathway. PAL, phenyl alanine ammonia lyase; CHS, chalcone synthase. Erythrose-4-phosphate (from the pentose phosphate pathway) and phosphoenolpyruvate combine to form C_{10} compounds. In the shikimic acid pathway the aromatic amino acid phenylalanine (C_9) is formed. The UV-B inducible enzyme PAL catalyzes a deamination reaction. Lignin is a complex phenolic macromolecule on the basis of C_9 units. Chalcone synthase catalyses the formation of C_{15} units, such as flavonoids. Tannins represent polyphenolics based on such C_{15} units. (Based on Beggs and Wellmann, 1994.)

at 280 nm). The production of these and other UV-B absorbing compounds has been assumed to be an important part of the evolution of land plants from aquatic plants (Kubitzki, 1987; Stafford, 1991). Complexity of polyphenolics tends to increase with evolutionary advancement from algae, charophycean algae, bryophytes, pteridophytes, and gymnosperms to angiosperms (Robinson, 1980; Rozema *et al.*, 1997d). A high degree of polymerization of phenolics is found in land plants such as flavonoids, tannins, and lignin (Rozema *et al.*, 1997d). These compounds serve many other functions apart from UV absorption such as signal transduction, plant hormones, defense against microorganisms, and herbivory and structural rigidity. It has been assumed that the filtering of UV-B radiation by polyphenolics provided land plants with an internal filter against damaging solar UV-B and allowed evolution of terrestrial plant life (Rozema *et al.*, 1997d).

The physical and chemical stability and the resistance to microbial breakdown of complex polyphenolics in plant litter contributes considerably to the long-term storage of organic carbon in terrestrial ecosystems (Berner, 1989, 1993; Rozema *et al.*, 1997d). The polyphenolic compound lignin determines to a large extent the resistance of dead organic plant matter to degradation by microorganisms (Berendse *et al.*, 1987). Lignin contributes to the persistence of peat and humus in soils. Thereby, complex polyphenolics help to establish large, long-lasting sinks of carbon in terrestrial ecosystems. The established relative homeostasis between oxygen and carbon dioxide in the current atmosphere (Berner, 1989, 1993) may partly be based on the presence of recalcitrant polyphenolics in terrestrial soil biota (Graham, 1993).

In some recent studies (Gehrke *et al.*, 1995; Rozema *et al.*, 1997c) enhanced solar UV-B, simulating about 15% stratospheric ozone depletion, results in an increased content of tannins and lignin in terrestrial plants. Decomposition of plant litter with increased content of tannins and lignin was retarded compared to litter from plants exposed to ambient solar UV-B.

VII. Ecosystem Processes

From the previous sections in this chapter it should be clear that limited data are available on the combined effects of enhanced atmospheric CO_2 and elevated solar UV-B (Rozema *et al.*, 1997c; Sullivan, 1997). Much of the $CO_2 \times$ UV-B research is based on laboratory growth chamber, and glasshouse UV-B experiments. Caldwell and Flint (1994) and Caldwell *et al.* (1995) have surveyed the limitations of UV-B effects found in growth chambers and greenhouse studies and discuss the difficulties in extrapolation to field and ecosystem-level situations. In particular, under relatively low PAR/UV-B ratios in indoor studies, plants are much less tolerant of UV-B (in part, perhaps because photorepair of UV-B damage may be limited). Generally growth and yield reduction by elevated UV-B in indoor studies is thought to overestimate effects occurring in the field (Caldwell and Flint, 1994). Short-term and long-term experiments of atmospheric CO_2 enrichment on natural and agro-ecosystems confirm many of the findings from indoor CO_2 enrichment experiments (Bazzaz, 1990; Rozema *et al.*, 1993). However, atmospheric CO_2 enrichment interacts with other environmental factors such as temperature, water, and nutrient in availability and this will alter the response to the increased CO_2 (Koch and Mooney, 1996; Körner and Bazzaz, 1996; Mark and Tevini, 1997). These interactions are also discussed in other chapters of this book.

Recent reviews indicate that, in contrast to elevated CO_2, enhancement of solar UV-B may not substantially affect ecosystem primary productivity (Caldwell *et al.*, 1995; Sullivan, 1997; Rozema *et al.*, 1997c,d,e). This does not, however, directly imply that CO_2 and UV-B may not interact in their influence on primary productivity of ecosystems, since enhanced UV-B may also affect plant morphogenesis and biomass allocation. Shifts in competitive relationships may occur in response to both CO_2 enrichment (Bazzaz, 1990) and elevated UV-B (Barnes *et al.*, 1988, 1995). These shifts may well relate to an altered plant and canopy architecture and subsequent changes in light interception of radiation (UV-B and PAR). Also altered root : shoot relationships may influence changes in competitive interactions (Gold and Caldwell, 1983, Ryel *et al.*, 1990). Remarkably, Barnes *et al.* (1988, 1995) reported changes in competitive plant relationships in the absence of UV-B effects on plant productivity.

CO_2 and UV-B experiments in Sweden and the Netherlands form part of a project funded by the European Union. Although there are some methodological differences, the general experimental design is similar for the Swedish and Dutch $CO_2 \times$ UV-B ecosystem research. Atmospheric CO_2 enrichment is achieved with open top chambers and enhanced UV-B is supplied by filtered fluorescent UV-B tubes (Fig. 3 and 4).

These two ecosystem-level studies of combined CO_2 and UV-B effects will be discussed in more detail.

A. Ecosystem Studies in a Subarctic Tundra Ecosystem at Abisko in Northern Sweden

Vegetation responses in a subarctic heathland (Fig. 3) at Abisko, North Sweden, to enhanced CO_2 (600 ppm) and elevated solar UV-B (simulating 15% stratosperic ozone reduction) are species specific. After 2 yr of enhanced UV-B, relative elongation growth of two evergreen species, *Vaccinium vitis-idaea* and *Empetrum hermaphroditum*, was reduced by 27 and 33%, respectively (Gwynn-Jones *et al.*, 1996). There was no reduced growth in two deciduous *Vaccinium* species. Leaf thickness of the evergreen *V. vitis-idaea* increased under elevated UV-B but did not in the deciduous species. Enhanced CO_2 increased growth of *V. myrtillus* but decreased growth of *E. hermaphroditum* and *V. vitis-idaea*. Tissue and Oechel (1987) reported only small plant responses to elevated CO_2 in a North American arctic ecosystem. Herbivory in the Swedish subarctic *Vaccinium* species was markedly affected by enhanced UV-B, but not by elevated CO_2. The latter is surprising, since the leaf C : N ratio increased under elevated CO_2.

Figure 3 UV-B supplementation and open top chambers for CO_2 enrichment in a subarctic heathland vegetation at the Abisko Scientific Research Station, North-Sweden. For more details see Johanson *et al.*, 1995; Gehrke *et al.*, 1996 and Gwynn-Jones *et al.*, 1996, 1997. (Photograph courtesy of J. Rozema.)

Effects of combined enhanced UV-B and elevated atmospheric CO_2 have been studied in the moss *Hylocomium splendens* (Sonesson *et al.*, 1996) and lichens (Sonesson *et al.*, 1995).

B. CO_2 and UV-B Ecosystem Experiments in a Dune Grassland, Heemskerk, The Netherlands

A similar $CO_2 \times$ UV-B ecosystem experiment is being conducted in a dune grassland in the Netherlands (Rozema *et al.*, 1995, 1998) (Fig. 4). The grass species *Calamagrostis epigeios,* along with *Holcus lanatus, Bromus hordeaceus,* and *Carex arenaria,* dominates the dune grassland. Photosynthesis, transpiration, and growth of *C. epigeios* was not adversely affected with a UV-B_{BE} dose simulating 30–40% stratospheric ozone reduction (Tosser-

Figure 4 Open top chambers for CO_2 enrichment (foreground) and UV-fluorescent tubes (background) for a UV-B supplementation system in the dune grassland in Heemskerk, The Netherlands. (Photograph courtesy of J. Rozema.)

ams and Rozema, 1995). Growth of *B. hordeaceus* at a rather low daily dose of UV-B (UV-B_{BE} of 4.6 kJ $m^{-2}d^{-1}$ exceeds that of plants that received no UV-B (Tosserams *et al.,* 1997). This is explained by more planophilic leaf inclination with the +UV-B and more erectophilic leaves with the no UV-B treatment (Fig. 5). With the more planophilic leaf inclination (at + UV-B), *Bromus* plants intercept more photosynthethic active radiation (PAR) and may therefore grow better than at 0 UV-B. This also emphasizes the relevance of plant and canopy architecture in UV-B and CO_2 research.

In outdoor experiments enhanced solar UV-B radiation appears to affect decomposition of plant litter in various direct and indirect ways. The indirect pathway is that under enhanced UV-B the chemical properties of the litter are changed and the rate of mass loss in litter bags (Rozema *et al.,* 1997c) is reduced. In *C. epigeios,* the reduced rate of decomposition of litter from plants exposed to enhanced UV-B is linked with an increased lignin content. In similar studies Gehrke *et al.* (1995) found tannins increased in +UV-B litter of a subarctic heathland vegetation. A consequence of this may be that enhanced UV-B affects the biogeochemical cycling of carbon (Schlesinger, 1993; Zepp *et al.,* 1995).

Enhanced solar UV-B may also promote the degradation of plant litter by a photodegradative effect. This assumes that the litter is at least partially

Figure 5 Leaf inclination of plants of *Bromus hordeaceus*, grown in pots in a greenhouse, exposed to no UV-B (left) and a daily dose of UV-B according to 35% stratospheric ozone depletion (right). The plant on the left has more erectophilic leaves and the plant on the right more planophilic leaves. For more detailed information see Tosserams *et al.*, 1997. (Photograph courtesy of M. Tosserams.)

exposed to solar UV-B under the canopy (Figs. 6 and 7). Obviously, photodegradative effects of solar UV-B on litter decomposition may be large in ecosystems with a low leaf area index, such as in desert and some tundra ecosystems. By affecting leaf inclination and canopy architecture, the enhanced solar UV-B influences its penetration into the vegetation canopy of terrestrial ecosystems (Moorhead and Callaghan, 1994). Penetration of part of the incident solar UV-B to the litter suggests that this UV-B radiation may also directly affect decomposer organisms (Fig. 8). In a laboratory study, negative effects of UV-B on the decomposer community have been reported by Gehrke *et al.* (1996).

Photodegradation by UV-B of dissolved organic matter (DOM) in aquatic ecosystems is well known (Herndl *et al.*, 1997). Photodegradation of the DOM by UV-B also increases the vertical penetration of solar UV-B into the water column (Häder, 1997).

It is known that mycorrhizal infection of the dune grassland species *C. epigeios* may be high (Ernst *et al.*, 1984; Rozema *et al.*, 1986). Currently, the degree of VAM infection of roots of dune grassland species that have been exposed to ambient or enhanced solar UV-B for 6 yr is being assessed.

Figure 6 Litter bags filled with senesced leaves of *Holcus lanatus* exposed to ambient solar UV-B or to enhanced UV-B in a dune grassland at Heemskerk. Litter bags are placed within the canopy of the dune grassland vegetation. For further details, see legend for Fig. 7. (Photograph courtesy of J. Rozema.)

Similarly the effects of elevated CO_2 on (mycorrhizal) fungal infection have been studied (Tingey *et al.*, 1995; Sanders, 1996).

A summary of UV-B and CO_2 effects on terrestrial ecosystems is presented in Fig. 9.

VIII. Conclusions

Atmospheric CO_2 enrichment is accompanied by increased solar UV-B radiation reaching the earth as a result of stratospheric ozone depletion. Effects of both climate change factors have been studied independently, but their combined effects have only been examined in a handful of studies.

Figure 7 Litter bags filled with senesced leaves of *Holcus lanatus* exposed to ambient or enhanced UV-B at an experimental field at the campus of the Vrije Universiteit, Amsterdam, The Netherlands. Litter bags are on the soil surface without vegetation to assess photodegrative effects of UV-B on the rate of mass loss of litter. Philips TL 12/40 fluorescent tubes have been installed in racks at a height of 1.40 m above the soil surface. Both for ambient UV-B and for enhanced UV-B, the fluorescent tubes were operating, but they have different filters to effect the appropriate UV-B supplementation levels. The enhanced UV-B treatment was designed to simulate about 15% stratospheric ozone depletion at this site. In this design, the ambient and enhanced UV-B treatment had the same flux of UV-A radiation. Litter in the litter bags originates from *Holcus lanatus* plants exposed to ambient (5 kJ $m^{-2}d^{-1}$) or enhanced (7.5 kJ $m^{-2}d^{-1}$ UV-B_{BE}) for one growing season (Photograph courtesy of J. Rozema.)

Enhanced solar UV-B may sometimes constitute a stress for plants leading to reduced plant elongation growth and occasionally primary production. However, probably of greater importance are other influences of enhanced solar UV-B such as the production of plant secondary metabolites, for example, flavonoids, tannins, and lignin. These complex polyphenolics are involved in plant–animal and plant–microorganism interactions and in resistance to litter decomposition. Atmospheric CO_2 enrichment tends pri-

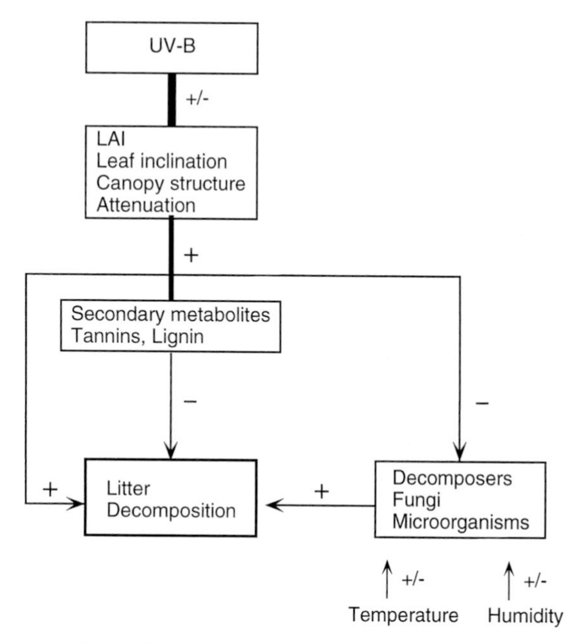

Figure 8 Structure of a model of the direct and indirect effects of UV-B radiation on the litter decomposition of a terrestrial ecosystem. Plant and canopy characteristics (LAI, leaf inclination) affect the attenuation of incident solar UV-B. By affecting leaf inclination, branching, and tillering, solar UV-B levels may, to some extent, influence the penetration of UV-B (and PAR) through the canopy (+/−). Intercepted UV-B radiation may induce production of various secondary metabolites (+), a.o. tannins, and lignin. Tannins and lignin in plant litter may retard litter decomposition (−). UV-B radiation penetrating to the litter layer may increase litter decomposition by photodegradation (+); activity of decomposers exposed to UV-B may be reduced (−). Temperature and humidity control the activity of decomposer organisms (+/−).

marily to promote photosynthesis and growth of terrestrial C_3 plant species (Fig. 9).

There is increasing evidence that enhanced UV-B in field-grown plants under natural spectral conditions does not depress photosynthesis. It can reduce elongation growth of stems and leaves, but does not often affect total primary production. (Caldwell *et al.*, 1995; Kim *et al.*, 1996; Day *et al.*, 1996). One reason for this may be that field-grown plants develop an effective UV-B screen of UV-B absorbing compounds in epidermal cells. In addition, field-grown plants are exposed to higher natural PAR/UV-B ratios than in many greenhouse and growth chamber studies and photoreactivation repair of UV-B damage also renders plants more tolerant of UV-B. In laboratory studies damage to PS II by UV-B may lead to reduced quantum efficiency and lower photosynthetic capacity, which may be due

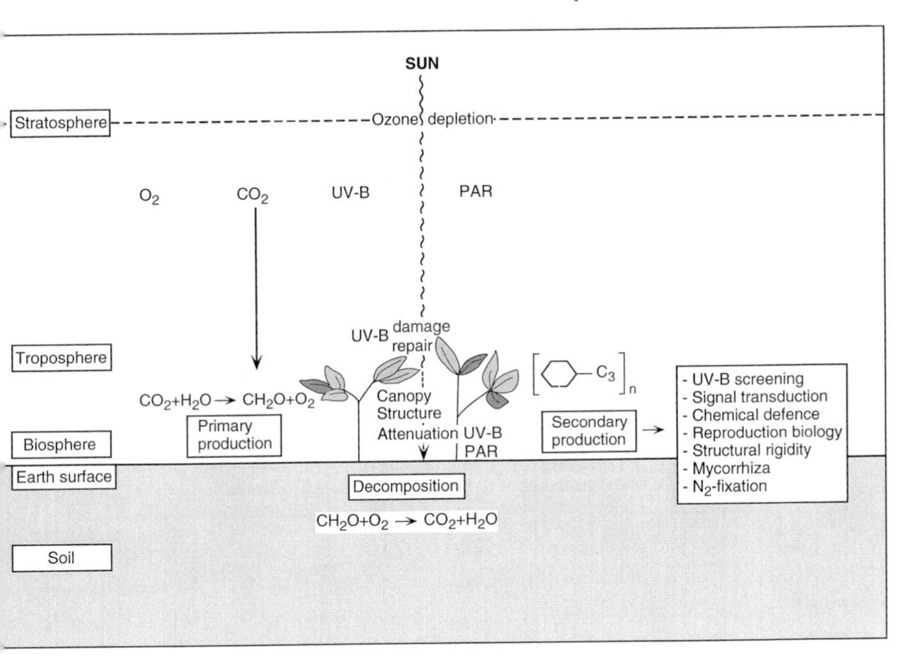

Figure 9 Diagram of CO_2 and solar UV-B radiation and its potential influence on the biosphere. The focus is on the role of UV-B. Depletion of stratospheric ozone allows an increased flux of solar UV-B to the earth's surface. Elevated UV-B and CO_2 may sometimes directly affect primary production. For enhanced UV-B, indirect effects are likely of much greater importance, such as influences on secondary chemistry, plant morphogenesis, and biomass allocation. Induction of secondary metabolites in plants may considerably affect litter decomposition, plant–animal, and plant–microorganism interactions. Changes in morphogenesis and biomass allocation can effect many ecosystem-level processes such as the competitive balance of plants.

to reduced regeneration of RuBP, the substrate for Rubisco. In addition, there may be direct UV-B damage to Rubisco by inhibiting RNA transcription for Rubisco (Jordan *et al.*, 1992).

For physiological processes, enhanced CO_2 and UV-B may interact in various ways. Damage to PS II and Rubisco may reduce the stimulation of elevated CO_2 to plant photosynthesis of C_3 plants. Alternatively, the effects of atmospheric CO_2 enrichment may contribute to protecting plants from UV-B damage to the photosynthetic system. At least in short-term experiments elevated CO_2 may increase net carboxylation by reduced oxygenase activity of Rubisco.

If, under outdoor conditions with a more natural spectral balance, negative effects of enhanced UV-B on photosynthesis and plant primary production do not occur, this does not rule out interactions between atmospheric

CO_2 enrichment and enhanced solar UV-B. However, these are to be demonstrated in field studies.

Although UV-B probably does not often affect plant primary productivity, it may influence biomass allocation and morphogenesis of plants. Significant interactions between CO_2 and UV-B on biomass partitioning have been previously reported (Sullivan and Teramura, 1994; Rozema *et al.*, 1997c; Sullivan, 1997; Tosserams *et al.*, 1997).

$CO_2 \times$ UV-B interactions on biomass allocation may have consequences for ecosystem-level processes such as plant competitive relationships. Plant morphogenesis is often altered under enhanced UV-B. Plant height, shoot and leaf length, and leaf area decrease under elevated UV-B, while leaf thickness and auxiliary branching are usually increased. Leaf angles and plant and canopy architecture may change under enhanced UV-B, as well as emergence, phenology, and senescence of plants. Limited published data are available concerning the influence of $CO_2 \times$ UV-B interaction on these characteristics.

The level of secondary metabolites in plants may be markedly changed under enhanced UV-B and atmospheric CO_2 (Beggs and Wellmann, 1994; Lambers, 1993). Secondary metabolites play an important role in the chemical or physical defense of plants against microorganisms, herbivores, and mechanical disturbance.

Some recent ecosystem studies indicate that enhanced UV-B affects decomposition of plant litter, herbivory (Paul *et al.*, 1997), and mycorrhizal relationshipships (Klironomos and Allen, 1995). It is likely that UV-B induced secondary metabolites play an important role. An increased content of tannins and/or lignin in plant growth under elevated UV-B is linked with a reduced rate of decomposition. Similarly, insect herbivory and mycorrhizal infection may be altered in plants in which UV-B has led to changes in the secondary chemistry of tissues. Effects of CO_2 enrichment on insect herbivory have been studied as well (Lincoln *et al.*, 1993) The impact of elevated atmospheric CO_2 and enhanced solar UV-B on the structure and processes in terrestrial ecosystems is summarized in Fig. 9. The number of studies on the combined effects of elevated CO_2 and UV-B is very limited and allows for few generalizations.

Acknowledgments

We are indebted to Karin Uyldert for word processing and to Henry Sion for drawing the diagrams. Mirella De Vries and Dr. Bob Kooi (Department of Mathemathical Biology, Vrije Universiteit) contributed to the development of the UV-B decomposition model (UVDECOM), which is highly appreciated. We thank Dr. Jos van de Staaij for critically reading the manuscript and Ir. Barbara Meijkamp for discussing the phenylpropanoid pathway. The $CO_2 \times$ UV-B dune grassland project in Heemskerk is being funded by the EU (contract ENV4-CT96-0208), which is greatly acknowledged. We thank Prof. J. A. Lee for permission to publish a photograph of the UV-B × CO_2 research at Abisko. The authors were grateful to be invited to attend the IGBP-GCTE workshop, Focus 1 and 3, May 14–18, 1995, Lake Tahoe. This paper is based on the presentations given by JR, AHT, and MMC.

References

Arp, W. J. (1991). Effects of source-sink relations on photosynthetic acclimation to elevated CO_2. *Plant Cell Environ.* **14,** 869–875.

Arp, W. J., and Drake, B. G. (1991). Increased photosynthetic capacity of *Scirpus olneyi* after four years of exposure to elevated CO_2. *Plant Cell Environ.* **14,** 1003–1006.

Ballaré, C. L., Barnes, P. W., and Flint, S. D. (1995). Inhibition of hypocotyl elongation by ultraviolet-B radiation in de-etiolating tomato seedlings. The photoreceptor. *Physiol. Plant.* **83,** 652–658.

Barnes, P. W., Jordan, P. W., Gold, W. G., Flint, S. D., and Caldwell, M. M. (1988). Competition, morphology and canopy structure in wheat (*Triticum aestivum* L.) and wild oat (*Avena fatua* L.) exposed to enhanced ultraviolet-B radiation. *Funct. Ecol.* **2,** 319–330.

Barnes, P. W., Flint, S. D., and Caldwell, M. M. (1995). Early season effects of supplemented solar UV-B radiation on seeding emergence, canopy structure, simulated stand photosynthesis and competition for light. *Global Change Biol.* **1,** 43–53.

Bazzaz, F. A. (1990). The response of natural ecosystems to rising global CO_2 levels. *Annu. Rev. Ecol. System.* **21,** 167–196.

Beggs, C. J., and Wellmann, E. (1994). Photocontrol of flavonoid biosynthesis. *In* "Photomorphogenesis in Plants," (R. E. Kendrick and G. H. M. Kronenberg, eds.), Vol. 2, pp. 733–750. Kluwer Academic Publishers, Dordrecht.

Berendse, F., Berg, B., and Bosatta, E. (1987). The effect of lignin and nitrogen on the decomposition of litter in nutrient poor ecosystems: A theoretical approach. *Can J. Bot.* **65,** 1116–1120.

Berner, R. A. (1989). Biogeochemical cycles of carbon and sulfur and their effects on atmospheric oxygen over Phanerozoic time. *Paleogeogr. Paleoclimatol. Paleoecol.* **75,** 97–122.

Berner, R. A. (1993). Paleozoic atmospheric CO_2: Importance of solar radiation and plant evolution. *Science* **261,** 68–70.

Björn, L. O. (1996). Effects of ozone depletion and increased UV-B on terrestrial ecosystems. *J. Environm. Science.* **51,** 217–243.

Bolker, B. M., Pacala, S. W., Bazzaz, F. A., Canham, C. D., and Levin, S. A. (1995). Species diversity and ecosystem response to carbon dioxide ferilization. Conclusions from a temperate forest model. *Global Change Biol.* **1,** 363–381.

Bothwell, M. L., Darren, M., Sherbot, J., and Pollock, M. (1994). Ecosystem response to solar ultraviolet-B radiation: Influence of trophic-level interactions. *Science* **265,** 97–100.

Braun, J., and Tevini, M. (1993). Regulation of UV-protection pigment synthesis in the epidermal layer of rye seedings (*Secale cereale L. cv Kustro*). *Photochem. Photobiol.* **57,** 318–323.

Britt, A. B. (1995). Repair of DNA damage induced by ultraviolet radiation. *Plant Physiol.* **108,** 891–896.

Caldwell, M. M. (1979). Plant life and ultraviolet radiation: Some perspective in the history of earth's UV-climate. *Bioscience* **29,** 520–525.

Caldwell, M. M., and Flint, S. D (1994). Stratospheric ozone reduction, solar UV-B radiation and terrestrial ecosystems. *Climatic Change* **28,** 375–394.

Caldwell, M. M., Teramura, A. H., and Tevini, M. (1989). The changing solar ultraviolet climate and the ecological consequences for higher plants. *Trends Ecol. Evol.* **4,** 363–367.

Caldwell, M. M., Teramura, A. H., Tevini, M. Bornman, J. F., Björn, L. O., and Kulandaivelu, G. (1995). Effects of increased solar ultraviolet radiation on terrestrial plants. *Ambio* **24,** 166–173.

Couteaux, M.-M., Mousseau, M., Céléries, M.-L., and Bottner, P. (1991). Increased atmospheric CO_2 and litter quality: Decomposition of sweet chestnut leaf litter with animal food webs of different complexities. *Oikos* **61,** 54–64.

Day, T. A. (1993). Relating UV-B radiation screening effectiveness of foliage to absorbing-compund concentration and anatomical characteristics in a diverse group of plants. *Oecologia* **95,** 542–550.

Day, T. A., Howells, B. W., and Ruhland, C. T. (1996). Changes in growth and pigment concentrations with leaf age in pea under modulated UV-B radiation field treatments. *Plant Cell Environ.* **19,** 101–108.

Ensminger, P. A., and Schäfer, E. (1992). Blue and ultraviolet-B light receptors in parsley cells. *Photochem. Photobiol.* **55,** 437–447.

Ernst, W. H. O., van Duin, W. E., and Oolbekking, G. T. (1984). Vesicular arbuscular mycorrhiza in dune vegetation. *Acta Bot. Neerlandica* **33,** 151–160.

Fiscus, E. L., and Booker, F. L. (1995). Is increased UV-B a threat to crop photosynthesis and productivity? *Photosynth. Res.* **43,** 81–92.

Gehrke, C., Johansson, U., Callaghan, T. V., Chadwick, D., and Robinson, C. H. (1995). The impact of enhanced ultraviolet-B radiation on litter quality and decomposition processes in *Vaccinium* leaves from the Subarctic. *Oikos* **72,** 213–222.

Gehrke, C., Johansson, U., Gwynn-Jones, D., Björn, L.-O., Callaghan, T. V., and Lee, J. A. (1996). Effects of enhanced ultraviolet-B radiation on terrestrial subarctic ecosystems and implications for interactions with increased atmospheric CO_2 *Ecol. Bull.* **45,** 192–203.

Gold, W. G., and Caldwell, M. M. (1983). The effects of ultraviolet-B radiation on plant competition in terrestrial ecosystems. *Physiol. Plant.* **58,** 435–444.

Graham, L. E. (1993). "Origin of Land Plants." Wiley, New York.

Gwynn-Jones, D., Björn, L. O., Callaghan, T. V., Gehrke, C., Johanson, U., Lee, J. A., and Sonesson, M. (1996). Effects of enhanced UV-B radiation and elevated concentrations of CO_2 on a subarctic heathland. *In* "Carbon Dioxide, Populations and Communities" (Ch. Körner and F. A. Bazzaz, eds.), pp. 197–207. Academic Press, San Diego.

Gwynn-Jones, D., Lee, J. A., Callaghan, T. V., and Sonesson, M. (1997). Effects of enhanced UV-B radiation and elevated carbon dioxide concentrations on subarctic forest heath ecosystems. *Plant Ecology* **128,** 242–249.

Häder, D-P. (1997). Penetration and effects of solar UV-B on phytoplankton and macroalgae. *Plant Ecol.* **128,** 4–13.

Herndl, G. J., Brugger, A., Hager, S., Kaiser, E., Overnosterer, I., Reitner, B., and Slezak, D. (1997). Role of ultraviolet-B radiation on bacterioplankton and the availability of dissolved organic matter, *Plant Ecol.* **128,** 42–51.

Jacob, J., Greitner, C., and Drake, B. G. (1995). Acclimation of photosynthesis in relation to Rubisco and non-structural carbohydrate content and *in situ* carboxylase activity in *Scirpus olneyi* grown in elevated CO_2 in the field. *Plant Cell Environ.* **18,** 875–885.

Johanson, U., Gehrke, C., Björn, L. O., Callaghan, T. V., and Sonesson, M. (1995). The effects of enhanced UV-B radiation on a subarctic heath ecosystem. *Ambio* **24,** 106–111.

Jordan, B. R., He, J., Chow, W. S., and Anderson, J. M. (1992). Changes in mRNA levels and polypeptide subunits of ribulose-1,5-bisphosphate carboxylase in response to supplementary ultraviolet-B radiation. *Plant Cell Environ.* **15,** 91–98.

Kim, H. Y., Kobayashi, K., Nouchi, I., and Yoneyama, L. (1996). Enhanced UV-B radiation has little effect on growth $\delta^{13}C$ values and pigments of pot-grown rice (*Oryza sativa*) in the field. *Physiol. Plant.* **96,** 1–5.

Klironomos, J. N., and Allen, M. F. (1995). UV-B mediated changes on below-ground communities associated with the roots of *Acer saccharum. Funct. Ecol.* **9,** 923–930.

Koch, G. W., and Mooney, H. A. (eds.). (1996). "Carbon Dioxide and Terrestrial Ecosystems." Academic Press, San Diego.

Körner, C., and Bazzaz, F. A. (eds.). (1996). "Carbon Dioxide, Populations and Communities," pp. 465. Academic Press, San Diego.

Kubitzki, K. (1987). Phenylpropanoid metabolism in land plant evolution. *J. Plant Physiol.* **131,** 17–24.

Lambers, H. (1993). Rising CO_2, secondary plant metabolism, plant herbivore interactions and litter decomposition. Theoretical considerations. *Vegetatio* **104/105,** 263–271.

Lenssen, G. M. (1993). "Response of C_3 and C_4 Species from Dutch Salt Marshes to Atmospheric CO_2 Enrichment," Doctorate thesis, pp. 113. Vrije Universiteit, Amsterdam.

Li, J., Ou-Lee, T. M. Raba, R., Amundson, R. G., and Last R. L. (1993). Arabidopsis flavonoid mutants are hypersensitive to UV-B radiation. *Plant Cell* **5,** 171–179.

Lincoln, D. E., Fajer, E. D., and Johnson, R. H. (1993). Plant insect herbivore interactions in elevated CO_2 environments. *Trends Ecol. Evol.* **8,** 64–72.

Lois, R., and Buchanan, B. B. (1994). Severe sensitivity to ultraviolet radiation in *Arabidopsis* mutant deficient in flavonoid accumulation. II. Mechanisms of UV-resistance in *Arabidopsis. Planta* **194,** 504–509.

Lowry, B., Lee, D., and Hébaut, C. (1980). The origin of land plants: A new look at an old problem. *Taxon* **29,** 183–197.

Margulis, L., Walker, J. C. G., and Rambler, M. (1976). Reassessment of roles of oxygen and ultraviolet light in Precambrium evolution. *Nature* **264,** 620–624.

Mark, U., and Tevini, M. (1997). Effects of elevated UV-B radiation temperature and CO_2 on growth and function of sunflower and maize seedlings. *Plant Ecol.* **128,** 224–234.

Melillo, J. M., Kicklighter, D. W., McGuire, A. D., Moore III, B., Voorosmart, C. J., and Grace, A. L. (1993). Global climate change and terrestrial net primary production *Nature* **363,** 234–240.

Mooney, H. A., Drake, B. G., Luxmoore, R. J., Oechel, W. C., and Pitelka, L. I. (1991). Predicting ecosystem responses to elevated CO_2 concentrations. *BioScience* **41,** 96–104.

Moorhead, D. L., and Callaghan, T. V. (1994). Effects of increasing UV-B radiation and soil organic matter dynamics: A synthesis and modelling study. *Biol. Fertil. Soil.* **18,** 19–26.

Norby, R., O'Neill, E. G. Hood, W. G., and Luxmoore, R. J. (1987). Carbon allocation, root exudation and mycorrhizal colonization of *Pinus echinata* seedlings grown under CO_2 enrichment. *Tree Physiol.* **3,** 203–210.

Paul, N. D., Rasanayagam, M. S., Moody, S. A., Hatcher, P. E., and Ayres, P. G. (1997). The role of interactions between trophic levels in determining the effects of UV-B on terrestrial ecosystems. *Plant Ecol.* **128,** 296–308.

Predieri, S., Krizek, D. T., Wang, C. Y., Mirecki, R. M., and Zimmerman, R. H. (1993). Influence of UV-B radiation on developmental changes, ethylene, CO_2 flux and polyamines in cv Doyenne d'hiver pear shoots grown *in vitro. Physiol. Plant* **87,** 109–117.

Robinson, J. M. (1980). Lignin, land plants and fungi: Biological evolution affecting Phanerozoic oxygen balance. *Geology* **15,** 607–610.

Rogers, H. H., and Dahlman, R. C. (1993). Crop responses to CO_2 enrichment. *Vegetatio* **104/105,** 117–132.

Ros, J., and Tevini, M. (1995). Interaction of UV-radiation and IAA during growth of seedlings and hypocotyl segments of sunflower. *J. Plant Physiol.* **146,** 295–302.

Rozema, J. (1993). Plant responses to atmospheric carbondioxide enrichment: Interactions with some soil and atmospheric conditions. *Vegetatio* **104/105,** 173–192.

Rozema, J. (1995). Plants and high CO_2. *In* "Encyclopaedia of Environmental Biology," Vol. 3, pp. 129–138. Academic Press, San Diego.

Rozema, J., Arp, W., van Diggelen, J. van Esbroek, M., Broekman, R., and Punte, H. (1986). Occurrence and significance of vesicular arbuscular mycorrhiza in the salt marsh environment. *Acta Bot. Neerlandica* **35,** 457–467.

Rozema, J., Lenssen, G. M., and van de Staaÿ, J. W. M. (1990). The combined effects of increased atmospheric CO_2 and UV-B radiation on some agricultural and salt marsh species. *In* "The Greenhouse Effect and Primary Production in European Agroecosystems" (J. Goudriaan, H. van Keulen, and H. H. van Laar, eds.), pp. 68–71. Pudoc Wageningen.

Rozema, J., Lambers, H., van de Geijn, S. C., and Cambridge, M. L. (1993). "CO_2 and Biosphere." Kluwer Academic Publishers, Dordrecht.

Rozema, J., Tosserams, M., and Magendans, M. G. M. (1995). Impact of enhanced solar UV-B radiation on plants from terrestrial ecosystems. *In* "Climate Change Research Evaluation and Policy Implications. pp. 997–1004. (S. Zwerver, R. S. A. R. van Rompaey, M. T. J. Kok and Berk, M. M. eds.). Elsevier, Amsterdam.

Rozema, J., Gieskes, W. W. C., van de Geijn, S. C., Nolan, C., and de Boois, H. (1997a). "UV-B and Biosphere." Kluwer Academic Publishers, Dordrecht.

Rozema, J., Lenssen, G. M., van de Staaij, J. W. M., Tosserams, M., Visser, A. J. and Broekman, R. A. (1997b). Effects of UV-B radiation on terrestrial plants: Interaction with CO_2-enrichment. *Plant Ecol.* **128,** 182–191.

Rozema, J., Tosserams, M., Nelissen, H. J. M. van Heerwaarden, L., Broekman, R. A., and Flierman, N. (1997c). Stratospheric ozone reduction and ecosystem processes: Enhanced UV-B radiation affects chemical quality and decomposition of leaves of the dune grassland species *Calamagrostis epigeios. Plant Ecol.* **128,** 284–294.

Rozema, J., van de Staaij, J., Björn, L.-O., and Caldwell, M. (1997d). UV-B and environmental factor in plant life: Stress and regulation. *Trends Ecol. Evol.* **12,** 22–28.

Rozema, J., van de Staaij, J. W. M., and Tosserams, M. (1997e). Effects on UV-B radiation on plants from agro-ecosystems and natural ecosystems. *In* "Plants and UV-B: Responses to Environmental Change" (P. J. Lumsden, ed.), pp. 213–232. Cambridge University Press, Cambridge.

Rozema, J., Oudejans, A. M. C., van de Staaij, J., van Beem, A., Stroetenga, M., Broekman, R., Meijkamp, B., Nelissen, H., Stoevelaar, R., and Van Marum, D. (1998). The effects of enhanced UV-B radiation on structure, processes and feedbacks in terrestrial ecosystems. *In* "Role of Solar UV-B Radiation on Ecosystems" (D. P. Häder and S. C., Nolan, eds.) in press.

Ryel, R. J., Barnes, P. W., Beyschlag, W., Caldwell, M. M., and Flint, S. D. (1990). Plant competition for light analyzed with a multispecies canopy model. 1. Model development and influence of enhanced UV-B conditions on photosynthesis in mixed wheat and wild oat canopies. *Oecologia* **82,** 304–310.

Sanders, I. A. (1996). Plant–fungal interactions in a CO_2-rich world. *In* "Carbon Dioxide, Populations and Communities" (Chr. Körner and F. A. Bazzaz, eds.), pp. 265–272. Academic Press, San Diego.

Schlesinger, W. H. (1993). Response of the terrestrial biosphere to global climate change and human perturbation. *Vegetatio* **104/105,** 295–306.

Sonesson, M., Callaghan, T. V., and Björn, L. O. (1995). Short-term effects of enhanced UV-B and CO_2 on lichens at different latitudes. *Lichenologist* 27, 547–557.

Sonesson, M., Callaghan, T. V., and Carlsson, B. A. (1996). Effects of enhanced ultraviolet radiation and carbondioxide concentration on the moss *Hylocomium splendens. Global Change Biol.* **2,** 67–73.

Stafford, H. E. (1991). Flavonoid evolution: An enzymic approach. *Plant Physiol.* **96,** 680–685.

Stapleton, A. E. (1992). Ultraviolet radiation and plants: Burning questions. *Plant Cell* **4,** 1353–1358.

Strain, B. R., and Cure, J. D. (eds.). (1985). "Direct Effects of Carbon Dioxide on Vegetation," DOE/ER-0238, pp. 286. Department of Energy, Washington, DC.

Sullivan, J. H. (1997). Effects of increasing UV-B radiation and atmospheric CO_2 on photosynthesis and growth: Implications for terrestrial ecosystems. *Plant Ecol.* **128,** 194–206.

Sullivan, J. H., and Teramura, A. H. (1994). The effects of ultra-violet-B radation in loblolly pine. 3. Interaction with CO_2 enhancement. *Plant Cell Environ.* **17,** 311–317.

Teramura, A. H., and Sullivan, J. H. (1989). UV-B effects on terrestrial plants. *Photochem. Photobiol.* **50,** 479–487.

Teramura, A. H., and Sullivan, J. H. (1994). Effects of UV-B radiation on photosynthesis and growth of terrestrial plants. *Photosynth. Res.* **39,** 463–473.

Teramura, A. H., and Sullivan, J. H., and Ziska, L. H. (1990). The interaction of elevated UV-B radiation and CO_2 on productivity in rice, wheat and soybean. *Plant Physiol.* **94,** 470–475.

Tevini, M. M., and Teramura, A. H. (1989). UV-B effects in terrestrial plants. *Photochem. Photobiol.* **50,** 479–487.

Tingey, D. T., Johnson, M. G., Phillips, D. L., and Storm, M. J. (1995). Effects of elevated CO_2 and nitrogen on ponderosa pine fine roots and associated fungal components. *J. Biogeogr.* **22,** 281–287.

Tissue, D. T., and Oechel, W. C. (1987). Response of *Eriophorum vaginatum* to elevated CO_2 and temperature in the Alaskan tussock tundra. *Ecology* **68,** 401–402.

Tosserams, M., and Rozema, J. (1995). Effects of ultraviolet-B radiation (UV-B) on growth and physiology of the dune grassland species *Calamagrostis epigeios. Environm. Pollution* **89,** 209–214.

Tosserams, M., Pais de Sa, A., and Rozema, J. (1996). The effect of solar UV radiation on four plant species occurring in a coastal grassland vegetation in The Netherlands. *Physiol. Plant.* **97,** 731–739.

Tosserams, M., Magendans, G. W. H., and Rozema, J. (1997). Differential effects of elevated ultraviolet-B radiation on plant species from a dune grassland ecosystem. *Plant Ecol.* **128,** 266–281.

van de Staaij, J. W. M., Lenssen, G. M., Stroetenga, M., and Rozema, J. (1993). The combined effects of elevated CO_2 levels and UV-B radiation on growth characteristics of *Elymus athericus. Vegetatio* **104/105,** 433–439.

van de Staaij, J. W. M., Ernst, W. H. O., Hakvoort, H. W., and Rozema, J. (1995). Ultraviolet-B (280–320 nm) absorbing pigments in the leaves of *Silene vulgaris:* Their role in UV-B tolerance. *J. Plant Physiol.* **147,** 75–80.

Visser, A. J., Tosserams, M., Groen, M. W., Kalis, G., Kwant, R., Magendans, G. W. H., and Rozema, J. (1997a). The combined effect of CO_2 and supplemental UV-B radiation on faba bean. 3. Leaf optical properties, pigments, stomatal index and epidermal cell density. *Plant Ecol.* **128,** 208–222.

Visser, A. J., Tosserams, M., Groen, M. W., Magendans, G. W. H., and Rozema, J. (1997b). The combined effects of CO_2 and solar UV-B radiation on faba bean grown in open top chambers. *Plant Cell Environ.* **120,** 189–199.

Zepp, R. G., Callaghan, T. V., and Erickson, D. J. (1995). Effects of increased solar ultraviolet radiation on biogeochemical cycles. *Ambio* **24,** 181–187.

Ziska, L. H., and Teramura, A. H. (1992). CO_2 enhancement of growth and photosynthesis in rice (*Oryza sativa*): Modification by increased ultraviolet-B radiation. *Plant Physiol.* **99,** 473–481.

7

Role of Carbon Dioxide in Modifying the Plant Response to Ozone

Andrea Polle and Eva J. Pell

I. Introduction

The tropospheric concentrations of CO_2 and ozone (O_3), have increased from preindustrial levels of about 280 ppm and 0.005–0.01 ppm, respectively, to current concentrations of about 365 ppm and 0.015–0.030 ppm, respectively (Enquete Commission, 1995). Most predictions suggest that present ambient CO_2 concentrations will rise further, reaching gas-phase concentrations of about 650–700 ppm at the end of the next century (Roeckner, 1992). The extent of further changes in the ground-level O_3 concentrations is more uncertain. Measurements at Mount Hohenpeisenberg (Bavaria, Germany) at 2000 m above sea level showed an increase in mean O_3 concentrations from about 0.025 to 0.035 ppm during the last 20 yr (Enquete Commission, 1995). The IPCC report (1992) predicts that the tropospheric O_3 concentration will increase by 1% per year in the northern hemisphere. However, this figure may vary largely because the production of O_3 depends on the availability of nitrogen oxides (NO_x), is strongly affected by the presence of volatile hydrocarbons, and is driven by energy-rich solar radiation ($\lambda < 400$ nm) (Mehlhorn and Wellburn, 1994). Thus, O_3 formation is promoted by anthropogenic and to some extent also biogenic emissions. Chameides *et al.* (1994) predicted significant increases in the tropospheric O_3 concentrations in northern mid-latitudes with significant impact for harvest in agricultural ecosystems if rising anthropogenic NO_x emissions were not abated.

Both CO_2 and O_3 have direct effects on the growth and vitality of plants. Carbon dioxide is a limiting factor to photosynthesis in C_3 plants. Therefore, increases in ambient CO_2 concentrations can result in increased rates of photosynthesis and stimulate growth and biomass production (Gunderson and Wullschleger, 1994). By contrast, O_3 is potentially damaging to plants and, therefore, may counteract or prevent positive responses to enhanced atmospheric CO_2 concentrations. With respect to the question of carbon fluxes, the interactive effects of elevated CO_2 and O_3 on growth and vitality of plants and plant communities are important, because the vegetation is a major sink for CO_2. Its *sink strength,* particularly the maintenance of enhanced growth under elevated CO_2 concentrations, depends on a wide range of edaphic, climatic, and species-inherent factors. These factors are discussed elsewhere in this volume (see also Stitt, 1991; Mousseau and Saugier, 1993; Gunderson and Wullschleger, 1994). Further limitations to the sink strength of the vegetation are imposed by man-made environmental constraints including O_3 as an important air pollutant. Primary reactions of O_3 occur in above-ground plant parts and initially cause subtle biochemical changes, which may then initiate a sequence of reactions leading to reductions in photosynthesis, changes in carbon allocation patterns, symptoms of visible injury on leaves, accelerated senescence, reduced growth, and loss of yield. The negative effects of O_3 on plant performance and biomass production have been reviewed in detail at regular intervals (Heck *et al.,* 1982; Darrell, 1989; Cowling *et al.,* 1990; Bytnerowicz and Grulke, 1991; Chameides *et al.,* 1994; Matyssek *et al.,* 1995b; Rennenberg *et al.,* 1996; Sandermann *et al.,* 1997; Pell *et al.,* 1997).

The objective of this contribution is to consider the potential responses of the plant to interacting effects of O_3 and elevated CO_2. Two issues are relevant to this discussion. First, each gas will elicit some distinctive responses; understanding the biochemistry and physiology of those responses will aid in developing predictions of potential interactions. Second, the exposure dynamics of the two gases in nature are quite different, and as such prediction of potential interactions must consider these differences. Carbon dioxide concentrations are relatively constant over an extended period of growth and will have a long-term influence on plant morphology, physiology, and biochemistry. Ozone, on the other hand, is under constant flux within the environment, frequently being present at relatively low concentrations, which may provide some low-level oxidative stress; occasionally concentrations of the pollutant will rise to higher levels, which will be more stressful to the plant. Thus, the level of CO_2 in which the plant grows may influence the relative sensitivity or tolerance which that plant will have to the more variable stress imposed by O_3.

II. Evidence of Interactions between Ozone and Carbon Dioxide

A. Measures of Interactions between Ozone and Carbon Dioxide

Many experiments have been conducted to consider whether plants grown in environments with elevated CO_2 exhibit any interaction when also treated with O_3. There are many indicators of plant response but the most frequent indicators include foliar injury, altered growth, and changes in net photosynthesis. Several examples of such interactions follow.

Barnes and Pfirrmann (1992) exposed radish to combinations of ambient/elevated CO_2 (350/750 ppm) and ambient/elevated O_3(0.020/0.080 ppm) and found a reduction in the extent of leaf injury from 28 to 18% when plants exposed to the higher O_3 level were grown under elevated atmospheric CO_2 concentrations. Balaguer *et al.*, (1995) provided further evidence that when plants are grown in an atmosphere with elevated CO_2, O_3-induced visible injury could be reduced. Wheat plants grown in the presence of elevated CO_2 (700 ppm) and elevated O_3 (0.075 ppm) exhibited reductions in visible injury from 30% when plants were treated with O_3 alone to 5% in the presence of both gases. But elevated CO_2 did not afford commensurate protection to physiological processes studied, since the extent of O_3-induced reduction in photosynthesis, carbohydrate availability, and growth observed at elevated CO_2 was similar to that induced by O_3 at ambient CO_2 levels (Balaguer *et al.*, 1995).

Elevated CO_2 does not always afford plants protection from O_3 induction of foliar injury. When young Norway spruce trees were continuously exposed to combinations of ambient/elevated O_3 (0.02/0.08 ppm) and ambient/elevated CO_2 (355/750 ppm) for one growth phase (Polle *et al.*, 1993; Barnes *et al.*, 1995), needles from trees exposed to elevated O_3 developed symptoms of O_3 injury (chlorotic mottling), although these symptoms appeared with a delay of approximately 2 weeks on needles from plants grown with elevated CO_2 (Barnes *et al.*, 1995). In the autumn, at the end of the experimental treatments, no difference in the degree of injury was apparent between needles from trees grown under ambient or elevated atmospheric CO_2 concentrations. One year after cessation of the experiment and postculture in the field, all trees that had been exposed to O_3 suffered from severe needle loss and chromosomal aberrations in root tips, irrespective of the CO_2 concentration applied (Pfirrmann, 1992; Müller *et al.*, 1994). Mortensen (1995) also found extensive leaf injury (28–34%) in birch (*Betula pubescens*) exposed to elevated O_3 (0.062 ppm, 8 h per day for 35 days) in both plants grown under ambient or elevated CO_2 concentrations (350 versus 560 ppm).

The interactive effects of O_3 and elevated CO_2 on plant growth were further explored by McKee *et al.* (1995). Wheat plants were exposed to 700 ppm CO_2 and 0.060 ppm O_3. Performance of young, mature, and early-senescent flag leaves of these plants were compared with those of plants grown under ambient CO_2 and O_3 concentrations. Growth under elevated atmospheric CO_2 concentrations resulted in protection from O_3-induced loss of Rubisco protein, but only in the last stage of development (McKee *et al.*, 1995). In the presence of elevated CO_2, the O_3-induced loss in shoot biomass was prevented but, more importantly, the increase in grain yield was not maintained (McKee *et al.*, 1997a). Fiscus *et al.* (1997) demonstrated that soybean plants grown in 1.5 times ambient O_3 exhibited reductions in seed yield; the effects of O_3 were completely ameliorated when plants were also grown in an environment with 700 μmol mol^{-1} CO_2.

There is some evidence that the interaction between the two gases can have antagonistic effects. Kull *et al.* (1996) examined the effects of growing aspen clones in an environment supplemented with 150 ppm CO_2, coupled with O_3 exposures at twice ambient concentrations. When photosynthetic CO_2 response curves (A/C_i) were measured in the leaves of clones that had been grown with episodic O_3 exposures, both carboxylation efficiency and maximum photosynthesis were depressed. Plants grown in an environment with elevated CO_2 did not show any amelioration of the O_3 response. The experiments were performed with two clones, one sensitive and the other tolerant to O_3. It is interesting that the sensitive clone exhibited similar photosynthetic responses when plants were grown with O_3 alone, or with CO_2 and O_3 together. Ozone and elevated CO_2 elicited a greater effect on photosynthetic rate of the tolerant clone when compared with the response of foliage to O_3 alone or to plants grown in charcoal-filtered air at ambient CO_2 levels. It is not entirely clear that CO_2 exacerbated the O_3 effect in the tolerant clone since A/C_i curves were not determined with elevated CO_2 alone. However, it is clear that elevated CO_2 did not suppress the O_3 response.

B. Stomatal Regulation of Ozone–Carbon Dioxide Interaction

From the preceding discussion it is apparent that exposure to O_3 and elevated CO_2 does not elicit the same type of interaction in all plant species studied and for all parameters measured. Exposure to both gases is limited by stomatal conductance. Plant cuticles are relatively impervious to O_3 (Kerstiens and Lendzian, 1989). Thus, stomatal closure provides the first line of defense against O_3. Because many plant species grown under approximately doubled CO_2 concentrations have ≈30–40% lower stomatal conductance than plants grown under ambient CO_2 concentrations (Field *et al.*, 1995), O_3 uptake will be diminished under these conditions. Thus, we ask the question "Is CO_2-induced reduction in stomatal conductance sufficient to predict changes in O_3 phytotoxicity?"

Barnes and Pfirrman (1992) found that the reduction in leaf injury of radish was accompanied by a significant (>50%) reduction in stomatal conductance in plants exposed to elevated concentrations of both $CO_2 \times O_3$, as compared with those exposed only to elevated O_3. When soybeans were grown in elevated CO_2 the reduction in leaf conductance reduced O_3 flux by 35% in an environment that was 1.5 times that of ambient, thus accounting for the suppression of O_3-induced yield reduction. Two groups reported that partial protection from O_3 toxicity afforded to wheat plants grown in the presence of elevated CO_2 could be ascribed to reduced stomatal conductance as well (Balaguer *et al.*, 1995; McKee *et al.*, 1995, 1997b). Balaguer *et al.* (1995) reported that wheat plants grown in the presence of elevated CO_2 (700 ppm) and/or elevated O_3 (0.075 ppm) had a 40% decrease in stomatal conductance, which was estimated to result in an approximately 30% reduction in the effective O_3 dose. As stated earlier, the protection reflected by decreased foliar injury did not transfer to other physiological responses. While Kull *et al.* (1996) report that elevated CO_2 and O_3 induce reductions in stomatal conductance of aspen foliage, elevated CO_2 did not eliminate effects of O_3 on net photosynthesis.

The relationship between O_3 and CO_2 becomes further complicated since reductions in stomatal conductance are not a general phenomenon in plants grown under elevated CO_2 concentrations. The stomatal conductance rates of many tree species grown under elevated CO_2 concentrations were not diminished as compared with that of trees grown under ambient CO_2 concentrations, even after exposure periods lasting for several years (Norby and O'Neill, 1991; Bunce, 1992; Petterson and McDonald, 1992; Barton *et al.*, 1993; Gunderson *et al.*, 1993; Idso *et al.*, 1995; Teskey; 1995; Tschaplinski *et al.*, 1995). When young Norway spruce trees were continuously exposed to combinations of ambient/elevated O_3 (0.02/0.08 ppm) and ambient/elevated CO_2 (355/750 ppm) for one growth phase (Polle *et al.*, 1993; Barnes *et al.*, 1995), elevated CO_2 had no effect on stomatal aperture of 1-yr-old needles, whereas enhanced O_3 resulted in significant reductions in stomatal conductance in current-years needles of seedlings grown under ambient (−31%) and elevated CO_2 concentrations (−12%) (Pfirrmann, 1992).

Data on $O_3 \times CO_2$ interactions in tree species and herbaceous plants are far too limited to allow general conclusions. Growth strategies of plants may play a role in how a plant utilizes elevated concentrations of CO_2. Short-lived, fast-growing species, with abundant resources, for example, may respond very differently to increased levels of CO_2, than would slow-growing, long-lived species in an environment with sparse resources. The effects of CO_2 and O_3 on foliar health, as measured by visible appearance or photosynthetic rate, may also be influenced by the stage of development when analytical assessments were performed.

III. Mechanisms for Ozone by Carbon Dioxide Interactions

When CO_2 did not induce an effect on leaf conductance, O_3 seemed to elicit similar responses regardless of the level of CO_2. Furthermore, CO_2-induced reduction in stomatal conductance did not always ensure reduced O_3 responses. In an elevated CO_2 environment plants may sustain increases in growth rates, shifts in above-and belowground carbon partitioning, and/or changes in carbon invested in defense compounds targeted at oxidative stress. Any of these strategies could influence the vulnerability of tissue to O_3. We will focus on effect on photosynthetic systems and antioxidant status because of data currently available in which O_3/CO_2 interactions have been explored.

A. Impact of Carbon Dioxide on Ozone as Mediated by Effects on Rubisco

When plants grow in an elevated CO_2 environment, there are often associated increases in the rate of net photosynthesis, at least in the short term (Ceulemans and Mousseau, 1994; Gunderson and Wullschleger, 1994). Despite the apparent CO_2-induced increases in net photosynthesis, the associated content of ribulose-1,5-bisphosphate carboxylase-oxygenase (Rubisco) often declines (Tissue *et al.*, 1993, 1996). There is recent evidence that when pea plants were grown in elevated CO_2 there was a reduction in the steady-state transcript level for *rbc*S, transcript for the small subunit of Rubisco (Majeau and Coleman, 1996). The decline in Rubisco is attributed to acclimation that sometimes occurs when the enzyme is used more efficiently; instead resources can be invested in synthesizing other compounds that are limiting to photosynthesis or to nonphotosynthetic processes (Sage, 1994). It is also possible that a decline in *rbc*S is caused by an acceleration in leaf ontogeny (U. Sonnewald, personal communication).

Ozone can also induce reductions in Rubisco (Dann and Pell, 1989; Pell *et al.*, 1992, 1994), but the response is dependent on a very different set of mechanisms. It has been observed by several laboratories that when plants received acute exposures to O_3, foliage responds with a reduction in *rbc*S and *rbc*L, mRNA transcripts for the small and large subunits of Rubisco, respectively (Reddy *et al.*, 1993; Glick *et al.*, 1995; Bahl and Kahl, 1995; Conklin and Last, 1995; Rao *et al.*, 1995a). The importance of this reduction is not clear but since O_3 acts at the cell entrance, either the wall or membrane, any change in nuclear or chloroplastic transcript levels must reflect a regulated response (Pell *et al.*, 1997). A second and more important mechanism to explain O_3-induced loss of Rubisco relates to enhanced proteolysis of the protein (Eckardt and Pell, 1994). A series of reports demonstrates that when plants are exposed to O_3 it is likely that Rubisco undergoes oxidative modification, which ultimately leads to greater susceptibility to proteolysis (reviewed by Pell *et al.*, 1997). The oxidation that occurs in the plastid is not the result of direct reactions with O_3, since it is unlikely that the gas penetrates to the interior of the cell (Laisk *et al.*, 1989). Rather, O_3 could induce greater oxidative stress in the plastid through the migration of secondary active oxygen species (e.g., hydrogen peroxide) or by inducing a reduction in chloroplastic antioxidants (Conklin and Last, 1995; Pell *et al.*, 1997). Another possibility is that O_3 impairs sucrose export, thereby leading to sugar and starch accumulation in chloroplasts and associated reduction in Rubisco (Meyer *et al.*, 1997).

If elevated CO_2 had no influence on stomatal conductance, then we might predict that since O_3 and elevated CO_2 have different mechanisms of action the effects on Rubisco content would be additive. Alternatively, elevated CO_2 could induce reduced stomatal conductance while also inducing a reduction in the concentration of Rubisco. The reduced O_3 flux might lead to a less than an additive effect on the remaining Rubisco. If elevated CO_2 were to induce a reduction in Rubisco, indirect effects of this response such as changes in plant morphology might also influence exposure to O_3. Evans *et al.* (1994) demonstrated that tobacco plants transformed with antisense for *rbc*S, which had lower concentrations of Rubisco protein, had greater intercellular space in contrast to cell volume. Wiese and Pell (1997), using similarly transformed tobacco plants, demonstrated an inverse relationship between O_3-induced visual injury and Rubisco concentration. They also determined that transgenic plants had greater relative intercellular air space, and concluded that the morphological modification could have explained the enhanced symptom expression of these plants. There is evidence that plants grown under elevated CO_2 may, in fact, have thicker or denser leaves, suggesting smaller intercelluar air spaces (Luo *et al.*, 1994). If this were the case then susceptibility to O_3 might actually decline.

This section focused on changes in Rubisco in response to elevated CO_2 and relationship to O_3 sensitivity. Because of the ubiquity and importance of Rubisco, it has been the focus of considerable effort. This does not in any way suggest that Rubisco is a unique target for the interaction of the two gases. However, at present we do not know enough about interactions between O_3 and CO_2 on other possible targets to develop potential scenarios.

B. Influence of Ozone and Elevated Carbon Dioxide on Antioxidant Defense Strategies

When carbon is acquired in the leaf it is partitioned into starch, sucrose, and precursors to the shikimate pathway. The ultimate fate of fixed carbon, into carbohydates or alternative metabolites, will depend on environmental stimuli. Oxidative stress is the outcome of normal cellular metabolism, for example, during photosynthesis, respiration, fatty acid metabolism, etc. (Winston, 1990; Scandalios, 1993; Asada, 1994), as well as many aberrant environmental stimuli including pathogens, paraquat, and the air pollutant O_3. Ozone is probably one of the most toxic of all of these entities, and has the potential to induce production of hydroxyl radicals, singlet oxygen, superoxide, and hydrogen peroxide species (reviewed by Pell *et al.*, 1997).

Plants have evolved a spectrum of metabolites with antioxidative function such as ascorbate, glutathione, α-tocopherol, polyamines, and phenolics that can either scavenge reactive oxygen species directly or serve as substrates of defense enzymes (Alscher, 1989; Bors *et al.*, 1989; Fryer, 1992; Foyer, 1993). The defense enzymes such as superoxide dismutase (SOD), catalase and peroxidases can only detoxify the long-lived intermediates of oxygen metabolism, H_2O_2 and O_2^-, respectively.

The relationship between presence and induction of antioxidants and relative sensitivity of plants to O_3 is complex. The level of antioxidants produced in plants growing with elevated CO_2 and associated changes in carbon balance may be influenced by the duration of the exposure and longevity of the plant species. Consequently, the interaction between O_3 and CO_2 as influenced by antioxidants may not be immediately predictable.

1. Fate of Antioxidants in Response to Ozone The pioneering studies of Freebairn (1960) and Menser (1964) have shown that spraying of tobacco leaves or feeding of detached tobacco leaves with ascorbate before O_3 exposure protects foliage from injury. Later, it was demonstrated that EDU, *N*-[2-(2-oxo-1-imidazolidinyl)ethyl]-*N*′-phenylurea, also protects plants against O_3 injury (Carnahan *et al.*, 1978). Lee and Bennett (1982) found that EDU protection of snap beans correlates with increased SOD activity. Since the SOD has been considered one of the most important defenses from ozone (Bennett *et al.*, 1984) and responses of antioxidative systems, particularly those of ascorbate and SOD, to O_3 have been studied intensively. A widely held view (e.g., Kangasjärvi *et al.*, 1994) is that plants generally respond to O_3 with an induction of SOD activity and/or that of other antioxidative defenses. However, compilation of literature data reveals conflicting results for all components of the antioxidative system so far studied. For example, in spruce, increases (Castillo *et al.*, 1987; Schittenhelm *et al.*, 1993) as well as decreases (Polle *et al.*, 1993; Hausladen *et al.*, 1990), or no effects of O_3 (Polle and Rennenberg, 1991; Nast *et al.*, 1993) on SOD activity were observed. Herbaceous species also showed contrasting effects of O_3 on both activity and expression of SOD. Willekens *et al.* (1994) reported significant increases in m-RNA levels for SOD in *Nicotiana tabacum* in association with the appearance of injury. By contrast, under similar exposure conditions mRNA levels of CuZn-SODs of *N. plumbaginifolia* remained unaffected and those of Fe-SOD decreased (Willekens *et al.*, 1994). Reasons for the observed conflicting results might be that antioxidants vary with developmental stage, leaf age, and nutritional stage of the plant, and that the response to oxidative stress might interact with the exposure regime (Bowler *et al.*, 1992; Polle and Rennenberg, 1994).

The following example illustrates the complex interactions between O_3, leaf age, nutrition and antioxidants in leaves of an O_3-susceptible birch clone. The trees were grown for one season in open top chambers in filtered air with or without addition of O_3 (0.09/0.04 ppm day/night). One-half of the trees of each set was supplied regularly with a complete fertilizer solution; the others received a 10-fold diluted fertilizer solution, which caused significant symptoms of nutrient deficiencies and severe growth reduction (for details of the experimental exposures, see Matyssek *et al.*, 1995a). In birch trees grown in filtered air, SOD activity declined with increasing leaf age but only in those plants that had been sufficiently supplied with nutrients and not in starved plants. In well-supplied birch trees exposed to O_3, low SOD activity was found in young leaves, which displayed no visible injury and an increase in SOD activity in older leaves with increasing injury. By contrast, nutrient-deficient plants maintained higher SOD activities than well-supplied birch trees, irrespective of leaf age and exposure to filtered air or O_3. Matyssek and coworkers (1997) showed that O_3 caused an enhanced leaf turnover in well-supplied plants, so that carbon gain depended mainly on the healthy leaves. By contrast, starved plants maintained damaged tissues and depended on injured leaves for biomass production (Matyssek *et al.*, 1997). It is possible that elevated antioxidative protection in leaves of starved plants afforded prolonged leaf maintenance.

It has also been found that some tobacco cultivars equipped with high activities of SOD and glutathione reductase were less O_3 susceptible than cultivars with lower antioxidative protection (Shaaltiel *et al.*, 1988). An O_3-tolerant clone of ponderosa pine contained constitutively higher activities of SOD and ascorbate peroxidase than an O_3-sensitive clone (Benes *et al.*, 1995). The susceptible clone developed needle injury ("flecks") during the first year, whereas the tolerant clone showed injury only in the second year of O_3 exposure (Benes *et al.*, 1995). Taken together these observations suggest that plants which possess high antioxidative capacity before the onset of O_3 stress are better protected from oxidative damage.

In crop improvement programs, it might therefore be a desirable aim to breed or construct plants with an elevated antioxidative protection. Genetically engineered plants are also useful to establish cause-and-effect relationships. To date, plants overexpressing SOD, ascorbate peroxidase, glutathione reductase, and glutathione synthetase activities in the cytosol or chloroplast are available. The stress responses of plants overexpressing antioxidative enzymes have been intensively investigated and have recently been reviewed (Foyer *et al.*, 1994; Rennenberg and Polle, 1994; Allen, 1995). In general, genetically engineered plants overexpressing either glutathione reductase or SOD conferred tolerance to photo-oxidative stresses such as high light in combination with low temperatures and herbicides, whereas a glutathione reductase underexpressing tobacco transformant was more stress susceptible (Aono *et al.*, 1995). The ability of plants overexpressing antioxidative enzymes to sustain O_3 stress without exhibiting injury has varied: Van Camp *et al.* (1994) found that tobacco overexpressing chloroplastic SOD isozymes exhibited a reduction in visible injury from 25 to 5% when exposed to near-ambient O_3. Investigation of transgenic poplars overexpressing glutathione synthetase activities showed that these plants did not contain elevated glutathione contents, although the enzyme activity was significantly increased (Strohm *et al.*, 1995). Employing this system, it was found that the availability of cysteine and the activity of γ-glutamylcysteine synthetase limited glutathione production in the first place (Strohm *et al.*, 1995). Even when this limitation was overcome, poplar plants containing elevated glutathione contents were not more protected from O_3 injury than their nontransformed counterparts (B. Will, H. Rennenberg, A. Polle, unpublished results).

The reason for the variable results linking antioxidants and O_3 toxicity may relate to localization of the antioxidants within the cell. Pitcher *et al.* (1992) reported that in tobacco, while O_3 induced increases in activity of cytosolic SOD, chloroplastic SOD activity was unaffected. Transgenic tobacco overexpressing SOD in the chloroplast afforded no protection from O_3-induced foliar injury (Pitcher *et al.*, 1991), while overexpression in the cytosol did reduce lesion formation (Pitcher *et al.*, 1996).

As discussed earlier, O_3 probably does not penetrate the cell. Model calculation of Chameides (1989) suggested that concentrations of >1 m*M* ascorbate present in cell walls would be sufficient to prevent diffusion of O_3 into the cells. Therefore, apoplastic ascorbate has been considered a primary defense from O_3. The role of apoplastic ascorbate in O_3 detoxification is still a matter of intensive research. Foyer *et al.* (1995) reported that the poplar hybrid, *Populus tremula x alba,* was particularly O_3 sensitive. In this species, only dehydroascorbate, but not ascorbate, was detected in the apoplastic space (A. Polle, unpublished results). In spinach, O_3 exposure led to a rapid decline of the apoplastic ascorbate pool, suggesting that the protective effect of ascorbate might be limited because the regeneration of ascorbate was too slow to keep pace with the influx of O_3 (Luwe *et al.*, 1993). However, since the apoplastic ascorbate pool was also not replenished on cessation of O_3 exposure, the O_3 dose employed (0.30 ppm for 6 h) might have already damaged sensitive proteins in the plasma membrane, thereby preventing regeneration of the apoplastic ascorbate pool. By contrast, mature spruce trees from a high altitude, which are acclimated to O_3 and other environmental stresses, contain higher antioxidative protection including elevated apoplastic ascorbate levels, than trees from lower altitudes (Polle and Rennenberg, 1992; Polle *et al.*, 1995; Madamanchi *et al.*, 1991). In this species, exposure to O_3 did not always result in a depletion of the apoplastic ascorbate pool but high effective O_3 doses defined as the product of O_3 taken up into the leaf and exposure time: $g_s \times C_{a \times} t$, resulted in an increase in apoplastic ascorbate (Polle *et al.*, 1995). Similar observations were reported for spinach and beech fumigated with moderate O_3 concentrations (Luwe and Heber, 1995). It has, therefore, been suggested that apoplastic ascorbate might be an inducible defense system in some species (Polle *et al.*, 1995).

It is now becoming clear that there may be an important restriction to the protection afforded by apoplastic ascorbate. The reaction of O_3 with ascorbate yields dehydroascorbate and oxygen (Giamalva *et al.*, 1985). But oxygen formed by this reaction is stoichiometrically released as $^1O_2$ (Kanofsky and Sima, 1991, 1995a,b) which is also highly toxic. Ascorbate itself can quench $^1O_2$ with an apparent rate constant similar to that of its reaction with O_3 [$k(^1O_2) = 0.84$ to 1.06×10^7 $M^{-1}s^{-1}$, Rooney, 1983; k $(O_3) = 2$ to 6×10^7 $M^{-1}s^{-1}$, Giamalva *et al.*, 1985]. Therefore, ascorbate can still be considered a protectant, but the magnitude of this protection might be more limited than initially anticipated and depends on the maintenance of a highly reduced pool.

It is not yet clear how ascorbate is exported into the apoplastic compartment and where the regeneration of its oxidized intermediates takes place. Apoplastic systems for the reduction of ascorbate comparable to those inside the cell have not been found (Castillo and Greppin, 1988; Polle *et al.*, 1990; Luwe *et al.*, 1993). Recently, Horeman *et al.* (1994) localized a Cyt_b-protein in the plasma membrane, which mediates the reduction of extracellular monodehydroascorbate radicals by concurrent oxidation of intracellular ascorbate. Furthermore, an active transport of dehydroascorbate into protoplasts has been documented (Rautenkrantz *et al.*, 1994). Both systems would afford a direct coupling between apoplastic and symplastic antioxidants and explain why an enhanced cellular antioxidative capacity protects from O_3.

2. Antioxidative Systems in Plants Grown under Elevated CO_2 and Alleviation of Injury Since elevated CO_2 increases the internal availability of carbohydrate, it has been suggested that plants grown under such conditions might be better supplied with substrates for repair and detoxification (Carlson and Bazzaz, 1982). For example, leaves of sour orange trees grown under elevated CO_2 contain significantly higher concentrations of soluble carbohydrates than leaves from trees grown under ambient CO_2 (Gries *et al.*, 1993), and these increases correlate with increased foliar asorbate concentrations (Schwanz *et al.*, 1996a). Apparently, a higher supply of glucose, which is a precursor for the synthesis of ascorbate (Foyer, 1993), permits enhanced synthesis of ascorbate.

However, increases in antioxidants or protective enzymes are not commonly observed in plants grown under elevated CO_2 concentrations. In tree species, most components of the antioxidative system remained unaffected by elevated CO_2, with two exceptions, SOD and catalase activities (spruce, pine, oak, beech, poplar; see Polle, 1996; Polle *et al.*, 1993, 1997; Schwanz *et al.*, 1996a,b). The activities of these two enzymes were generally significantly reduced in comparison with controls grown under ambient CO_2 concentrations. The reductions in SOD activity were found in deciduous as well as in coniferous trees grown in growth chambers, glasshouses, and open top chambers. Decreased SOD activities were not the result of a transient response to high CO_2 concentrations because they were also observed in leaves of sour orange trees that had been grown for more than 7 yr in an atmosphere enriched with CO_2 (Schwanz *et al.*, 1996a). We have previously suggested that the decrease in SOD activity might be an acclimation to decreased oxidative stress encountered in plants grown under elevated CO_2 (Polle, 1996).

There is now increasing evidence that nutrient availability affects SOD activity. It is well known that growth of plants under enhanced CO_2 frequently results in a drop in the N:C ratio of the plant (see elsewhere in this volume). Therefore, we considered whether the nutrient supply affected foliar SOD activities in plants grown at ambient and elevated CO_2 concentrations. In beech seedlings grown with high, intermediate, and low nutrient supply rates, significant effects of nutrition and atmospheric CO_2 concentration on SOD activity were found. Although plants raised in elevated CO_2 generally had lower SOD activities than plants grown at ambient CO_2 concentrations, the differences were most pronounced in the intermediate nutritional situation and negligible under a high nutrient supply rate (Polle *et al.*, 1997). Hence, agricultural plants with relatively high N requirements fertilized at regular intervals may not show a CO_2-induced reduction in SOD activity. Actually, in tobacco and wheat plants grown under elevated CO_2 concentrations, the activity of SOD remained unaffected compared with control plants grown under ambient conditions (Havir and McHale, 1989; Rao *et al.*, 1995b). On the other hand, it is likely that tree species in their natural—normally N-limited—environments would respond with decreases in SOD activities to increasing CO_2 concentrations. This assumption is supported by the observation that leaves from mature *Quercus pubescens* trees grown along a transect at natural CO_2 springs in Italy show decreasing SOD activities with decreasing distance from the CO_2 source (Schwanz and Polle, 1998; for a description of CO_2 springs, see Miglietta *et al.*, 1993).

From the observation that plants with high antioxidative capacity should be better protected from oxidative stresses, the question arises of whether tree species grown at high CO_2 with reduced SOD activity are more susceptible to oxidative injury. Our group has employed a range of photo-oxidative stress factors such as drought stress, paraquat, and high light intensities with low temperature to test the stress tolerance of plants grown under ambient and elevated atmospheric CO_2 concentrations, respectively (Schwanz *et al.*, 1996b). We have consistently observed decreases in SOD activity in stressed plants from ambient conditions and increases in SOD activity in stressed plants from elevated CO_2 concentrations (e.g., Schwanz *et al.*, 1996b). It seems that plants grown under elevated CO_2 have an enhanced metabolic flexibility to encounter intrinsic oxidative stresses (Schwanz *et al.*, 1996b).

One example supports the notion that elevated CO_2 will provide protection from O_3-induced foliar injury. Rao *et al.* (1995b) investigated the effect of O_3 (0.12 ppm for 5 h per day) on the time course of SOD, ascorbate peroxidase. and glutathione reductase activities during the development of wheat plants grown for 5 weeks under ambient (390 ppm) and elevated (880 ppm) CO_2 concentrations. In the absence of O_3, differences in the activities of the antioxidative enzymes between plants grown under ambient or elevated CO_2 concentrations were not observed. When elevated O_3 was also present, the activities of antioxidative enzymes were initially increased compared to plants grown at ambient O_3 concentrations, irrespective of the CO_2 concentration (Rao *et al.*, 1995b). But after 2–3 weeks of O_3 exposure, the activities of antioxidative enxymes decreased in plants grown at ambient CO_2, whereas those of plants grown at elevated CO_2 continued to rise (Rao *et al.*, 1995b). Notably, O_3-induced leaf injury was prevented to a significant extent (36% in ambient CO_2 versus 10% in elevated CO_2), and the increase in biomass afforded by growth under elevated CO_2 was not adversely affected by the presence of O_3. Rao *et al.* (1995b) ascribed this improved protection of wheat plants to the observed increases in symplastic antioxidants. Since primary reactions of O_3 occur presumably in the apoplast, it is possible that increased symplastic protection also strengthens the capacity of apoplastic antioxidants by redox coupling. On the other hand, potential contributions of reduced O_3 fluxes to improved plant performance under elevated CO_2 have not been considered by Rao *et al.* (1995b). In contrast to the results with wheat, Polle *et al.* (1993) found significant reduction in SOD activity in spruce needles exposed for one growing season to elevated CO_2 (750 ppm) and/or O_3 (0.08 ppm) with only a slight delay in visible injury (see earlier discussion).

IV. Summary

Decline in O_3 sensitivity due to elevated CO_2 is probably an integration of factors including reduction in stomatal conductance, possible anatomical changes in the leaf, and flexibility to induce antioxidants when needed (Fig. 1). Developing a unified explanation for how O_3 and elevated CO_2 interact to elicit plant response is difficult because of the variable responses

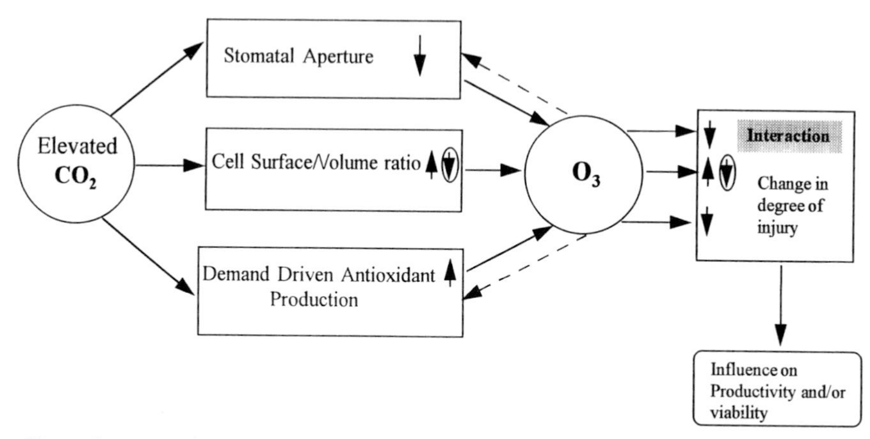

Figure 1 Influence of elevated CO_2 on stomatal aperture, leaf anatomy, and carbon availability for antioxidant production, as these parameters influence O_3 sensitivity.

reported in the literature. There is considerable evidence that when plants are grown in elevated CO_2, O_3-induced foliar symptoms decline; there are similar reports related to plant growth and yield. However, this type of interaction is not universal. There is a relationship between CO_2-induced reduction in stomatal conductance and diminished leaf response to O_3; however, once again, the interactions are not observed in all situations. In addition to the role that CO_2 plays in controlling stomatal conductance, and thereby O_3 flux, there are other mechanistic explanations for how these two gases may elicit an interacting plant response. Plants grown in elevated CO_2 induce changes in sizes of metabolite pools, for example, increases in carbohydrates and decreases in associated enzymes like Rubisco. In cases where elevated CO_2 does not reduce O_3 flux by closing stomates, any O_3 target that is in lower concenteration in elevated CO_2, like Rubisco, could become the subject of an adverse interaction between the gases.

Limited experiments have actually been performed to determine how elevated CO_2 influences status of antioxidants, and what role that might play in relative O_3 sensitivity. From data presented in this paper it appears that plants grown with elevated levels CO_2 may have greater flexibility to shift carbon to increase pools of antioxidants as needed. Thus, if plants grow in an environment with elevated CO_2 and chronic O_3 exposure, the foliage may be better able to limit long-term increases in oxidative stress.

Most of the discussion in this chapter is based on the premise that the plants are growing in an environment that is optimal for all factors including nutrient status. A three-way interaction among CO_2-O_3-nutrient stress is beyond the scope of this paper. However, it is well known that responses of plants to elevated CO_2 or O_3 are influenced substantially by nutrient availability. There is every reason to predict that if nutrients are not limiting, as in agricultural situations, CO_2 could have an ameliorating affect on O_3 toxicity. In natural ecosystems, where nutrients are more limiting, elevated CO_2 may be less likely to be able to allow the plant the flexibility to synthesize protective compounds and as a result offer less protection from O_3.

Acknowledgments

The collaboration in various projects and stimulating discussions with Drs. H. Rennenberg, C. Foyer, B. Kimball, S. Long, R. Matyssek, and A. Raschi are gratefully acknowledged. Financial support for these projects was provied by the Bayerisches Staatsministerium für Landesentwicklung und Umweltfragen, EUROSILVA, the German National Science Foundation (DFG), and the Commission of the European Community (EV5V-CT94-0432, EV5V-CT94-0438). Additional support was provided in part by the Pennsylvania Agricultural Experiment Station and the Environmental Resources Research Institute, The Pennsylvania State University.

References

Allen, R. D. (1995). Dissection of oxidatives stress tolerance using transgenic plants. *Plant Physiol.* **107,** 1049–1054.

Alscher, R. G. (1989). Biosynthesis and antioxidant function of glutathione in plants. *Physiol. Plant.* **77,** 457–464.

Aono, M., Saji, H., Fuijama, K., Sugita, M., Kondo, N., and Tanaka, K. (1995). Decrease in activity of glutathione reductase enhances paraquat sensitivity in transgenic *Nicotiana tabacum. Plant Physiol.* **197,** 645–48.

Asada, K. (1994). Production and action of reactive oxygen species in photosynthetic tissues. *In* "Causes of Photo-Oxidative Stress and Amelioration of Defense Systems in Plants" (C. Foyer and P. Mullineaux, eds.), pp. 77–105. CRC Press, Boca Raton, FL.

Bahl, A., and Kahl, G. (1995). Air pollutant stress changes the steady-state transcript levels of three photosynthesis genes. *Environ. Pollut.* **88,** 57–65.

Balaguer, L., Barnes, J. D., Panucucci, A., and Borland, A. M. (1995). Production and utilisation of assimilates in wheat (*Triticum aestivum* L.) leaves exposed to elevated O_3 and/or CO_2. *New Phytol.* **129,** 557–568.

Barnes, J., and Pfirrmann, T. (1992). The influence of CO_2 and O_3, singly and in combination, on gas exchange, growth and nutrient status of radish (*Raphanus sativus,* L.). *New Phytol.* **121,** 403–412.

Barnes, J. D., Pfirrmann, T., Steiner, K., Lütz, C., Busch, U., Küchenhoff, H., and Payer, H. D. (1995). Effects of elevated CO_2, elevated O_3, and potassium deficiency on Norway spruce [*Picea abies* (L.) Karst.]: Seasonal changes in photosynthesis and non-structural carbohydrate content. *Plant Cell Environ.* **18,** 1345–1357.

Barton, C. V., Lee, H. S., and Jarvis, P. G. (1993). A branch bag and CO_2 control system for long-term CO_2 enrichment of mature Sitka spruce (*Picea sitchensis* (Bong.) Carr.). *Plant Cell Environ.* **16,** 1139–1148.

Benes, S. E., Murphy, T. M., Anderson, P. D., and Houpis, J. L. (1995). Relationship of antioxidant enzymes to ozone tolerance in branches of mature ponderosa pine (*Pinus ponderosa*) trees exposed to long-term, low concentration, ozone fumigation and acid precipitation. *Physiol. Plant.* **94,** 124–134.

Bennett, J., Lee, E., and Heggestad H. (1984). Biochemical aspects of plant tolerance to ozone and oxyradicals: superoxide dismutase. *In* "Gaseous Air Pollutants and Plant Metabolism" (M. Koziol and F. Whatley, eds.), pp. 413–423. Butterworths, New York.

Bors, W., Langebartels, C., Michael, C., and Sandermann, H. (1989). Polyamines as radical scavengers and protectants against ozone damage. *Phytochem.* **28,** 1589–1595.

Bowler, C., Van Montagu, M., and Inzé, D. (1992). Superoxide dismutase and stress tolerance. *Annu. Rev. Plant Physiol. Plant Mol. Biol.* **43,** 83–116.

Bunce, J. A. (1992). Stomatal conductance, photosynthesis and respiration of temperate deciduous tree seedlings grown outdoors at an elevated concentration of carbon dioxide. *Plant Cell Environ.* **15,** 541–549.

Bytnerowicz, A., and Grulke, N. E. (1991). Physiological effects of air pollutants on western trees. *In* "The Response of Western Forests to Air Pollution" (R. K. Olson, D. Binkley and M. Böhm, eds.), pp. 183–232. Springer Verlag, Heidelberg.

Carlson, R. W., and Bazzaz, F. A. (1982). Photosynthetic and growth response to fumigation with SO_2 and elevated CO_2 in C_3 and C_4 plants. *Oecologia* **54,** 50–54.

Carnahan, J. E., Jenner, E. L., and Wat, E. K. (1978). Prevention of ozone injury to plants by a new protectant chemical. *Disease Contr. Pest Manag.* **68,** 1225–1229.

Castillo, F., and Greppin, H. (1988). Extracellular ascorbic acid and enzyme activities related to ascorbic acid metabolism in *Sedum album* leaves after ozone exposure. *Exp. Environ. Bot.* **28,** 231–238.

Castillo, F. J., Miller, P. R., and Greppin, H. (1987). 'Waldsterben' extracellular biochemical markers of photochemical air pollution damage to Norway spruce. *Experientia* **43,** 111–115.

Ceulemans, R., and Mousseau, M. (1994). Tansley Review No. 71, Effects of elevated atmospheric CO_2 on woody plants. *New Phytol.* **127,** 425–446.

Chameides, W. (1989). The chemistry of ozone deposition to plant leaves: Role of ascorbic acid. *Environ. Sci. Technol.* **23,** 595–600.

Chameides, W. L., Kasibhlata, P. S., Yienger, J., and Levy II, H. (1994). Growth of continental-scale metro-agroplexes, regional ozone pollution, and world food production. *Science* **264,** 74–77.

Conklin, P. L., and Last, R. L. 1995. Differential accumulation of antioxidant mRNAs in *Arabidopsis thaliana* exposed to ozone. *Plant Physiol.* **109,** 203–212.

Cowling, E., Shriner, J., Barnard, A., Lucier, A., Johnson, A., and Kiester, A. (1990). Air borne chemicals and forest health in the United States. *In* "Report from 19th IUFRO World Congress," Vienna, pp. 25–36.

Dann, M. S., and Pell, E. J. (1989). Decline of activity and quantity of ribulose bisphosphate carboxylase/oxygenase and net photosynthesis in ozone-treated foliage. *Plant Physiol.* **91,** 427–432.

Darrell, N. M. (1989). The effect on air pollutants on physiological processes in plants. *Plant Cell Environ.* **12,** 1–30.

Eckardt, N. A., and Pell, E. J. 1994. O_3-induced degradation of Rubisco protein and loss of Rubisco protein and loss of Rubiscoo mRNA in relation of leaf age in potato. *New Phytol.* **127,** 741–748.

Enquete Commission. (1995). "Protection in the Earth's Atmosphere," pp. 22–36. Economia Verlag, Bonn.

Evans, J. R. v. Caemmerer, S., Setchell, B. A., and Hudson, G. S. (1994). The relationship between CO_2 transfer conductance and leaf antatomy in transgenic tobacco with a reduced content of rubisco. *Austrl. J. Plant Physiol.* **21,** 475–495.

Field, C. B., Jackson, R. B., and Mooney, H. A. (1995). Stomatal responses to increased CO_2: Implications from the plant to the global scale. *Plant Cell Environ.* **18,** 1214–1225.

Fiscus, E. L., Reid, C. D., Miller, J. E., and Heagle, A. S. (1997). Elevated CO_2 reduces O_3 flux and O_3-induced yield losses in soybeans: Possible implications for elevated CO_2 studies. *J. Exp. Bot.* **48,** 307–313.

Foyer, C. (1993). Ascorbic acid. *In* "Antioxidants in Higher Plants" (Alscher, R. G. and Hess, J. L., eds.), pp. 31–58. CRC Press, Boca Raton, FL.

Foyer, C., Descourvières, P., and Kunert, K. J. (1994). Protection against oxygen radicals: An important defence mechanism studied in transgenic plants. *Plant Cell Environ.* **17,** 507–523.

Foyer, C., Jouanin, L., Souriau, N., Perret, S., Lelandais, M., Kunert, K.-J., Noctor, G., Pruvost, C., Strohm, M., Mehlhorn, H., Polle, A., and Rennenberg, H. (1995). The molecular, biochemical and physiological function of glutathione and its action in poplar. *In* "EUROSILVA Contribution to Forest Tree Physiology" (H. Sandermann and M. Bonnet-Masimbert, eds.), pp. 141–170. INRA Edition, Paris.

Freebairn, H. T. (1960). The prevention of air pollution damage to plants by use of vitamin C sprays. *J. Air Poll. Control. Assoc.* **10,** 314–137.

Fryer, M. J. (1992). The antioxidant effects of thylakoid vitamin E (a-tocopherol). *Plant Cell Environ.* **15,** 381–392.

Giamalva, D., Church, D., and Pryor, W. (1985). A comparison of the rate of ozonization of biological antioxidants and oleate and linoleate esters. *Biochem. Biophys. Res. Comm.* **133,** 773–779.

Glick, R. E., Schlagnhaufer, C. D., Arteca, R. N., and Pell, E. J. (1995). Ozone-induced ethylene emission accelerates the loss of ribulose-1,5-bisphosphate carboxylate/oxygenase and nuclear-encoded mRNAs in senescing potato leaves. *Plant Physiol.* **109,** 891–898.

Gries, C., Kimball, B. A., and Idso, S. B. (1993). Nutrient uptake during the course of a year by sour orange trees growing in ambient and elevated atmospheric carbon dioxide concentrations. *J. Plant Nutr.* **16,** 129–147.

Gunderson, C. A., and Wullschleger, S. D. (1994). Photosynthetic acclimation in trees to rising atmospheric CO_2: A broader perspective. *Photosynth. Res.* **39,** 369–388.

Gunderson, C. A., Norby, R. J., and Wullschleger, S. D. (1993). Foliar gas exchange responses of two deciduous hardwoods during 3 years of growth in elevated CO_2: No loss of photosynthetic enhancement. *Plant Cell Environ.* **16,** 797–807.

Hausladen, A., Madamanchi N., Fellows, S., Alscher, R., and Amundson, R. (1990). Seasonal changes in antioxidants in red spruce as affected by ozone. *New Phytol.* **115,** 447–456.

Havir, E. A., and McHale, N. A. (1989). Regulation of catalase activity in leaves of *Nicotiana sylvestris* by high CO_2. *Plant Physiol.* **89,** 952–957.

Heck, W. W., Taylor, O. C., Adams, R., Bingham, G., Miller, J., Preston, E., and Weinstein, L. (1982). Assessment of crop loss from ozone. *J. Air. Poll. Contr. Assoc.* **32,** 353–361.

Horeman, N., Asard, H., and Caubergs, R. (1994). The role of ascorbate free radical as an electron acceptor to cytochrome *b*-mediated trans-plasma membrane electron transport in higher plants. *Plant Physiol.* **104,** 1455–1458.

IPCC. (1992). Climate change 1992). *In* "The Supplementary Report to the IPCC Scientific Assessment" (J. T. Houghton, B. A. Callander, and S. K. Varney, eds.). IPCC, Cambridge University Press, Cambridge.

Kangasjärvi, J., Talvinen J., Utriainen, M., and Karjalainen, R. (1994). Plant defence systems induced by ozone. *Plant Cell Environ.* **17,** 783–794.

Kanofsky, J., and Sima, P. (1991). Singlet oxygen production from the reactions of ozone with biological molecules. *J. Biol. Chem.* **266,** 9039–9042.

Kanofsky, J. R., and Sima, P. D. (1995a). Reactive absorption of ozone by aqueous biomolecule solutions: Implications for the role of sulfhydryl compounds as targets for ozone. *Arch. Biochem. Biophys.* **316,** 52–62.

Kanofsky, J. R., and Sima, P. D. (1995b). Singlet oxygen generation from the reaction of ozone with plant leaves. *J. Biol. Chem.* **270,** 7850–7852.

Kerstiens, G., and Lendzian, K. (1989). Interactions between ozone and plant cuticles. I. Ozone deposition and permeability. *New Phytol.* **112,** 13–19.

Kull, O., Sober, A., Coleman, M. D., Dickson, R. E., Isebrands, J. G., Gagnon, Z., and Karnosky, D. F. (1996). Photosynthetic responses of aspen clones to simultaneous exposures of ozone and CO_2. *Can. J. For. Res.* **26,** 639–648.

Laisk, A., Kull, O., and Moldau, H. (1989). Ozone concentration in leaf intercellular air spaces is close to close to zero. *Plant Physiol.* **90,** 1163–1167.

Lee, E. H., and Bennett, J. H. (1982). Superoxide dismutase. A possible protective enzymes against ozone injury in snap beans (*Phaseolus vulgaris* L.). *Plant Physiol.* **69,** 1444–1449.

Luo, Y., Field, C. B., and Mooney, H. A. (1994). Predicting responses of photosynthesis and root fractions to elevated CO_2: Interactions among carbon, nitrogen, and growth, theoretical paper. *Plant Cell Environ.* **17,** 1195–1204.

Luwe, M., and Heber, U. (1995). Ozone detoxification in the apoplast and symplast of spinach, broad bean and beech leaves at ambient and elevated concentrations of ozone in air. *Planta* **197,** 448–455.

Luwe, M., Takahama, U., and Heber, U. (1993). Role of ascorbate in detoxifying ozone in the apoplast of spinach (*Spinacia oleracea*) leaves. *Plant Physiol.* **101,** 969–976.

Madamanchi, N., Hausladen, A., Alscher, R., Amundson, J., and Fellows, S. (1991). Seasonal changes in antioxidants in red spruce (*Picea rubens* Sarg.) from three field sites in the northeastern United States. *New Phytol.* **118,** 331–338.

Majeau, N., and Coleman, J. R. (1996). Effect of CO_2 concentration on carbonic anhydrase and ribulose-1,5-bisphosphate carboxylase/oxygenase expression in pea. *Plant Physiol.* **112,** 569–574.

Matyssek, R., Pfanz, H., and Lomsky, B. (1995a). Nutrition and the ecophysiology of spruce and birch under SO_2 and O_3 impact. *In* "EUROSILVA Contribution to Forest Tree Physiology" (H. Sandermann and M. Bonnet-Masimbert, eds.), pp. 119–140. INRA Edition, Paris.

Matyssek, R., Reich, P., Oren, R., and Winner, W. E. (1995b). Response mechanisms of conifers to air pollutants. *In* "Ecophysiology of Coniferous Forests" (W. K. Smith, and T. M. Hinckley, eds.), pp. 255–308. Academic Press, New York.

Matyssek, R., Maurer, S., Gunthardt-Goerg, M., Landoldt, W., Saurer, M., and Polle, A. (1997). Nutrition determines the 'strategy' of *Betula pendula* for coping with ozone stress. *Phyton* **37,** 157–168.

McKee, I. F., Farage, P. K., and Long, S. P. (1995). The interactive effect of elevated CO_2 and O_3 concentration on photosynthesis in spring wheat. *Photosynth. Res.* **45,** 111–119.

McKee, I. F., Bullimore, J. F., and Long, S. P. (1997a). Will elevated CO_2 protect the yield of wheat from O_3 damage? *Plant Cell Environ.* **20,** 77–84.

McKee, I. F., Eiblmeier, M., and Polle, A. (1997b). Enhanced ozone-tolerance in wheat grown at an elevated CO_2 concentration: Ozone exclusion and ozone detoxification. *New Phytol.* **137,** 275–284.

Mehlhorn, H., and Wellburn, A. (1994). Man-induced causes of free radical damage: O_3 and other gaseous air pollutants. *In* "Causes of Photo-Oxidative Stress and Amelioration of Defense Sysems in Plants" (C. Foyer, and P. Mullineaux, eds.), pp. 155–175. CRC Press, Boca Raton, FL.

Menser, A. (1964). Response of plants to air pollutants: III. A relation between ascorbic acid levels and ozone susceptibility of light pre-conditioned tobacco plants. *Plant Physiol.* **39,** 564–567.

Meyer, U., Köllner, B., Willenbrink, J., and Krause, G. H. M. (1997). Physiological changes on agricultural crops induced by different ambient ozone expossure regimes. I. Effects on photosynthesis and assimilate allocation in spring wheat. *New Phytol.* **136,** 645–652.

Miglietta, F., Raschi, A., Bettarini, I., Resti, R., and Salvi, F. (1993). Natural CO_2 springs in Italy: A Resource of examining long-term response of vegetation to rising atmospheric CO_2 concentrations. *Plant Cell Environ.* **16,** 873–878.

Montensen, L. M. (1995). Effect of carbon dioxide concentration on biomass production and partitioning in Betula pubescens Ehrh. seedlings at different ozone and temperature regimes. *Environ. Poll.* **87,** 557–343.

Mousseau, M., and Saugier, B. (1993). The direct effect of increased CO_2 on gas exchange and growth of forest tree species. *J. Exp. Bot.* **43,** 1121–1130.

Müller, M., Köhler, B., Grill, D., Guttenberger, H., and Lütz, C. (1994). The effect of various soils, different provenances and air pollution on root tip chromosomes in Norway spruce. *Trees* **9,** 73–79.

Nast, W., Mortensen, L., Fischer, K., and Fitting, I. (1993). Effects of air pollutants on the growth and antioxidative system of Norway spruce exposed in open-top chambers. *Environ. Poll.* **80,** 85–90.

Norby, R. J., and O'Neil, E. G. (1991). Leaf area compensation and nutrient interactions in CO_2-enriched seedlings of yellow-poplar (*Liriodendron tulipifera* L.) *New Phytol.* **117,** 515–528.

Pell, E. J., Eckardt, N., and Enyedi, A. J. (1992). Timing of ozone stress resulting status of ribulose bisphosphate carboxylase/oxygenase and associated net photosynthesis. *New Phytol.* **120,** 397–405.

Pell, E. J., Eckardt, N. A., and Glick, R. e. (1994). Biochemical and molecular basis for impairment of photosynthetic potential. *Photosyn. Research* **39,** 453–462.

Pell, E. J., Schlagnhaufer, C. D, and Arteca R. N. (1997). Ozone-induced oxidative stress: Mechanisms of action and reaction. *Physiol. Plant.* **100,** 264–273.

Petterson, R., and McDonald, J. S. (1992). Effects of elevated carbon dioxide concentration on photosynthesis and growth of small birch plants (*Betula pendula* Roth.) at optimal nutrition. *Plant Cell Environ.* **15,** 911–919.

Pfirrmann, T. (1992). Wechselwirkugen von Ozon, Kohlendioxid und Wassermangel bei zwei Klonen unterschiedlich mit Kalium ernährter Fichten. *Giessener Bodenkundliche Abhandlungen,* Giessen.

Pitcher, L. H., and Zilinskas, B. A. (1996). Overexpression of copper/zinc superoxide dismutase in the cytosol of transgenic tobacco confers partial resistance to ozone-induced foliar necrosis. *Plant Physiol.* **110,** 583–588.

Pitcher, L. H., Brennan, E., Hurley, A., Dunsmuir, P., Tepperman, J. M., and Zilinskas, B. A. (1991). Overproduction of petunia chloroplastic copper/zinc superoxide dismutase does not confer ozone tolerance in transgenic tobacco. *Plant Physiol.* **97,** 452–455.

Pitcher, L. H., Brennan, E., and Zilinskas, B. A. (1992). The antiozonant ethylene diurea does not act via superoxide dismutase induction in bean. *Plant Physiol.* **99,** 1388–1392.

Polle, A. (1996). Protection from oxidative stress in trees as affected by elevated CO_2 and environmental stress. *In* "Terrestrial Ecosystem Response to Elevated CO_2 "(G. Koch, and H. Mooney, eds.), pp. 299–315. Academic, Press, New York.

Polle, A., and Rennenberg, H. (1991). Superoxide dismutase activity in needles of Norway spruce (*Picea abies* L., Karst.) and in Scots pine (*Pinus sylvestris,* L.) under field and chamber conditions: Lack of ozone effects. *New Phytol.* **117,** 335–343.

Polle, A., and Rennenberg, H. (1992). Field studies on Norway spruce trees at high altitudes. II. Defence systems against oxidative stress in needles. *New Phytol.* **121,** 635–642.

Polle, A., and Rennenberg, H. (1994). Photooxidative stress in trees. *In* "Causes of Photo-Oxidative Stress and Amelioration of Defense Systems in Plants" (Foyer, C. and Mullineaux, P., eds.) pp. 199–218. CRC Press, Boca Raton, FL.

Polle, A., Chakrabarti, K., Schürmann, W., and Rennenberg, H. (1990). Composition and and properties of hydrogen peroxide decomposing systems in extracellular and total extracts from needles of Norway spruce (*Picea abies,* L. Karst.). *Plant Physiol.* **94,** 312–319.

Polle, A., Pfirrmann, T., Chakrabarti, S., and Rennenberg, H. (1993). The effects of enhanced ozone and enhanced carbon dioxide concentrations on biomass, pigments and antioxidative enzymes in spruce needles (*Picea abies,* L.). *Plant Cell Environ.* **16,** 311–316.

Polle, A., Wieser, G., and Havranek, W. M. (1995). Quantification of ozone influx and apoplastic ascorbate content in needles of Norway spruce trees (*Picea abies* L. Karst) at high altitude. *Plant Cell Environ.* **18,** 681–688.

Polle, A., Eiblmeir, M., Sheppard, L., and Murray, M. (1997). Respones of antioxidative enzymes to elevated CO_2 in leaves of beech (*Fagus sylvativa,* L.) seedlings grown under a range of nutrient regimes. *Plant Cell Environ.* **20,** 1317–1321.

Rao, M. V., Paliyath, G., and Ormrod, D. P. (1995a). Differential response of photosynthetic pigments, Rubisco activity and Rubisco protein of *Arabidopsis thaliana* exposed to UVB and ozone. *Photochem. Photobiol.* **62,** 727–735.

Rao, M. V., Hale, B. A., and Ormrod, D. P. (1995b). Amelioration of ozone-induced oxidative damage in wheat plants grown under high carbon dioxide. Role of antioxidant enzymes. *Plant Physiol.* **109,** 421–432.

Rautenkrantz, A., Li, L., Mächler E., Martinoia, E., and Oertli, J. (1994). Transport of ascorbic and dehydroascorbic acids across protoplast and vacuole membranes isolated from barley (*Hordeum vulgare* L. cv Gerbel) leaves. *Plant Physiol.* **106,** 187–193.

Reddy, G. N., Arteca, R. N., Dai, Y.-R., Flores, H. F., Negm, F. B., and Pell, E. J. (1993). Changes in ethylene and polyamines in relation to mRNA levels of the large and small subunits of ribulose bisphosphate carboxylase/oxygenase in O_3-stressed potato foliage. *Plant Cell Environ.* **16,** 819–826.

Rennenberg, H., and Polle, A. (1994). Protection from oxidative stress in transgenic plants. *Biochem. Soc. Trans.* **22,** 936–940.

Rennenberg, H., Herschbach, C., and Polle, A. (1996). Consequences of air pollution on shoot–root interaction. *J. Plant Physiol.* **148,** 296–301.

Roeckner, E. (1992). Past, present and future levels of greenhouse gases in the atmosphere and model projections of related climatic changes. *J. Exp. Bot.* **43,** 1097–1109.

Rooney, M. (1983). Ascorbic acid as photooxidation inhibitor. *Photochem. Photobiol.* **38,** 619–621.

Sage, R. F. (1994). Acclimation of photosynthesis to increasing atmospheric CO_2: The gas exchange perspective. *Photosyn. Res.* **39,** 351–368.

Sandermann H., Wellburn, A., and Heath, R. (eds.). "Forest decline and ozoone: A comparison of controlled chamber and field experiments, Ecological Studies." Springer Verlag, Berlin,

Scandalios, J. (1993). Oxygen stress and superoxide dismutases. *Plant Physiol.* **101,** 7–12.

Schittenhelm, J., Westphal, S., Toder, S., and Wagner, E. (1993). Das antioxidative System der Fichte: Einfluß von verschiedenen Streßfaktoren. Forstw. Cbl. **112,** 240–250.

Schwanz, P., Kimball, B. A., Idso, S. B., Hendrix, D., and Polle, A. (1996a). Antioxidants in sun- and shade leaves of sour orange trees (*Citrus aurantinum*) after long-term acclimation to elevated CO_2. *J. Exp. Bot.* **47,** 1941–1950.

Schwanz, P., Picon, C., Vivin, P., Dreyer, E., Guehl, J. M., and Polle, A. (1996b). Responses of antioxidative systems to drought stress in pendunculate oak and maritime pine as modulated by elevated CO_2. Plant Physiol. **110,** 393–402.

Schwanz, P., and Polle, A. (1998). Antioxidative systems, pigment and protein contents in leaves of adult mediterranean oak species (*Quercus pubescens* and *Quercus ilex*) with lifetime exposure to elevated CO_2. *New Phytologist* **140,** 411–423.

Shaaltiel, Y., Glazer, A., Bozion, P., and Gressel, J. (1988). Cross tolerance to herbicidal and environmental oxidants of plant biotypes tolerant to paraquat, sulfur dioxide and ozone. *Pest Biochem. Physiol.* **31,** 13–33.

Stitt, M. (1991). Rising CO_2 levels and their potential significance for carbon flow in photosynthetic cells. *Plant Cell Environ.* **14,** 741–742.

Strohm, M., Jouanin, L., Kunert, K.-J., Pruvost, C., Polle, A., Foyer, C. H., and Rennenberg, H. (1995). Regulation of glutathione synthesis in leaves of transgenic poplar (*Populus tremula x P. alba*) overexpressing glutathione synthetase. *Plant J.* **7,** 141–145.

Teskey, R. O. (1995). A field study of the effects of elevated CO_2 on carbon assimilation, stomatal conductance and leaf and branch growth of *Pinus taeda* trees. *Plant Cell Environ.* **18,** 565–573.

Tissue, D. T., Thomas, R. B., and Strain, B. R. (1993). Long-term effects of elevated CO_2 and nutrients on photosynthesis and Rubisco in loblolly pine seedlings. *Plant Cell Environ.* **16,** 859–865.

Tissue, D. T., Thomas, R. B., and Strain, B. R. (1996). Growth and photosynthesis of loblolly pine (*Pinus taeda*) after exposure to elevated CO_2 for 19 months in the field. *Tree Physiol.* **16,** 49–59.

Tschaplinski, T. J., Stewart, D. B., Hanson, P. J., and Norby, R. J. (1995). Interactions between drought and elevated CO_2 on growth and gas exchange of seedlings of deciduous tree species. *New Phytol.* **129,** 63–71.

Van Camp, W., Willekens, H., Bowler, C., Van Montagu, M., Reupold-Popp, P., Sandermann, H., and Langebartels, C. (1994). Elevated levels of superoxide dismutase protect transgenic plants against ozone damage. *Bio/Technol.* **12,** 165–168.

Wiese, C. B., and Pell, E. J. (1997). Influence of ozone on transgenic tobacco plants expressing reduced quantities of Rubisco. *Plant Cell and Environ* **20,** 1283–1291.

Willekens, H., Van Camp, Van Montagu, M., Inzé, D., Langebartels, C., and Sandermann, H. (1994). Ozone, sulfur dioxide, ultraviolet B have similar effects on mRNA accumulation of antioxidant genes in *Nicotiana plumbaginifolia* L. *Plant Physiol.* **106,** 1007–1014.

Winston, G. (1990). Physiochemical basis for free radical formation in cells: production and defenses. *In* "Stress Responses in Plants: Adaptation and Acclimation Mechanisms" (R. Alscher and J. Cumming, eds.), pp. 57–86. Wiley Liss, New York.

8

Response of Plants to Elevated Atmospheric CO_2: Root Growth, Mineral Nutrition, and Soil Carbon

Hugo H. Rogers, G. Brett Runion, Stephen A. Prior, and H. Allen Torbert

I. CO_2 Response

The rise in atmospheric CO_2, due mainly to fossil fuel combustion and land use change, is an undisputed fact. This ongoing CO_2 increase has important implications for vegetation. Plant growth is typically enhanced by elevated CO_2. Carbon dioxide is the substrate for photosynthesis and, when elevated, both carbon assimilation and water use efficiency generally increase. Stimulation of root system development associated with increased growth implies more rooting, which, in turn, implies the possibility of increased water and nutrient capture. Microbes mediate C and nutrient flows within the soil, and CO_2-induced changes in the structure and function of plant root systems may lead to changes in the microbiology of both rhizosphere and soil. Enhanced plant growth further suggests greater delivery of C to soil, and thus potentially greater soil C storage. Soil is a vital reservoir in the global C cycle. Sequestration of soil C is closely linked to nutrient cycling. Root growth, rhizosphere microbiology, nutrient cycling and availability, and C storage in soils are integrally linked (Fig. 1; Zak *et al.*, 1993) and have important implications for plant health. Predicting how belowground processes respond to rising CO_2 will be necessary for the management of future crop and forest systems.

II. Roots

Effects of atmospheric CO_2 concentration on belowground processes have recently received increased attention within the research community,

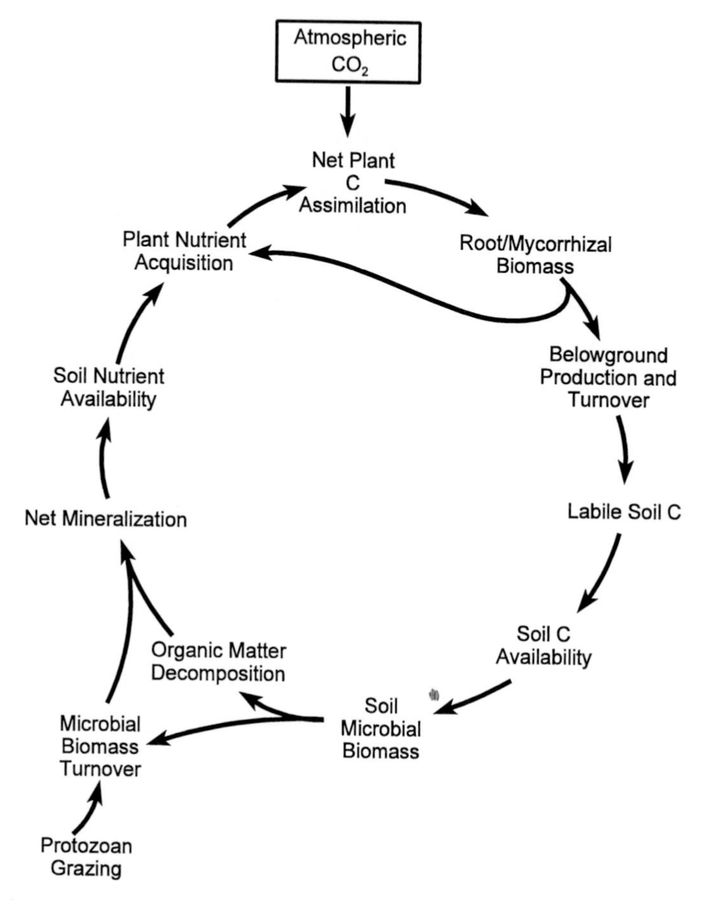

Figure 1 A conceptual model depicting the influence of elevated atmospheric CO_2 on plant production, microbial activity, and the cycling of C and nutrients. The model is characterized by a series of positive feedbacks in which increased net carbon assimilation under elevated CO_2 results in increased fine root/mycorrhizal growth, microbial biomass, and rate of nutrient mineralization. (Adapted with permission from Zak *et al.*, 1993.)

particularly with regard to root growth. Luo *et al.* (1994) have developed an excellent framework for predicting response of photosynthesis and root fraction to CO_2 enrichment which includes the interaction of C, N, and growth. They found that the response of the root fraction tends to rise or fall if photosynthesis is more sensitive to changes in CO_2 concentration than to relative growth rate. Apparently, this is often the case, because raising the CO_2 level frequently leads to increased root growth.

Rogers *et al.* (1992a) demonstrated enhanced root growth in soybean [*Glycine max* (L.) Merr.]. Root dry weight, length, diameter, and volume increased when CO_2 was elevated; however, total root number exhibited no response. Del Castillo *et al.* (1989) found that soybean root weight was approximately 28% higher in CO_2-enriched chambers and that cumulative root length correspondingly increased; no effect on rate of root elongation was seen. However, there was a linear increase in number of actively growing roots, that is, root systems were more branched. They concluded that roots of soybean plants growing under high concentrations of CO_2 would not explore a greater volume of soil, but would explore a given volume more thoroughly. These results contrast with Rogers *et al.* (1992a), who found a 110% increase in root length of soybean with no change in number of lateral roots.

Chaudhuri *et al.* (1990) found that winter wheat (*Triticum aestivum* L.) under elevated CO_2 achieved maximum rooting depth faster than in ambient air. Differences in root growth occurred primarily in the upper 10 cm of the soil profile, and they concluded that high CO_2 could compensate for restricted root growth brought about by drought, particularly in this top 10 cm of soil. With [*Sorghum bicolor* (L.) Moench], Chaudhuri *et al.* (1986) found that numbers and dry weights of roots were higher at all soil profile depths (to 150 cm) under elevated CO_2.

Belowground responses have been observed in cotton (*Gossypium hirsutum* L.) under free-air CO_2 enrichment (Rogers *et al.*, 1992b; Prior *et al.*, 1994a,b, 1995). Dry weights, lengths, and volumes of taproots, lateral roots, and fine roots were often higher for CO_2-enriched cotton plants. Although the numbers of lateral roots per unit length of taproot tended not to be increased by elevated CO_2, the overall greater taproot lengths under CO_2 enrichment usually had increased total numbers of laterals. Fine root patterns within the soil profile were also investigated; vertical and horizontal distributions of fine root density per unit volume of soil (expressed as length or dry weight m^{-3}) were measured. Although rooting was most pronounced in the upper 45 cm of the soil profile, the density of fine roots was seen to increase under CO_2 enrichment at most depths to 90 cm. Root length and dry weight densities also tended to exhibit greater differences between ambient and elevated CO_2 treatments with horizontal distance from row center, indicating a faster and more prolific spread of cotton roots. The location and density of roots within the soil profile is particularly important in that it determines nutrient acquisition, especially for ions with low diffusivity such as phosphate (Barber and Silberbush, 1984; Ikram, 1990; Caldwell *et al.*, 1992).

Highly significant increases in the growth of corms and roots of the C_3 tropical konjak (*Amorphophallus konjac* K. Koch) have been observed (Imai and Coleman, 1983); doubling CO_2 concentration doubled corm yield. In a separate study (Imai *et al.*, 1984), a large increase in growth (150%) was measured for the tropical root crop cassava (*Manihot esculenta* Crantz), with greater partitioning to roots. Carbon dioxide also causes large gains in root dry weight in rice (*Oryza sativa* L.) (Ziska and Teramura, 1992).

Adding CO_2 to greenhouses with mist systems raised the percentage of cuttings that formed roots in numerous horticultural species (Lin and Molnar, 1981; French, 1989). Elevating CO_2 during propagation also increased root number and length in sweet potato [*Ipomoea batatas* (L.) Lam.] (Bhattacharya *et al.*, 1985). Laforge *et al.* (1991) saw increases in root dry weight and root number of raspberry (*Rubus idaeus* L.) plantlets under high CO_2. Davis and Potter (1983) reported increases in root length and dry weight for several ornamentals, but root number increased only for *Peperomia glabella* A. Dietr. 'Variegata.' In further work with leafy pea (*Pisum sativum* L.) cuttings, elevated CO_2 increased carbohydrates, water potential, and root system size, but not root number (Davis and Potter, 1989). However, Grant *et al.* (1992) suggested that improved rooting was due to better water relations rather than increased carbohydrate levels.

There have been few studies on natural ecosystems. In a field study of elevated CO_2 effects on a Chesapeake Bay salt marsh, Drake (1992) noted increased numbers of roots and rhizomes (with increased allocation of C to them) for the C_3 sedge, *Scirpus olneyi* A. Gray. Investigations of a tallgrass prairie ecosystem (Owensby *et al.*, 1993b) have demonstrated increased root biomass for some species that may have led, at least in part, to shifts in species composition. Billings *et al.* (1984) concluded from vegetation/soil core studies that net production in tundra ecosystems was unlikely to be directly affected by CO_2 even at twice the ambient value. Indirect effects of CO_2 that could possibly become important in the tundra were temperature, water table, peat decomposition, and soil nutrient availability.

Berntson and Woodward (1992) reported that elevated CO_2 and water stress resulted in root foraging, branching patterns, and root lengths that resembled adequately watered, ambient CO_2-grown *Senecio vulgaris* L. plants, implying possible alleviation of stress. Prior *et al.* (1997) reported similar findings for longleaf pine (*Pinus palustris* Mill.). Root biomass of water-stressed seedlings increased (110%) when grown under elevated CO_2 conditions; elevated CO_2 also increased total fine root length for seedlings under water-stressed conditions.

Root response to CO_2 is influenced by interacting treatment conditions and, in their recent review, Stulen and den Hertog (1993) attributed experimental variability in root response (growth and function) to differential treatment of plants (i.e., water, nutrients, and pot size). They concluded that more CO_2 research on belowground effects is required and that the root to shoot ratio (R:S) effect needs critical reexamination. We have recently reviewed both belowground responses to CO_2 enrichment and the effects of elevated CO_2 on R:S.

In an examination of 167 studies of root response to CO_2 enrichment (Rogers *et al.*, 1994), root dry weight was the most frequently examined root measure and was included in about 50% of the studies. Most of these investigations (≈87%) found increases in root dry weight under elevated atmospheric CO_2, regardless of species or study conditions. Further, roots often exhibited the greatest relative dry weight gain, among tissues, for plants exposed to high CO_2 (Imai and Murata, 1976; Wittwer, 1978; Rogers *et al.*, 1983; Imai *et al.*, 1985; Hocking and Meyer, 1991b; Norby *et al.*, 1992). A majority (77%) of the studies in our survey (Rogers *et al.*, 1994) found that elevated CO_2 resulted in more and/or longer plant roots, possibly leading to increased spread (Idso and Kimball, 1991, 1992) and/or penetration of the soil profile (Baker *et al.*, 1990; Rogers *et al.*, 1992b).

Tognoni *et al.* (1967) reported the general promotion of root growth by higher levels of CO_2 with accompanying increases in R:S. However, despite fairly consistent results with most root measures, R:S responses have been more variable. For example, Laforge *et al.* (1991) reported an approximate doubling of R:S for raspberry plantlets and Rogers *et al.* (1992a) reported an increase in soybean R:S under high CO_2; while Chu *et al.* (1992) studying wild radish (*Raphanus sativus x raphanistrum* L.) and Prior *et al.* (1994a) working with cotton grown under field conditions reported that R:S was unaffected by CO_2 treatment. Further, in one study of 27 herbaceous species, R:S decreased in 14 species, increased in 6 species, and was unaffected in 7 species under elevated CO_2 (Hunt *et al.*, 1991).

In our examination of root response to CO_2 enrichment (Rogers *et al.*, 1994), R:S increased in agronomic (88% of studies), forest (86%), and natural community (81%) species; however, a wide range was seen in R:S among species and among study conditions. More recently, we have identified 264 determinations of R:S response in crop species under elevated atmospheric CO_2 (Rogers *et al.*, 1996), which adds further support to the contention that the response of R:S to elevated atmospheric CO_2 is highly variable among crop species and experimental conditions. Positive responses in R:S to elevated CO_2 occurred in 59.5% of these reports, while negative responses occurred in 37.5%, and no response occurred 3.0% of the time. We, in fact, found that the response of R:S to increased concentrations of atmospheric CO_2 approximates a normal distribution (Rogers *et al.*, 1996). Further analysis demonstrated that, while most of these observations were clustered close to zero (75.4% occurred between ±30%), there was a mean positive response (+11.1%), which was statistically greater than zero. This increase in R:S was found to be in general agreement with other reviews in the CO_2 literature (Acock and Allen, 1985; Enoch and Zieslin, 1988; Norby *et al.*, 1995; Wullschleger *et al.*, 1995).

In addition to the four primary variables (root dry weight, R:S, root length, and root number), other root responses to elevated atmospheric

CO_2 have been observed. Examination of structural aspects of roots (i.e., diameter, volume, branching, and relative growth rate) have usually shown positive effects of high CO_2. Tubers (number, dry weight, and diameter) and nodulation (number, dry weight, and activity) also benefit from elevated CO_2 in most cases (Rogers *et al.*, 1994). In a summary of available data, Porter and Grodzinski (1985) reported a mean yield ratio (for high CO_2) of 1.40 for mature root crops and of 1.77 for immature root crops.

Other root factors seen to increase with CO_2 (though infrequently examined) include parenchyma cell division and expansion, mycorrhizae, and carbohydrate levels (Rogers *et al.*, 1994). Results of Ferris and Taylor (1994), using a biophysical analysis of root cell elongation as influenced by elevated CO_2 suggested that root growth is stimulated after increased cell expansion. Indications were that increased P and cell wall tensiometric extensibility are probably both key to enhanced root growth of plants in elevated CO_2. Time to harvest (though seldom reported) was found to be shortened for root and tuber crops (Cummings and Jones, 1918).

Available literature demonstrates that, while responses may vary among species and study conditions, more CO_2 tends to result in positive, sometimes dramatic, responses in the belowground portion of plants. Increased rooting has the potential to alter significantly the edaphic environment through increased C deposition and/or nutrient uptake by plants; this may have important consequences for managed and natural terrestrial ecosystems, particularly under potentially changing climates. However, increased C storage in soil depends not only on the quantity and quality of C deposited, but also on the manner and speed with which it is recycled once on or in the soil (Fig. 1). Similarly, increased resource acquisition from soil depends not only on the ability of roots to forage for nutrients, but also on the accessibility and availability of these nutrients. Effects of CO_2 on these processes will largely depend on how rhizosphere and soil microorganisms respond.

III. The Rhizosphere

Carbon dioxide-induced changes in plants will affect the structure (species populations) and function (activity) of rhizosphere and soil microorganisms. As pathogens, symbionts, and decomposers, microbes exert a strong influence on C and nutrient cycling in plant/soil systems (Fig. 1). Changes in plant structure (Thomas and Harvey, 1983; Prior *et al.*, 1995), physiology (Amthor, 1991; Rogers and Dahlman, 1993; Amthor *et al.*, 1994; Rogers *et al.*, 1994), and phytochemistry (Melillo, 1983; Lekkerkerk *et al.*, 1990; Liljeroth *et al.*, 1994; Pritchard *et al.*, 1997) brought about by elevated atmospheric CO_2 may alter plant–microbe interactions and, thus, plant health and C and nutrient cycling in the soil.

Despite the potential for increasing CO_2 (through changes in growth, exudation, and tissue chemistry) to alter interactions of plant roots with pathogenic microbes, this area remains virtually unstudied. Runion *et al.* (1994) reported a trend for increased infestation of root-zone soil of cotton by *Rhizoctonia* spp., under free-air CO_2 enrichment (FACE). Although this suggested the potential for elevated CO_2 to increase root disease, a bioassay using this soil demonstrated no increase in damping-off potential. Additionally, they reported a trend for lower numbers of parasitic nematodes under FACE, which they related to increased competition from a larger population of saprophagous nematodes (Runion *et al.*, 1994).

It has been hypothesized that, due to additional C entering plant rhizospheres, atmospheric CO_2 enrichment could result in greater mycorrhizal colonization of roots (Luxmoore, 1981; Lamborg *et al.*, 1983). This potential increase in mycorrhizal colonization could benefit host plants via increased nutrient (Bowen, 1973; Tinker, 1984) or water (Bowen, 1973; Augé *et al.*, 1987) uptake and by protecting roots from adverse edaphic conditions such as temperature extremes, high salinity, or pathogenic microorganisms (Marx, 1973; Duchesne, 1994). The fact that mycorrhizae can provide additional water to plants through hyphal proliferation in soil (Luxmoore, 1981) might explain, in part, observed increases in biomass of CO_2-enriched plants under drought stress. These beneficial effects of mycorrhizae could, then, increase plant productivity, leading to greater amounts of organic matter supplied to the soil.

Carbon dioxide enrichment has been reported to increase mycorrhizal colonization in roots of several plant species (Norby *et al.*, 1987; O'Neill *et al.*, 1987b; Monz *et al.*, 1994; Runion *et al.*, 1997) or, even in the absence of an effect on a unit root length basis, to increase numbers of mycorrhizae on a whole-plant basis due to significant increases in total root length (Norby *et al.*, 1986a; O'Neill *et al.*, 1991; Runion *et al.*, 1994). Runion *et al.* (1997) suggested that sink/source relationships may be a major factor regulating mycorrhizal development in high-CO_2 environments. That is, when either photosynthate supply is not source limited (elevated CO_2) or when tissue N concentrations are low (elevated CO_2 or low soil N), plants alter allocation to soil resource acquisition by investing in roots and mycorrhizae. Lewis *et al.* (1994b) found that, for loblolly pine (*Pinus taeda* L.) from different geographic sources, mycorrhizal colonization responded differentially to CO_2 enrichment. For two pines (*Pinus radiata* D. Don and *P. caribaea* P.M. Morelet var. *hondurensis*), elevated CO_2 boosted P uptake when soil P availability was low (Conroy *et al.*, 1990b); changes in mycorrhizae appeared to be the most probable cause. It was concluded that high leaf P concentrations would be necessary if the full potential for more atmospheric CO_2 on pine forests were to be realized.

Study of N fixation (the incorporation of atmospheric N into nitrogenous compounds which can be utilized by living organisms) as influenced by atmospheric CO_2 has generally shown that legume/bacterial symbiosis is favored by elevated CO_2 (Reddy *et al.*, 1989; Reardon *et al.*, 1990). Acock (1990) concluded that the positive effect of CO_2 on N fixation appeared to be simply the result of more biomass. This is supported by work showing that, when atmospheric CO_2 is high, nodule dry weight increases but specific nodule activity (g N g^{-1} nodule) does not (Williams *et al.*, 1981; Finn and Brun, 1982). However, others have found increases in nodule activity under CO_2 enrichment (Hardy and Havelka, 1973; Whiting *et al.*, 1986; Norby and Sigal, 1989) and relate the increased activity to photosynthetic stimulation. These differing results may be due to differences in plant species (Masuda *et al.*, 1989a,b), experimental conditions, or duration of CO_2 exposure. Phillips *et al.* (1976) obtained results indicating that long-term enrichment promoted fixation by enhancing nodule development in peas, while short-term high-CO_2 exposures increased fixation by affecting nodule function. In their work with white clover (*Trifolium repens* L.), Masterson and Sherwood (1978) found that the expected decrease in N fixation at high levels of soil N did not occur under elevated CO_2. Allen *et al.* (1991) observed a positive effect of high CO_2 on N-fixing soil bacteria, which improved N nutrition and yield of rice.

In addition to their effects as symbionts, soil microorganisms are responsible for the decomposition of organic matter; through the processes of decomposition and mineralization, nutrients become available to plants (Grant and Long, 1981; Paoletti *et al.*, 1993). Not only can soil and rhizosphere microbes make nutrients available to plants, they can also make them unavailable through immobilization. In an experiment with labeled plant residues, Jenkinson (1966) found that about 35% of labeled C remaining in the soil after 1 yr was present as microbial biomass. Soil microfauna (e.g., nematodes and microarthropods) are also important in nutrient cycling (Coleman *et al.*, 1984; Freckman, 1988; Ingham, 1988; Rabatin and Stinner, 1988; Runion *et al.*, 1994).

In many terrestrial ecosystems, nutrients are a primary factor limiting plant growth; however, nutrient uptake and plant growth are not usually related to the total nutrient content of the soil, but rather to the quantity of nutrients mineralized by microbes and available for plant growth (Grove *et al.*, 1988). For example, microbial transformations of organic P often produce the majority of the plant available solution P (Paul and Voroney, 1980).

Zak *et al.* (1993) saw a positive feedback between C and N soil dynamics and elevated CO_2 (Fig. 1) using *Populus grandidentata* Michx. grown on nutrient-poor soil. They reported significant increases in microbial biomass C in the rhizosphere and in bulk soil associated with plants grown under elevated CO_2. They also observed greater N mineralization, which was possibly related to both increased turnover of microbial N and release of N from soil organic matter. O'Neill *et al.* (1987a) found that, while nitrite-oxidizing and phosphate-dissolving bacteria in the rhizosphere of yellow poplar (*Liriodendron tulipifera* L.) seedlings were reduced at the final harvest, total N and P uptake were increased under high CO_2. They speculated that the decline in bacteria populations was a function of decreased nutrient availability due to increased competition with seedling roots as the growing season progressed. Runion *et al.* (1994) found that, while dehydrogenase activity (a measure of microbial respiration) was significantly higher in soils from CO_2-enriched cotton plants, no appreciable differences in microbial populations (fungi, bacteria, and actinomycetes) were observed.

Elevated CO_2-induced changes in plant tissue quantity and quality can affect the composition and activity of rhizosphere and soil microbes and thus impact C turnover and storage in soils (Goudriaan and de Ruiter, 1983; Lamborg *et al.*, 1983). Overdieck and Reining (1986) raised the possibility of slowing decomposition rates by increased C:N ratios in high-CO_2 grown plants. Such an effect could slow the cycling of both C and nutrients, and thus reverse the positive CO_2 effect on vegetation in the long run. Melillo (1983) reported higher C:N ratios and higher levels of phenolics in sweetgum leaves exposed to high CO_2 and hypothesized that this would result in reduced decomposition rates and decreased soil fertility. Ball (1992) found decreased growth of lignocellulose-degrading actinomycetes on elevated CO_2-grown wheat material, resulting in reduced degradation, which was related to increased lignification and changes in the C:N ratio. Lekkerkerk *et al.* (1990) found that the input of easily decomposable root-derived material in soil supporting wheat plants was increased and, due to microbial preference for these materials, turnover of more resistant soil organic matter was reduced under elevated CO_2. Coûteaux *et al.* (1991) demonstrated similar results for an initial decomposition period and related the reduction in decomposition rate to lower N concentration and higher C:N ratios of CO_2-enriched plants. However, when they allowed decomposition to continue, changes in the composition of the decomposer population (increase in microfauna and introduction of white-rot fungi) resulted in an increased decomposition rate of material (produced under CO_2 enrichment) while the rate for control materials declined. These CO_2-induced shifts in decomposer composition led to an overall 30% rise in C mineralization. An increase in C turnover was also observed in soils where CO_2-enriched cotton plants had grown for three seasons (Wood *et al.*, 1994) and could be related to increases in soil microfauna and saprophagous nematode populations (Runion *et al.*, 1994).

Elevated atmospheric CO_2, acting through plant-mediated changes in the quantity and quality of organic materials entering the rhizosphere, may alter the composition and activity of soil microbes. Increasing atmospheric CO_2 may negatively affect pathogenic soil microbes and, thus, decrease root diseases, but this possibility has not been studied. In contrast, CO_2 effects on beneficial microorganisms (mycorrhizal fungi and N-fixing bacteria) have received more attention and enhanced plant health through increased interactions with microbial symbionts under CO_2 enrichment appears likely. Stimulation of N fixation and nutrient release by increased microbial (including mycorrhizal) and root activity are possible mechanisms that could raise fertility levels (Brinkman and Sombroek, 1996). Thus, rhizosphere and soil microbes may also improve plant health under CO_2 enrichment by increasing nutrient cycling and availability and by improving soil quality through increased C storage; however, generalizations regarding these effects cannot yet be made and further study is required.

IV. Mineral Nutrition

Mineral nutrition is essential to plant growth (Shuman, 1994; Marschner, 1995) and is therefore an important aspect of the CO_2 response. As Wolfe and Erickson (1993) have stated, "An accurate assessment of the effects of CO_2-doubling on crop productivity will require more information regarding complex interactions between CO_2 and nutrient availability, and a better understanding of how specific nutrient deficiencies influence plant response to CO_2." There is uncertainty associated with interactions between atmospheric CO_2 concentration and mineral nutrition (Sinclair, 1992) since complex interactions between roots and nutrients depend on the specific nutrient, its concentration, form when applied, and environmental conditions (Sattelmacher *et al.*, 1993).

A review of the literature for the past decade revealed that, while the absolute magnitude of plant growth is greater under adequate resource availability, the relative response to elevated CO_2 is generally enhanced most when resource limitations and environmental stresses are greatest (Idso and Idso, 1994). Although higher CO_2 levels may contribute to the plant's capacity to grow under nutrient deficient conditions, it is important to note this is not always the case. It has been suggested that larger root systems, higher photosynthetic rates, and greater activity in the rhizosphere may help overcome nutrient stress (Rogers *et al.*, 1994). In addition, root exudation, which can increase in plants grown under CO_2 enrichment (Norby *et al.*, 1987), might enhance nutrient acquisition, especially under stress conditions (Uren and Reisenauer, 1988).

Nitrogen and P are the nutrients most likely to be influenced by rising CO_2 since relatively high quantities of both are needed in the photoreductive C cycle and the photo-oxidative cycle (Rogers *et al.*, 1993; Gifford, 1992). Research has revealed that CO_2 enrichment will induce the greatest productivity in C_3 plants when soil N and P availability are high; that low N does not always eliminate the CO_2 growth effect; and that some species do not respond when P is insufficient, owing to a lessening of photosynthetic activity (Conroy *et al.*, 1986a,b; 1988; 1990a,b; 1992; Conroy, 1992; Conroy and Hocking, 1993). In general, maximum productivity under CO_2 enrichment requires higher tissue P concentrations, but the N requirement is reduced (Conroy, 1992). Rogers *et al.* (1993) reported that critical leaf N concentration was decreased while critical P level was increased in wheat and cotton grown under high CO_2. Duchein *et al.* (1993) demonstrated that supplying supraoptimal levels of P to clover (*Trifolium subterraneum* L.) exposed to high CO_2 greatly improved growth. However, Seneweera *et al.* (1994) reported that P level for maximum yield of rice was the same at both ambient and twice ambient CO_2. Lewis *et al.* (1994a) showed that P availability and photosynthetic capacity of loblolly pine were influenced by CO_2 level. Although long-term exposure to elevated CO_2 may reduce P stress, the authors point out that, in the field where P is already only marginally sufficient, plants could soon become P limited as global CO_2 level rises. Collectively, the above results imply that critical nutrient concentrations (those that promote maximum productivity) may need to be reconsidered in light of the rising level of atmospheric CO_2 (Hocking and Meyer, 1991a; Conroy *et al.*, 1992).

The essential need for nitrogen has made it a frequent target of research. Kimball and Mauney (1993) saw no interaction between CO_2 and soil N in cotton. Similarly, Wong (1979) reported reduced dry weight and leaf area for both maize (*Zea mays* L.) and cotton under low N supply at both ambient and elevated CO_2. Recently compiled literature on forest responses (Wullschleger *et al.*, 1997) also suggests that limited supplies of N, P, and water will only slightly inhibit tree growth response to CO_2 enrichment. This is supported by the work of Samuelson and Seiler (1993) who concluded that response of red spruce (*Picea rubens* Sarg.) to CO_2 rise may proceed even if water and nutrients are in short supply, and by Cure *et al.* (1988) who showed that atmospheric CO_2 enrichment can enhance yield of non-nodulating soybean even if N availability is limiting. In contrast, Bowler and Press (1993) showed no growth effect of elevated CO_2 in *Nardis stricta* L. at low N, but an increase in growth at high N; similar results have also been reported for loblolly (Griffin *et al.*, 1993) and longleaf pine (Prior *et al.*, 1997). Likewise, Wong *et al.* (1992) saw a positive interaction of CO_2 and N on dry weight and leaf area in four species of eucalyptus. Brown and Higginbotham (1986) reported that spruce [*Picea glauca* (Moench)

Voss] biomass was increased by high CO_2 at high N; aspen (*Populus tremuloides* Michx.) biomass also increased initially but this effect did not persist, suggesting development of a nutrient limitation as the plants grew. In contrast, Wong and Osmond (1991) found that wheat shoot growth was stimulated more in low than in high nitrate when subjected to elevated CO_2. Bowler and Press (1993) showed a proportionally greater growth response to high CO_2 at low compared with high N (78 and 58%, respectively) for the upland grass species, *Agrostis capillaris* L. These variable results suggest a need for further research.

Interacting effects of CO_2 and N on plant growth may be related to effects on photosynthesis, and changes in N supply will influence the photosynthetic response to CO_2 in a complex fashion (Woodward, 1992). In loblolly pine, photosynthetic rate showed similar responses to CO_2 and N as did growth; that is, rates of photosynthesis were higher at elevated CO_2 only when supplemental N was provided (Tissue *et al.*, 1993). However, increased photosynthesis in longleaf pine tended to occur under elevated CO_2 regardless of N fertility level (G. B. Runion, unpublished data). Bunce (1992) also showed that photosynthetic adjustment to elevated CO_2 in soybean and sugar beet (*Beta vulgaris* L.) was unaltered by increasing nutrient supply.

In addition to variability from differing levels of N, the source of this N can also influence response to CO_2. Carbon dioxide enhancement of photosynthetic rate in carob (*Ceratonia siliqua* L.) was proportionally higher than biomass increase when nitrate, rather than ammonium was the N source (Cruz *et al.*, 1993); the preferred N source was ammonium at both CO_2 levels. Also, carob grown with nitrate had higher sucrose content, while those on ammonium had greater starch storage. BassiriRad *et al.* (1996) observed that elevated CO_2 enhanced root uptake capacity for nitrate, but not for ammonium in loblolly pine. Substantially higher carbohydrate levels in roots were noted without a significant change in root N concentration. Elucidating the preferred form of nutrients will contribute to understanding the relationship of CO_2 to mineral cycling in the field. Clarifying this relationship will allow the development of better models and more accurate predictions. One forest modeling effort has suggested that decreases in foliar concentration of N could slow rates of N cycling between vegetation and the soil, and lower rates of N loss through gaseous emission, fire, and leaching (Comins and McMurtrie, 1993). It was revealed that in the long term the equilibrium elevated CO_2 response would be sensitive to the rate of gaseous N loss due to mineralization. Another interesting aspect of greater nutrient capture by plant root systems is the possibility of a shift in the N cycle where more N is held in organic form within CO_2-enriched agro-ecosystems, leading to improved groundwater quality (Torbert *et al.*, 1996).

Some research has examined responses to elevated CO_2 under generalized nutrient regimes, and again results have been variable. Sionit (1983) found that limited nutrient supply in soybean could be at least partially ameliorated by CO_2 enrichment and reported that yield response to high CO_2 rose significantly at the highest nutrient level, suggesting a need for more fertilization. Sionit *et al.* (1981) reported that total dry matter of wheat increased with elevated CO_2 regardless of nutrition level (1/16, 1/8, 1/2, and full-strength Hoagland's solution); for ambient CO_2, yield increased up to the 1/2 strength level, but fell at full strength. Patterson and Flint (1982) reported growth enhancement by CO_2 in soybean, sicklepod (*Cassia obtusifolia* L.), and showy crotalaria (*Crotalaria spectabilis* Roth) at both 1/8 and 1/2 strength Hoagland's solution. Newbery (1994) observed no interaction between nutrient supply and elevated CO_2 for *Agrostis capillaris.* Norby and O'Neill (1991) also reported whole plant and root dry weight increases in yellow poplar with and without added mineral nutrients. Growth of bean (*Phaseolus vulgaris* L.) was also positively affected with or without adequate nutrition (Radoglou and Jarvis, 1992). In contrast to this lack of response to nutritional level, McKee and Woodward (1994) found that CO_2 enhancement of spring wheat was limited by low nutrient supplies. Similarly, Curtis *et al.* (1995) observed greater dry weight gain in hybrid poplar [*Populus x euramericana* (Dode) Guinier cv. Eugenia] exposed to twice ambient CO_2 at high (49%) compared with low (25%) fertility levels. They also found that photosynthetic capacity in the high-CO_2 treatment fell off more rapidly for the low fertility plants (100 days) than for those at high fertility (135 days). Bazzaz and Miao (1993) reported that the greatest stimulation in growth by CO_2 enrichment for six tree species (*Betula alleghaniensis* Britton, *B. populifolia* Marsh., *Fraxinus americana* L., *Acer rubrum* L., *A. pennsylvanicum* L., and *Quercus rubra* L.,) occurred under high nutrient conditions; further, the three early successional species (*B. populifolia, F. americana,* and *A. rubrum*) showed significant growth increases only in the high nutrient regime. In sweet chestnut (*Castanea sativa* Mill), soil nutrient availability caused a shift in partitioning with CO_2 enrichment (El Kohen *et al.*, 1992); root dry weight increased in unfertilized soil, while only stem dry weight increased with fertilization.

Interactions of CO_2 and nutrients have primarily focused on effects on biomass (production and allocation) and/or photosynthesis. However, other physiological parameters of interest have been examined. Radoglou *et al.* (1992) observed that, while water use efficiency (WUE) of bean plants showed a linear increase over the CO_2 concentration range, WUE was doubled at a high nutrient and tripled at a low nutrient supply. Masle *et al.* (1992) also observed a positive correlation between WUE and leaf mineral content in several species. Coleman *et al.* (1993) suggested that decreases in plant N concentration may not be the result of a physiological change in N use efficiency, but rather a phenomenon dependent on plant size increase that occurs because CO_2 accelerates growth (which could affect water use). Conroy and Hocking (1993) suggest that, besides the diluting effect of extra C compounds, reduction in nutrient absorption might result from slower transpiration rates. Experiments by Ito (1970) raised the possibility that CO_2-induced reduction in transpiration would lower the uptake of P. In light of these findings, and the often reported increase in WUE by elevated CO_2, there is good reason to explore the interaction of CO_2 and mineral nutrition.

Results on nutrient uptake and concentration are variable due to differences in nutrient application during the course of the experiments (Linder and McDonald, 1993). For example, Israel *et al.* (1990) reported that, if plants growing under high CO_2 are supplied with higher levels of nutrients, tissue nutrient concentrations and nutrient uptake efficiency are generally not significantly affected by CO_2 concentration. On the other hand, it has been suggested that when plants are grown under nutrient levels considered poor to adequate for ambient conditions, high CO_2 results in larger plants with lower tissue nutrient concentrations (Norby *et al.,* 1986a,b; Yelle *et al.,* 1987; Silvola and Ahlholm, 1995). In fact, nutrient concentrations in plant parts are often lower under high CO_2 regardless of soil nutrient availability. For example, Reeves *et al.* (1994) observed lower tissue N concentrations in soybean (N-fixing) grown under elevated CO_2. Johnson *et al.* (1995) also saw dilution of tissue N concentration in ponderosa pine (*Pinus ponderosa* Douglas ex P. Laws. & C. Laws.) exposed to elevated CO_2, even when growing in high N soil; however, a more efficient use of internal N reserves was noted. El Kohen *et al.* (1992) found that, while the total N pool was unaffected by extra CO_2 tissue N concentration was reduced in sweet chestnut. A decline (approximately 25%) in leaf mineral content (N, P, K, Ca, and Mg) under high CO_2 has also been observed in bean (Porter and Grodzinski, 1984); carbohydrate dilution was suggested as the cause (Porter and Grodzinski, 1985). McKee and Woodward (1994) reported that high-CO_2 treatment reduced both the shoot N concentration and the proportional allocation of N to the uppermost leaves of spring wheat, which could impact photosynthesis. Thompson and Woodward (1994) found that, while CO_2 enrichment increased grain yield of spring wheat and spring barley (*Hordeum vulgare* L.), the grain N content was reduced, enough so that current bread-making procedures would be affected. Hocking and Meyer (1991b) saw higher nitrate reductase activity in wheat grown in high CO_2 with an accompanying increase in N use efficiency. Total N accumulated was greater but the increase did not rise in proportion to the CO_2-induced dry weight gain and, therefore, N concentration fell; similar results were reported for cocklebur (*Xanthium occidentale* Bertol.) (Hocking and Meyer, 1985).

Few studies have explored nutrient interactions *in situ,* within CO_2-enriched plant communities. In a 3-yr study of tallgrass prairie response to elevated CO_2, Owensby *et al.* (1993a) saw increased biomass of C_4 grasses. Total N content went up while tissue N concentration came down. The total N rise was attributed to enhanced root exploration of the soil profile by larger root systems. Enhanced P nutrition was also observed. Growth of C_3 species was not substantially increased, but tissue concentration of N was reduced as in the C_4 grasses. In a salt marsh community of three species growing under an approximate CO_2 doubling, Curtis *et al.* (1989) noted a substantial rise in C:N ratio for *Scirpus olneyi* (C_3), but no effect on two C_4 plants, [*Spartina patens* (Aiton) Muhl.] and [*Distichlis spicata* (L.) Greene]. Total aboveground N for *S. olneyi* remained unchanged, suggesting that the observed growth stimulation was dependent on reallocation of stored N. A lowering of decomposition rates as well as forage quality appeared to be probable consequences of the reduced N concentration. Woodin *et al.* (1992), investigating the effect of CO_2 on heather (*Culluna vulgaris* L.), observed that the total nutrients absorbed did not increase, thus tissue concentration fell. Their results suggested that heather, in its usually nutrient-poor habitats, would become nutrient limited by even small rises in atmospheric CO_2 level. Nutrient cycling is integrally linked to competition within natural communities. The effect of CO_2 on this cycling may alter competitive relationships, thereby changing community structure and function.

Although a decline in nutrient concentration of plant tissue under high CO_2 is the general rule, nutrient uptake is often increased by elevated CO_2; however, results have varied among species and among elements. Peñuelas and Matamala (1990) examined the N and S content of leaves in herbarium specimens of 14 species collected from 1450 to 1985. A significant overall decrease in N was seen. The S content showed no long-term trend, but a sharp upturn in the 1940s was attributed to the burning of high-sulfur coal during those years. In a similar study (Peñuelas and Matamala, 1993) of mineral content of 12 species (trees, shrubs, and herbs) during the last 250 yr, present levels of Al, Ca, Cu, Sr, Fe, P, Mg, Mn, K, Na, S, and Zn were found to be lower than at any other period. Larigauderie *et al.* (1994) noted greater N uptake rates for loblolly pine grown with high N supply, but lower uptake with low N supply, under high CO_2. Newbery (1994) observed that the uptake of N, P, and K did not increase proportionally with CO_2-induced increases in growth; K concentration declined, while N and P concentrations were similar for *Agrostis capillaris* plants grown with elevated CO_2 compared with those grown in ambient air. The large rise in K demand was suggested to have been related to shifts in osmoregulation. Seneweera *et al.* (1994) reported that leaf P content of rice was unaffected by CO_2 treatment over a range of added P (0–480 mg kg^{-1} as $CaHPO_4 \cdot 2H_2O$).

Luxmoore *et al.* (1986) saw increases in dry weight and uptake of N, Ca, Al, Fe, Zn, and Sr in Virginia pine (*Pinus virginiana* Mill.) under CO_2 enrichment. Greater uptake was associated with increased root weight. Specific absorption rates (uptake per unit root dry weight) were generally unaltered by high CO_2. Uptake of P and K did not increase with elevated CO_2, but had greater nutrient use efficiency. Nitrogen and Ca use efficiency were not affected. The rise in Zn uptake may have resulted from increased pH in the rhizosphere since, under elevated CO_2, cations were more readily absorbed than anions. Results suggested increased nutrient retention under high CO_2. O'Neill *et al.* (1987a) reported that, for yellow poplar, absorption of N, S, and B were lower at elevated CO_2, while uptake of Ca, Mg, Sr, Ba, Zn, and Mn remained uninfluenced by enrichment. Uptake of P, K, Cu, Al, and Fe was proportional to growth at both ambient and high CO_2. Wong *et al.* (1992), working with four eucalypts, saw increased N use efficiency under elevated CO_2 for both leaf and whole plant at low N. In the case of high N, leaf N use efficiency went up in two species, but went down in the other two species. Kuehny *et al.* (1991) noted nutrient dilution by CO_2 enrichment in *Chrysanthemum x morifolium* Ramat. cv. Fiesta. When corrected for leaf starch content, differences in N, P, K, Ca, Mg, S, B, Fe, and Cu disappeared, but not differences in Mn and Zn. Carbon dioxide enrichment of celery (*Apium graveolens* L.) decreased shoot levels of N, P, K, Mg, and B from 4 to 12% (Tremblay *et al.*, 1988). In roots, concentrations of N and K were reduced, Mg was increased (34%), and P, Ca, and B were unchanged. Nitrogen addition increased the N concentration of both roots and shoots, but caused reductions of P, K, and Ca levels in roots. Fertilization with P increased root and shoot P concentration, but lowered K in the shoot. In a study of three grass species (*Agrostis capillaris, Poa alpina* L., and *Festuca vivipara* L.), Baxter *et al.* (1994) reported that total nutrient absorption was greatest in *A. capillaris,* while N and K remained unchanged in *P. alpina*. In *A. capillaris* neither nutrient use efficiency (dry matter accumulated per unit of nutrient) nor nutrient productivity (dry weight gained per day over a defined period per mean weight of nutrient over the same period) was significantly affected by a doubling of CO_2. Depression of tissue nutrient content arose from increased growth instead of differences in nutrient use efficiency. The productivities of K, Mg, and Ca in *P. alpina* were lowered by elevated CO_2, while photosynthetic N and P use efficiencies were doubled. Poor growth of *F. vivipara* was accompanied by a decline in photosynthetic N use efficiency and photosynthetic P use efficiency and a large rise in the ratio of nonstructural carbohydrate to N content.

Plant tissue nutrient concentration is largely determined by plant roots, because these are the primary means of extracting nutrients from the soil profile. Therefore, effects of atmospheric CO_2 on roots, and belowground processes, will affect plant nutrition. Variability in the literature predicates that generalizations regarding the interacting effects of nutrients and CO_2, both in regard to growth and tissue nutrient content, be made with caution. In some cases, limited nutrient supply appears to inhibit only slightly plant growth response to CO_2 enrichment, and elevated CO_2 appears to aid plants in partially overcoming nutrient stress; in other experiments, positive growth responses to increased CO_2 are only observed when nutrients are not limiting. Nonetheless, it appears certain that critical nutrient concentrations will need to be reconsidered in light of the rising level of atmospheric CO_2. In addition, elevated CO_2 usually increases the overall size of plants, thus increasing whole-plant nutrient uptake for many plant species, but these nutrients are distributed throughout larger plants and, thus, concentration per unit weight of tissue is diminished. Also, nutrient utilization efficiency (unit of biomass produced per unit of nutrient) generally rises under elevated CO_2 (Gifford, 1992), while nutrient uptake efficiency (unit of nutrient per unit weight of root) declines in most studies. Plants under high CO_2 are able to produce more biomass with available nutrients; however, their larger root systems appear unable to gather proportionally more nutrients.

Nutrient supplies are critical to plant systems and attenuate the dynamic flows (Fig. 1) of essential materials (e.g., carbon and water) through them. Perhaps the principal difference between agricultural and natural ecosystems is that in the former, we seek, by whatever means that are economically feasible and environmentally sound, to minimize stresses on plant growth. Part of this "agri-culture" is to fertilize, to reduce nutrient stress. But in the wilderness meadow and backcountry forest, soil fertility is fixed. Here nutrients are in short supply and must be competed for, used sparingly and efficiently, and then recycled for yet another wave of hungry critters. The key role of nutrients in the changing global C cycle must be elucidated if we are to predict how these changes will impact ecosystem structure, function, and stability.

V. Soil Carbon Storage

Brinkman and Sombroek (1996), in discussing the reaction of soil conditions in relation to plant growth and food production in the context of global change, suggested that soil quality could be affected by changes in organic matter, soil biology, fertility, temperature, and hydrology; heightened soil resilience to degradation was also seen as a possible outcome. Increases in plant growth above and below the ground under increased atmospheric CO_2 have been well substantiated (Rogers and Dahlman, 1993; Rogers *et al.*, 1994; Goudriaan and Zadoks, 1995). Post *et al.* (1992) underscore the complexity of C fluxes as they dynamically interact with the global biogeochemical-climate system. They go on to say that it is "essential that we understand how terrestrial vegetation and ocean processes respond to changes in CO_2 and climate."

Woodward *et al.* (1991) have pointed out that the enormous propensity for CO_2 to stimulate plant growth has great potential for C sequestration in terrestrial ecosystems. However, Schlesinger (1986, 1990) found little evidence for soil C storage. Schlesinger (1995) further suggests that a lack of response of natural vegetation to higher CO_2 levels is likely to result from nutrient limitations.

Not only is soil fertility essential to terrestrial C sequestration, but as Tiessen *et al.* (1994) have argued for tropical soils, the recycling of nutrients from soil organic matter is necessary to maintain fertility (Fig. 1). Nepstad *et al.* (1994) have demonstrated the vital role of deep rooting in C deposition to soil in Amazonian forests. In these forests, there is more C below 1 m than above the ground. Pregitzer (1993) suggested the possibility that increased C flux to the soil may enhance available N and further noted that warmer temperatures could lead to faster decomposition and hence a more rapid return of CO_2 to the atmosphere. Pregitzer *et al.* (1995) also reported that production and mortality of fine roots, which can be the greatest source of C input to forest soils, were both increased under elevated CO_2 and as soil N availability increased. Nitrogen plays a key role in C storage in both plants and soils, and will therefore be pivotal in C processes affected by elevated CO_2 (Davidson, 1995).

Rastetter *et al.* (1992) used a general model of ecosystem biogeochemistry to examine C sequestration in two systems (arctic tundra and temperate hardwood forest). Increasing CO_2 increased C storage in both systems due to gains in C:N ratio in plants and soil. In a modeling study of soil organic C storage, Kirschbaum (1993) showed that equilibrium soil C content would increase with increasing rates of N addition (from atmospheric deposition and biological fixation). Hudson *et al.* (1994), employing a global C cycle model, have found that it is likely that a sizable part of the missing CO_2 sink is attributable to fertilization by N emitted from anthropogenic sources.

Huettl and Zoettl (1992), in discussing forest fertilization for enhanced C storage capacity, concluded that proper nutrient management may increase C in forest systems by stimulating biomass production both above and below the ground. Nilsson (1993), in fact, has observed an increase in C sequestration rate after fertilization of a Norway spruce [*Picea abies* (L.) H. Karst.] forest.

Leavitt *et al.* (1994), using $\delta^{13}C$ analysis of free-air CO_2-enriched cotton, observed that about 10% of the soil C had been replaced with "fresh" C, including the more recalcitrant fractions, in 3 yr of elevated CO_2 exposure. They suggest that the soils may be acting as an enhanced C sink under high CO_2. Follett (1993) states that, through improved management, agriculture (including forestry) has a great opportunity to help mitigate potential climate change by "stashing" CO_2 as C in soil and vegetation. He concludes that these practices can offset not only the CO_2 emissions from U.S. agriculture, but also part of that from U.S. sources outside of agriculture.

Stewart *et al.* (1992) have considered the interactive nature of global C and nutrient element cycles; they underscore the critical need for an integrated understanding if our actions to mitigate change are to be effective. For agricultural ecosystems, the management of crop residues is key to soil retention of C (Stewart, 1993).

VI. Conclusion

Predicting how belowground processes respond to rising CO_2 will be necessary for the management of future crop and forest systems. Among the most important of these processes is the absorption of mineral nutrients from the soil by roots. The following statements summarize what we know of the relationships of elevated CO_2 and plant nutrition:

- More CO_2 tends to result in positive responses in the belowground portion of plants and this has the potential to alter significantly the edaphic environment through increased carbon deposition and/or nutrient uptake by plants.
- Enhanced plant health and higher soil fertility levels through increased interactions with microbial symbionts under CO_2 enrichment appear likely.
- Rhizosphere and soil microbes may improve plant growth under CO_2 enrichment by increasing nutrient cycling and availability, and by improving soil quality through increased C storage.
- Critical nutrient concentrations (fertilizer recommendations for crops) will need to be reconsidered in light of the rising level of atmospheric CO_2.
- Elevated CO_2, through increased plant size, increases whole-plant nutrient uptake but concentration per unit weight of tissue is diminished.
- Plants under high CO_2 are able to produce more biomass with available nutrients, so nutrient utilization efficiency is generally increased.
- Larger plant root systems produced under elevated CO_2 appear unable to gather proportionally more nutrients, so nutrient uptake efficiency declines.
- Plant growth stimulation by CO_2 and higher C:N ratios suggest the potential for increased C sequestration in terrestrial ecosystems.

The relationship of carbon to factors such as mineral nutrition must be understood as we seek to mitigate and adapt to global environmental change. Knowledge of how plants and soils will respond to and sequester carbon is essential. We recognize the importance of carbon's edaphic interactions, and it is imperative that we develop a thorough understanding through further research.

Acknowledgment

The authors wish to express their sincere appreciation for support provided by the Terrestrial Carbon Processes Program of the Environmental Sciences Division, U.S. Department of Energy (Interagency Agreement DE-AI05-95ER62088).

References

Acock, B. (1990). Effects of CO_2 on photosynthesis, plant growth and other processes. *In* "Impact of CO_2, Trace Gases, and Climate Change on Global Agriculture" (B. A. Kimball, N. J. Rosenberg, and L. H. Allen, Jr., eds.), ASA Special Pub. No. 53, pp. 45–60. ASA, CSSA, and SSSA, Madison, WI.

Acock, B., and Allen Jr., L. H. (1985). Crop responses to elevated carbon dioxide concentrations. *In* "Direct Effects of Increasing Carbon Dioxide on Vegetation" (B. R. Strain and J. D. Cure, eds.), Report DOE/ER-0238, pp. 53–97. U.S. Department of Energy, Office of Energy Research, Washington, DC.

Allen Jr., L. H., Boote, K. J., Jones, J. W., Jones, P. H., Baker, J. T., Albrecht, S. L., Wasehmann, R. S., and Kamuru, F. (1991). Carbon dioxide effects on growth, photosynthesis and evapotranspiration of rice at three nitrogen fertilizer levels. Series No. 062, "Response of Vegetation to Carbon Dioxide." U.S. Department of Energy, Carbon Dioxide Research Division, and U.S. Department of Agriculture, Agricultural Research Service, Washington, DC.

Amthor, J. S. (1991). Respiration in a future, higher-CO_2 world. *Plant Cell Environ.* **14,** 13–20.

Amthor, J. S., Mitchell, R. J., Runion, G. B., Rogers, H. H., Prior, S. A., and Wood, C. W. (1994). Energy content and construction costs of plants grown in elevated CO_2. *New Phytol.* **128,** 443–450.

Augé, R. M., Schekel, K. A., and Wample, R. L. (1987). Leaf water and carbohydrate status of VA mycorrhizal rose exposed to drought stress. *Plant Soil* **99,** 291–302.

Baker, J. T., Allen Jr., L. H., and Boote, K. J. (1990). Growth and yield responses of rice to carbon dioxide concentration. *J. Agric. Sci.* **115,** 313–20.

Ball, A. S. (1992). Degradation of plant material grown under elevated CO_2 conditions by *Streptomyces viridosporus. In* "Progress in Biotechnology, Vol. 7, Xylans and Xylanes" (J. Visser *et al.,* eds.), pp. 379–382. Elsevier, New York.

Barber, S. A., and Silberbush, M. (1984). Plant root morphology and nutrient uptake. *In* "Roots, Nutrients and Water Influx, and Plant Growth" (S. A. Barber and D. R. Bouldin, eds.), ASA Special Pub. No. 49, pp. 65–87. ASA, CSSA, and SSSA, Madison, WI.

BassiriRad, H., Thomas, R. B., Reynolds, J. F., and Strain, B. R. (1996). Differential responses of root uptake kinetics of NH^+_4 and NO^-_3 to enriched atmospheric CO_2 in field grown loblolly pine. *Plant Cell Environ.* **19,** 367–371.

Baxter, R., Gantley, M., Ashenden, T. W., and Farrar, J. F. (1994). Effects of elevated carbon dioxide on three grass species from montane pasture. II. Nutrient uptake, allocation and efficiency of use. *J. Exp. Bot.* **45,** 1267–1278.

Bazzaz, F. A., and Miao, S. L. (1993). Successional status, seed size, and responses of tree seedlings to CO_2, light, and nutrients. *Ecology* **74,** 104–112.

Berntson, G. M., and Woodward, F. I. (1992). The root system architecture and development of *Senecio vulgaris* in elevated CO_2 and drought. *Funct. Ecol.* **6,** 324–333.

Bhattacharya, S., Bhattacharya, N. C., and Strain, B. R. (1985). Rooting of sweet potato stem cuttings under CO_2-enriched environment and with IAA treatment. *Hort. Sci.* **20,** 1109–1110.

Billings, W. D., Peterson, K. M., Luken, J. O., and Mortensen, D. A. (1984). Interaction of increasing atmospheric carbon dioxide and soil nitrogen on the carbon balance of tundra microcosms. *Oecologia* **65,** 26–29.

Bowen, G. D. (1973). Mineral nutrition of ectomycorrhizae. *In* "Ectomycorrhizae: Their Ecology and Physiology" (G. C. Marks and T. T. Kozlowski, eds.), pp. 151–205. Academic Press, New York.

Bowler, J. M., and Press, M. C. (1993). Growth responses of two contrasting upland grass species to elevated CO_2 and nitrogen concentration. *New Phytol.* **124,** 515–522.

Brinkman, R., and Sombroek, W. G. (1996). The effects of global change on soil conditions in relation to plant growth and food production. *In* "Global Climatic Change and Agricultural Production. Direct and Indirect Effects of Changing Hydrological, Pedological, and Plant Physiological Processes" (F. Bazzaz and W. Sombroek, eds.), pp. 49–63. John Wiley & Sons, New York.

Brown, K., and Higginbotham, K. O. (1986). Effects of carbon dioxide enrichment and nitrogen supply on growth of boreal tree seedlings. *Tree Physiol.* **2,** 223–232.

Bunce, J. A. (1992). Light, temperature and nutrients as factors in photosynthetic adjustment to an elevated concentration of carbon dioxide. *Physiol. Plant.* **86,** 173–179.

Caldwell, M. M., Dudley, L. M., and Lilieholm, B. (1992). Soil solution phosphate, root uptake kinetics and nutrient acquisition: Implications for a patchy soil environment. *Oecologia* **89,** 305–309.

Chaudhuri, U. N., Burnett, R. B., Kirkham, M. B., and Kanemasu, E. T. (1986). Effect of carbon dioxide on sorghum yield, root growth, and water use. *Agric. For. Meteorol.* **37,** 109–122.

Chaudhuri, U. N., Kirkham, M. B., and Kanemasu, E. T. (1990). Root growth of winter wheat under elevated carbon dioxide and drought. *Crop. Sci.* **30,** 853–857.

Chu, C. C., Coleman, J. S., and Mooney, H. A. (1992). Controls of biomass partitioning between roots and shoots: Atmospheric CO_2 enrichment and the acquisition and allocation of carbon and nitrogen in wild radish. *Oecologia* **89,** 580–587.

Coleman, D. C., Ingham, R. E., McClellan, J. F., and Trofymow, J. A. (1984). Soil nutrient transformations in the rhizosphere via animal–microbial interactions. *In* "Invertebrate–Microbial Interactions" (J. M. Anderson *et al.*, eds.), pp. 35–58. Cambridge University Press, New York.

Coleman, J. S., McConnaughay, K. D. M., and Bazzaz, F. A. (1993). Elevated CO_2 and plant nitrogen use: Is reduced tissue nitrogen concentration size dependent? *Oecologia* **93,** 195–200.

Comins, H. N., and McMurtrie, R. E. (1993). Long-term response of nutrient limited forests to CO_2 enrichment: Equilibrium behavior of plant–soil models. *Ecol. Appl.* **3,** 666–681.

Conroy, J. P. (1992). Influence of elevated atmospheric CO_2 concentrations on plant nutrition. *Aust. J. Bot.* **40,** 445–456.

Conroy, J., and Hocking, P. (1993). Nitrogen nutrition of C_3 plants at elevated atmospheric CO_2 concentrations. *Physiol. Plant.* **89,** 570–576.

Conroy, J., Barlow, E. W., and Bevege, D. I. (1986a). Response of *Pinus radiata* seedlings to carbon dioxide enrichment at different levels of water and phosphorus: Growth, morphology and anatomy. *Ann. Bot.* **57,** 165–177.

Conroy, J. P., Smillie, R. M., Küppers, M., Bevege, D. I., and Barlow, E. W. (1986b). Chlorophyll *a* fluorescence and photosynthetic and growth responses of *Pinus radiata* to phosphorus deficiency, drought stress, and high CO_2. *Plant Physiol.* **81,** 423–429.

Conroy, J. P., Küppers, M., Küppers, B., Virgona, J., and Barlow, E. W. R. (1988). The influence of CO_2 enrichment, phosphorus deficiency and water stress on the growth, conductance and water use of *Pinus radiata* D. Don. *Plant Cell Environ.* **11,** 91–98.

Conroy, J. P., Milham, P. J., Bevege, D. I., and Barlow, E. W. R. (1990a). Influence of phosphorus deficiency on the growth response of four families of *Pinus radiata* seedlings to CO_2-enriched atmospheres. *For. Ecol. Manag.* **30,** 175–188.

Conroy, J. P., Milham, P. J., Reed, M. L., and Barlow, E. W. (1990b). Increases in phosphorus requirements for CO_2-enriched pine species. *Plant Physiol.* **92,** 977–982.

Conroy, J. P., Milham, P. J., and Barlow, E. W. R. (1992). Effect of nitrogen and phosphorus availability on the growth response of *Eucalyptus grandis* to high CO_2. *Plant Cell Environ.* **15,** 843–847.

Coûteaux, M.-M., Mousseau, M., Célérier, M.-L., and Bottner, P. (1991). Increased atmospheric CO_2 and litter quality: Decomposition of sweet chestnut leaf litter with animal food webs of different complexities. *Oikos* **61,** 54–64.

Cruz, C., Lips, S. H., and Martins-Loução, M. A. (1993). The effect of nitrogen source on photosynthesis of carob at high CO_2 concentrations. *Physiol. Plant.* **89,** 552–556.

Cummings, M. B., and Jones, C. H. (1918). The aerial fertilization of plants with carbon dioxide. Vermont Agric. Exp. Sta. Bull. No. 211.

Cure, J. D., Israel, D. W., and Rufty Jr., T. W. (1988). Nitrogen stress effects on growth and seed yield of nonnodulated soybean exposed to elevated carbon dioxide. *Crop Sci.* **28,** 671–677.

Curtis, P. S., Drake, B. G., and Whigham, D. F. (1989). Nitrogen and carbon dynamics in C_3 and C_4 estuarine marsh plants grown under elevated CO_2 *in situ*. *Oecologia* **78,** 297–301.

Curtis, P. S., Vogel, C. S., Pregitzer, K. S., Zak, D. R., and Teeri, J. A. (1995). Interacting effects of soil fertility and atmospheric CO_2 on leaf area growth and carbon gain physiology in *Populus* x *euramericana* (Dode) Guinier. *New Phytol.* **129,** 253–263.

Davidson, E. A. (1995). Linkages between carbon and nitrogen cycling and their implications for storage of carbon in terrestrial ecosystems. *In* "Biotic Feedbacks in the Global Climatic System: Will the Warming Feed the Warming?" (G. M. Woodwell and F. T. Mackenzie, eds.), pp. 219–230. Oxford University Press, New York.

Davis, T. D., and Potter, J. R. (1983). High CO_2 applied to cuttings: Effects on rooting and subsequent growth in ornamental species. *Hort. Sci.* **18,** 194–196.

Davis, T. D., and Potter, J. R. (1989). Relations between carbohydrate, water status and adventitious root formation in leafy pea cuttings rooted under various levels of atmospheric CO_2 and relative humdity. *Physiol. Plant.* **77,** 185–190.

Del Castillo, D., Acock, B., Reddy, V. R., and Acock, M. C. (1989). Elongation and branching of roots on soybean plants in a carbon dioxide-enriched aerial environment. *Agron. J.* **81,** 692–695.

Drake, B. G. (1992). A field study of the effects of elevated CO_2 on ecosystem processes in a Chesapeake Bay wetland. *Aust. J. Bot.* **40,** 579–595.

Duchein, M.-C., Bonicel, A., and Betsche, T. (1993). Photosynthetic net CO_2 uptake and leaf phosphate concentrations in CO_2 enriched clover (*Trifolium subterraneum* L.) at three levels of phosphate nutrition. *J. Exp. Bot.* **44,** 17–22.

Duchesne, L. C. (1994). Role of ectomycorrhizal fungi in biocontrol. *In* "Mycorrhizae and Plant Health" (F. L. Pfleger and R. G. Linderman, eds.), pp. 27–45. American Phytopathological Society Press, St. Paul, MN.

El Kohen, A., Rouhier, H., and Mousseau, M. (1992). Changes in dry weight and nitrogen partitioning induced by elevated CO_2 depend on soil nutrient availability in sweet chestnut (*Castanea sativa* Mill). *Ann. Sci. For.* **49,** 83–90.

Enoch, H. Z., and Zieslin, N. (1988). Growth and development of plants in response to carbon dioxide concentrations. *Appl. Agric. Res.* **3,** 248–256.

Ferris, D. M., and Taylor, G. (1994). Increased root growth in elevated CO_2: A biophysical analysis of root cell elongation. *J. Exp. Bot.* **45,** 1603–1612.

Finn, G. A., and Brun, W. A. (1982). Effect of atmospheric CO_2 enrichment on growth, nonstructural carbohydrate content, and root nodule activity in soybean. *Plant Physiol.* **69,** 327–331.

Follett, R. F. (1993). Global climate change, U.S. agriculture, and carbon dioxide. *J. Prod. Agric.* **6,** 181–190.

Freckman, D. W. (1988). Bacterivorous nematodes and organic-matter decomposition. *In* "Biological Interactions in Soil" (C. A. Edwards, B. R. Stinner, D. Stinner, and S. Rabatin, eds.), pp. 195–217. Elsevier, New York.

French, C. J. (1989). Propagation and subsequent growth of *Rhododendron* cuttings: Varied response to CO_2 enrichment and supplementary lighting. *J. Am. Soc. Hort. Sci.* **114,** 251–259.

Gifford, M. (1992). Interaction of carbon dioxide with growth-limiting environmental factors in vegetation productivity: Implications for the global carbon cycle. *In* "Advances in Bioclimatology" (R. L. Desjardins, R. M. Gifford, T. Nilson, and E. A. N. Greenwood, eds.), Vol. I, pp. 24–58. Springer Verlag, Berlin.

Goudriaan, J., and de Ruiter, H. E. (1983). Plant growth in response to CO_2 enrichment, at two levels of nitrogen and phosphorus supply. I. Dry matter, leaf area and development. *Neth. J. Agric. Sci.* **31,** 157–169.

Goudriaan, J., and Zadoks, J. C. (1995). Global climate change: Modelling the potential responses of agro-ecosystems with special reference to crop protection. *Environ. Poll.* **87,** 215–224.

Grant, W. D., and Long, P. E. (1981). "Environmental Microbiology." Halsted Press, John Wiley & Sons, New York.

Grant, W. J. R., Fan, H. M., Downton, W. J. S., and Loveys, B. R. (1992). Effects of CO_2 enrichment on the physiology and propagation of two Australian ornamental plants, *Chamelaucium uncinatum* (Schauer) x *Chamelaucium floriferum* (MS) and *Correa schlechtendalii* (Behr). *Sci. Hort.* **52,** 337–342.

Griffin, K. L., Thomas, R. B., and Strain, B. R. (1993). Effects of nitrogen supply and elevated carbon dioxide on construction cost in leaves of *Pinus taeda* (L.) seedlings. *Oecologia* **95,** 575–580.

Grove, T. L., Riha, S. J., and Boulding, D. R. (1988). Relating nutrient and water uptake models to biotic interactions, nutrient cycling and plant growth. *In* "Biological Interactions in Soil" (C. A. Edwards, B. R. Stinner, D. Stinner, and S. Rabatin, eds.), pp. 361–368. Elsevier, New York.

Hardy, R. W. F., and Havelka, U. D. (1973). Symbiotic N_2 fixation: Multifold enhancement by CO_2-enrichment of field-grown soybeans. *Plant Physiol.,* **48** (Suppl.), 35 (Abstr.).

Hocking, P. J., and Meyer, C. P. (1985). Responses of Noogoora burr (*Xanthium occidentale* Bertol.) to nitrogen supply and carbon dioxide enrichment. *Ann. Bot.* **55,** 835–844.

Hocking, P. J., and Meyer, C. P. (1991a). Carbon dioxide enrichment decreases critical nitrate and nitrogen concentrations in wheat. *J. Plant Nutr.* **14,** 571–584.

Hocking, P. J., and Meyer, C. P. (1991b). Effects of CO_2 enrichment and nitrogen stress on growth, and partitioning of dry matter and nitrogen in wheat and maize. *Aust. J. Plant Physiol.* **18,** 339–356.

Hudson, R. J. M., Gherini, S. A., and Goldstein, R. A. (1994). Modeling the global carbon cycle: Nitrogen fertilization of the terrestrial biosphere and the "missing" CO_2 sink. *Global Biogeochem. Cycles* **8,** 307–333.

Huettl, R. F., and Zoettl, H. W. (1992). Forest fertilization: Its potential to increase the CO_2 storage capacity and to alleviate the decline of the global forests. *Water Air Soil Poll.* **64,** 229–249.

Hunt, R., Hand, D. W., Hannah, M. A., and Neal, A. M. (1991). Response to CO_2 enrichment in 27 herbaceous species. *Funct. Ecol.* **5,** 410–421.

Idso, K. E., and Idso, S. B. (1994). Plant responses to atmospheric CO_2 enrichment in the face of environmental constraints: A review of the past 10 years' research. *Agric. For. Meteorol.* **69,** 153–203.

Idso, S. B., and Kimball, B. A. (1991). Effects of two and a half years of atmospheric CO_2 enrichment on the root density distribution of three-year-old sour orange trees. *Agric. For. Meteorol.* **55,** 345–349.

Idso, S. B., and Kimball, B. A. (1992). Seasonal fine-root biomass development of sour orange trees grown in atmospheres of ambient and elevated CO_2 concentration. *Plant Cell Environ.* **15,** 337–341.

Ikram, A. (1990). Rhizosphere microorganisms and crop growth. *The Planter, Kuala Lumpur* **66,** 630–631.

Imai, K., and Coleman, D. F. (1983). Elevated atmospheric partial pressure of carbon dioxide and dry matter production of konjak (*Amorphophallus konjac* K. Koch). *Photosynth. Res.* **4,** 331–336.

Imai, K., and Murata, Y. (1976). Effect of carbon dioxide concentration on growth and dry matter production in crop plants. *Proc. Crop. Sci. Soc. Jap.* **45,** 598–606.

Imai, K., Coleman, D. F., and Yanagisawa, T. (1984). Elevated atmospheric partial pressure of carbon dioxide and dry matter production of cassava (*Manihot esculenta* Crantz). *Jap. J. Crop Sci.* **53,** 479–485.

Imai, K., Coleman, D. F., and Yanagisawa, T. (1985). Increase in atmospheric partial pressure of carbon dioxide and growth and yield of rice (*Oryza sativa* L.). *Jap. J. Crop Sci.* **54,** 413–418.

Ingham, R. E. (1988). Interactions between nematodes and vesicular-arbuscular mycorrhizae. *In* "Biological Interactions in Soil" (C. A. Edwards, B. R. Stinner, D. Stinner, and S. Rabatin, eds.), pp. 169–182. Elsevier, New York.

Israel, D. W., Rufty Jr., T. W., and Cure, J. D. (1990). Nitrogen and phosphorus nutritional interactions in a CO_2 enriched environment. *J. Plant Nutr.* **13,** 1419–1433.

Ito, T. (1970). Absorption and distribution of radioactive phosphorus in tomato plant with respect to the carbon dioxide concentration in the atmosphere. Tech. Bull. No. 18, pp. 21–28. Faculty of Hort., Chiba Univ., Japan.

Jenkinson, D. S. (1966). Studies on the decomposition of plant material in soil. II. Partial sterilization of soil and the soil biomass. *J. Soil Sci.* **17,** 280–302.

Johnson, D. W., Ball, T., and Walker, R. F. (1995). Effects of elevated CO_2 and nitrogen on nutrient uptake in ponderosa pine seedlings. *Plant Soil* **168/169,** 535–545.

Kimball, B. A., and Mauney, J. R. (1993). Response of cotton to varying CO_2, irrigation, and nitrogen: Yield and growth. *Agron. J.* **85,** 706–712.

Kirschbaum, M. U. F. (1993). A modelling study of the effects of changes in atmospheric CO_2 concentration, temperature and atmospheric nitrogen input on soil organic carbon storage. *Tellus* **45B,** 321–334.

Kuehny, J. S., Peet, M. M., Nelson, P. V., and Willits, D. H. (1991). Nutrient dilution by starch in CO_2-enriched chrysanthemum. *J. Exp. Bot.* **42,** 711–716.

Laforge, F., Lussier, C., Desjardins, Y., and Gosselin, A. (1991). Effect of light intensity and CO_2 enrichment during *in vitro* rooting on subsequent growth of plantlets of strawberry, raspberry and asparagus in acclimatization. *Sci. Hort.* **47,** 259–269.

Lamborg, M. R., Hardy, R. W. F., and Paul, E. A. (1983). Microbial effects. *In* "CO_2 and Plants: The Response of Plants to Rising Levels of Atmospheric CO_2" (E. R. Lemon, ed.), pp. 131–176. Am. Assoc. Adv. Sci., Washington, DC.

Larigauderie, A., Reynolds, J. F., and Strain, B. R. (1994). Root responses to CO_2 enrichment and nitrogen supply in loblolly pine. *Plant Soil* **165,** 21–32.

Leavitt, S. W., Paul, E. A., Kimball, B. A., Hendrey, G. R., Mauney, J., Rauschkolb, R., Rogers, H., Lewin, K., Nagy, J., Pintor, P., and Johnson H. B. (1994). Carbon isotope systematics of FACE cotton and soils. *Agric. For. Meteorol.* **70,** 87–101.

Lekkerkerk, L. J. A., van de Geijn, S. C., and Van Veen, J. A. (1990). Effects of elevated atmospheric CO_2-levels on the carbon economy of a soil planted with wheat. *In* "Soils and the Greenhouse Effect" (A. F. Bouwman, ed.), pp. 423–429. John Wiley, New York.

Lewis, J. D., Griffin, K. L., Thomas, R. B., and Strain, B. R. (1994a). Phosphorus supply affects the photosynthetic capacity of loblolly pine grown in elevated carbon dioxide. *Tree Physiol.* **14,** 1229–1244.

Lewis, J. D., Thomas, R. B., and Strain, B. R. (1994b). Effect of elevated CO_2 on mycorrhizal colonization of loblolly pine (*Pinus taeda* L.) seedlings. *Plant Soil* **165,** 81–88.

Liljeroth, E., Kuikman, P., and Van Veen, J. A. (1994). Carbon translocation to the rhizosphere of maize and wheat and influence on the turnover of native soil organic matter at different soil nitrogen levels. *Plant Soil* **161,** 233–240.

Lin, W. C., and Molnar, J. M. (1981). Effects of CO_2 mist and high intensity supplementary lighting on propagation of selected woody ornamentals. *Can. J. Plant Sci.* **61,** 965–969.

Linder, S., and McDonald, A. J. S. (1993). Plant nutrition and the interpretation of growth response to elevated concentrations of atmospheric carbon dioxide. *In* "Design and Execution of Experiments on CO_2 Enrichment" (E.-D. Schulze and H. A. Mooney, eds.), pp. 73–82. Office for Official Publications of the European Communities, Luxembourg, Belgium.

Luo, Y., Field, C. B., and Mooney, H. A. (1994). Predicting responses of photosynthesis and root fraction to elevated $[CO_2]_a$: Interactions among carbon, nitrogen, and growth. *Plant Cell Environ.* **17,** 1195–1204.

Luxmoore, R. J. (1981). CO_2 and phytomass. *BioScience* **31,** 626.

Luxmoore, R. J., O'Neill, E. G., Ells, J. M., and Rogers, H. H. (1986). Nutrient uptake and growth responses of Virginia pine to elevated atmospheric CO_2. *J. Environ. Qual.* **15,** 244–251.

Marschner, H. (1995). "Mineral nutrition in higher plants," 2nd ed. Academic Press, Orlando, FL.

Marx, D. H. (1973). Mycorrhizae and feeder root diseases. *In* "Ectomycorrhizae: Their Ecology and Physiology" (G. C. Marks and T. T. Kozlowski, eds.), pp. 351–382. Academic Press, New York.

Masle, J., Farquhar, G. D., and Wong, S. C. (1992). Transpiration ratio and plant mineral content are related among genotypes of a range of species. *Aust. J. Plant Physiol.* **19,** 709–721.

Masterson, C. L., and Sherwood, M. T. (1978). Some effects of increased atmospheric carbon dioxide on white clover (*Trifolium repens*) and pea (*Pisum sativum*). *Plant Soil* **49,** 421–426.

Masuda, T., Fujita, K., Kogure, K., and Ogata, S. (1989a). Effect of CO_2 enrichment and nitrate application on vegetative growth and dinitrogen fixation of wild and cultivated soybean varieties. *Soil Sci. Plant Nutr.* **35,** 357–366.

Masuda, T., Fujita, K., and Ogata, S. (1989b). Effect of CO_2 enrichment and nitrate application on growth and dinitrogen fixation of wild and cultivated soybean plants during pod-filling stage. *Soil Sci. Plant Nutr.* **35,** 405–416.

McKee, I. F., and Woodward, F. I. (1994). CO_2 enrichment responses of wheat: Interactions with temperature, nitrate and phosphate. *New Phytol.* **127,** 447–453.

Melillo, J. M. (1983). Will increases in atmospheric CO_2 concentrations affect decay processes? *In* "Ecosystems Center Annual Report," pp. 10–11. Marine Biology Laboratory, Woods Hole, MA.

Monz, C. A., Hunt, H. W., Reeves, F. B., and Elliott, E. T. (1994). The response of mycorrhizal colonization to elevated CO_2 and climate change in *Pascopyrum smithii* and *Bouteloua gracilis*. *Plant Soil* **165,** 75–80.

Nepstad, D. C., de Carvalho, C. R., Davidson, E. A., Jipp, P. H., Lefebvre, P. A., Negreiros, G. H., da Silva, E. D., Stone, T. A., Trumbore, S. E., and Vieira, S. (1994). The role of deep roots in the hydrological and carbon cycles of Amazonian forests and pastures. *Nature* **372,** 666–669.

Newbery, R. M. (1994). Influence of elevated CO_2 and nutrient supply on growth and nutrient uptake of *Agrostis capillaris. Biol. Plant.* **36** (Suppl.), S285. (Abstr.).

Nilsson, L. O. (1993). Carbon sequestration in Norway spruce in south Sweden as influenced by air pollution, water availability, and fertilization. *Water Air Soil Poll.* **70,** 177–186.

Norby, R. J., and O'Neill, E. G. (1991). Leaf area compensation and nutrient interactions in CO_2-enriched seedlings of yellow-poplar (*Liriodendron tulipifera* L.). *New Phytol.* **117,** 515–528.

Norby, R. J., and Sigal, L. L. (1989). Nitrogen fixation in the lichen *Lobaria pulmonaria* in elevated atmospheric carbon dioxide. *Oecologia* **79,** 566–568.

Norby, R. J., O'Neill, E. G., and Luxmoore, R. J. (1986a). Effects of atmospheric CO_2 enrichment on the growth and mineral nutrition of *Quercus alba* seedlings in nutrient-poor soil. *Plant Physiol.* **82,** 83–89.

Norby, R. J., Pastor, J., and Melillo, J. M. (1986b). Carbon–nitrogen interactions in CO_2-enriched white oak: Physiological and long-term perspectives. *Tree Physiol.* **2,** 233–241.

Norby, R. J., O'Neill, E. G., Hood, W. G., and Luxmoore, R. J. (1987). Carbon allocation, root exudation and mycorrhizal colonization of *Pinus echinata* seedlings grown under CO_2 enrichment. *Tree Physiol.* **3,** 203–210.

Norby, R. J., Gunderson, C. A., Wullschleger, S. D., O'Neill, E. G., and McCracken, M. K. (1992). Productivity and compensatory responses of yellow poplar trees in elevated CO_2. *Nature* **357,** 322–324.

Norby, R. J., O'Neill, E. G., and Wullschleger, S. D. (1995). Belowground responses to atmospheric carbon dioxide in forests. *In* "Carbon forms and Functions in Forest Soils" (W. W. McFee and J. M. Kelly, eds.), pp. 397–418. Soil Science Society of America, Madison, WI.

O'Neill, E. G., Luxmoore, R. J., and Norby, R. J. (1987a). Elevated atmospheric CO_2 effects on seedling growth, nutrient uptake, and rhizosphere bacterial populations of *Liriodendron tulipifera* L. *Plant Soil* **104,** 3–11.

O'Neill, E. G., Luxmoore, R. J., and Norby, R. J. (1987b). Increases in mycorrhizal colonization and seedling growth in *Pinus echinata* and *Quercus alba* in an enriched CO_2 atmosphere. *Can. J. For. Res.* **17,** 878–883.

O'Neill, E. G., O'Neill, R. V., and Norby, R. J. (1991). Hierarchy theory as a guide to mycorrhizal research on large-scale problems. *Environ. Poll.* **73,** 271–284.

Overdieck, D., and Reining, E. (1986). Effect of atmospheric CO_2 enrichment on perennial ryegrass (*Lolium perenne* L.) and white clover (*Trifolium repens* L.) competing in managed model ecosystems. II. Nutrient uptake. *Acta Ecologica/Ecol. Plant.* **7,** 367–378.

Owensby, C. E., Coyne, P. I., and Auen, L. M. (1993a). Nitrogen and phosphorus dynamics of a tallgrass prairie ecosystem exposed to elevated carbon dioxide. *Plant Cell Environ.* **16,** 843–850.

Owensby, C. E., Coyne, P. I., Ham, J. M., Auen, L. M., and Knapp, A. K. (1993b). Biomass production in a tallgrass prairie ecosystem exposed to ambient and elevated levels of CO_2. *Ecol. Appl.* **3,** 644–653.

Paoletti, M. G., Foissner, W., and Coleman, D. (eds.). (1993). "Soil Biota, Nutrient Cycling, and Farming Systems." Lewis Publishers, Boca Raton, FL.

Patterson, D. T., and Flint, E. P. (1982). Interacting effects of CO_2 and nutrient concentration. *Weed Sci.* **30,** 389–394.

Paul, E. A., and Voroney, R. P. (1980). Nutrient and energy flows through soil microbial biomass. *In* "Contemporary Microbial Ecology" (D. C. Ellwood, ed.), pp. 215–237. Academic Press, New York.

Peñuelas, J., and Matamala, R. (1990). Changes in N and S leaf content, stomatal density and specific leaf area of 14 plant species during the last three centuries of CO_2 increase. *J. Exp. Bot.* **41,** 1119–1124.

Peñuelas, J., and Matamala, R. (1993). Variations in the mineral composition of herbarium plant species collected the last three centuries. *J. Exp. Bot.* **44,** 1523–1525.

Phillips, D. A., Newell, K. D., Hassell, S. A., and Felling, C. E. (1976). The effect of CO_2 enrichment on root nodule development and symbiotic N_2 reduction in *Pisum sativum* L. *Am. J. Bot.* **63,** 356–362.

Porter, M. A., and Grodzinski, B. (1984). Acclimation to high CO_2 in bean. *Plant Physiol.* **74,** 413–416.

Porter, M. A., and Grodzinski, B. (1985). CO_2 enrichment of protected crops. *Hort. Rev.* **7,** 345–398.

Post, W. M., Emanuel, W. R., and King, A. W. (1992). Soil organic matter dynamics and the global carbon cycle. *In* "World Inventory of Soil Emission Potentials" (N. H. Batjes and E. M. Bridges, eds.), pp. 107–119. Int. Soil Ref. Information Centre, Wageningen, The Netherlands.

Pregitzer, K. S. (1993). Impact of climate change on soil processes and soil biological activity. *In* "Global Climatic Change: Its Implications for Crop Protection" (D. Atkinson, ed.), BCPC Monograph No. 56, pp. 71–82. British Crop Protection Council, Farnham, Surrey, UK.

Pregitzer, K. S., Zak, D. R., Curtis, P. S., Kubiske, M. E., Teeri, J. A., and Vogel, C. S. (1995). Atmospheric CO_2, soil nitrogen, and turnover of fine roots. *New Phytol.* **129,** 579–585.

Prior, S. A., Rogers, H. H., Runion, G. B., and Hendrey, G. R. (1994a). Free-air CO_2 enrichment of cotton: Vertical and lateral root distribution patterns. *Plant Soil* **165,** 33–44.

Prior, S. A., Rogers, H. H., Runion, G. B., and Mauney, J. R. (1994b). Effects of free-air CO_2 enrichment on cotton root growth. *Agric. For. Meteorol.* **70,** 69–86.

Prior, S. A., Rogers, H. H., Runion, G. B., Kimball, B. A., Mauney, J. R., Lewin, K. F., Nagy, J., and Hendrey, G. R. (1995). Free-air CO_2 enrichment of cotton: Root morphological characteristics. *J. Environ. Qual.* **24,** 678–683.

Prior, S. A., Runion, G. B., Mitchell, R. J., Rogers, H. H., and Amthor, J. S. (1997). Effects of atmospheric CO_2 on longleaf pine: Productivity and allocation as influenced by nitrogen and water. *Tree Physiol.* **17,** 397–405.

Pritchard, S., Peterson, C., Runion, G. B., Prior, S., and Rogers, H. (1997). Atmospheric CO_2 concentration, N availability, and water status affect patterns of ergastic substance deposition in longleaf pine (*Pinus palustris* Mill.) foliage. *Trees* **11,** 494–503.

Rabatin, S. C., and Stinner, B. R. (1988). Indirect effects of interactions between VAM fungi and soil-inhabiting invertebrates on plant processes. *In* "Biological Interactions in Soil" (C. A. Edwards, B. R. Stinner, D. Stinner, and S. Rabatin, eds.), pp. 135–146. Elsevier, New York.

Radoglou, K. M., and Jarvis, P. G. (1992). The effects of CO_2 enrichment and nutrient supply on growth morphology and anatomy of *Phaseolus vulgaris* L. seedlings. *Ann. Bot.* **70,** 245–256.

Radoglou, K. M., Aphalo, P., and Jarvis, P. G. (1992). Response of photosynthesis, stomatal conductance and water use efficiency to elevated CO_2 and nutrient supply in acclimated seedlings of *Phaseolus vulgaris* L. *Ann. Bot.* **70,** 257–264.

Rastetter, E. B., McKane, R. B., Shaver, G. R., and Melillo, J. M. (1992). Changes in C storage by terrestrial ecosystems: How C–N interactions restrict responses to CO_2 and temperature. *Water Air Soil Poll.* **64,** 327–344.

Reardon, J. C., Lambert, J. R., and Acock, B. (1990). The influence of carbon dioxide enrichment on the seasonal patterns of nitrogen fixation in soybeans. Series No. 016, "Response of Vegetation to Carbon Dioxide." U.S. Department of Energy, Carbon Dioxide Research Division, and U.S. Department of Agriculture, Agricultural Research Service, Washington, DC.

Reddy, V. R., Acock, B., and Acock, M. C. (1989). Seasonal carbon and nitrogen accumulation in relation to net carbon dioxide exchange in a carbon dioxide-enriched soybean canopy. *Agron. J.* **81,** 78–83.

Reeves, D. W., Rogers, H. H., Prior, S. A., Wood, C. W., and Runion, G. B. (1994). Elevated atmospheric carbon dioxide effects on sorghum and soybean nutrient status. *J. Plant Nutr.* **17,** 1939–1954.

Rogers, G. S., Payne, L., Milham, P., and Conroy, J. (1993). Nitrogen and phosphorus requirements of cotton and wheat under changing atmospheric CO_2 concentrations. *In* "Plant Nutrition—From Genetic Engineering to Field Practice" (N. J. Barrow, ed.), Proceedings of the Twelfth Int. Plant Nutrition Colloquium, September 21–26, 1993, Perth, Western Australia, pp. 257–260. Kluwer Academic Publishers, Dordrecht, The Netherlands.

Rogers, H. H., and Dahlman, R. C. (1993). Crop responses to CO_2 enrichment. *Vegetatio* **104/105,** 117–131.

Rogers, H. H., Bingham, G. E., Cure, J. D., Smith, J. M., and Surano, K. A. (1983). Responses of selected plant species to elevated carbon dioxide in the field. *J. Environ. Qual.* **12,** 569–574.

Rogers, H. H., Peterson, C. M., McCrimmon, J. N., and Cure, J. D. (1992a). Response of plant roots to elevated atmospheric carbon dioxide. *Plant Cell Environ.* **15,** 749–752.

Rogers, H. H., Prior, S. A., and O'Neill, E. G. (1992b). Cotton root and rhizosphere responses to free-air CO_2 enrichment. *Crit. Rev. Plant Sci.* **11,** 251–263.

Rogers, H. H., Runion, G. B., and Krupa, S. V. (1994). Plant responses to atmospheric CO_2 enrichment with emphasis on roots and the rhizosphere. *Environ. Poll.* **83,** 155–189.

Rogers, H. H., Prior, S. A., Runion, G. B., and Mitchell, R. J. (1996). Root to shoot ratio of crops as influenced by CO_2. *Plant Soil* **187,** 229–248.

Runion, G. B., Curl, E. A., Rogers, H. H., Backman, P. A., Rodríguez-Kábana, R., and Helms, B. E. (1994). Effects of free-air CO_2 enrichment on microbial populations in the rhizosphere and phyllosphere of cotton. *Agric. For. Meteorol.* **70,** 117–130.

Runion, G. B., Mitchell, R. J., Rogers, H. H., Prior, S. A., and Counts, T. K. (1997). Effects of resource limitations and elevated atmospheric CO_2 on ectomycorrhizae of longleaf pine. *New Phytol.* **137,** 681–689.

Samuelson, L. J., and Seiler, J. R. (1993). Interactive role of elevated CO_2, nutrient limitations, and water stress in the growth responses of red spruce seedlings. *For. Sci.* **39,** 348–358.

Sattelmacher, B., Gerendas, J., Thoms, K., Bruck, H., and Bagdady, N. H. (1993). Interaction between root growth and mineral nutrition. *Environ. Exp. Bot.* **33,** 63–73.

Schlesinger, W. H. (1986). Changes in soil carbon storage and associated properties with disturbance and recovery. *In* "The Changing Carbon Cycle: A Global Analysis" (J. R. Trabalka and D. E. Reichle, eds.), pp. 194–220. Springer Verlag, Berlin.

Schlesinger, W. H. (1990). Evidence from chronosequence studies for a low carbon-storage potential of soils. *Nature* **348,** 232–234.

Schlesinger, W. H. (1995). Soil respiration and changes in soil carbon stocks. *In* "Biotic Feedbacks in the Global Climatic System: Will the Warming Feed the Warming?" (G. M. Woodwell and F. T. Mackenzie, eds.), pp. 159–168. Oxford University Press, New York.

Seneweera, S., Milham, P., and Conroy, J. (1994). Influence of elevated CO_2 and phosphorus nutrition on the growth and yield of a short-duration rice (*Oryza sativa* L. cv. Jarrah). *Aust. J. Plant Physiol.* **21,** 281–292.

Shuman, L. M. (1994). Mineral nutrition. *In* "Plant–Environment Interactions" (R. E. Wilkinson, ed.), pp. 149–182. Marcel Dekker, New York.

Silvola, J., and Ahlholm, U. (1995). Combined effects of CO_2 concentration and nutrient status on the biomass production and nutrient uptake of birch seedlings (*Betula pendula*). *Plant Soil* **168/169,** 537–553.

Sinclair, T. R. (1992). Mineral nutrition and plant growth response to climate change. *J. Exp. Bot.* **43,** 1141–1146.

Sionit, N. (1983). Response of soybean to two levels of mineral nutrition in CO_2-enriched atmosphere. *Crop Sci.* **23,** 329–33.

Sionit, N., Mortensen, D. A., Strain, B. R., and Hellmers, H. (1981). Growth response of wheat to CO_2 enrichment and different levels of mineral nutrition. *Agron. J.* **73,** 1023–7.

Stewart, B. A. (1993). Managing crop residues for the retention of carbon. *Water Air Soil Poll.* **70,** 373–380.

Stewart, J. W. B., Victoria, R. L., and Wolman, G. (1992). Global cycles. *In* "An Agenda of Science for Environment and Development into the 21st Century" (J. C. I. Dodge *et al.*, eds.), pp. 129–140. Cambridge University Press, Cambridge, UK.

Stulen, I., and den Hertog, J. (1993). Root growth and functioning under atmospheric CO_2 enrichment. *Vegetatio* **104/105,** 99–115.

Thomas, J. F., and Harvey, C. N. (1983). Leaf anatomy of four species grown under continuous CO_2 enrichment. *Bot. Gaz.* **144,** 303–309.

Thompson, G. B., and Woodward, F. I. (1994). Some influences of CO_2 enrichment, nitrogen nutrition and competition on grain yield and quality in spring wheat and barley. *J. Exp. Bot.* **45,** 937–942.

Tiessen, H., Cuevas, E., and Chacon, P. (1994). The role of soil organic matter in sustaining soil fertility. *Nature* **371,** 783–785.

Tinker, P. B. (1984). The role of microorganisms in mediating and facilitating the uptake of plant nutrients from soil. *Plant Soil* **76,** 77–91.

Tissue, D. T., Thomas, R. B., and Strain, B. R. (1993). Long-term effects of elevated CO_2 and nutrients on photosynthesis and Rubisco in loblolly pine seedlings. *Plant Cell Environ.* **16,** 859–865.

Tognoni, F., Halevy, A. H., and Wittwer, S. H. (1967). Growth of bean and tomato plants as affected by root absorbed growth substances and atmospheric carbon dioxide. *Planta* **72,** 43–52.

Torbert, H. A., Prior, S. A., Rogers, H. H., Schlesinger, W. H., Mullins, G. L., and Runion, G. B. (1996). Elevated atmospheric CO_2 in agro-ecosystems affects groundwater quality. *J. Environ. Qual.* **25,** 720–726.

Tremblay, N., Yelle, S., and Gosselin, A. (1988). Effects of CO_2 enrichment, nitrogen and phosphorus fertilization during the nursery period on mineral composition of celery. *J. Plant Nutr.* **11,** 37–49.

Uren, N. C., and Reisenauer, H. M. (1988). The role of root exudates in nutrient acquisition. *Adv. Plant Nutr.* **3,** 79–114.

Whiting, G. J., Gandy, E. L., and Yoch, D. C. (1986). Tight coupling of root-associated nitrogen fixation and plant photosynthesis in the salt marsh grass *Spartina alterniflora* and carbon dioxide enhancement of nitrogenase activity. *Appl. Environ. Microbiol.* **52,** 108–113.

Williams, L. E., DeJong, T. M., and Phillips, D. A. (1981). Carbon and nitrogen limitations on soybean seedling development. *Plant Physiol.* **68,** 1206–1209.

Wittwer, S. H. (1978). Carbon dioxide fertilization of crop plants. *In* "Problems in Crop Physiology" (U. S. Gupta, ed.), pp. 310–333. Haryana Agric. University, Hissar, India.

Wolfe, D. W., and Erickson, J. D. (1993). Carbon dioxide effects on plants: Uncertainties and implications for modeling crop response to climate change. *In* "Agricultural Dimensions of Global Climate Change" (H. M. Kaiser and T. E. Drennen, eds.), pp. 153–178. St. Lucie Press, Delray Beach, FL.

Wong, S. C. (1979). Elevated atmospheric partial pressure of CO_2 and plant growth. I. Interactions of nitrogen nutrition and photosynthetic capacity in C_3 and C_4 plants. *Oecologia* **44,** 68–74.

Wong, S. C., and Osmond, C. B. (1991). Elevated atmospheric partial pressure of CO_2 and plant growth. III. Interactions between *Triticum aestivum* (C_3) and *Echinochloa frumentacea* (C_4) during growth in mixed culture under different CO_2, N nutrition and irradiance treatments, with emphasis on below-ground responses estimated using the $\delta^{13}C$ value of root biomass. *Aust. J. Plant Physiol.* **18,** 137–152.

Wong, S. C., Kriedemann, P. E., and Farquhar, G. D. (1992). $CO_2 \times$ nitrogen interaction on seedling growth of four species of eucalypt. *Aust. J. Bot.* **40,** 457–472.

Wood, C. W., Torbert, H. A., Rogers, H. H., Runion, G. B., and Prior, S. A. (1994). Free-air CO_2 enrichment effects on soil carbon and nitrogen. *Agric. For. Meteorol.* **70,** 103–116.

Woodin, S., Graham, B., Killick, A., Skiba, U., and Cresser, M. (1992). Nutrient limitation of the long term response of heather [*Calluna vulgaris* (L.) Hull] to CO_2 enrichment. *New Phytol.* **122,** 635–642.

Woodward, F. I. (1992). Predicting plant responses to global environmental change. *New Phytol.* **122,** 239–251.

Woodward, F. I., Thompson, G. B., and McKee, I. F. (1991). The effects of elevated concentrations of carbon dioxide on individual plants, populations, communities and ecosystems. *Ann. Bot.* **67,** 23–38.

Wullschleger, S. D., Norby, R. J., and Gunderson, C. A. (1997). Forest trees and their response to atmospheric CO_2 enrichment—a compilation of results. *In* "Advances in CO_2 Effects Research" (L. H. Allen, Jr., M. B. Kirkham, D. M. Olszyk, and C. E. Whitman, eds.), pp. 79–100. ASA, CSSA, and SSSA, Madison, WI.

Wullschleger, S. D., Post, W. M., and King, A. W. (1995). On the potential for a CO_2 fertilization effect in forests: Estimates of the biotic growth factor based on 58 controlled-exposure studies. *In* "Biotic Feedbacks in the Global Climatic System: Will the Warming Feed the Warming?" (G. M. Woodwell and F. T. Mackenzie, eds.), pp. 85–107. Oxford University Press, New York.

Yelle, S., Gosselin, A., and Trudel, M. J. (1987). Effect of atmospheric CO_2 concentration and root zone temperature on growth, mineral nutrition, and nitrate reductase activity of greenhouse tomato. *J. Am. Soc. Hort. Sci.* **112,** 1036–1040.

Zak, D. R., Pregitzer, K. S., Curtis, P. S., Teeri, J. A., Fogel, R., and Randlett, D. L. (1993). Elevated atmospheric CO_2 and feedback between carbon and nitrogen cycles. *Plant Soil* **151,** 105–117.

Ziska, L. H., and Teramura, A. H. (1992). Intraspecific variation in the response of rice (*Oryza sativa*) to increased CO_2—photosynthetic, biomass and reproductive characteristics. *Physiol. Plant.* **84,** 269–276.

9

Rhizosphere Processes under Elevated CO_2

Weixin Cheng

I. Introduction

Atmospheric concentration of CO_2 is predicted to double in the next century (Keeling *et al.*, 1989). An increase in CO_2 may result in an increase in primary production (Schimel, 1995) and the subsequent allocation and utilization of this photosynthetically fixed carbon in the rhizosphere has been predicted to be important for soil organic matter and nutrient dynamics (Van Veen *et al.*, 1991). Many important aspects of plant–soil interactions are mediated by rhizosphere processes, such as plant nutrient acquisition (Uren and Reisennauer, 1988), colonization of rhizosphere microorganisms (Baker, 1991; Miller, 1990), and soil organic matter decomposition (Sallih and Bottner, 1988; Cheng and Coleman, 1990). Understanding rhizosphere processes in relation to the effect of increased atmospheric CO_2 concentration is important for predicting the response of soil nutrients and organic matter to global environmental changes because of the unique role that the rhizosphere plays in plant–soil interactions.

The physiological response of plants to elevated atmospheric CO_2 has received considerable attention because CO_2 is a substrate for photosynthesis and its atmospheric concentration is predicted to double in the next century if the current trend continues (Keeling *et al.*, 1989). What is probably most challenging is to predict functional changes in belowground systems in response to CO_2 increase (Mooney, 1991). Compared to aboveground components, considerably less attention has been given to belowground components. This is especially true for rhizosphere processes such as root exudation, root respiration, rhizosphere microbial respiration, and root-induced or root-suppressed soil organic matter decomposition.

Rhizosphere processes play an important role in carbon sequestration and nutrient cycling in terrestrial ecosystems (Helal and Sauerbeck, 1989; Van Veen *et al.*, 1991). The rhizosphere has been identified as one of the key fine-scale components in the overall research of global carbon cycles (Coleman *et al.*, 1992). This chapter focuses on the interactions between atmospheric CO_2 levels and rhizosphere processes and discusses the feedback between rhizosphere processes and other ecosystem components.

II. Rhizodeposition

Rhizodeposition was first defined by Whipps and Lynch (1985) as all material loss from plant roots, including water-soluble exudates, secretions of insoluble materials, lysates, dead fine roots, and gases such as CO_2 and ethylene. It is generally believed that carbon input from the roots is the driving force for most of the rhizosphere processes. Many experiments have been carried out to quantify rhizodeposition. Using nutrient solution cultures under gnotobiotic conditions, the amount of rhizodeposition has been quantified to be less than 0.6 g g^{-1} of root dry weight (Newman, 1985; Whipps, 1990) for seedlings a few weeks old. However, addition of microorganisms may increase the amount of rhizodeposition (Martin 1977, Barber and Martin 1976). Due to these highly artificial conditions, this value must represent considerable underestimation. Using ^{14}C-labeling techniques, rhizodeposition has been quantified under more realistic conditions. Values of rhizodeposition measured by this labeling technique may range between 30 and 90% of the carbon transferred to belowground components of various plant–soil systems. An in-depth review of this topic has been given by Whipps (1990).

Plants grown under elevated CO_2 conditions often exhibit increased growth and often a disproportional increase in C allocation to belowground components (Curtis *et al.*, 1990; Norby *et al.*, 1986; Rogers *et al.*, 1992). The effect of elevated CO_2 on plant root growth has been covered extensively by Rogers *et al.* (1994) in a recent review and is, therefore, not included in this chapter.

In a growth chamber study using continuous ^{14}C-labeling techniques, Kuikman *et al.* (1991) reported that total rhizodeposition of wheat (*Triticum aestivum*) plants was significantly higher when plants were grown under 700 ppm CO_2 than 350 ppm CO_2, mostly due to increased root growth and belowground $^{14}CO_2$ output under elevated CO_2. By the end of the experiment ^{14}C-labeled plant C residue left in the soil was also higher under elevated CO_2 than ambient CO_2. They also reported an increase in rhizodeposition per unit of root dry weight under elevated CO_2, suggesting a changed root activity in response to elevated CO_2. In another study using similar techniques, Billes *et al.* (1993) also reported that rhizodeposition of wheat plants was significantly higher when plants were grown under 700 ppm CO_2 than 350 ppm CO_2, due to the increased size of the root system under elevated CO_2. There was no difference in rhizodeposition per unit of root dry weight between ambient and elevated CO_2 treatments, which indicated that root growth was the overall driving variable for rhizosphere processes, and other root-associated processes were in proportion to total root growth. Similar to the results of Billes *et al.* (1993), Rouhier *et al.* (1996), using a ^{14}C pulse-labeling technique, also reported significant increase in total rhizodeposition when sweet chestnut seedlings (*Castanea sativa*) were grown under elevated CO_2. Using ^{14}C pulse labeling in a growth chamber study with wheat and perennial ryegrass (*Lolium perenne*), Paterson *et al.* (1996) reported that elevated CO_2 significantly increased carbon allocation of current photosynthate to the rhizosphere, possibly as soluble deposition. In a growth chamber experiment using ^{13}C natural tracers, rhizodeposition of wheat plants was significantly higher under elevated CO_2 than under ambient CO_2, due to the increased root growth under elevated CO_2 (Cheng and Johnson, 1998). In a pot study using ^{13}C natural tracers and grown birch (*Betula pendula*) seedlings in free-air CO_2 enrichment rings, Ineson *et al.* (1996) reported that rhizodeposition was significantly higher under elevated CO_2 than under ambient CO_2 as measured by the plant-derived carbon input from the seedlings at the end of the experiment. By analyzing data from both an open-top chamber experiment and a small pot experiment, Hungate *et al.* (1997) reported that elevated CO_2 significantly increased rhizodeposition as indicated by the increased carbon partitioning to rapidly cycling carbon pools belowground.

Based on limited numbers of reports, elevated CO_2 tends to increase rhizodeposition either due to increased root growth or increased deposition per unit of root mass, or due to both. However, because of methodological limitations, available information on this topic so far comes from either growth chamber studies or pot studies. The disturbance and the artificial conditions inherent with these growth chamber studies may limit the application of these results to undisturbed natural ecosystems. How rhizodeposition changes in response to an increased atmospheric CO_2 concentration in natural ecosystems awaits investigation.

III. Root Exudate Quality and Quantity

It is likely that the amount of root exudates will increase if root growth is enhanced when plants are grown under elevated atmospheric CO_2. To predict the response of plant–soil systems to higher atmospheric CO_2, it is essential to investigate the effects of elevated CO_2 concentrations on the quality and quantity of root exudates. However, virtually no report on this topic is available. Results from a growth chamber experiment with wheat indicated that water-soluble carbon concentration in the rhizosphere of plants grown under elevated CO_2 was much higher than plants under ambient CO_2, even though overall carbon allocation patterns between the two CO_2 treatments were unchanged (Cheng and Johnson, 1998).

Root exudation and associated processes may act as one of the key links between environmental changes and soil organic matter dynamics in terrestrial ecosystems (Elliott *et al.*, 1979; Ingham *et al.*, 1985; Norby *et al.*, 1987; Curtis *et al.*, 1989; Kuikman *et al.*, 1990; Abbadie *et al.*, 1992; Bormann *et al.*, 1993; Zak *et al.*, 1993). This key link may invoke a positive feedback in total ecosystem carbon sequestration under increased atmospheric CO_2 conditions (Zak *et al.*, 1993; Billes *et al.*, 1993) because increased exudation may exacerbate the so-called "priming" effect (stimulation of soil organic matter decomposition caused by the addition of labile substrates) on soil nitrogen mineralization, which may, as a consequence, increase plant growth due to the increased nitrogen availability. Microbial activities in the rhizophere may be largely controlled by the quality and quantity of root exudates.

IV. Rhizosphere Respiration

Rhizosphere respiration consists of root respiration and rhizosphere microbial respiration, which uses current plant materials as its substrate. Rhizosphere respiration is a component of rhizodeposition. Total rhizosphere respiration of annual plants grown under ambient CO_2 concentration may use 20–80% of the plant C transfer to the belowground system via roots (Martin and Kemp, 1986; Lambers, 1987; Lambers *et al.*, 1991; Whipps, 1990). For carbon budgeting purposes, rhizosphere respiration is a significant source of carbon in all plant–soil studies.

Continuous ^{14}C-labeling studies have shown that plants grown in elevated CO_2 concentration allocate more carbon to total rhizosphere respiration either proportionally (Whipps, 1985; Billes *et al.*, 1993) or disproportionally (Kuikman *et al.*, 1991) relative to root growth. The range of this net increase in rhizosphere respiration is from about 20% to as high as 82% above the ambient CO_2 treatment. In all three experiments, $^{14}CO_2$ evolved from the root–soil system has been treated as one category—rhizosphere respiration. Functionally, rhizosphere respiration consists of root respiration and microbial respiration utilizing root-derived materials (rhizomicrobial respiration). These two functional categories of rhizosphere respiration have rarely been separated *in situ* because of technical difficulties in partitioning respiration between roots and their rhizosphere associates (Helal and Sauerbeck,

1991). The ecological and biogeochemical implications of root respiration are different from those of rhizosphere microbial respiration. Root respiration is a direct release of photosynthetically fixed C, whereas rhizosphere microbial respiration is a process through which rhizosphere microorganisms utilize photosynthetically fixed C. This latter process may have a profound impact on soil carbon and nitrogen dynamics.

Although total rhizosphere respiration seems very sensitive to any changes in environmental conditions and plant species (Whipps, 1984, 1990), very little is known about which portion of rhizosphere respiration varies more in response to environmental changes. Some reports suggest that root exudation varies substantially depending on the soil type and environmental conditions, such as anoxia, mechanical force, water stress, nutrient status, temperature, pH, and length of day (e.g., Barber and Gunn, 1974; Hale and Moore, 1979; Wiedenroth and Poskuta, 1981; Lee and Gaskins, 1982; Martin and Foster, 1985; Merckx *et al.*, 1987; Smucker and Erickson, 1987; Meharg and Killham, 1990). Other reports indicate that root respiration and exudation may co-vary in response to different environmental conditions (Martin, 1977; Hemrika-Wagner *et al.*, 1982; Bloom *et al.*, 1992; Barneix *et al.*, 1984; Kuiper 1983; Lambers *et al.*, 1981; Kuiper and Smid, 1985; Palta and Nobel, 1989). Very little information is available about the effects of elevated atmospheric CO_2 on the relative partitioning between root respiration and rhizosphere microbial respiration, possibly due to the limited availability of usable methods.

The isotopic trapping method of Cheng *et al.* (1993, 1994) may potentially be used to partition root respiration from rhizosphere microbial respiration under different atmospheric CO_2 concentrations. Briefly, isotopic trapping is achieved through substrate competition. If ^{14}C of low weight molecules is exuded by roots after a pulse labeling of the shoot and simultaneously taken up by the rhizosphere microorganisms, then adding glucose to the rhizosphere will reduce the microbial uptake of ^{14}C exudates (or trap the ^{14}C exudates) since the microorganisms will use both the added glucose and the exudates instead of relying solely on the root exudates. Therefore, rhizosphere microbial respiration ($^{14}CO_2$ evolved from microbial utilization of exudates) must be inversely proportional to the glucose–^{12}C concentration in the rhizosphere, whereas root respiration is independent of the glucose–^{12}C concentration in the rhizosphere. The following equation can be written for each concentration of glucose and a family of two equations can be established for two glucose concentrations added to the same experiment:

$$\frac{^{14}CO_2 \text{ with glucose–root resp.}}{^{14}CO_2 \text{ without glucose–root resp.}} = \frac{\text{soluble C (g liter}^{-1})}{\text{soluble C (g liter}^{-1}) + \text{glucose C (g liter}^{-1})}.$$

The two unknown variables in the equation are the $^{14}CO_2$ evolution rate that is due solely to root respiration and the soluble C concentration in the rhizosphere of intact plants. By solving two equations in a family simultaneously, the values of the two variables can be determined.

Results from a growth chamber experiment with wheat using this isotopic trapping method indicated that both root respiration and rhizosphere microbial respiration under elevated CO_2 were much higher than under ambient CO_2, even though overall carbon allocation patterns between the two CO_2 treatments were unchanged (Cheng and Johnson, 1998). One of the limitations of this method is that it can only give relative values of root respiration and rhizomicrobial respiration during the short period of about 4–6 h.

A natural ^{13}C tracer method for measuring total rhizosphere respiration and soil organic carbon decomposition has been introduced (Cheng, 1996). This method can be used to study the effect of elevated CO_2 on total rhizosphere respiration and soil organic matter decomposition under growth chamber settings. The principle of this ^{13}C natural tracer method is based on the difference in $^{13}C:^{12}C$ ratio (often reported as a $\delta^{13}C$ value) between plants with the C_3 photosynthetic pathway, whose mean $\delta^{13}C$ is $-27‰$, and plants with the C_4 pathway, whose mean $\delta^{13}C$ is $-12‰$ (Smith and Epstein, 1971), and on the subsequent difference between soil organic matter derived from the two types of plants. Soil organic matter derived from C_4 plants (C_4-derived soil) such as tropical grasslands has $\delta^{13}C$ values ranging from -12 to $-20‰$, whereas $\delta^{13}C$ values of soil organic matter derived from cold and temperate forest (C_3-derived soil) range from -24 to $-29‰$. If one grows C_3 plants in a C_4-derived soil, or grows C_4 plants in a C_3-derived soil, the carbon entering the soil via roots will have a different $\delta^{13}C$ value than the $\delta^{13}C$ value of the soil. Based on Cerri *et al.* (1985), the following equation can be used to partition soil-derived C_4-carbon from plant-derived C_3-carbon:

$$C_3 = C_t\ (\delta_t - \delta_4)\ /\ (\delta_3 - \delta_4), \qquad (1)$$

where $C_t = C_3 + C_4$, is the total carbon from belowground CO_2, C_3 is the amount of carbon derived from C_3 plants, C_4 is the amount of carbon derived from C_4 soil, δ_t is the $\delta^{13}C$ value of the C_t carbon, δ_3 is the $\delta^{13}C$ value of the C_3 plant carbon, and δ_4 is the $\delta^{13}C$ value of the C_4 soil carbon.

This method has the following advantages compared to ^{14}C labeling of either the plants or the soil:

1. Both the plants and the soil are uniformly and naturally labeled, so that the problem of nonuniformity in the ^{14}C-labeling approach is eliminated.
2. No radiation safety problems are involved.
3. Much effort is saved in labeling because it happens naturally.

But there are these disadvantages:

1. Low sensitivity and low resolution due to both the natural variation of $\delta^{13}C$ values in plant or soil samples and the relatively small difference in the $\delta^{13}C$ value between C_3 and C_4 plants.
2. It is limited to C_3 plants on C_4-derived soils or C_4 plants on C_3 soils only.
3. Sample analysis is relatively slow and expensive.

In a growth chamber experiment using this ^{13}C natural tracer method, total rhizosphere respiration of wheat plants was significantly higher under elevated CO_2 than under ambient CO_2 (Cheng and Johnson, 1998). Based on data from a pot study using a ^{13}C natural tracer method, Hungate *et al.* (1997) reported that total rhizosphere respiration of grass species significantly increased under elevated CO_2 compared to ambient CO_2. Using dual natural tracers of ^{13}C and ^{18}O in a sun-lit growth chamber study, Lin *et al.* (in press) also reported that total rhizosphere respiration of 4-year old Douglas fir seedlings was significantly higher when tree seedlings were grown under elevated CO_2 than under ambient CO_2.

V. Rhizosphere Effects on Soil Organic Matter Decomposition

An elevated CO_2 concentration may indirectly influence soil organic matter decomposition. Indirect effects such as increased plant photosynthesis, altered litter quality (C:N ratio), the projected rise in temperature, as well as changes in soil moisture have received considerable attention (Anderson, 1992; Post *et al.*, 1992; Nie *et al.*, 1992; Peterjohn *et al.*, 1993). However, very little attention has been given to the effect of elevated CO_2 concentration on original soil organic matter decomposition via plant roots. It is this process that has the greatest potential to link the increased CO_2 concentration with soil carbon sequestration/loss and soil nitrogen cycling.

The effects of elevated CO_2 on soil N availability are of paramount importance to effects on ecosystem C accumulation because (1) N is the most frequently limiting nutrient in the northern hemisphere, and (2) soil N pools are very large relative to vegetation N requirements, with only a small fraction (typically 1% or less) being available for uptake at any given time. There is some evidence that elevated CO_2 can affect soil carbon and nitrogen mineralization through rhizosphere effects. Körner and Arnone (1992) found a reduction in soil C and increases in soil respiration and nitrate leaching in an artificial tropical ecosystem subjected to elevated CO_2. They attributed these results to an increased rate of soil organic matter decomposition in the rhizosphere. Similarly, Zak *et al.* (1993) found increased microbial biomass C and N mineralization in the rhizosphere soils of *Populus grandidentata* seedlings subjected to elevated CO_2. These findings have considerable implications for the ability of rapidly growing forests to acquire more N from the soil in times of high N demand. On the other hand, if elevated CO_2 causes rhizodeposition of labile C with a high C:N ratio, it may increase microbial N demand and the immobilization rather than the mineralization of available N (Diaz *et al.*, 1993), potentially causing a nutrient feedback in the opposite direction of that posed by Zak *et al.* (1993).

Contradictory data exist in the literature about the effects of living roots on soil organic matter decomposition. Decomposition of labeled plant material was markedly lowered in the presence of cultivated plant cover or in natural grasslands when compared to bare soil controls. This result was due to the difference of the physical environment between the plant-covered soil and the fallow soil (Fuhr and Sauerbeck, 1968; Shields and Paul, 1973; Jenkinson, 1977). During laboratory experiments under controlled conditions, Reid and Goss (1982, 1983) and Sparling *et al.* (1982) observed that when ^{14}C-labeled plant material was decomposed in soil planted with maize, ryegrass, wheat, or barley, $^{14}CO_2$ release from the soil was reduced compared to bare soil controls. They surmised that this negative effect of living roots on soil organic matter decomposition was due to both root uptake of organic C and the competition between the roots and the rhizosphere microflora for substrates. In contrast, a stimulatory effect of living roots on soil organic matter decomposition was reported based on laboratory experiments (Helal and Sauerbeck, 1984, 1985, 1986, 1987; Cheng and Coleman, 1990). These authors suggested that the breakdown of soil aggregates and the stimulation of rhizosphere microflora were the cause of this phenomenon. Sallih and Bottner (1988) demonstrated in a 2-yr study that the presence of plants suppressed the decomposition of newly incorporated ^{14}C-labeled plant material during the first 200 days of decomposition and stimulated the mineralization of the ^{14}C in the soil during the latter stage compared to bare soil. It is likely that the response of total microbial metabolism is what determines the effect of roots on soil organic matter decomposition (Cheng and Coleman, 1990). If the presence of roots induces greater microbial growth, it will stimulate loss of soil organic matter and nitrogen mineralization. If the presence of plants reduces microbial growth, it will also reduce the loss of soil organic matter and tend to immobilize nitrogen.

Increasing atmospheric CO_2 may change the effect of rhizosphere processes on soil organic matter decomposition. Alternatively, it could (1) exacerbate the stimulatory effect of living roots on soil organic matter decomposition, (2) reduce the stimulatory effect, (3) aggravate the suppressive effect, (4) reduce the suppressive effect, or (5) have no effect. The first outcome has been assumed to be more likely (Luxmoore, 1981; Körner and Arnone, 1992; Zak *et al.*, 1993; Billes *et al.*, 1993). However, results from the limited number of studies are inconsistent. Stimulatory (Billes *et al.*, 1993; Zak *et al.*, 1993), suppressive (Kuikman *et al.*, 1991; Rouhier *et al.*, 1994), and neutral (Liljeroth *et al.*, 1990; Lin *et al.*, in press) results have been reported. Results from a growth chamber experiment using ^{13}C natural tracers showed that soil nitrogen status was important in determining the directions of the effect of elevated CO_2 on soil organic matter decomposition (Cheng and Johnson, 1998). Elevated CO_2 increased soil organic matter decomposition in the nitrogen-added treatment but decreased soil organic matter decomposition without nitrogen addition. In a microcosm study using yellow birch (*Betula alleghaniensis*), Berntson and Bazzaz (1997) found that elevated CO_2 increased soil organic matter decomposition (as indirectly indicated by nitrogen mineralization) during the initial period, but decreased soil organic matter decomposition during the later period, suggesting that temporal dynamics might play an important role. The issue of elevated CO_2 effects on soil organic matter decomposition remains an open one, and more research is needed to understand these important processes and the effects of CO_2 on them.

VI. Rhizosphere Associations

A. Mycorrhizae

Mycorrhizal symbioses between plant roots and soil fungi are widespread in agricultural and natural ecosystems (Allen, 1994). It is widely known that under many circumstances mycorrhizal fungi can give the plant hosts a number of advantages, including (1) enhanced nutrient uptake of phosphorus (Jayachandran *et al.*, 1992; Pearson and Jakobsen, 1993), nitrogen (Finlay *et al.*, 1992), and other nutrients (Leake and Read, 1989; Marschner and Dell, 1994); (2) increased plant resistance to pathogens (Perrin, 1990); and (3) reduced plant drought stress (Davies *et al.*, 1992). Plant roots can be significantly affected by elevated CO_2 (Norby, 1994). It has been frequently reported that root dry weight increased in response to elevated CO_2 (Rogers *et al.*, 1994). Fine roots are the major sites of mycorrhizal infection, and often respond to elevated CO to a greater extent that coarse roots (Norby *et al.*, 1986). Under elevated CO_2 concentrations, there are entry points from mycorrhizae because of more fine roots. Therefore, whole plant mycorrhizal influence may increase in response to elevated CO_2 even if percent infection is similar under both elevated and ambient CO_2 conditions. Mycorrhizal fungi can be a major sink for plant photosynthate (Rygiwicz *et al.*, 1994) and may, therefore, affect plant carbon allocation.

Many reports presented data in percent mycorrhizal infection of fine roots (Table I). Elevated CO_2 increased percent infection of vesicular-

Table I Effect of Elevated CO_2 on Mycorrhizal Colonization of Roots

Plant species	Mycorrhizae	Growth conditions	Percent infection	Reference
Pascopyrum smithii	VAM	Growth chamber	↑	Monz *et al.* (1994)
Trifolium repens	VAM	Growth chamber	↑	Jongen *et al.* (1996)
Liriodendron tulipifera	VAM	Open-top chamber	NE	O. Neill *et al.* (1991)
Beilschmiedia pendula	VAM	Open-top chamber	↑	Lovelock *et al.* (1996)
Pinus taeda	EM	Green house	NE	Lewis *et al.* (1994)
Pinus echinata	EM	Growth chamber	↑	O'Neill *et al.* (1987)
Pinus radiata	EM	Growth chamber	NE	Conroy *et al.* (1990)
Pinus caribaea	EM	Growth chamber	NE	Conroy *et al.* (1990)
Pinus silvestris	EM	Growth chamber	↑	Ineichen *et al.* (1995)
Pinus silvestris	EM	Growth chamber	NE	Perez-Soba *et al.* (1995)
Quercus alba	EM	Growth chamber	↑	O'Neill *et al.* (1987)

[a] VAM, vesicular-arbuscular mycorrhizae; EM, ectomycorrhizae; NE, no significant effect; ↑, increased.

arbuscular mycorrhizae (VAM) in a study using *Pascopyrum smithii* in growth chambers (Monz *et al.*, 1994), and had no effect on percent VAM infection in another study using *Liriodendron tulipifera* in open-top chambers (O'Neill *et al.*, 1991, O'Neill, 1994). For ectomycorrhizae, elevated CO_2 either increased percent infection (O'Neill *et al.*, 1987; O'Neill, 1994; Norby *et al.*, 1987; Ineichen *et al.*, 1995) or had no effect on percent infection (Lewis *et al.*, 1994; Conroy *et al.*, 1990; Perez-Soba *et al.*, 1995). However, we need to be cautious about the interpretation of percent infection data. We can only use percent infection data to assess the degree of mycorrhizal colonization, not the total function of the fungal symbiont, which is more important in term of the interactions between plants and the fungal component. Higher percent infection does not necessarily translate into higher amount of total mycorrhizal hyphae, and vice versa. Measurements of total mycorrhizal hyphal length or biomass are definitely needed to solve this problem.

B. Biotic Nitrogen Fixation

Biotic fixation of atmospheric N_2 is one of the important ways for ecosystem N input, which bears critical implications for ecosystem feedback to elevated CO_2 because N is often limiting ecosystem primary production. There are two groups of biotic N_2 fixation, symbiotic associations such as root nodulating rhizobia and actinomycetes and asymbiotic N fixers (sometimes called free living). Symbiotic N fixers get all carbon from their plant hosts and provide available N to their hosts. Biological dinitrogen fixation requires high amounts of energy, especially those of an associative nature. At least 16 ATP molecules are consumed to convert one N_2 to two NH_3 molecules (Postgate, 1987) in addition to other processes required for associative N_2 fixation. If one of the primary effects of elevated CO_2 on plants is increased photosynthesis, N_2 fixers may benefit the most from an increased carbon availability because very often biotic nitrogen fixation is limited by carbon availability. So far, based on a limited database, elevated CO_2 tends to increase symbiotic N_2 fixation across several types of associations (Arnone and Gordan, 1990; Hardy and Havelka, 1976; Hibbs *et al.*, 1995; Masterson and Sherwood, 1978; Norby, 1987; Philips *et al.*, 1976; Thomas *et al.*, 1991; Tissue *et al.*, 1997) (Table II). This increased symbiotic N_2 fixation is primarily seen in total nitrogenase activity and in total plant nitrogen. However, we need to be cautious about this general indication because all the data are from growth chamber experiments under highly disturbed and highly controlled conditions (Table I). The significance of the potential increased symbiotic N_2 fixation under elevated CO_2 to ecosystem-level processes may not be directly extrapolated from these growth chamber results. Studies of biotic N_2 fixation in more realistic systems are definitely needed for our further understanding of this issue.

It is widely known that the rhizosphere is one of the important sites for potential associative free-living nitrogen fixation due to the favorable conditions in the rhizosphere (supply of carbon source, mainly root exudates, and the relatively low oxygen potential caused by root and microbial respiration in the rhizosphere). The list of free-living nitrogen-fixing bacteria continues to grow as more genera and species are described. It seems that most plant species in natural environments are colonized to some degree by free-living diazotrophs (Silvester and Musgrave, 1991; Kapulnik, 1991). Recent reports have indicated that the contribution of biologically fixed N_2 by free-living diazotrophs can be substantial in some ecosystems, such as savanna grasslands (Abbadie *et al.*, 1992) and pine forests (Bormann *et al.*, 1993). However, this subject remains controversial in broader perspectives. Some reported values of fixed N_2 by free-living diazotrophs exceed that which can possibly be supported by the estimated amount of carbon available to the diazotrophs. Much of the controversy stems from the energetic requirement of the nitrogen fixation process and the estimated amount of carbon available to the rhizosphere diazotrophs (Zuberer, 1990). If plants grown under elevated CO_2 increase their carbon input to the rhizosphere as discussed in the previous sections, asymbiotic diazotrophs in the rhizosphere may respond to this increased carbon source and fix more nitrogen, which can feed back to the plant primary production by increasing nitrogen availability. Whiting *et al.* (1986) reported a five- to sixfold increase in rhizosphere nitrogenase activity when the C_4 grass *Spartina alterniflora* was exposed to elevated CO_2. Crush (1994) reported that rhizosphere nitrogenase activity was not affected by growing several grasses

Table II Effect of Elevated CO_2 on Symbiotic N_2 Fixation[a]

Plant species/symbiont	Conditions	NM	SNA	TNA	TPN	Reference
Herbaceous plants						
Trifolium repens/Rhizobium	GC	NE	↑	↑	↑	Masterson and Sherwood (1978)
Trifolium repens/Rhizobium	GC	?	↑	↑	↑	Crush (1993)
Trifolium repens/Rhizobium	GC	↑	?	NE	?	Ryle *et al.* (1992)
Medicago sativa	GC	?	↑	↑	↑	Crush (1993)
Pisum sativum/Rhizobium	GC	↑	NE	↑	↑	Materson and Sherwood (1978)
Pisum sativum/Rhizobium	GC	↑	NE	↑	↑	Philips *et al.* (1976)
Glycine max/Bradyrhizobium	GC	↑	↑	↑	↑	Hardy and Havelka (1976)
Glycine max/Bradyrhizobium	GC	↑	NE	↑	?	Finn and Brun (1982)
Woody plants						
Robinia pseudoacadia/Rhizobium	GC	↑	NE	↑	NE	Norby (1987)
Gliricidia sepium/Rhizobium	GC	↑	?	?	↑	Thomas *et al.* (1991)
Gliricidia sepium/Rhizobium	GC	NE	↑	↑	↑	Tissue *et al.* (1997)
Alnus glutinosa/Frankia	GC	↑	NE	↑	NE	Norby (1987)
Alnus glutinosa/Frankia	OTC	?	↑	?	↑	Vogel and Curtis (1995)
Alnus rubra/Frankia	GC	NE	NE	↑	↑	Arnone and Gordon (1990)
Alnus rubra/Frankia	GC	NE	↑	NE	↑	Hibbs *et al.* (1995)
Eleagnus angustifolia/Frankia	GC	NE	NE	NE	NE	Norby (1987)

[a] GC, growth chamber; OTP, open-top chamber; NM, nodule mass; SNA, specific nitrogenase activity; TNA, total nitrogenase activity; TPN, total plant nitrogen; NE, no significant effect; ↑, increased; ?, not determined.

(*Lolium boucheanum, Plantago lanceolata, Pennisetum clandestinum*) under elevated CO_2 conditions in a growth chamber study. More studies on this issue are urgently needed.

VII. Rhizosphere-Based Communities

Microbial communities in the rhizosphere are largely controlled by root processes such as exudation, cell sloughing, and fine root turnover. The quality of these substrates differs even under existing CO_2 concentrations and may change differentially with CO_2 enrichment (Schwab *et al.*, 1983; Paterson *et al.*, 1996). Mostly due to methodological inadequacies, we have very limited information on rhizosphere microbial community structure and how it changes with season, phenology, maturity of the ecosystem, and in response to perturbation such as elevated CO_2. Elevated CO_2 concentrations may indirectly influence the composition of microbial communities and their function through changes in the quality and quantity of microbial substrates from plant roots (O'Neill, 1994). Changes in root-derived substrates may serve either to enhance or suppress the microbial activity and change the composition of microbial communities (Van Veen *et al.*, 1991).

The hypothesis that an increased input of plant-derived materials under elevated CO_2 may alter the composition of rhizosphere microbial communities was not supported by the results of Zak *et al.* (1996) using phospholipid fatty acid analysis. In a short (one growing season) open-top chamber experiment with *Populus grandidentata* in sandy subsurface soil boxes, they reported no significant changes in microbial community structures either in the rhizosphere or in the bulk soil. They suggested that one growing season might not be long enough for such changes to occur. The use of subsurface soil in their experiment might have also conditioned the results differently from the situation of surface soils because rhizosphere microbial communities were dually controlled by root-related processes and soil-related processes (Bachmann and Kinzel, 1992). The study by Ringelberg *et al.* (1997) did support this hypothesis. Using ester-linked polar lipid fatty acid technology, they found significant changes in rhizosphere microbial community composition in response to elevated CO_2 in a 4-year-long open-top chamber experiment with white oak (*Quercus alba* L.) growing in field soils at Oak Ridge, Tennessee. They mainly detected community changes in prokaryotic microorganisms. It is hoped a better understanding of this topic will come when more studies are carried out in the near future.

VIII. Summary

Rhizosphere processes are important in carbon sequestration and nutrient cycling in terrestrial ecosystems. It is critical to understand rhizosphere processes in relation to increased atmospheric CO_2 concentrations for predicting the potential changes of nutrient cycling and carbon sequestration in terrestrial ecosystems. It is generally believed that rhizodeposition is the driving force for most of the rhizosphere processes. Based on available results from either growth chamber or pot studies, elevated CO_2 tends to increase rhizodeposition either due to increased total root growth or increased deposition per unit of root mass, or both. However, due to the disturbance and the artificial conditions inherent with these studies, how rhizodeposition will change in response to an increased atmospheric CO_2 concentration in natural ecosystems awaits to be fully explored when suitable methods become available. The amount of root exudates may increase if root growth is enhanced when plants are grown under elevated atmospheric CO_2. However, virtually no report on this topic is available. Continuous ^{14}C-labeling studies have shown that plants grown in elevated CO_2 concentration allocate more carbon to total rhizosphere respiration. The range of this net increase in rhizosphere respiration is from about 20% to as high as 82% above the ambient CO_2 treatment.

Elevated CO_2 may alter rhizosphere processes, which, in turn, may affect soil organic matter decomposition. Alternatively, it may (1) exacerbate the stimulatory effect of living roots on soil organic matter decomposition, (2) reduce the stimulatory effect, (3) exacerbate the suppressive effect, (4) reduce the suppressive effect, or (5) have no effect. Results from the limited number of studies are inconsistent. Stimulatory, suppressive, and neutral results have been reported. The issue of elevated CO_2 effects on soil organic matter decomposition remains to be further investigated.

Elevated CO_2 tends to increase rhizosphere symbiants such as mycorrhizae and rhizobia across several types of associations. However, we need to be cautious about this general indication because all the data are from growth chamber experiments under highly disturbed and highly controlled conditions. The potential effect of elevated CO_2 on asymbiotic N_2 fixation in the rhizosphere needs urgent attention. Based on two reports, elevated CO_2 may or may not alter the structure of rhizosphere communities.

References

Abbadie, L., Mariotti, A., and Menaut, J. (1992). Independence of savanna grass from soil organic matter for their nitrogen supply. *Ecology* **73,** 608–613.

Allen, M. F. (1994). "The Ecology of Mycorrhizae," p. 184. Cambridge University Press, New York.

Anderson, J. M. (1992). Responses of soils to climate change. *Adv. Ecol. Res.* **22,** 163–210.

Arnone III, J. A., and Gordon, J. C. (1990). Effect of nodulation, nitrogen fixation and CO_2 enrichment on the physiology, growth and dry mass allocation of seedlings of *Alnus rubra* Bong. *New Phytol.* **116,** 55–66.

Bachmann, G., and Kinzel, H. (1992). Physiological and ecological aspects of the interactions between plant roots and the rhizosphere soil. *Soil Biol. Biochem.* **24,** 543–552.

Baker, R. (1991). Induction of rhizosphere competence in the biocontrol fungus *Trichoderma*. *In* "The Rhizosphere and Plant Growth" (D. L. Keister and P. B. Cregan, eds.) pp. 221–228. Kluwer Academic, Dordrecht, The Netherlands.

Barber, D. A., and Gunn, K. B. (1974). The effect of mechanical forces on the exudation of organic substances by the roots of cereal plants grown under sterile conditions. *New Phytol.* **73,** 39–45.

Barber, D. A., and Martin, J. K. (1976). The release of organic substances by cereal roots in soil. *New Phytol.* **76,** 69–80.

Barneix, A. J., Breteler, H., and Van de Geijn, S. C. (1984). Gas and ion exchanges in wheat roots after nitrogen supply. *Physiol. Plant.* **61,** 357–362.

Berntson, G. M., and Bazzaz, F. A. (1997). Nitrogen cycling in microcosms of yellow birch exposed to elevated CO_2: Simultaneous positive and negative below-ground feedbacks. *Global Change Biol.* **3,** 247–258.

Billes, G., Rouhier, H., and Bottner, P. (1993). Modifications of the carbon and nitrogen allocations in the plant (*Triticum aestivum* L.) soil system in response to increased atmospheric CO_2 concentration. *Plant Soil* **157,** 215–225.

Bloom, A. J., Sukrapanna, S. S., and Warner, R. L. (1992). Root respiration associated with ammonium and nitrate absorption and assimilation by barley. *Plant Physiol.* **99,** 1294–1301.

Bormann, B. T., Brmann, F. H., Bowden, W. B., Pierce, R. S., Hamburg, S. P., Wang, D., Snyder, M. C., Li, C. Y., and Ingersoll, R. C. (1993). Rapid N_2 fixation in pines, alder, and locust: Evidence from the sandbox ecosystem study. *Ecology* **74,** 583–598.

Cerri, C. C., Feller, C., Balesdent, J., Victoria, R., and Plenecassagne, A. (1985). Application du tracage isotopique naturel en ^{13}C a letude de la bynamique de la matiere oranique dans les sols. *Comptes Rend. Acad. Sci. Paris* **300,** 423–428.

Cheng, W. (1996). Measurement of rhizosphere respiration and organic matter decomposition using natural ^{13}C. *Plant Soil* **183,** 263–268.

Cheng, W., and Coleman, D. C. (1990). Effect of living roots on soil organic matter decomposition. *Soil Biol. Biochem.* **22,** 781–787.

Cheng, W., and Johnson, D. W. (1998). Effect of elevated CO_2 on rhizosphere processes and soil organic matter decomposition. *Plant Soil* **202,** 167–174.

Cheng, W., Coleman, D. C., Carroll, C. R., and Hoffman, C. A. (1993). *In situ* measurement of root respiration and soluble carbon concentrations in the rhizosphere. *Soil Biol. Biochem.* **25,** 1189–1196.

Cheng, W., Coleman, D. C., Carroll, C. R., and Hoffman, C. A. (1994). Investigating short-term carbon flows in the rhizospheres of different plant species using isotopic trapping. *Agron. J.* **86,** 782–791.

Coleman, D. C., Odum, E. P., and Crossley, D. A. (1992). Soil biology, soil ecology, and global change. *Biol. Fertil. Soils* **14,** 104–111.

Conroy, J. P., Milham, P. J., Reed, M. L., and Barlow, E. W. (1990). Increases in phosphorus requirements of CO_2-enriched pine species. *Plant Physiol.* **92,** 977–982.

Crush, J. R. (1993). Hydrogen evolution from root nodules of *Trifolium repens* and *Medicago sativa* plants grown under elevated atmospheric CO_2. *N. Z. J. Agric. Res.* **36,** 177–183.

Crush, J. R. (1994). Elevated CO_2 concentration and rhizosphere nitrogen fixation in four forage plants. *N. Z. J. Agric. Res.* **37,** 455–463.

Curtis, P. S., Drake, B. G., and Whigman, D. F. (1989). Nitrogen and carbon dynamics in C_3 and C_4 estuarine marsh plants grown under elevated CO_2 in situ. *Oecologia* **78,** 297–301.

Curtis, P. S., Baldmann, L. M., Drake, B. G., and Whigman, D. F. (1990). Elevated atmospheric CO_2 effects on belowground processes in C3 and C4 estuarine communities. *Ecology* **71,** 2001–2006.

Davies, F. T., Potter, J. R., and Linderman, R. G. (1992). Mycorrhiza and repeated drought exposure affect drought resistance and extra-radical hyphae development of pepper plants independent of plant size and nutrient content. *J. Plant Physiol.* **139,** 289–294.

Diaz, S., Grime, J. P., Harris, J., and McPherson, E. (1993). Evidence of feedback mechanism limiting plant response to elevated carbon dioxide. *Nature* **364,** 616–617.

Elliott, E. T., Coleman, D. C., and Cole, C. V. (1979). The influence of amoebae on the uptake of nitrogen by plants in gnotobiotic soil. *In* "The Soil-Root Interface" (J. L. Harley and R. S. Russel, eds.), pp. 221–229. Academic Press, London.

Finley, R. D., Frostegard, A., and Sonnerfeldt, A. M. (1992). Utilization of organic and inorganic nitrogen sources by ectomycorrhizal fungi in pure culture and in symbiosis with *Pinus contorta* Dougl. ex Loud. *New Phytol.* **120,** 105–115.

Finn, G. A., and Brun, W. A. (1982). Effect of atmospheric CO_2 enrichment on growth, nonstructural carbohydrate content and root nodule activity in soybean. *Plant Physiol.* **69,** 327–331.

Fuhr, F., and Sauerbeck, D. R. (1968). Decomposition of wheat straw in the field as influenced by cropping and rotation. *In* "Isotopes and Radiation in Soil Organic Matter Studies," pp. 241–250. IAEA, Vienna.

Hale, H. G., and Moore, L. D. (1979). Factors affecting root exudation. II: 1970–1978. *Adv. Agron.* **31,** 93–124.

Hardy, R. W. F., and Havelka, U. D. (1976). Photosynthate as a major limiting factor limiting nitrogen fixation by field grown legumes with emphasis on soybeans. *In* "Symbiotic Nitrogen Fixation in Plants" (P. S. Nutman, ed.), pp. 421–439. Cambridge University Press, London.

Helal, H. M., and Sauerbeck, D. R. (1984). Influence of plant roots on C and P metabolism in soil. *Plant Soil* **76,** 175–182.

Helal, H. M., and Sauerbeck, D. R. (1985). Transformation of ^{14}C labelled plant residues and development of microbial biomass in soil as affected by maize roots. *Landwirtsch. Forsch.* **38,** 104–109.

Helal, H. M., and Sauerbeck, D. R. (1986). Effect of plant roots on carbon metabolism of soil microbial biomass. *Z. Pflanzen. Bodenk.* **149,** 181–188.

Helal, H. M., and Sauerbeck, D. R. (1987). Direct and indirect influences of plant roots on organic matter and phosphorus turnover in soil. *INTECOL Bull.* **15,** 49–58.

Helal, H. M., and Sauerbeck, D. (1989). Carbon turnover in the rhizosphere. *Z. Pflanzen. Bodenk.* **152,** 211–216.

Helal, H. M., and Sauerbeck, D. (1991). Short term determination of the actual respiration rate of intact plant roots. *In* "Plant Roots and Their Environment" (B. L. Michael and H. Persson, eds.), pp. 88–92. Elsevier Science Publishers, Amsterdam.

Hemrika-Wagner, A. M., Kreuk, K. C. M., and Van derPlas, L. H. W. (1982). Influence of growth temperature on respiratory chraracteristics of mitochondria from callus-forming potato tuber discs. *Plant Physiol.* **70,** 602–605.

Hibbs, D. E., Chan, S. S., Castellano, M., and Niu, C. H. (1995). Response of red alder seedlings to CO_2 enrichment and water stress. *New Phytol.* **129,** 569–577.

Hungate, B. A., Holland, E. A., Jackson, R. B., Chapin III, F. S., Mooney, H. A., and Field, C. B. (1997). The fate of carbon in grasslands under carbon dioxide enrichment. *Nature* **388,** 576–579.

Ineichen, K., Wiemken, V., and Wiemkne, A. (1995). Shoots, roots and ectomycorrhiza formation of pine seedlings at elevated atmospheric carbon dioxide. *Plant Cell Env.* **18,** 703–707.

Ineson, P., Cotrufo, M. F., Bol, R., Harkness, D. D., and Blum, H. (1996). Quantification of soil carbon inputs under elevated CO_2: C_3 plants in a C_4 soil. *Plant Soil* **187,** 345–350.

Ingham, R. E., Trofymow, J. A., Ingham, E. R., and Coleman, D. C. (1985). Interactions of bacteria, fungi, and their nematode grazers: Effects of nutrient cycling and plant growth. *Ecol. Monog.* **55,** 119–140.

Jayachandran, K., Schwab, A. P., and Hentrick, B. A. D. (1992). Mineralization of organic phosphorus by vesicular-arbuscular mycorrhizal fungi. *Soil. Biol. Biochem.* **24,** 897–903.

Jenkinson, D. S. (1977). Studies on decomposition of plant material in soil: V. The effect of plant cover and soil type on the loss of carbon from ^{14}C labelled ryegrass decomposing under field conditions. *J. Soil Sci.* **28,** 424–434.

Jongen, M., Fay, P., and Jones, M. B. (1996). Effects of elevated carbon dioxide and arbuscular mycorrhizal infection on *Trifolium repens. New Phytol.* **132,** 413–423.

Kapulnik, Y. (1991). Nonsymbiotic nitrogen-fixing microorganisms. *In* "Plant Roots: The Hidden Half" (Y. Waisel, A. Eshel, and U. Kafkafi, eds.), pp. 703–716. Marcel Dekker, New York.

Keeling, C. D., Bacastow, R. B., Carter, A. F., Piper, S. C., Whorf, T. P., Heimann, M., Mook, W. G., and Roeloffzen, H. (1989). A three dimensional model of atmospheric CO_2 transport based on observed winds: 1. Analysis of observational data. *In* "Aspects of Climate Variability in the Pacific and the Western Americas" (D. H. Peterson, ed.), pp. 165–236. *Geophys. Monog.* **55,** Washington (USA).

Körner, C., and Arnone, J. A. (1992). Responses to elevated carbon dioxide in artificial tropical ecosystems. *Science* **257,** 1672–1675.

Kuikman, P. J., Jansen, A. G., Van Veen, J. A., and Zehnder, A. J. B. (1990). Protozoan predation and the turnover of soil organic carbon and nitrogen in the presence of plants. *Biol. Fert. Soils* **10,** 22–28.

Kuikman, P. J., Lekkerkerk, L. J. A., and Van Veen, J. A. (1991). Carbon dynamics od a soil planted with wheat under elevated CO_2 concentration. *In* "Advances in Soil Organic Matter Research: The Impact on Agriculture and the Environment" (W. S. Wilson, ed.), Special Publication 90, pp. 267–274. The Royal Society of Chemistry, Cambridge.

Kuiper, D. (1983). Genetic differentiation in *Plantago major:* Growth and root respiration and their role in phenotypic adaptation. *Physiologia. Plantarum* **57,** 222–230.

Kuiper, D., and Smid, A. (1985). Genetic differentiation and phenotypic plasticity in *Plantago major* ssp. major: 1. The effect of differences in level of irradiance on growth, photosynthesis, respiration and chlorophyll content. *Physiol. Plant.* **65,** 520–528.

Lambers, H. (1987). Growth, respiration, exudation and symbiotic associations: The fate of carbon translocated to the roots: *In* "Root Development and Function" (P. J. Gregory, J. V. Lake, and D. A. Rose, eds.), pp. 125–145. Cambridge University Press, London.

Lambers, H., Posthumus, F., Stulen, L., Lanting, L., Van de Dijk, S. J. and Hofstra, R. (1981). Energy metabolism of *Plantago major* var. major as dependent on the supply of nutrients. *Physiol. Plant.* **51,** 245–252.

Lambers, H., Van der Werf, A., and Konings, H. (1991). Respiratory patterns in roots in relation to their functioning. *In* "Plant Roots: The Hidden Half" (Y. Waisel, A. Eshel, and U. Kafkafi, eds.), pp. 229–263. Marcel Dekker, New York.

Leake, J. R., and Read, D. J. (1989). The biology of mycorrhiza in the Ericaceae XV. The effect of mycorrhizal infection on calcium uptake by *Calluna vulgaris* (L.) Hull. *New Phytol.* **113,** 535–545.

Lee, D. H., and Gaskins, M. H. (1982). Increased root exudation of ^{14}C-compounds by sorghum seedlings inoculated with nitrogen-fixing bacteria. *Plant Soil* **69,** 391–399.

Lewis, J. D., Thomas, R. B., and Strain, B. R. (1994). Effect of elevated CO_2 on mycorrhizal colonization of loblolly pine (*Pinus taeda* L.) seedlings. *Plant Soil* **165,** 81–88.

Liljeroth, J. A., Van Veen, J. A., and Miller, H. J. (1990). Assimilate translocation to the rhizosphere of two wheat lines and subsequent utilization by rhizosphere microorganisms at two nitrogen concentrations. *Soil Biol. Biochem.* **22,** 1015–1021.

Lin, G., Ehleringer, J. R., Rygiewicz, P. T., Johnson, M. G., and Tingey, D. T. (in press). Elevated CO_2 and temperature impacts on different components of soil CO_2 efflux in Douglas-fir Terracosms. *Global Chang. Biology.*

Lovelock, C. E., Kyllo, D., and Winter, K. (1996). Growth responses to vesicular-arbuscular mycorrhizae and elevated CO_2 in seedlings of a tropical tree, *Beilschmiedia pendula*. *Funct. Ecol.* **10,** 662–667.

Luxmoore, R. J. (1981). CO_2 and phytomass. *BioScience* **31,** 626.

Marschner, H., and Dell, B. (1994). Nutrient uptake in mycorrhizal symbiosis. *Plant Soil* **159,** 89–102.

Martin, J. K. (1977). Factors influencing the loss of organic carbon from wheat roots. *Soil Biol. Biochem.* **9,** 1–7.

Martin, J. K., and Foster, R. C. (1985). A model system for studying the biochemistry and biology of the root–soil interface. *Soil Biol. Biochem.* **17,** 261–269.

Martin, J. K., and Kemp, J. R. (1986). The measurement of C transfers within the rhizosphere of wheat grown in field plots. *Soil Biol. Biochem.* **18,** 103–107.

Masterson, C. L., and Sherwood, M. T. (1978). Some effects of increased atmospheric carbon dioxide on white clover (*Trifolium repens*) and pea (*Pisum sativum*). *Plant Soil* **49,** 421–426.

Meharg, A. A., and Killham, K. (1990). The effect of soil pH on rhizosphere carbon flow of *Lolium perenne*. *Plant Soil* **123,** 1–7.

Merckx, R. A., Dijkstra, A., den Hartog, A., and van Veen, J. A. (1987). Production of root-derived materials and associated microbial growth in soil at different nutrient levels. *Biol. Fertil. Soils* **5,** 126–132.

Miller, R. H. (1990). Soil microbiological inputs for sustainable agricultural systems. *In* "Sustainable Agricultural Systems" (C. A. Edwards, R. Lal, P. Madden, R. H. Miller, and G. House, eds.), pp. 614–623. Elsevier, Amsterdam, The Netherlands.

Monz, C. A., Hunt, H. W., Reeves, F. B., and Elliott, E. T. (1994). The response of mycorrhizal colonization to elevated CO_2 and climate change in *Pascopyrum smithii* and *Bouteloua gracilis*. *Plant Soil* **165,** 75–80.

Mooney, H. A. (1991). Biological response to climate change: An agenda for research. *Ecol. Appl.* **1,** 112–117.

Newman, E. I. (1985). The rhizosphere: Carbon sources and microbial populations. *In* "Ecological Interactions in Soil: Plants, Microbes and Animals" (A. H. Fitter, D. Atkinson, D. J. Read and M. B. Usher, eds.), pp. 107–121. Blackwell Scientific Publications, Oxford, UK.

Nie, D., Kirkham, M. B., Ballou, L. K., Lawlor, D. J., and Kanemasu, E. T. (1992). Changes in prairie vegetation under elevated carbon dioxide levels and two soil moisture regimes. *J. Vegetation Sci.* **3,** 673–678.

Norby, R. J. (1987). Nodulation and nitrogenase activity in nitrogen-fixing woody plants stimulated by CO_2 enrichment of the atmosphere. *Physiol. Plant.* **71,** 77–82.

Norby, R. J. (1994). Issues and perspectives for investigating root responses to elevated atmospheric carbon dioxide. *Plant Soil* **165,** 9–20.

Norby, R. J., O'Neill, E. G., and Luxmoore, R. J. (1986). Effects of CO_2 enrichment on the growth and mineral nutrition of Quercus alba seedlings in nutrient-poor soil. *Plant Physiol.* **82,** 83–89.

Norby, R. J., O'Neill, E. G., Hood, W. G., and Luxmoore, R. J. (1987). Carbon allocation, root exudation, and mycorrhizal colonization of *Pinus echinata* seedlings grown under CO_2 enrichment. *Tree Physiol.* **3,** 203–210.

O'Neill, E. G. (1994). Responses of soil biota to elevated atmospheric carbon dioxide. *Plant Soil* **165,** 55–65.

O'Neill, E. G., Luxmoore, R. J., and Norby, R. J. (1987). Increases in mycorrhizal colonization and seedling growth in *Pinus echinata* and *Quercus alba* in an enriched CO_2 atmosphere. *Can. J. For. Res.* **17,** 878–883.

O'Neill, E. G., O'Neill, R. V., and Norby, R. J. (1991). Hierarchy theory as a guide to mycorrhizal research on large-scale problems. *Env. Pollut.* **73,** 271–284.

Palta, J. A., and Nobel, P. S. (1989). Influence of soil O_2 and CO_2 on root respiration for *Agave deserti*. *Physiol. Plant.* **76,** 187–192.

Paterson, E., Rattray, E. A. S., and Killham, K. (1996). Effect of elevated atmospheric CO_2 concentration on C-partitioning and rhizosphere C-flow for three plant species. *Soil Biol. Biochem.* **28,** 195–201.

Pearson, J. N., and Jakobsen, I. (1993). The relative contribution of hyphae and roots to phosphorus uptake by arbuscylar mycorrhizal plants, measured by dual labeling with ^{32}P and ^{33}P. *New Phytol.* **124,** 481–488.

Perez-Soba, M., Dueck, T. A., Puppi, G., and Kuiper, P. J. C. (1995). Interactions of elevated CO_2, NH_3 and O_3 on mycorrhizal infection, gas exchange and N metabolism in saplings of Scots pine. *Plant Soil* **176,** 107–116.

Perrin, R. (1990). Interactions between mycorrhizae and diseases caused by soil-born fungi. *Soil Use Manag.* **6,** 189–195.

Peterjohn, W. T., Melillo, J. M., Bowles, F. P., and Steudle, P. A. (1993). Soil warming and trace gas fluxes: Experimental design and preliminary flux results. *Oecologia* **93,** 18–24.

Phillips, D. A., Newell, K. D., Hassell, S. A., and Felling, C. E. (1976). The effect of CO_2 enrichment on root nodule development and symbiotic N_2 fixation in *Pisum sativum* L. *Am. J. Bot.* **63,** 356–362.

Post, W. M., Pastor, J., King, A. W., and Emanuel, W. R. (1992). Aspects of the interaction between vegetation and soil under global change. *Water Air Soil Pollut.* **64,** 345–363.

Postgate, J. (1987). "Nitrogen Fixation," p. 13. Edward Arnold, London.

Reid, J. B., and Goss, M. J. (1982). Suppression of decomposition of ^{14}C-labelled plant roots in the presence of living roots of maize and perennial ryegrass. *J. Soil Sci.* **33,** 387–395.

Reid, J. B., and Goss, M. J. (1983). Growing crops and transformations of ^{14}C-labelled soil organic matter. *Soil Biol. Biochem.* **15,** 687–691.

Ringelberg, D. B., Stair, J. O., Almeida, J., Norby, R. J., O'Neill, E. G., and White, D. C. (1997). Consequences of rising atmospheric carbon dioxide levels for the belowground microbiota associated with white oak. *J. Environ. Qual.* **26,** 495–503.

Rogers, H. H., Peterson, C. M., McCrimmon, J. N., and Cure, J. D. (1992). Response of plant roots to elevated atmospheric carbon dioxide. *Plant Cell Env.* **15,** 749–752.

Rogers, H. H., and Runion, G. B. (1994). Plant responses to atmospheric CO_2 enrichment with emphasis on roots and the rhizosphere. *Env. Pollut.* **83,** 155–189.

Rouhier, H., Billes, G., Elkohen, A., Mousseau, M., and Bottner, P. (1994). Effect of elevated CO_2 on carbon and nitrogen distribution within a tree (*Castanea-sativa* Mill) soil system. *Plant Soil* **162,** 281–292.

Rouhier, H., Billes, G., Billes, L., and Bottner, P. (1996). Carbon fluxes in the rhizosphere of sweet chestnut seedlings (*Castanea sativa*) grown under two atmospheric CO_2 concentrations: ^{14}C partitioning after pulse labelling. *Plant Soil* **180,** 101–111.

Rygiewicz, P. T., and Andersen, C. P. (1994). Mycorrhizae alter quality and quantity of carbon allocated balow ground. *Nature (London)* **369,** 58–60.

Ryle, G. J. A., Powell, C. E., and Davidson, I. A. (1992). Growth of white clover, dependent on N_2 fixation, in elevated CO_2 and temperature. *Ann. Bot.* **70,** 221–228.

Sallih, Z., and Bottner, P. (1988). Effect of wheat (*Triticum aestivum*) roots on mineralization rates of soil organic matter. *Biol. Fertil. Soils* **7,** 67–70.

Schimel, D. S. (1995). Terrestrial ecosystems and the carbon cycle. *Global Change Biol.* **1,** 77–91.

Schwab, S. M., Menge, J. A., and Leonard, R. T. (1983). Quantitative and qualitative effects of phosporus on extracts and exudates of sudangrass roots in relation to vesicular-arbuscular mycorrhiza formation. *Plant Physiol.* **73,** 761–765.

Shields, J. A., and Paul, E. A. (1973). Decomposition of ^{14}C-labelled plant material under field conditions. *Can. J. Soil Sci.* **53,** 297–306.

Silvester, W. B., and Musgrave, D. R. (1991). Free-living diazotrophs. *In* "Biology and Biochemistry of Nitrogen Fixation" (M. J. Dilworth and A. R. Glenn, eds.), pp. 162–186. Elsevier, Amsterdam.

Smith, B. N., and Epstein, S. (1971). Two categories of $^{13}C/^{12}C$ ratios for higher plants. *Plant Physiol.* **47,** 380–384.

Smucker, A. J. M., and Erickson, A. E. (1987). Anaerobic stimulation of root exudates and disease of peas. *Plant Soil* **99,** 423–433.

Sparling, G. S., Cheshire, M. V., and Mundie, C. M. (1982). Effect of barley plants on the decomposition of ^{14}C-labelled soil organic matter. *J. Soil Sci.* **33,** 89–100.

Thomas, R. B., Richter, D. D., Ye, H., Heine, P. R., and Strain, B. R. (1991). Nitrogen dynamics and growth of seedlings of an N-fixing tree (*Gliricidia sepium*) exposed to elevated atmospheric carbon dioxide. *Oecologia* **8,** 415–421.

Tissue, D. T., Megonigal, J. P., and Thomas, R. B. (1997). Nitrogenase activity and N_2 fixation are stimulated by elevated CO_2 in a tropical N_2-fixing tree. *Oecologia* **109,** 28–33.

Uren, N. C., and Reisennauer, H. M. (1988). The role of root exudates in nutrient acquisition. *Adv. Plant Nutr.* **3,** 79–114.

Van Veen, J. A., Liljeroth, E., Lekkerkerk, L. J. A., and Van de Geijn, S. C. (1991). Carbon fluxes in plant–soil systems at elevated atmospheric CO_2 levels. *Ecol. Appl.* **1,** 175–181.

Vogel, C. S., and Curtis, P. S. (1995). Leaf gas exchange and nitrogen dynamics of N_2-fixing field-grown *Alnus glutinosa* under elevated atmospheric CO_2. *Global Change Biol.* **1,** 55–61.

Whipps, J. M. (1984). Environmental factors affecting the loss of carbon from the roots of wheat and barley seedlings. *J. Exp. Bot.* **35,** 767–773.

Whipps, J. M. (1985). Effect of CO_2-concentration on grow, carbon distribution and loss of carbon from the roots of maize. *J. Exp. Bot.* **36,** 644–651.

Whipps, J. M. (1990). Carbon economy. *In* "The Rhizosphere" (J. M. Lynch, ed.), pp. 59–97. John Wiley, New York.

Whipps, J. M., and Lynch, J. M. (1985). Energy losses by the plant in rhizodeposition. *Ann. Proc. Phytochem. Soc Eur.* **26,** 59–71.

Whiting, G. J., Gandy, E. L., and Yoch, D. C. (1986). Tight coupling of root-associated nitrogen-fixation and plant photosynthesis in the salt marsh grass *Spartina alterniflora* and carbon dioxide enhancement of nitrogenase activity. *Appl. Environ. Microbiol.* **52,** 108–113.

Wiedenroth, E., and Poskuta, J. (1981). The influence of oxygen deficiency in roots on CO_2 exchange rates of shoots and distribution of ^{14}C-photoassimilates of wheat seedlings. *Z. Pflanzenphysiol.* **103,** 459–467.

Zak, D. R., Pregitger, K. S., Curtis, P. S., Teeri, J. A., Fogel, R., and Randlett, D. L. (1993). Elevated atmospheric CO_2 and feedback between carbon and nitrogen cycles. *Plant Soil* **151,** 105–117.

Zak, D. R., Ringelberg, D. B., Pregitger, K. S., Randlett, D. L., White, D. C., and Curtis, P. S. (1996). Soil microbial communities beneath Populus grandidentata grown under elevated atmospheric CO_2. *Ecol. Appl.* **6,** 257–262.

Zuberer, D. A. (1990). Soil and rhizosphere aspects of N_2-fixing plant-microbe associations. *In* "The Rhizosphere" (J. M. Lynch, ed.), pp. 317–353. John Wiley & Sons, Chichester, UK.

10

Ecosystem Responses to Rising Atmospheric CO_2: Feedbacks through the Nitrogen Cycle

Bruce A. Hungate

I. Introduction

Rising atmospheric CO_2 could alter soil nitrogen (N) cycling, shaping the responses of many terrestrial ecosystems to elevated CO_2. Increased carbon input to soil through increased root growth, altered litter quality, and increased soil water content through decreased plant water use in elevated CO_2 can all affect soil N transformations and thus N availability to plants. Nitrogen limits net primary productivity (NPP) in many terrestrial ecosystems, so changes in N availability to plants will influence NPP in an elevated CO_2 environment. Furthermore, changes in NPP will alter carbon uptake by the terrestrial biosphere, and thus feed back to rising atmospheric CO_2. Elevated CO_2 could also influence the processes that regulate N inputs to and losses from ecosystems—N fixation, gaseous N losses (N_2, N_2O, NO_x), and N leaching. Such changes could alter ecosystem nitrogen stocks and thus nitrogen available to support NPP. Additionally, soil emissions of N_2O contribute to the greenhouse effect and stratospheric ozone destruction, and emissions of NO_x contribute to photochemical smog and acid rain. Thus, by altering soil nitrogen cycling, elevated CO_2 could cause other changes in atmospheric chemistry. Predicting these feedbacks requires that we understand what changes in soil nitrogen cycling caused by elevated CO_2 are likely, as well as how such changes might vary among terrestrial ecosystems. Here, I discuss the mechanisms through which elevated CO_2 can cause changes in soil nitrogen cycling, and review what changes have been observed and what mechanisms implicated in studies to date.

II. Soil Nitrogen Cycle

The soil nitrogen cycle comprises plant, microbial, and abiotic nitrogen transformations (Fig. 1A). Nitrogen mineralization converts organic nitrogen to ammonium, making it available for uptake by plants, immobilization by microbes, nitrification, and volatilization as ammonia. In most terrestrial ecosystems, ammonium and nitrate are the dominant forms of nitrogen that plants take up, so mineralization is an important bottleneck regulating nitrogen availability to plants (Chapin, 1995). Some plants acquire much of their nitrogen through associations with nitrogen-fixing bacteria (which convert atmospheric N_2 to organic nitrogen) or through uptake of dissolved organic nitrogen. Both of these processes "short circuit" the mineralization bottleneck (Chapin *et al.*, 1993), adding qualitatively different control points to the nitrogen cycle (Eviner and Chapin, 1997). A common view holds that the balance of mineralization and NH_4^+ immobilization determines how much NH_4^+ is left over for plant NH_4^+ uptake and nitrification (Rosswall, 1982; Tiedje *et al.*, 1981; Myrold and Tiedje, 1986). However, some recent results show that plant NH_4^+ uptake can limit microbial immobilization (Norton and Firestone, 1996), and that nitrifiers may not be restricted to heterotrophic leftovers (Davidson *et al.*, 1990). Nitrification (the autotrophic oxidation of ammonia to nitrate) is the major biological nitrogen transformation controlling ecosystem losses of nitrogen because it produces N_2O and NO_x, and also provides substrate for denitrification (conversion of nitrate to N_2, N_2O, and some NO_x) and for nitrate leaching.

III. Mechanisms through Which Elevated CO_2 Alters Soil Nitrogen Cycling

The responses of plants to elevated CO_2 provide the starting point for considering potential changes in soil nitrogen cycling. First, elevated CO_2 usually increases photosynthesis, often increasing plant growth. To the extent that greater plant growth increases their demand for belowground resources, carbon allocation to roots may increase (Rogers *et al.*, 1994). Increased root allocation can be manifested as increased standing root mass, but also as increased root turnover, respiration, and exudation (Norby, 1994; Day *et al.*, 1996; Cardon, 1996; Berntson and Bazzaz, 1996a), in all cases enhancing the flux of carbon from plants to soil (Van Veen *et al.*, 1991; van de Geijn *et al.*, 1993; Gorissen, 1996; Hungate *et al.*, 1997b). Greater availability of carbon to soil microorganisms is likely to alter soil nitrogen transformations (Fig. 1B).

Second, elevated CO_2 often decreases stomatal conductance and plant transpiration (Morrison, 1987). In ecosystems where plant canopies strongly

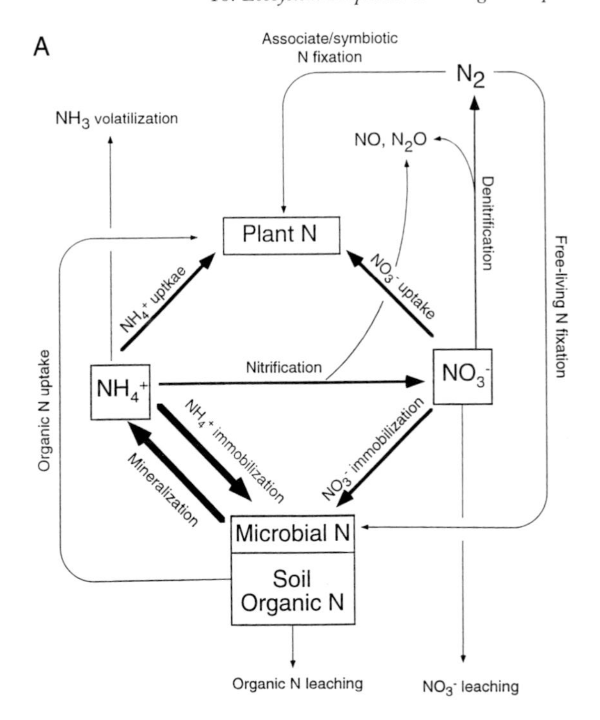

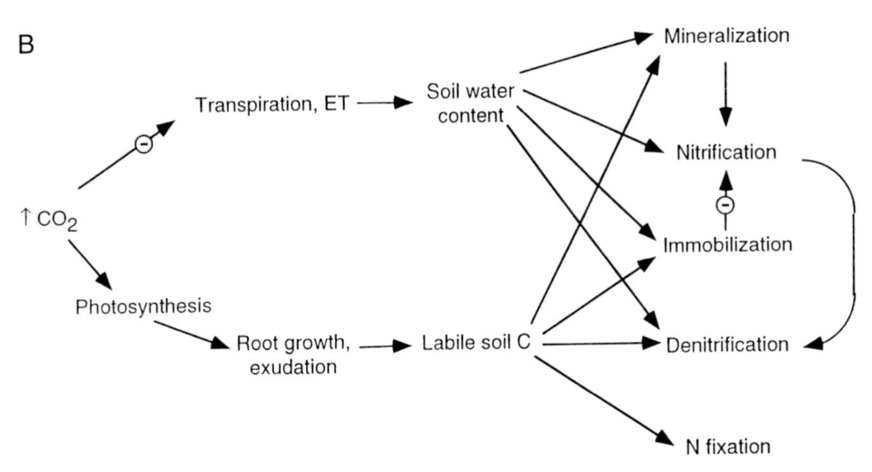

Figure 1 (A) Major nitrogen transformations in terrestrial ecosystems. Boxes show pools of nitrogen in the plant–soil system; arrows show fluxes between these pools and exchange with the atmosphere and groundwater. The magnitude of the flux is proportional to the thickness of the arrow. (B) Mechanisms through which increased carbon input to soil and increased soil water content alter microbial nitrogen transformations. Arrows marked with a minus indicate a negative effect; unmarked arrows indicate a positive effect.

influence evapotranspiration, reduced plant transpiration in elevated CO_2 can increase soil water content (Morgan *et al.*, 1994; Fredeen *et al.*, 1997; Hungate *et al.*, 1997a; Lutze, 1996; Bremer *et al.*, 1996), as long as there is no compensatory increase in leaf area (Field *et al.*, 1995). It is also possible that CO_2-induced increases in leaf area could more than compensate for lower transpiration per unit leaf area, resulting in greater ET in elevated CO_2, and thus drier soils. Changes in soil water content could alter soil nitrogen transformations in many ecosystems (Fig. 1B).

Third, elevated CO_2 usually decreases the nitrogen concentration of plant tissues, especially leaves, apparently due to reduced concentrations of Rubisco and other enzymes, as well as to greater accumulation of starch and other carbon storage compounds (Drake *et al.*, 1997). Reduced nitrogen concentrations in live plant tissue can lead to an increased litter carbon : nitrogen ratio (e.g., Cortrufo *et al.*, 1994), but apparently nitrogen retranslocation and carbon metabolism during senescence often eliminate the CO_2 effect on C : N (O'Neill and Norby, 1996; Canadell *et al.*, 1996; Hirschel *et al.*, 1997). This mechanism through which elevated CO_2 could alter soil nitrogen cycling has received considerable attention and is not discussed in detail here.

IV. Nitrogen Mineralization and Immobilization and Increased Carbon Input to Soil

The balance of nitrogen mineralization and immobilization depends on the carbon and nitrogen contents of microbial substrates. When carbon content is relatively high, immobilization will dominate and, conversely, when nitrogen content is relatively high, mineralization will dominate. Thus, the simplest prediction of the consequences of increased carbon input to soil is increased N immobilization by soil microorganisms. However, carbon input to soil can stimulate nitrogen mineralization. A burst of bacterial and fungal growth (caused by increased carbon availability) followed by protozoan and nematode grazing can cause nitrogen mineralization, because the nitrogen content of bacteria and fungi is relatively high compared to the nitrogen demand of their protozoan and nematode predators (Clarholm, 1985; Ingham *et al.*, 1985). The mineralization associated with the consumption of bacteria and fungi will be larger than the immobilization associated with their initial growth if the net effect of increased carbon input to soil was to transfer some of the total amount of organic nitrogen held in the total microbial N pool to the soil ammonium pool, or if the increase in carbon input to soil enhanced bacterial and fungal access to other sources of organic nitrogen (i.e., the "priming effect"). Although several studies have demonstrated that the "priming effect" occurs (Clarholm, 1985, 1989; Helal and Sauerbeck, 1986), its quantitative significance may be negligible (Griffiths and Robinson, 1992). Understanding which pool(s) of soil organic nitrogen are susceptible to release via priming should assume a high priority for future research.

Lambourgh *et al.* (1983) and Luxmoore (1981) predicted that increased carbon input to soil in elevated CO_2 would alter the balance of nitrogen mineralization and immobilization, but Lambourgh (1983) predicted a relatively stronger increase in mineralization, while Luxmoore (1981) predicted a larger stimulation of immobilization. The two earliest tests of these ideas came to opposite conclusions, thus providing empirical support for both. In these experiments, elevated CO_2 stimulated root growth, thereby increasing the pool of carbon in the soil microbial biomass in poplar monocultures (Zak *et al.*, 1993) and in grass and herb communities (Díaz *et al.*, 1993). In the poplar study, increased carbon input to soil stimulated net nitrogen mineralization (measured by laboratory incubations), presumably increasing nitrogen availability to plants (Zak *et al.*, 1993). In contrast, carbon input to soil enhanced nitrogen immobilization in the soil microbial biomass in the grass and herb communities, apparently decreasing nitrogen availability to plants (Díaz *et al.*, 1993).

Subsequent experiments have also yielded conflicting results (summarized in Table I), though none since Zak *et al.* (1993) has found direct evidence that increased carbon input to soil stimulates nitrogen mineralization via priming. Billès *et al.* (1993) and Rouhier *et al.* (1994, 1996) used $^{14}CO_2$ pulse labeling to investigate CO_2 effects on carbon flow and nitrogen pools in wheat and sweet chestnut. In both experiments, elevated CO_2 increased the flux of carbon from plant roots to soil. However, the effects of elevated CO_2 on plant and soil inorganic nitrogen in the two experiments were in direct opposition: In wheat (Billès *et al.*, 1993), elevated CO_2 increased the sum of plant and soil inorganic nitrogen, but in sweet chestnut (Rouhier *et al.*, 1994), elevated CO_2 decreased the sum of plant and soil inorganic nitrogen. In the wheat experiment, and in similar experiments where elevated CO_2 increases the total mass of plant nitrogen, because the stimulation of photosynthesis in elevated CO_2 may directly enhance root growth, and thereby increase the volume of soil mined for nitrogen (Berntson and Woodward, 1992), it is not possible to separate this direct effect of elevated CO_2 from an indirect effect on plant growth caused by increased mineralization. Indeed, the former is a simpler explanation than the priming effect for the observation that elevated CO_2 increases total plant nitrogen. Explicitly testing the effects of increased carbon input to soil on the balance of mineralization and immobilization requires measuring these processes directly, preferably *in situ* (Berntson and Bazzaz, 1996b).

Other experiments support the idea that carbon input to soil stimulates nitrogen immobilization, thereby decreasing nitrogen availability to plants.

Table I Summary of Studies Investigating Feedbacks to Plant Nitrogen Availability in Elevated CO_2[a]

Reference	System	↑ C input?	↑ soil H_2O?	Δ N avail	Δ MBN	Δ Plant N
Diaz *et al.* (1993)	Grassland microcosms (GC)	√	N.R.		↑	↓[b]
Zak *et al.* (1993)	Poplar (OTC)	√	N.R.	↑		
Billes *et al.* (1993)	Wheat (GC)	√	N.R.	(↑)		↑
Rouhier *et al.* (1996)	Sweet chestnut (GC)	√	N.R.	↓		↓
Morgan *et al.* (1994)	C_4 grassland (GC)	√	√	↓		↓
Johnson *et al.* (1997)	Pondersa pine (OTC)	√	N.R.	↓		
Matamala (1997)	Salt marsh (OTC)	√	0	↓		
Rice *et al.* (1994), Owensby *et al.* (1993)	C_4 grassland (OTC)	√	√		↑	↑
Hungate *et al.* (1996a)	Serpentine annuals (OTC)	√	0		↓ ↑	↓ ↑[c]
Hungate *et al.* (1996b)	Annual grasslands (OTC)	√	√	↑	0	↑
Berntson and Bazzaz (1996)	Yellow birch, deciduous forest (GC)	√	N.R.	↑	0	↑

[a] √ Indicates empirical support that elevated CO_2 caused increased C input to soil (support includes increased root growth, increased microbial biomass carbon, as well as increased C flow according to ^{14}C measurements), and that elevated CO_2 caused increased soil moisture. Arrows under "Δ N avail," "Δ MBN," and "Δ Plant N" indicate whether CO_2 decreased (↓), increased (↑), or caused no change (0) in nitrogen availability, microbial biomass nitrogen, and total plant nitrogen, respectively. Measures of nitrogen availability include mineralization rates (net or gross, *in situ* or laboratory assays) and extractable soil nitrogen. Microbial biomass nitrogen is included as a separate category, because changes in it do not necessarily indicate changes in nitrogen availability to plants. N.R., no response.

[b] Plant shoots only.

[c] Depended on plant species.

In monocultures of *Bouteloua gracilis,* a C_4 grass, elevated CO_2 increased root growth and VAM infection, indicating greater carbon input to soil, and resulting in less extractable soil nitrate and less total plant nitrogen uptake (Morgan *et al.,* 1994). Increased immobilization resulting from increased carbon input to soil could explain this reduction in nitrogen availability. Elevated CO_2 reduced extractable NH_4^+ and NO_3^- in soils planted with ponderosa pine after 3 yr, apparently due to increased immobilization in the soil microbial biomass (Henderson and Johnson, 1996). After 10 yr of exposure to elevated CO_2, extractable ammonium in soil was consistently lower throughout the summer growing season in a C_3 salt marsh, perhaps due to increased microbial immobilization of ammonium (Matamala, 1997). In this case, however, total plant nitrogen pools were not significantly different between CO_2 treatments (Matamala, 1997).

In some cases, though elevated CO_2 increases nitrogen immobilization in the soil microbial biomass, plant nitrogen pools also increase. For example, in an unfertilized C_4 grassland exposed to elevated CO_2 for 3 yr, elevated CO_2 increased nitrogen in the soil microbial biomass by 8–9% (Rice *et al.,* 1994) and total nitrogen pools in plants by 24% (calculated from Owensby *et al.,* 1993). Similarly, in six annual grasses and forbs under conditions where nutrients strongly limited plant growth, CO_2-induced changes in microbial nitrogen pools were parallel to CO_2-induced changes in plant nitrogen pools (Hungate *et al.,* 1996). Parallel increases in plant and microbial nitrogen pools in these experiments could be explained by increased mineralization via the priming effect (Zak *et al.,* 1993), though increased root growth and thus mining of nitrogen in a larger volume of soil may be a simpler explanation (Berntson and Woodward, 1992).

Berntson and Bazzaz (1996b) investigated the effects of elevated CO_2 on soil nitrogen cycling in deciduous forest microcosms. In these experiments, elevated CO_2 decreased gross nitrogen mineralization as well as both plant and microbial ammonium uptake, measured during a 48-h period after 14 months of treatment. The decrease in plant ammonium uptake was relatively larger than the decrease in microbial ammonium uptake. Sustaining these altered rates of microbial ammonium production and consumption over longer periods would cause reduced ammonium availability to plants, first because less ammonium is produced (i.e., gross mineralization decreases), but also because as a fraction of mineralized (i.e., available) ammonium, microbes immobilized relatively more in elevated compared to ambient CO_2 in this experiment. A reduction in the carbon quality of root-derived substrates could decrease both mineralization and immobilization (Berntson and Bazzaz, 1996b). Alternatively, decreased nitrogen cycling rates in this experiment could indicate a shift in the seasonal pattern of belowground activity under elevated CO_2. If, for example, plants were phenologically more advanced in elevated compared to ambient CO_2, and if in this more advanced stage of growth they allocated less carbon to roots and more to aboveground sinks, this could explain the reduced plant ammonium uptake. Reduced carbon allocation to roots would also slow the carbon input that stimulates microbial growth, and thus production and consumption of ammonium.

Several possible reasons could account for this wide range of results. First, the methods for determining nitrogen availability differ in these experiments, ranging from instantaneous availability of inorganic nitrogen measured by soil extractions, to rates of mineralization and immobilization measured over 1 or 2 d, to the net balance of mineralization and immobilization measured by the amount of nitrogen in the soil microbial biomass, to the size of the potentially mineralizable pool of soil nitrogen. It is likely that some of the variation among these studies is due to differences in methodology. For example, if elevated CO_2 stimulated nitrogen immobilization *in situ,* thereby increasing the size of the microbial nitrogen pool, one might expect a laboratory incubation to show increased nitrogen mineralization (Zak *et al.,* 1993), because nitrogen sequestered in the microbial biomass *in situ* is released once cut off from a continual supply of root-derived carbon.

Second, the timescales of the experiments differ, ranging from several weeks to several years. Even in systems where the responses of the plant–soil system to elevated CO_2 are very similar, it is very likely that those responses change through time as the system "equilibrates" after the step change in atmospheric CO_2 concentration used in these experiments. For example, Berntson and Bazzaz (1997) found that elevated CO_2 decreased mineralization, ammonium immobilization, and plant ammonium uptake over a 48-h period after 14 months of growth. Despite this, elevated CO_2 increased the total mass of nitrogen in plants, an integrated measure of nitrogen uptake over the entire 14-month period. Thus, net feedbacks to the availability of nitrogen to plants strongly depended on the timescale at which the question was addressed. In this case, elevated CO_2 apparently stimulated nitrogen acquisition early in the experiment, but negative feedbacks to nitrogen availability through reduced mineralization eventually caused a decline in plant ammonium uptake, a pattern that, if sustained, could limit plant growth responses to elevated CO_2.

Third, the plant species differ among these experiments. Plant responses to elevated CO_2 vary among plant species (Poorter, 1993), so it is not surprising that the effects of elevated CO_2 on soil nitrogen cycling might also vary among species. For example, the effects of elevated CO_2 on microbial nitrogen pools varied among six serpentine annuals, with some species showing increased and others decreased microbial nitrogen pools (Hungate *et al.,* 1996).

Finally, the effects of increased carbon input to soil on microbial nitrogen transformations may depend on the nutrient status of the soil. Just as litter decomposition depends on the relative availability of carbon and nutrients in the decomposing substrate (Melillo and Aber, 1982), the decomposition of soil organic matter in the rhizosphere may depend on soil nutrient status. Carbon sources in the rhizosphere vary simultaneously in carbon quality and nutrient content. Root-derived carbon substrates are lower in nitrogen content than older soil organic matter constituents, but root-derived carbon (excepting, perhaps, suberized cell walls) is more readily degraded (i.e., higher carbon quality) than older soil organic matter. Changes in nutrient availability can alter the relative rates at which microorganisms decompose these different carbon sources (Merckx *et al.*, 1987; Van Veen *et al.*, 1989; Liljeroth *et al.*, 1990), as described by the concept of "preferential substrate utilization" (Kuikman and Gorissen, 1993): Given adequate nutrient supply, microbes preferentially consume root-derived carbon (Van Veen *et al.*, 1989; Liljeroth *et al.*, 1990). If inorganic nutrients are scarce, root-derived carbon compounds accumulate in the soil as microbes preferentially degrade more nutrient-rich soil organic matter (Liljeroth *et al.*, 1990).

When some external factor (e.g., CO_2 concentration) stimulates root growth and causes greater input of root-derived carbon to the rhizosphere, changes in nitrogen availability to plants may thus depend on the nutrient status of the soil (Fig. 2). If nutrient availability is high, increased rhizodeposition in elevated CO_2 will drive nitrogen immobilization, because microbes preferentially consume the more labile root-derived carbon compounds and immobilize inorganic soil nitrogen to meet their nitrogen requirement. In this case, nitrogen availability to plants would decrease. If, on the other hand, nutrient availability is low, increased rhizodeposition may exacerbate microbial nutrient limitation. In this case, rhizodeposits may accumulate in the soil, causing little change in decomposition of soil organic matter; or, to the extent the priming effect occurs, increased rhizodeposition under strongly nutrient-limiting conditions may enhance microbial degradation of nutrient-rich soil organic matter. In this case, nitrogen availability to plants would increase.

Several experiments support the idea that decomposition in the rhizosphere in response to an increase in the supply of root-derived carbon depends on soil nutrient status. In wheat grown with high nitrogen supply, elevated CO_2 increased rhizodeposition, increased microbial use of rhizodeposits, and decreased decomposition of older soil organic matter (Lekkerkerk *et al.*, 1990; Kuikman *et al.*, 1991). Cardon (1996) also found that elevated CO_2 decreased decomposition of native soil organic matter, but only when soil nutrient availability was high. Under nutrient-limiting conditions, elevated CO_2 decreased microbial use of root-derived carbon

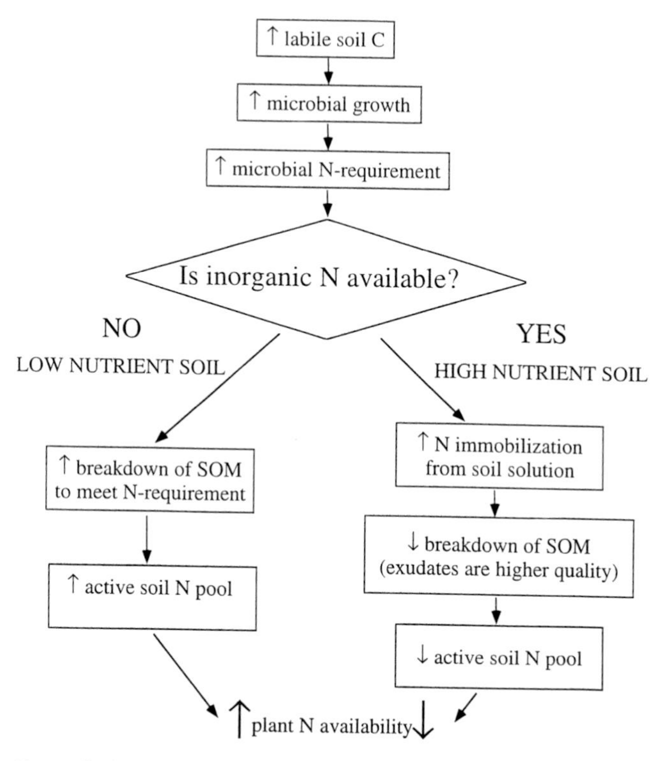

Figure 2 How soil nitrogen status may influence changes in nitrogen availability to plants that occur in response to increased carbon input to soil. See text for explanation.

compounds (Paterson *et al.*, 1996). In a recent experiment, however, elevated CO_2 had no effect on the decomposition of native soil organic matter at either low or high soil nutrient levels (van Ginkel *et al.*, 1996).

The nutrient dependence of decomposition in the rhizosphere may explain some of the variation in the direction of changes in nitrogen cycling observed in the studies cited. For example, Zak *et al.* (1993) grew their poplar trees in a "nutrient-poor soil," whereas Díaz *et al.* (1993) described their tall herb community as "high fertility." Given these albeit qualitative descriptions of soil nutrient status, the changes in nitrogen availability to plants observed in these studies match well the predictions of this framework.

V. Increased Carbon Flux to Soil and Nitrogen Inputs and Losses

Rates of transfer between organic and inorganic soil nitrogen pools (mineralization and immobilization) are the most important controls of nitrogen availability to plants on the timescale of months to several years. As the timescale of inquiry expands to decades and centuries (the timescale of CO_2-induced global change), the processes that regulate ecosystem nitrogen inputs and losses—biological nitrogen fixation, atmospheric deposition, gaseous nitrogen emissions (N_2, N_2O, and NO_x), and nitrogen leaching—become increasingly important in determining nitrogen availability and nitrogen limitation of plant growth. In addition to altering nitrogen availability to plants in the short term, elevated CO_2 could also affect the processes that regulate nitrogen inputs to and losses from ecosystems. A greater supply of fixed carbon in elevated CO_2 can enhance nitrogen fixation by energy-limited nitrogen-fixing bacteria (Hardy and Havalka, 1976; Poorter, 1993; Soussana and Hartwig, 1996). Less often considered, elevated CO_2 could also affect nitrification and denitrification, the microbial nitrogen transformations that return mineral nitrogen to the atmosphere, as dinitrogen, nitrous oxide, or nitric oxide. Recent work in this area focuses on the idea that increased carbon input to soil is likely the major mechanism altering ecosystem nitrogen inputs and losses.

Several modeling studies show that small changes in annual rates of nitrogen inputs or losses can appreciably affect ecosystem carbon uptake over the timescale of centuries (Comins and McMurtrie, 1993; Gifford *et al.*, 1996). Thus, although very small changes in nitrogen inputs or losses would have negligible effects on the nitrogen budget in short-term CO_2 experiments, if such changes were sustained over several decades or centuries as CO_2 accumulates in the atmosphere, they could substantially affect ecosystem carbon cycling.

Empirical studies demonstrate that elevated CO_2 can also alter ecosystem nitrogen losses. By increasing labile carbon availability to carbon-limited denitrifiers, elevated CO_2 stimulated nitrogen losses through denitrification in the rhizosphere of wheat grown in hydroponic systems (Smart *et al.*, 1997). Similarly, elevated CO_2 enhanced nitrogen losses, presumably via denitrification, from soil planted with *Quercus agrifolia* seedlings (Hinkson, 1996). Though elevated CO_2 also increased asymbiotic nitrogen fixation in this study, the total mass of nitrogen in the system was lower in elevated CO_2, indicating that the stimulation of nitrogen losses was larger than the increase in nitrogen inputs (Hinkson, 1996). By contrast, in a C_3 salt marsh, elevated CO_2 decreased denitrification potential, possibly because increased labile carbon availability in elevated CO_2 enhanced nitrogen immobilization (Matamala, 1997), thereby limiting nitrate supply to denitrifiers.

In seasonally dry ecosystems, the first rains after a long drought stimulate microbial nitrogen transformations, including gaseous soil efflux of nitrous oxide and nitric oxide (Davidson, 1991). Such "wet-up" events can contribute substantially to the annual losses of nitrogen from seasonally dry ecosystems (Davidson, 1991). In a California annual grassland, elevated CO_2 increased root growth and the C:N ratio of root detritus after plant senes-

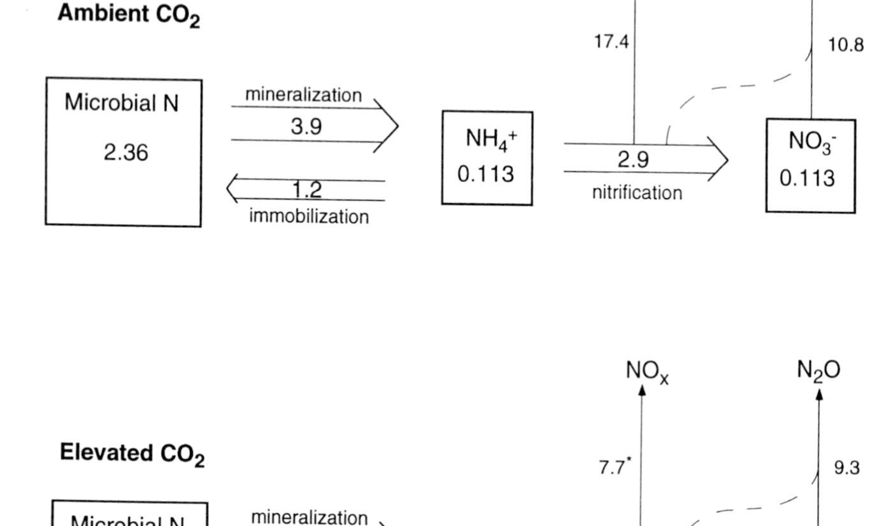

Figure 3 Nitrogen pools and fluxes under ambient and elevated CO_2 and nutrient-enriched soil during a 9-d period after the first autumn rains in a mediterranean annual grassland in California (data from Hungate *et al.,* 1997c). Plants were grown during the previous growing season under ambient and elevated CO_2, and high and low nutrients. Values for gross mineralization, immobilization, and nitrification are in g N m^{-2} for the 9-d period. Values for net NO_x and N_2O flux are in mg N m^{-2} $9d^{-1}$. Values for microbial N are in gN m^{-2} (at the end of the 9-d period), for ammonium and nitrate in g m^{-2} (weighted average for the 9-d period). Values significantly affected by elevated CO_2 are indicated by an asterisk, * ($P < 0.05$).

cence (Hungate *et al.,* 1997c). At the end of the summer drought, this increase in root litter mass and C:N ratio stimulated nitrogen immobilization in the soil microbial biomass (Fig. 3). Under nutrient enrichment, higher immobilization decreased ammonium availability to nitrifiers, and thereby decreased nitrification and associated NO efflux from soil (Fig. 3). Total nitrogen losses due to NO efflux from soil were substantial in this experiment, so the reduction in NO efflux by elevated CO_2 could be quantitatively significant for nitrogen retention.

VI. Altered Nitrogen Cycling and Soil Water Content

Decreased plant transpiration in elevated CO_2 will increase soil moisture in ecosystems where plant transpiration is a large component of evapotranspiration, and if increases in soil evaporation, leaf area, or canopy temperature (Field *et al.*, 1995) do not offset reduced transpiration at the leaf level. Based on such considerations, Field *et al.* (1995) predicted that elevated CO_2 would increase soil water content in a broad range of grassland, shrubland, and forest ecosystems. Empirical work shows that elevated CO_2 increases soil water content in agricultural forb ecosystems (Clifford *et al.*, 1993) and in annual (Fredeen *et al.*, 1997; Field *et al.*, 1995) and perennial grasslands (Rice *et al.*, 1994; Ham *et al.*, 1995; Bremer *et al.*, 1996). Elevated CO_2 also increased soil water content in several grassland microcosm experiments (Lutze, 1996; Morgan *et al.*, 1994; Ross *et al.*, 1995).

Water content influences microbial nitrogen transformations in soil, so increased soil water content is another potentially important mechanism through which elevated CO_2 could alter soil nitrogen transformations. In two annual grassland ecosystems, elevated CO_2 increased gross nitrogen mineralization and plant nitrogen uptake (Hungate *et al.*, 1997c). Elevated CO_2 increased soil water content and labile carbon availability (measured as increased microbial biomass carbon) in this experiment, either of which could have been the mechanism increasing nitrogen mineralization. As discussed earlier, increased labile carbon availability could stimulate mineralization by increasing microbial growth, nitrogen uptake, and subsequent turnover of microbial cells (Zak *et al.*, 1993; Clarholm, 1985). Wetter soils could also enhance nitrogen mineralization by increasing substrate diffusion, by increasing motility of microorganisms, or by directly relieving physiological stress due to dry soils (Stark and Firestone, 1995). Thus, in this grassland experiment, there was evidence for two possible mechanisms that could have caused the observed increase in nitrogen mineralization. Soil water content was more strongly correlated to gross nitrogen mineralization than was microbial biomass carbon, suggesting that increased soil moisture was the more likely mechanism causing increased nitrogen mineralization (Hungate *et al.*, 1997a).

Rice *et al.* (1994) investigated the effects of elevated CO_2 on soil microbial respiration in a perennial C_4 grassland. In this experiment also, increased soil moisture may be the simplest explanation for increased microbial respiration in elevated CO_2 (Rice *et al.*, 1994). In this experiment, there was a strong correlation between the CO_2 effect on soil moisture and the CO_2 effect on microbial respiration over the 8-month period during which they were measured during both dry ($r = 0.74$) and wet ($r = 0.72$) years (Fig. 4). Total rainfall during these years was 669 cm in 1991 and 1028 cm

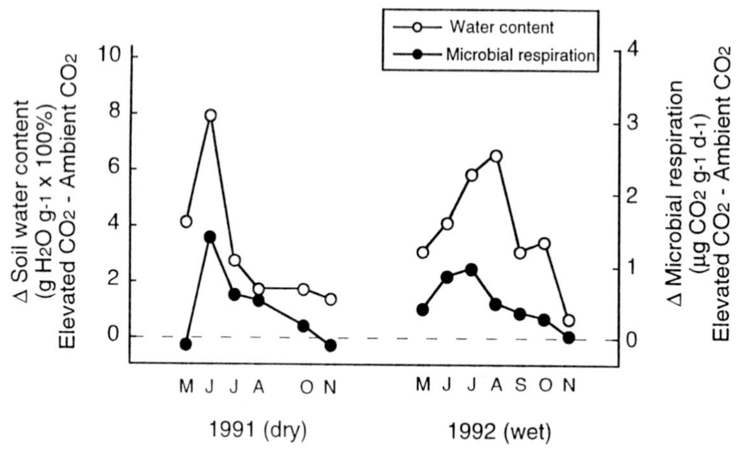

Figure 4 The relationship between the effects of elevated CO_2 on soil water content and on soil microbial respiration in an intact C_4 grassland for 2 yr (data replotted from Rice *et al.*, 1994). Values are the mean differences between the elevated and ambient CO_2 treatments in soil water content and microbial respiration rate. These absolute CO_2 effects on soil water content and microbial respiration rate are linearly correlated ($r^2 = 0.53$, $P = 0.005$).

in 1992, and mean soil water contents were 22.7 and 30.8%, respectively (Rice *et al.*, 1994). Despite these considerable differences in precipitation and soil moisture, the magnitude of the CO_2 stimulation of soil water content and of microbial respiration was comparable for the 2 yr, with water content increasing by 15 and 14% and microbial activity by 28 and 29% during 1991 and 1992, respectively (Fig. 4). Increased soil moisture in elevated CO_2 is a simple and plausible explanation for the observed increase in microbial respiration in this experiment given the strong correlation between CO_2 effects on soil moisture and soil microbial respiration and the known causal relationship between these in this grassland (Garcia, 1992).

VII. Relative Importance of Increased Carbon Input versus Altered Soil Water Content

The relative importance of the mechanisms through which elevated CO_2 alters nitrogen cycling will likely vary among ecosystems. Understanding this variation will aid in the effort to incorporate changes in soil processes in simulation models of the feedbacks of terrestrial ecosystem to rising CO_2. In part, such variation may involve a simple trade-off in which the magnitude of increased plant growth, and thus carbon input to soil, is inversely related to the reduction in evapotranspiration (ET), and thus increased soil moisture. Elevated CO_2 is likely to cause the largest reductions in ET in ecosystems where aboveground growth responses to elevated CO_2 are smallest, for the simple reason that increased leaf area provides more transpiring surface, offsetting decreased transpiration per unit leaf area (Fig. 5). Thus, factors that favor large increases in aboveground growth in response to elevated CO_2 will limit reductions in ET, while factors that constrain aboveground growth responses to elevated CO_2 will allow reductions in ET. To the extent that aboveground growth is directly proportional to net ecosystem carbon uptake, the relative importance of feedbacks through increased carbon input to soil and altered soil moisture will have a simple, predictable relationship (Fig. 5).

For example, decreased evapotranspiration under elevated CO_2 is likely to be larger in a system dominated by C_4 plants than one dominated by C_3 plants, due to generally larger aboveground growth responses in C_3 plants (Poorter, 1993), whereas the stimulation of net ecosystem carbon uptake should be relatively stronger in C_3-dominated systems. Consistent with this, elevated CO_2 decreased evapotranspiration by 20% in a salt marsh dominated by C_3 plants and by 29% in one dominated by C_4 plants, whereas the stimulation of net ecosystem carbon uptake was larger in the C_3 marsh (56%) than in the C_4 marsh (24%) (Arp, 1991). Similarly, as growth and photosynthetic responses to CO_2 increase with nutrient availability (McGuire *et al.*, 1995) and temperature (Long and Drake, 1992), decreased

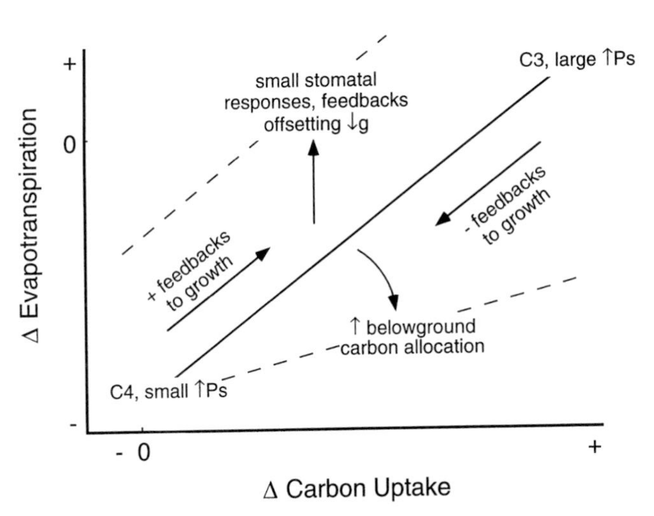

Figure 5 Predicted relationship between changes in soil water content and changes in carbon input to soil, based on a positive relationship between evapotranspiration and leaf area. See text for a discussion of this relationship, and of the factors that modify it.

ET in elevated CO_2 is likely to be more pronounced where low nutrient availability and low temperatures constrain aboveground plant growth responses to elevated CO_2.

Several factors will modify this relationship. First, increased carbon allocation belowground allows greater carbon uptake with minimal increase in leaf area, weakening the trade-off between reduced ET and increased carbon input to soil (Fig. 5). Second, the CO_2 reduction in stomatal conductance is quite variable among species (Ellsworth *et al.*, 1995; Tissue and Oechel, 1987), apparently more variable than the stimulation of photosynthesis among C_3 species. Third, canopy roughness affects the translation of decreased conductance to decreased ET: In systems where the canopy is aerodynamically rough, CO_2 will have larger effects on ET than in systems where the canopy is aerodynamically smooth (Field *et al.*, 1995). In ecosystems dominated by species in which stomatal conductance is unresponsive to elevated CO_2, or in systems where aerodynamically smooth canopies limit stomatal control of ET, the relationship between feedbacks through ET and through increased soil moisture will fall above the line shown in Fig. 5.

It is also likely that the relative importance of these feedbacks will change through time within ecosystems as they adjust to rising CO_2 concentrations (or to an experimental step change in CO_2). For example, if plant nitrogen acquisition increases as a result of increased nitrogen mineralization in wetter soils (Hungate *et al.*, 1997a), leaf area should increase, reducing the CO_2 effect on ET and limiting further increases in nitrogen cycling. On the other hand, if increased soil moisture enhances nitrogen losses or otherwise reduces nitrogen availability, this could reduce leaf area and amplify the CO_2 effect on ET, creating a strong negative feedback to increased plant growth in elevated CO_2. For example, plant transpiration influences the height of the water table in coastal salt marshes and enhances sediment oxidation (Dacey and Howes, 1984). In these marshes, reduced ET in elevated CO_2 (Arp, 1991) could increase the height of the water table, thereby reducing sediment oxidation and associated nutrient mineralization. (Observing this phenomenon on the scale of a 1-m^2 plot would be difficult, because rapid lateral water flow in saturated soil would buffer any change in water table height over such a small area.) Similarly, a negative feedback to plant growth through nutrient accumulation in microbes and soil organic matter, by limiting increases in leaf area, would favor larger feedbacks through reduced ET.

VIII. Conclusions

Elevated CO_2 can alter soil nitrogen cycling by increasing carbon input to soil, a consequence of greater photosynthesis and increased belowground allocation in response to elevated CO_2. Increased carbon input to soil in elevated CO_2 has been reported to increase, decrease, and have no effect on nitrogen availability to plants. Increased soil moisture is another important mechanism through which elevated CO_2 can alter nitrogen cycling in soil and N availability to plants, but the degree to which CO_2 increases soil moisture and the subsequent effects on nitrogen cycling are likely to vary among ecosystems. Ecosystem inputs and losses of nitrogen, though not quantitatively significant in the short term, merit attention for their effects on nitrogen stocks in the long term. Over the timescale of CO_2 doubling, changes in nitrogen inputs and losses could affect carbon storage in the terrestrial biosphere. Again, however, while some experiments indicate that elevated CO_2 will decrease N losses (e.g., Hungate *et al.*, 1997c), others suggest that N losses will increase (e.g., Arnone and Bohlen, 1998). On the surface, qualitatively different results from different studies suggest that soil responses to elevated CO_2 are largely idiosyncratic, overriding the utility of any general patterns of response across ecosystems, if indeed any general patterns exist. However, some of the observed variation may be predictable (according to differences in plant species and in soil nutrient status, for example), and some is likely due to differences in techniques for measuring nitrogen cycling and in the timescale of the various experiments. In the future, researchers in this area should consider greater coordination, using the same techniques across many experiments. By doing so, we will be able to assess whether changes in soil nitrogen cycling in response to elevated CO_2 are general across many different systems, or whether they are idiosyncratic.

Acknowledgments

Thanks to Glenn Berntson, Terry Chapin, Zoe Cardon, Bert Drake, Chris Field, Dale Johnson, Yiqi Luo, Roser Matamala, Miquel Gonzalez-Meler, and an anonymous reviewer for helpful discussions. This work was supported in part by the Alexander Hollaender Distinguished Postdoctoral Fellowship Program sponsored by the U.S. Department of Energy, Office of Health and Environmental Research, and administered by the Oak Ridge Institute for Science and Education.

References

Arnone III, J. A., and Bohlen, P. J. (1998). Stimulated N_2O flux from intact grassland monoliths after two growing seasons under elevated atmospheric CO_2. *Oecologia* (in press).

Arp, W. J. (1991). Vegetation of a North American salt marsh and elevated atmospheric carbon dioxide. PhD thesis. Free University of Amsterdam.

Berntson, G. M., and Bazzaz, F. A. (1996a). The influence of elevated CO_2 on the allometry of root production and root loss in Acer rubrum and Betula papyerifera. *Am. J. Bot.* **83,** 608–616.

Berntson, G. M., and Bazzaz, F. A. (1996b). Belowground positive and negative feedbacks on CO_2 growth enhancement. *Plant Soil* **187,** 119–134.

Berntson, G. M., and Bazzaz, F. A. (1997). Nitrogen cycling in microcosms of yellow birch exposed to elevated CO_2: Simultaneous positive and negative feedbacks. *Global Change Biol.* **3,** 247–258.

Berntson, G. M., and Woodward, F. I. (1992). The root system architecture and development of Senecio vulgaris in elevated carbon dioxide and drought. *Funct. Ecol.* **6,** 324–333.

Billès, G., Rouhier, H., and Bottner, P. (1993). Modifications of the carbon and nitrogen allocations in the plant (*Triticum aestivum* L.) soil system in response to increased atmospheric CO_2 concentration. *Plant Soil* **157,** 215–225.

Bremer, D. J., Ham, J. M., and Owensby, C. E. (1996). Effect of elevated atmospheric carbon dioxide and open-top chambers on transpiration in a tallgrass prairie. *J. Env. Qual.* **25,** 691–704.

Canadell, J. G., Pitelka, L. F., and Ingram, J. S. I. (1996). The effects of elevated CO_2 on plant–soil carbon belowground: A synthesis. *Plant Soil* **187,** 391–400.

Cardon, Z. G. (1996). Effects of root exudation and rhizodeposition on ecosystem carbon storage under elevated CO_2. *Plant Soil* **187,** 277–288.

Chapin III, F. S. (1995). New cog in the nitrogen cycle. *Nature* **377,** 199–200.

Chapin III, F. S., Moilanen, L., and Kielland, K. (1993). Preferential use of organic nitrogen for growth by a non-mycorrhizal arctic sedge. *Nature* **361,** 150–153.

Clarholm, M. (1985). Interactions of bacteria, protozoa and plants leading to mineralization of soil nitrogen. *Soil Biol. Biochem.* **17,** 181–187.

Clarholm, M. (1989). Effects of plant–bacterial–amoebal interactions on plant uptake of nitrogen under field conditions. *Biol. Fertil. Soils* **8,** 373–378.

Clifford, S. C., Stronach, I. M., Mohamed, A. D., Azam-Ali, S. N., and Crout, N. M. J. (1993). The effects of elevated atmospheric carbon dioxide and water stress on light interception, dry matter production and yields in stands of groundnut (*Arachis hypogaea* L.) *J. Exp. Bot.* **44,** 1763–1770.

Comins, H. N., and McMurtrie, R. E. (1993). Long-term response of nutrient-limited forests to CO_2 enrichment; equilibrium behavior of plant–soil models. *Ecol. Appl.* **3,** 666–681.

Cortrufo, M. F., Ineson, P., and Rowland, A. P. (1994). Decomposition of tree leaf litters grown under elevated CO_2: Effect of litter quality. *Plant Soil* **163,** 121–130.

Dacey, J. W. H., and Howes, B. L. (1984). Water uptake by roots controls water table movement and sediment oxidation in short *Spartina* marsh. *Science* **224,** 487–489.

Davidson, E. A. (1991). Fluxes of nitrous oxide and nitric oxide from terrestrial ecosystems. *In* "Microbial Production and Consumption of Greenhouse Gases: Methane, Nitrous Oxides, and Halomethanes" (J. E. Rogers and W. E. Whitman, eds.), pp. 219–235. American Society of Microbiology, Washington, DC.

Davidson, E. A., Stark, J. M., and Firestone, M. K. (1990). Microbial production and consumption of nitrate in an annual grassland. *Ecology* **71,** 1968–1975.

Day, F. P., Weber, E. P., Hinkle, C. R., and Drake, B. G. (1996). Effects of elevated atmospheric CO_2 on fine root length and distribution in an oak-palmetto scrub ecosystem in central Florida. *Global Change Biology* **2,** 143–148.

Díaz, S., Grime, J. P., Harris, J., and McPherson, E. (1993). Evidence of a feedback mechanism limiting plant response to elevated carbon dioxide. *Nature* **364,** 616–617.

Drake, B. G., Long, S. P., and Gonzelez-Meler, M. (1997). Increased plant efficiency: A consequence of elevated atmospheric CO_2? *Ann. Rev. Plant Physiol.* **48,** 609–639.

Ellsworth, D. S., Oren, R., Huang, C., Phillips, N., and Hendrey, G. R. (1995). Leaf and canopy responses to elevated CO_2 in a pine forest under free-air CO_2 enrichment. *Oecologia* **104,** 139–146.

Eviner, V. T., and Chapin III, F. S. (1997). Plant-microbial interactions. *Nature* **385,** 26–27.

Field, C. B., Jackson, R. B., and Mooney, H. A. (1995). Stomatal responses to increased CO_2: Implications from the plant to the global scale. *Plant Cell Env.* **18,** 1214–1225.

Fredeen, A. L., Randerson, J. T., Holbrook, N. M., and Field, C. B. (1997). Elevated atmospheric CO_2 increases water availability in a water-limited grassland ecosystem. *J. Am. Water Res. Assoc.* **33,** 1033–1039.

Garcia, F. O. (1992). Carbon and nitrogen dynamics and microbial ecology in tallgrass prairie. Ph.D. Dissertation. Kansas State University, Manhattan.

Garnier, E. (1991). Resource capture, biomass allocation and growth in herbaceous plants. *Trends Ecol. Evol.* **6,** 126–131.

Gifford, R. M., Lutze, J. L., and Barrett, D. (1996). Global integration of atmospheric change effects on plant–soil interactions. *Plant Soil* **187,** 369–387.

Gorissen, T. (1996). Elevated CO_2 evokes quantitative and qualitative changes in carbon dynamics in a plant/soil system: Mechanisms and implications. *Plant Soil* **187,** 289–298.

Griffiths, B., and Robinson, D. (1992). Root-induced nitrogen mineralisation: A nitrogen balance model. *Plant Soil* **139,** 253–263.

Ham, J. M., Owensby, C. E., Coyne, P. I., and Bremer, D. J. (1995). Fluxes of CO_2 and water vapor from a prairie ecosystem exposed to ambient and elevated atmospheric CO_2. *Agric. For. Meteor.* **77,** 73–93.

Hardy, R. W. F., and Havelka, H. D. (1976). Photosynthate as a major factor limiting nitrogen fixation by field-grown legumes with emphasis on soybeans. *In* "Symbitoic Nitrogen Fixation" (P. S. Nutman, ed.), pp. 421–439. Cambridge University Press, Cambridge.

Helal, H. M., and Sauerbeck, D. (1986). Effect of plant roots on carbon metabolism of soil microbial biomass. *Z. Planzenern. Bodenk.* **149,** 181–188.

Hinkson, C. L. (1996). Gas exchange, growth response, and nitrogen dynamics of Quercus agrifolia under elevated atmospheric CO_2 and water stress. MS Thesis, San Diego State University.

Hirschel, G., Korner, Ch., Arnone III, J. A. (1997). Will rising atmospheric CO_2 affect leaf litter quality and in situ decomposition rates in native plant communities? *Oecologia* **110,** 387–392.

Hungate, B. A., Canadell, J., and Chapin III, F. S. (1996). Plant species mediate changes in soil microbial N in response to elevated CO_2. *Ecology* **77,** 2505–2515.

Hungate, B. A., Chapin III, F. S., Zhong, H., Holland, E. A., and Field, C. B. (1997a). Stimulation of grassland nitrogen cycling under carbon dioxide enrichment. *Oecologia* **109,** 149–153.

Hungate, B. A., Holland, E. A., Jackson, R. B., Chapin III, F. S., Field, C. B., and Mooney, H. A. (1997b). The fate of carbon in grasslands under carbon dioxide enrichment. *Nature* **388,** 576–579.

Hungate, B. A., Lund, C. P., Pearson, H. L., and Chapin III, F. C. (1997c). Elevated CO_2 and nutrient addition alter soil N cycling and N trace gas fluxes with early season wet-up in a California annual grassland. *Biogeochemistry* **37,** 89–109.

Ingham, R. E., Trofymow, J. A., Ingham, E. R., and Coleman, D. C. (1985). Interactions of bacteria, fungi, and their nematode grazers: Effects on nutrient cycling and plant growth. *Ecol. Monogr.* **55,** 119–140.

Johnson, D. W., Ball, J. T., and Walker, R. F. (1997). Effects of CO_2 and nitrogen fertilization on vegetation and soil nutrient content in juvenile ponderosa pine. *Plant Soil* **190,** 20–40.

Kuikman, P. J., and Gorissen, A. (1993). Carbon fluxes and organic matter transformations in plant–soil systems. *In* "Climate Change: Crops and Terrestrial Ecosystems" (S. C. van de Geijn, J. Goudriaan, and F. Berendse, eds.), pp. 97–108. CABO-DLO, Wageningen.

Kuikman, P. J., Lekkerkerk, L. J. A., and Van Veen, J. A. (1991). Carbon dynamics of a soil planted with wheath under an elevated atmospheric CO_2 concentration. *In* "Advances in Soil Organic Matter Research," pp. 267–274. The Royal Society of Chemistry, Special Publication 90, Cambridge.

Lamborgh, M. R., Hardy, R. W. F., and Paul, E. A. (1983). Microbial effects. *In* "CO_2 and Plants: The Response of Plants to Rising Levels of Atmospheric CO_2" (E. R. Lemon, ed.), pp. 131–176. Westview Press, Boulder, CO.

Lekkerkerk, L. J. A., Van de Geijn, S. C., and Van Veen, J. A. (1990). Effects of elevated atmospheric CO_2 levels on the carbon economy of a soil planted with wheat. *In* "Soils and the Greenhouse Effect" (A. F. Bouwman, ed.), pp. 423–429. John Wiley and Sons, Chichester.

Liljeroth, E., Van Veen, J. A., and Miller, H. J. (1990). Assimilate translocation to the rhizosphere of two wheat lines and subsequent utilization by rhizosphere microorganisms at two soil nitrogen concentrations. *Soil Biol. Biochem.* **22,** 1015–1021.

Long, S. P., and Drake, B. G. (1992). Photosynthetic CO_2 assimilation and rising atmospheric C CO_2 O2 concentration. Commissioned review. *In* "Topics in Photosynthesis" (N. R. Baker and H. Thomas, eds.), Vol. II, pp. 69–107. Elsevier Science Publishers, Amsterdam.

Lutze, J. L. (1996). Carbon and nitrogen relationships in swards of *Danthonia richardsonii* in response to CO_2 enrichment and nitrogen supply Ph.D. thesis. Australian National University.

Luxmoore, R. (1981). CO_2 and phytomass. *BioScience* **31,** 626.

Matamala, R. (1997). Carbon and nitrogen cycling in a salt marsh under elevated atmospheric CO_2. Ph.D. dissertation, University of Barcelona.

Melillo, J. M., and Aber, J. D. (1982). Nitrogen and lignin control of hardwood leaf litter decomposition dynamics. *Ecology* **63,** 621–626.

Merckx, R., Dijkstra, A., den Hartog, A., and Van Veen, J. A. (1987). Production of root-derived material and associated microbial growth in soil at different nutrient levels. *Biol. Fert. Soils* **5,** 126–132.

McGuire, A. D., Melillo, J. M., and Joyce, L. A. (1995). The role of nitrogen in the response of forest net primary production to elevated atmospheric carbon dioxide. *Ann. Rev. Ecol. Syst.* **26,** 473–503.

Morgan, J. A., Knight, W. G., Dudley, L. M., and Hunt, H. W. (1994). Enhanced root system C-sink activity, water relations and aspects of nutrient acquisition in mycotrophic *Bouteloua gracilis* subjected to CO_2 enrichment. *Plant Soil* **165,** 139–146.

Morrison, J. I. L. (1987). Intercellular CO_2 concentration and stomatal response to CO_2. *In* "Stomatal Function" (E. Zeiger, G. D. Farquhar, and I. R. Cowan, eds.), pp. 229–252. Stanford University Press, Stanford, CA.

Myrold, D. D., and Tiedje, J. M. (1986). Simultaneous estimation of several nitrogen cycle rates using ^{15}N: theory and application. *Soil Biol. Biochem.* **18,** 559–568.

Norby, R. J. (1994). Issues and perspectives for investigating root response to elevated atmospheric carbon dioxide. *Plant Soil* **164,** 9–20.

Norton, J. M., and Firestone, M. K. (1996). Nitrogen dynamics in the rhizosphere of Pinus ponderosa seedlings. *Soil Biol. Biochem.* **28,** 351–362.

O'Neill, E. G., and Norby, R. J. (1996). Litter quality and decomposition rates of foliar litter produced under CO_2 enrichment. *In* "Carbon Dioxide and Terrestrial Ecosystems" (G. W. Koch and H. A. Mooney, eds.), pp. 87–103. Academic Press, San Diego.

Owensby, C. E., Coyne, P. I., and Auen, L. M. (1993). Nitrogen and phosphorus dynamics of a tallgrass prairie ecosystem exposed to elevated carbon dioxide. *Plant Cell Env.* **16,** 843–850.

Paterson, E., Rattray, E. A. S., and Kollham, K. (1996). Effect of elevated atmospheric CO_2 concentration on C-partitioning and rhizosphere C-flow for three plant species. *Soil Biol. Biochem.* **28,** 195–201.

Poorter, H. (1993). Interspecific variation in the growth response of plants to an elevated ambient CO_2 concentration. *Vegetatio* **104/105,** 77–97.

Rice, C. W., Garcia, F. O., Hampton, C. O., and Owensby, C. E. (1994). Soil microbial response in tallgrass prairie to elevated CO_2. *Plant Soil* **165,** 67–75.

Rogers, H. H., Runion, G. B., and Krupa, S. V. (1994). Plant responses to atmospheric CO_2 enrichment with emphasis on roots and the rhizosphere. *Env. Pollut.* **83,** 155–189.

Ross, D. J., Tate, K. R., and Newton, P. C. D. (1995). Elevated CO_2 and temperature effects on soil carbon and nitrogen cycling in ryegrass/white clover turves of an Endoaquept soil. *Plant Soil* **176,** 37–49.

Rosswall, T. (1982). Microbiological regulation of the biogeochemical nitrogen cycle. *Plant Soil* **67,** 15–34.

Rouhier, H., Billès, G., El Kohen, A., Mousseau, M., and Bottner, P. (1994). Effect of elevated CO_2 on carbon and nitrogen distribution within a tree (*Castanea sativa*)-soil system. *Plant Soil* **162,** 281–292.

Rouhier, H., Billès, G., Billès, L., and Bottner, P. (1996). Carbon fluxes in the rhizosphere of sweet chestnut seedlings (*Castanea sativa*) grown under two atmospheric CO_2 concentrations: ^{14}C partitioning after pulse labeling. *Plant Soil* **180,** 101–111.

Smart, D. R., Ritchie, K., Bugbee, B., and Stark, J. M. (1997). The influence of elevated CO_2 on rhizosphere denitrifier activity. *Appl. Env. Microbiol.* **63,** 4621–4624.

Soussana, J. F., and Hartwig, U. A. (1996). The effects of elevated CO_2 on symbiotic N2 fixation: a link between the carbon and nitrogen cycles in grassland ecosystems. *Plant Soil* **187,** 321–332.

Stark, J. M., and Firestone, M. K. (1995). Mechanisms for soil moisture effects on nitrifying bacteria. *Appl. Env. Microbiol.* **61,** 218–221.

Tiedje, J. M., Sorenson, J., and Chang, Y. Y. L. (1981). *In* "Terrestrial Nitrogen Cycles" (F. E. Clark and T. Rosswall, eds.), pp. 331–342. Ecological Bulletins, Stockholm.

Tissue, D. T., and Oechel, W. C. (1987). Response of *Eriophorum vaginatum* to elevated CO_2 and temperature in the Alaskan tussock tundra. *Ecology* **68,** 401–410.

Van de Geijn, S. C., and Van Veen, J. A. (1993). Implications of increased carbon dioxide levels for carbon input and turnover in soils. *Vegetatio* **104/105,** 283–292.

Van Ginkel, J. H., Gorissen, A., and Van Veen, J. A. (1996). Carbon and nitrogen allocation in Lolium perenne in response to elevated atmospheric CO_2 with emphasis on soil carbon dynamics. *Plant Soil* **188,** 299–308.

Van Veen, J. A., Mercks, R., and Van de Geijn, S. C. (1989). Plant and soil related controls of the flow of carbon from roots through the soil microbial biomass. *Plant Soil* **115,** 179–188.

Van Veen, J. A., Liljeroth, E., Lekkerkerk, L. J. A., and Van de Geijn, S. C. (1991). Carbon fluxes in plant–soil systems at elevated atmospheric CO_2 levels. *Ecol. Appl.* **1,** 175–181.

Zak, D. R., Pregitzeer, K. S., Curtis, P. S., Teeri, J. A., Fogel, R., and Randlett, D. A. (1993). Elevated atmospheric CO_2 and feedback between carbon and nitrogen cycles. *Pland Soil* **151,** 105–117.

II

Evolutionary, Scaling, and Modeling Studies of CO_2 and Stress Interactions

11

Implications of Stress in Low CO_2 Atmospheres of the Past: Are Today's Plants Too Conservative for a High CO_2 World?

Rowan F. Sage and Sharon A. Cowling

I. Introduction: The Case for Studying Responses to Low CO_2

Research addressing biological responses to variation in atmospheric CO_2 has focused on responses to a doubling of the present CO_2 level (Körner *et al.*, 1995). Although this approach has obvious merit, a drawback is that the current level of atmospheric CO_2 is already 33% enriched above the average level of 270 μmol CO_2 mol^{-1} air that persisted between 10,000 yr B.P. (before present) and 1870 A.D. (Neftel *et al.*, 1988; Jasper and Hayes, 1990). Moreover, during recent geological history, atmospheric CO_2 level oscillated between approximately 180 and 300 μmol mol^{-1} in concert with the advance and retreat of major glacial events (Fig. 1). The pattern of low CO_2 glacials of 100,000 yr interspersed with relatively short, CO_2-enriched interglacials lasting 15,000 yr is known to have persisted for at least the past 250,000 yr and probably extends back 1–3 million years (Trabalka, 1985; Raymo, 1992). Ice core records show that CO_2 concentrations during glacial episodes averaged 225 μmol mol^{-1}, and during the coldest extremes of the last two glaciations (18,000 and 150,000 yr), CO_2 fell below 200 μmol mol^{-1} for periods longer than 10,000 yr (Barnola *et al.*, 1987; Jouzel *et al.*, 1993). Not only does the current CO_2 level of 365 μmol mol^{-1} represent a 30–50% rise over CO_2 levels predominant in the late Pleistocene, but the current period of CO_2 enrichment is but a small fraction of recent geological time.

Preindustrial CO_2 levels are important because they may represent the baseline CO_2 to which much of the world's flora is adapted. Except for

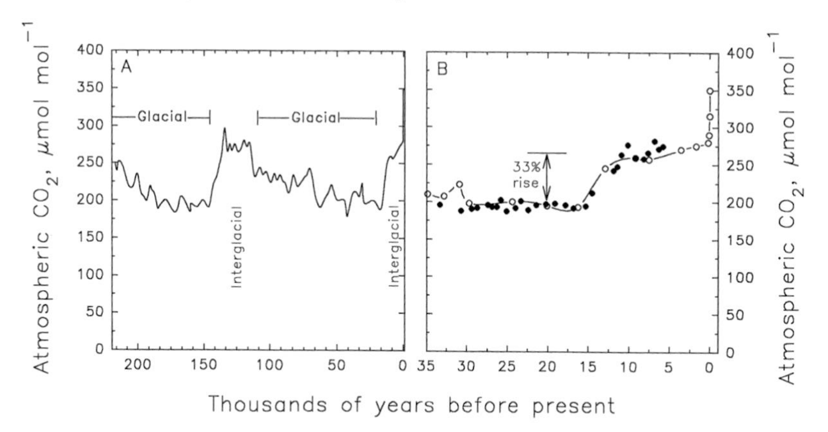

Figure 1 Antarctic ice core data for the partial pressure of CO_2 in the atmosphere over the past 220,000 yr. Data of Barnola *et al.* (1987; for time 0 to 160,000 yr B.P. in panel A, open circles in panel B), Jouzel *et al.* (1993; for 160,000 to 220,000 yr B.P. in panel A), and Neftal *et al.* (1988; filled circles panel B). (Adapted from Sage, 1995.)

rapidly evolving annuals and short-lived perennials, the past century of anthropogenic CO_2 enrichment may be an insufficient period of time for evolutionary adaptation to rising CO_2. Even for annual species, this period may be inadequate for CO_2-dependent evolution because most of the atmospheric CO_2 rise has occurred in the past 50 yr. Evolution over longer time periods is also questionable because atmospheric CO_2 enrichment is not a lethal selection agent nor does it result in strong, consistent directional selection (Kingsolver, 1996; Thomas and Jasienski, 1996). In addition, species with long-lived individuals could still be adapted to the low CO_2 levels of the Pleistocene because (1) there may have been relatively few generations on which selection could act and (2) genes contributed to the breeding pool by old individuals could dilute novel, high CO_2 genes for long periods after CO_2 enrichment has occurred. In numerous clonal species, individual genotypes have been identified with ages exceeding 10,000 yr (Cook, 1985; Silander, 1985; Vasek, 1980). Individual clones of aspen (*Populus tremuloides*) may even be 1 million years old (Mitton and Grant, 1996). Thus, it appears likely that clones dating from the Pleistocene still persist in modern ecosystems.

If plants have indeed adapted to the low CO_2 levels of the past, we need to consider the significance this may have for responses to future atmospheric change. In particular, how might low CO_2 adaptations influence the response of plants to CO_2 enrichment? Could the traits conferring survival to low CO_2 constrain potential responses to CO_2 enrichment, creating situations where potential benefits of high CO_2 are poorly exploited?

In the context of plant stress, strong interactions are recognized between stress intensity and carbohydrate availability. One important mode of stress action is an inhibition of carbon gain potential (Chapter 1, this volume; Osmond *et al.*, 1987), which in a CO_2-limiting atmosphere could aggravate carbohydrate deficiency and, in turn, enhance the degree to which the stress is internalized. Given the potential for strong interaction between CO_2 deficiency and intensity of environmental stress, one could envision strong selection pressure for mechanisms conferring stress tolerance in low CO_2 environments. Such mechanisms could lead to plants being overly conservative in terms of growth potential, allocation patterns, and storage investment. Moreover, the environmental thresholds plants use to trigger stress responses may also be too conservative in CO_2-enriched environments. Understanding how plants adjusted their stress reactions for CO_2-deficient atmospheres of the past may be the key to predicting how they will respond to future CO_2 enrichment.

In this chapter, we evaluate various C_3-plant processes which are responsive to subambient CO_2, particularly those which influence carbon gain and plant growth. We then discuss stress × low CO_2 interactions, highlighting how subambient CO_2 may increase stress frequency and intensity. Finally, we discuss how selection for stress survival at low CO_2 (<200 μmol mol^{-1}) may constrain plant responses to CO_2 enrichment above the current ambient of 365 μmol mol^{-1}, possibly creating novel stresses associated with excessive carbohydrate accumulation.

II. Plant Responses to Subambient CO_2

A. Photosynthesis

The primary effect of CO_2 on photosynthesis is to modulate the activity of ribulose-bisphosphate carboxylase/oxygenase (Rubisco), the enzyme catalyzing the fixation of CO_2 into organic compounds. As a substrate, CO_2 availability substantially limits Rubisco activity at atmospheric concentrations below 360 μmol mol^{-1} (Fig. 2A; Farquhar and von Caemmerer, 1982; Sage and Reid, 1994). CO_2 fixation by Rubisco is also inhibited at CO_2 levels below 360 μmol mol^{-1} by the oxygenation reaction of Rubisco (Fig. 2B; Brooks and Farquahar, 1985). Rubisco oxygenation leads to photorespiratory inhibition of photosynthesis, with the degree of inhibition rising with increasing temperature and declining CO_2. In a historical context, predicted rates of C_3 photosynthesis at 30°C were 20–30% less in 1870 (270 μmol mol^{-1}) than today, with about 20% of this reduction arising from increased photorespiration (predictions modeled as described in the legend of Fig. 3). At the height of the last glaciation 20,000 yr B.P. (CO_2 level ≈ 180 μmol mol^{-1}), photosynthesis at 30°C (a common temperature

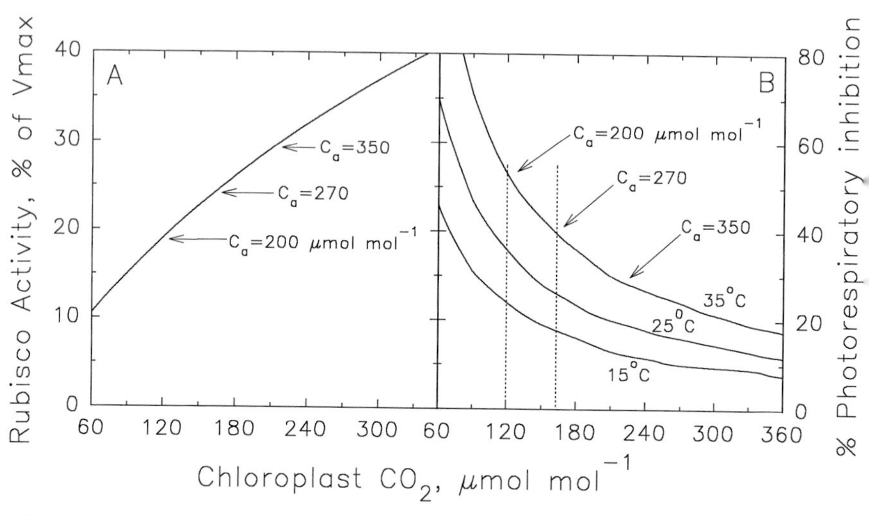

Figure 2 Modeled relationships between (A) the activity of Rubisco relative to its maximum capacity (*V*max), and chloroplast CO_2 levels and (B) percent photorespiratory inhibition (expressed as the ratio of photorespiration to photosynthesis × 100%) and chloroplast CO_2 at 15, 25, and 35°C. Relationships modeled according to Farquhar and von Caemmerer (1982) as modified by Sage (1990). Atmospheric CO_2 levels (C_a) corresponding to chloroplast CO_2 levels indicated by arrows and vertical dashed lines. (From Sage, 1995.)

in the tropics and subtropics at this time; Kutzbach and Ruddiman, 1993) would have been 40–60% less, with about one-third of this decline arising from increased photorespiratory activity. Individual studies with modern taxa grown at a range of subambient CO_2 support the modeled estimates. In 18 studies of the effect of low CO_2 on crops, native shrubs, wild annuals, and herbaceous perennials, reducing CO_2 from 350 ± 10 μmol m^{-2} s^{-1} to 180 μmol m^{-2} s^{-1} reduced both photosynthesis and yield by approximately 50% (Fig. 3; Sage, 1995). These studies were conducted under favorable thermal, light, and nutrient conditions, and not under conditions of drought, thermal extremes, low humidity, or nutrient deficiency, which can alter the relationship between photosynthesis, productivity, and atmospheric CO_2.

B. Interactions between Low CO_2 and Stress

As described elsewhere in this volume, CO_2 enrichment attenuates the inhibitory effects of moderate drought, humidity, salinity, and temperature stress. By the same logic, CO_2 depletion should aggravate inhibitory effects of stress; however, few have studied the relationship.

During the past 2 yr, we have studied interactive effects between CO_2 depletion (from the current ambient CO_2 to 200 μmol mol^{-1}) and tempera-

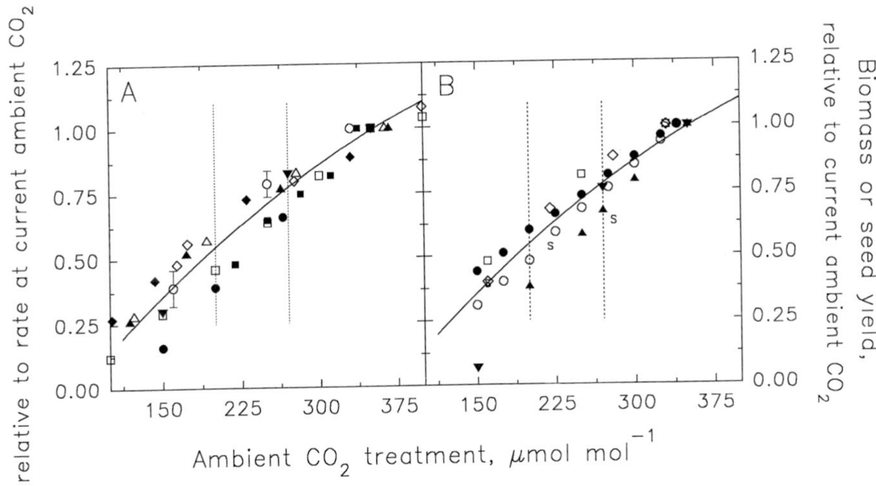

Figure 3 The response of (A) net photosynthesis and (B) plant productivity to ambient CO_2 in a variety of C_3 plants grown over a range of atmospheric CO_2 concentrations. Data relative to the value at current levels of CO_2. Vertical dashed lines indicate atmospheric CO_2 levels existing at 18,000 yr B.P. ($CO_2 \cong 180\ \mu mol\ mol^{-1}$) and 200 yr B.P. ($CO_2 \cong 270\ \mu mol\ mol^{-1}$). Solid curves in each panel are photosynthetic responses modeled according to Farquhar and von Caemmerer as modified by Sage (1990) with these assumptions: Rubisco V_{max} = 100 μmol $m^{-2}\ s^{-1}$, typical of C_3 crops; mesophyll transfer conductance = 1 mol $m^{-2}\ s^{-1}$; respiration = 1.2 μmol $m^{-2}\ s^{-1}$. Intercellular CO_2 was corrected to ambient CO_2 using an intercellular to ambient CO_2 ratio of 0.71 above 230 μmol CO_2 mol^{-1} air, and 0.72–0.9 between 230 and 80 μmol mol^{-1}. Photosynthetic data for *Abutilon theophrasti* (▼; Tissue *et al.*, 1995); oats, *Avena sativa* (◇ grown at 166 μmol mol^{-1}, ◆ grown at 330 μmol mol^{-1}; Johnson *et al.*, 1993); *Brassica kaber* (●; Polley *et al.*, 1992); beans, *Phaseolus vulgaris* (△ grown at 200 μmol mol^{-1}, ▲ grown at 350 μmol mol^{-1}; Sage and Reid, 1992); *Prosopis glandulosa* (□; Johnson *et al.*, 1993); rice, *Oryza sativa* (○; Baker *et al.*, 1990); wheat, *Triticum aestivum* (■; Polley *et al.*, 1993b). Biomass data from *Abutilon theophrasti* (▼; Dippery *et al.*, 1995); soybean (◇, plant mass, S, seed mass; Allen *et al.*, 1991); rice (□; Baker *et al.*, 1990); wheat (▲), oats (○), and *Brassica kaber* (●) Polley *et al.*, 1993a. (From Sage, 1995.)

ture enhancement (from 25°/20°C day/night to 35°/29°C) in *Phaseolus vulgaris* (*cv.* Black Turtle), emmer wheat (*Triticum dicoccum*), and tobacco (*Nicotiana tabaccum*). In each species, reduction from the current CO_2 level (380 μmol mol^{-1} in Toronto) to 200 μmol mol^{-1} aggravated the inhibition of growth associated with high temperature (Figs. 4 and 5; Table I). At a growth C_a of 380 μmol mol^{-1}, plants were 40–70% smaller at 35°/29°C than 25°/20°C. Reducing CO_2 at 25°/20°C similarly reduced growth rate so that plants were 40–60% smaller at 200 μmol mol^{-1} than at 380 μmol mol^{-1}. Increasing temperature from 25°/20°C to 35°/29°C and reducing CO_2 to 200 μmol mol^{-1} produced a strong interactive effect of CO_2 and heat such that plants size was reduced 75–95% relative to plants grown at 25°/20°C and 380 μmol mol^{-1}.

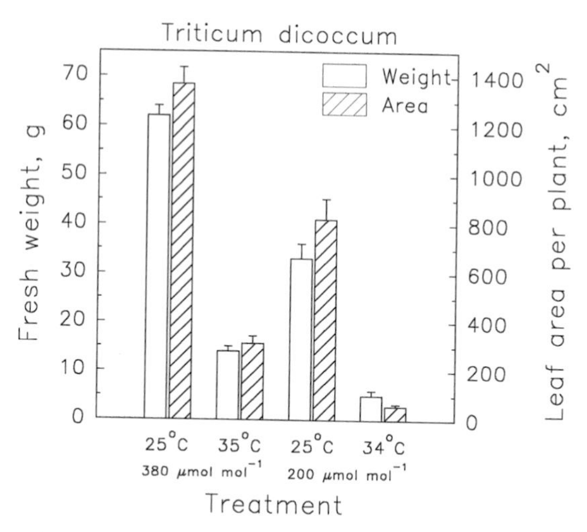

Figure 4 Shoot biomass yield and leaf area per plant of emmer wheat (*Triticum dicoccum*) grown for 41 d at the indicated day temperature and a CO_2 level of 380 or 200 μmol mol^{-1}. See Table I for other growth conditions. Mean ± SE, $N = 3$.

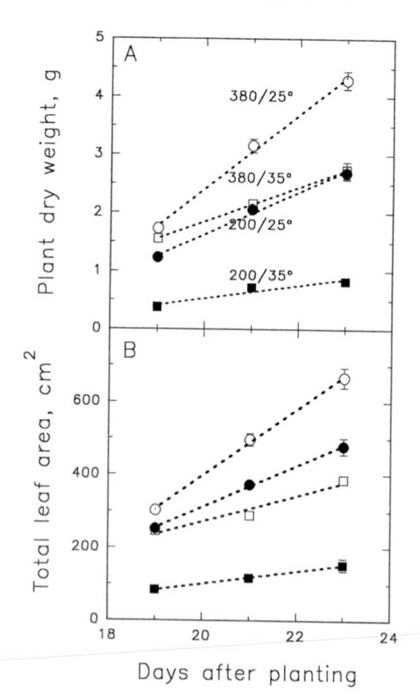

Figure 5 (A) Whole plant dry weight and (B) leaf area growth response to day temperature and subambient CO_2 in black turtle beans (*Phaseolus vulgaris* var. Black Turtle). CO_2 partial pressures and temperatures beside curves indicate daytime treatment conditions. Night temperatures were 5–6°C below day temperature. Error bars ±SE. $N = 15$. See Table I for other growth conditions. (From Cowling and Sage, 1998.)

Table I Growth Parameters of Tobacco (*Nicotiana tabacum*) Grown for 21 Days at the Indicated Daytime Temperature and CO_2 Levels[a,b]

	Growth treatment			
	380 μmol mol^{-1} CO_2		200 μmol mol^{-1} CO_2	
	25°C	35°C	25°C	34°C
Shoot dry weight, g	1.20[a] ±0.06	0.84[b] ±0.06	0.51[c] ±0.06	0.09[d] ±0.01
Leaf area per plant, cm^2	390[a] ±14	287[b] ±19	208[b] ±29	34[c] ±2
Largest leaf size, cm^2	92.7[a] ±4.4	62.8[b] ±2.2	50.3[c] ±6.4	5.9[d] ±0.5
Average area per leaf, cm^2	24.4[a] ±28.6	25.5[a] ±22.0	15.4[b] ±17.0	3.3[c] ±1.8
Leaves per plant	16 ± 1[a]	12 ± 1[c]	14 ± 1[a]	10 ± 0.4[c]

[a] Means ± SE. $N = 6$. Superscripted letters in the table body indicate statistically different groups at $p = 0.05$.

[b] *Methods:* Plants were grown in 2-L plots of soil in Conviron growth chambers set to deliver the indicated conditions. Photon flux density = 500 ± 50 μmol m^{-2} s^{-1} for 12-h photoperiods. Night temperature was 20 and 29°C for cool and warm treatments, respectively.

The greatest impact of the low CO_2–high temperature interaction was on leaf area. In all three species, leaf area production was reduced in similar proportion as biomass growth (Figs. 4–6; Table I). This falloff in area was because of reduced leaf expansion and less frequent leaf initiation. For example, in bean at 200 μmol mol^{-1} and 35°/29°C, few axillary leaves were produced, whereas other treatments produced multiple sets of axillary leaves (Fig. 6). Photosynthesis rate per unit leaf area in beans grown at 200 μmol mol^{-1} and 35°/29°C was not affected relative to plants at 200 μmol mol^{-1} air and 25°/20°C CO_2 (Table II), in part because plants at higher temperature had greater leaf nitrogen content. This further implicates leaf area reduction as a primary mechanism controlling the interaction between CO_2 supply and elevated temperature. Bean leaf respiration was not affected by temperature alone, but was by the combination of heat and low CO_2, being twice as high at 200 μmol mol^{-1} and 35°/29°C as in the two 380 μmol mol^{-1} treatments (Table II). At 35°/29°C and 200 μmol mol^{-1}, plants failed to flower in contrast to other treatments, in-

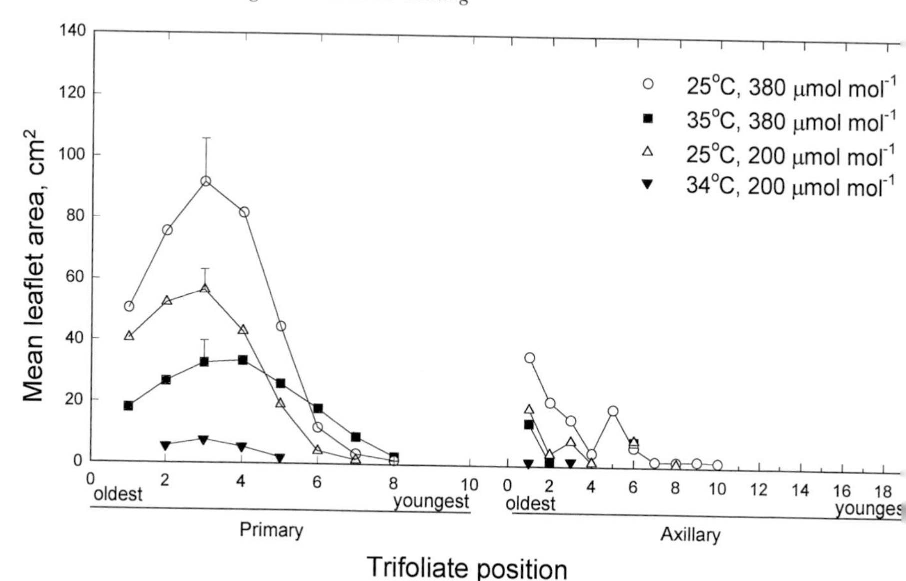

Figure 6 The response of leaflet area and leaflet initiation to growth temperature and subambient CO_2 exposure in *Phaseolus vulgaris* cv. Black Turtle grown for 21 d at the indicated daytime growth conditions. Night temperatures were 5–6°C below day temperature. (From S. Dexter, T. Sage, and R. Sage, unpublished.)

dicating they would not complete their life cycle. Interestingly, in bean, carbohydrate starvation was not evident in the 200 μmol mol^{-1}/35°C treatment, because the levels of nonstructural carbohydrate in leaves of this treatment were equivalent to ambient CO_2 controls (Table II). This indicates that despite low potential for CO_2 assimilation, plants regulate leaf carbohydrate levels to ensure reserves are maintained, presumably to deal with nights and low-light episodes.

These results demonstrate a strong interactive relationship between CO_2 depletion and the degree to which other unfavorable conditions inhibit growth. Thermal conditions that are moderately inhibitory at current CO_2 levels become severely inhibitory at 200 ppm CO_2. From these results, we hypothesize that a relationship exists between CO_2 and moderate stress as described in Fig. 7. At optimal conditions, the reduction in plant performance with declining CO_2 is relatively modest. As conditions deteriorate, the drop in performance with CO_2 reduction is proportionally greater because of stress impacts on carbon acquisition (which aggravate carbohydrate deficiency), and reduced carbon availability within the plant to compensate for deleterious effects of the stress. The exact magnitude of the performance reduction will depend on (1) how limiting carbon is for growth, (2) the degree to which carbon is involved in the mechanisms that

Table II Effects of CO_2 Reduction and Daytime Temperature During Growth on Leaf Parameters in *Phaseolus vulgaris* L[a]

Treatment (μmol/mol^{-1}/°C)	A μmol m^{-2} s^{-1}	R_d μmol m^{-2} s^{-1}	g_s mmol m^{-2} s^{-1}	C_i/C_a μbar bar^{-1}	N g m^{-2}	TNC g m^{-2}
380/25	14.9 ± 0.7[a]	0.9 ± 0.2[b]	0.19 ± 0.1[c]	0.66 ± .03[c]	1.9 ± 0.1[ab]	4.8 ± 0.4[a]
380/35	18.0 ± 0.9[a]	1.0 ± 0.1[b]	0.47 ± 0.1[a]	0.83 ± .01[a]	1.8 ± 0.1[b]	4.1 ± 0.2[a]
200/25	9.4 ± 0.6[b]	0.6 ± 0.1[c]	0.37 ± 0.1[b]	0.78 ± .01[b]	1.6 ± 0.1[c]	2.3 ± 0.1[b]
200/35	10.4 ± 1.4[b]	1.9 ± 0.1[a]	0.65 ± 0.2[a]	0.85 ± .02[a]	2.0 ± 0.1[a]	4.0 ± 0.2[a]

[a] Treatment effects on rates of net CO_2 Assimilation (A) and respiration (R_d), stomatal conductance (g_s), the ratio of intercellular CO_2 concentration to ambient CO_2 concentration (C_i/C_a), foliar nitrogen content (N), and total nonstructural carbohydrate content (TNC). Leaves were sampled for nitrogen 22 ± 2 d after planting ($N = 20$). Leaf gas exchange was measured on mature leaves 27–42 d after planting ($N = 3$). Dissimilar superscripts denote significant differences (ANOVA, $P < 0.05$). Night growth temperatures were 5 to 6°C below daytime growth temperature. From Cowling and Sage (1998).

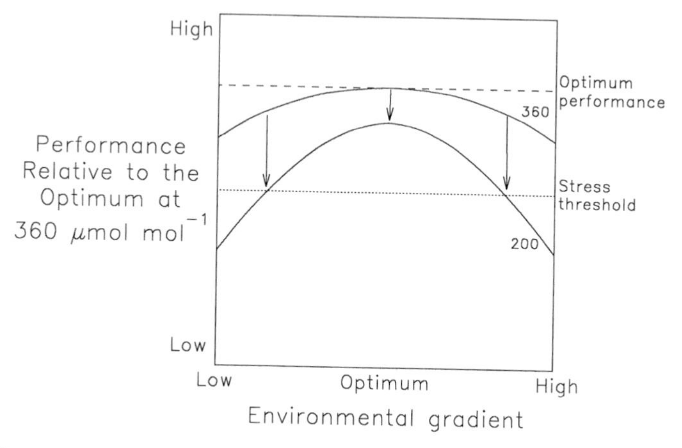

Figure 7 Hypothesized relationships between performance (for example, shoot growth) and the interaction of environmental stress and CO_2 reduction from 360 to 200 μmol mol^{-1}. Dashed lines refer to the magnitude of response under optimal conditions. Dotted lines refer to a hypothetical stress threshold, which depending on stress criteria used, could be (1) a level where the performance response is substantially less than the optimum, (2) a point where tissue injury may occur, or (3) where the probability of survival is significantly reduced by the environmental perturbation.

compensate for stressful conditions, and (3) the sensitivity of photosynthesis and leaf area production to the shift from optimal to nonoptimal conditions. Inhibition due to drought, for example, increases with CO_2 depletion. Drought reduces carbon gain potential by restricting leaf expansion and stomatal conductance, and the mechanisms conferring drought tolerance such as osmotic adjustment and root elongation are dependent on carbohydrate supply (Chapter 1, this volume; Chaves and Pereira, 1992; Rogers *et al.*, 1994). Performance reduction from heat exposure at low CO_2 arises from reduced leaf area and higher photo- and dark respiration (Chapter 3, this volume; Tables I and II). In contrast, interactions between soil infertility and carbon supply are less responsive to CO_2 variation because carbohydrate demand is curtailed by nutrient deficiency (Chapter 13, this volume; Field *et al.*, 1992).

III. Adaptations to Environmental Stress under Low CO_2

The relationship in Fig. 7 indicates that as CO_2 levels decline, the inhibitory range of environmental conditions broadens substantially. As a consequence, many conditions that are currently nonstressful may have been stressful during past episodes of low CO_2, and conditions that are now moderately stressful may have been lethal at low CO_2. If true, then strong selection pressures may have favored mechanisms that compensated for consequences of interactions between low CO_2 and carbon-depleting stresses. Exactly what adaptations confer fitness to CO_2 is a matter of conjecture; however, based on models proposed for other resource limitations—nutrients in particular—it is possible to hypothesize which traits confer fitness in low CO_2 environments.

In plants, strategies for dealing with nutrient limitation often involve modulating growth and allocation patterns (1) to enhance acquisition of the limiting resources (for example, high root allocation is associated with soil infertility); (2) to store resources as a buffer against fluctuations in resource supply, particularly fluctuations that lead to resource starvation; and (3) to limit growth rates to avoid depleting resources during stressful episodes (Chapin, 1988; Grime and Campbell, 1991). An example of a species adapted to low nutrient conditions is *Streptanthus morrisonii* (Jewel flower), a biennial herb that grows on highly infertile serpentine soils of Northern California (Sage *et al.*, 1989). In comparison to *Chenopodium album,* a common weed of fertile soils, *S. morrisonii* has 10% the growth capacity, one-fourth the allocation to leaves (as measured by leaf area partitioning), and three times the allocation to roots (Table III). Taken together, the characteristics of plants adapted to infertile soils lead to a condition of "luxury consumption," in which plants acquire and store resources in amounts exceeding their ability to use them for immediate growth (Chapin, 1980). Strategies for dealing with nutrient stress have a substantial cost in terms of reduced carbon gain and growth potential, largely because developmental controls on meristematic activity restrict the maximum growth rate of plants adapted to infertile soils, or because

Table III Growth Characteristics of Two Herbaceous Species Adapted to Nutrient-Poor or Nutrient-Rich Soils[a]

	Streptanthus morrisonii	*Chenopodium album*
Relative growth rate, %/d	3.5	34
Net CO_2 assimilation rate, $\mu mol\ m^{-2}\ s^{-1}$ at saturating light	25	42
Net assimilation rate, $g\ m^{-2}\ d^{-1}$	7	22
Leaf area partitioning,[b] $cm^{-2}\ d^{-1}/g\ d^{-1}$	37	149
Root weight partitioning $g\ d^{-1}/g\ d^{-1}$	39	13
Root/shoot	0.59	0.23

[a] Both species were grown under nutrient rich conditions (sand culture with daily watering of Johnson-Hoaglands solution). From Sage *et al.*, 1989 (*S. morrisonii*) and Sage and Pearcy, 1987 (*Chenopodium album*).

[b] Leaf area partitioning is the ratio of absolute leaf area growth rate to absolute whole plant growth rate. Root weight partitioning is the ratio of root growth rate to whole plant growth rate.

resources may be diverted into storage rather than growth during favorable periods (Chapin, 1980; Grime and Campbell, 1991). Traits associated with luxury consumption are successful, however, because they prevent depletion of internal resource levels below what is required to support minimum levels of metabolism (Chapin *et al.*, 1990). Plants that lack these adaptations tend to outgrow low levels of nutrient supply, and thus internalize the stress and increase the probability of premature death (Chapin, 1988). Sage and Pearcy (1987), for example, observed that the high-growth-capacity weed *Amaranthus retroflexus* developed extensive necrosis under severe nitrogen deficiency, which they attributed to insufficient nitrogen reserves to maintain leaf viability.

A. Plants as Luxury Consumers of Carbon

Using the luxury consumption model for nutrient deficiency as a conceptual guide, it is possible to envision how plants may have adapted to the low CO_2 conditions of the past. To survive low CO_2 availability, plants would need higher carbon acquisition capacity than would be required in a high CO_2 environment. This would occur as a result of producing more leaves and photosynthetic enzymes within leaves than needed at high CO_2. In particular, the high levels of Rubisco commonly observed in leaves may occur because the activity of this enzyme tends to be more limiting under low CO_2 conditions (von Caemmerer and Farquhar, 1981; Sage and Reid, 1994). The optimal fraction of carbon allocated to storage would depend on the degree of harshness encountered in any given situation, but compared to conditions in today's atmosphere, an identical habitat at low CO_2 could require a greater degree of storage because stressful conditions would be more frequent and intense than under higher CO_2 levels. Conservative growth patterns would restrict competition between storage sinks and growing regions, thus promoting enough carbohydrate accumulation to ensure survival during unfavorable periods. Conservative growth patterns could arise from either a genetically fixed limit on growth rate, or alternatively because a strong stress reaction is initiated in response to a given set of nonoptimal conditions. In plants, the response to stress often involves sensing unfavorable environmental conditions by roots, which then release ABA and modulate other signals that inhibit shoot growth (Chapin, 1991; Beck, 1994; Schulze, 1994). Because low CO_2 expands the range of stressful conditions, the mechanisms that sense unfavorable conditions and initiate growth adjustments may be more reactive than needed in current and future CO_2 conditions. Consequently, the degree of growth inhibition associated with a set of given nonoptimal conditions may be unnecessarily large in CO_2 enriched conditions.

In an evolutionary context, fitness is enhanced when multiple finite resources equally share control over productivity and survival, rather than a single resource, such as carbon, dominating the control (Mooney and Chiariello, 1984). Allocation patterns that compensate for resource deficiencies through enhanced acquisition and storage lead to a condition of shared control (that is, a balance) between the supply potential of various limiting factors in the environment (Bloom *et al.*, 1985). However, as demonstrated by adaptation to variation in light, nutrients, and water supply (Chapin, 1988; Schulze *et al.*, 1987; Givinish, 1988), allocation strategies that work for one suite of resource availabilities are often maladapted for others, and much of the competitive interactions in nature involve an interplay between resource availabilities and resource acquisition strategies (Grime, 1988; Tilman, 1993; Tilman and Wedin, 1991). Just as noncarbon resources are recognized as producing adaptive syndromes in response to varying resource availability, we suggest plants have evolved adaptive syndromes to low carbon availability of the past, such that during the Pleistocene, control over productivity and fitness was equally controlled by carbon supply and other finite resources in the plants habitat. If low carbon adaptations still predominate, they may restrict the ability of plants to exploit the increased growth potential CO_2 enrichment provides, and contribute to situations where nutrients, rather than carbon, dominate control of biomass productivity.

B. Are Plants Too Photosynthetic and Conservative for a High CO_2 World?

In support of the hypothesis that low CO_2 adaptations restrict plant responses to high CO_2, we cite the many reports which observe little effect of CO_2 enrichment on growth processes in natural or simulated plant communities, despite large increases in the level of stored carbohydrate (for review see Bazzaz, 1990; Koch and Mooney, 1996). Although CO_2 enrichment often produces large biomass responses in isolated plants grown with high soil fertility, the same species show far less response to CO_2 enrichment when grown in mixed communities or under natural nutrient regimes (Ackerly and Bazzaz, 1995). In field plants, stimulation of photosynthesis often remains following long-term CO_2 enrichment, yet is typically associated with growth enhancement of a much lower magnitude (Luo *et al.*, 1997). In a recent review of major ecosystem-level studies of CO_2 enrichment, Koch and Mooney (1996) concluded "there is often little or no measurable increases in plant biomass despite large increases in photosynthetic carbon uptake." They further note that when increases in biomass are observed, they are generally associated with drought, high temperature stress, or enriched soil fertility—conditions that normally increase the degree to which carbon is limiting. In a particularly strong example, a doubling of atmospheric CO_2 above herbaceous communities in low elevation Swiss pastures, and alpine tundra in the Swiss Alps, produced sustained increases in photosynthesis of more than 25% (Diemer, 1994; Körner, 1995; Körner *et al.*, 1996). In contrast, associated growth responses of individual species generally ranged between 0 and 20%, while overall community production was little affected (Körner, 1995; Schäppi and Körner, 1996; Leadley and Körner, 1996; Schäppi, 1996).

Where growth stimulation lags behind photosynthetic stimulation, the excess carbon is thought to be secreted to the soil or stored in the plant (Diaz *et al.*, 1993; Schäppi and Körner, 1996; Stitt, 1991). A near-universal observation of plants grown at elevated CO_2 is that nonstructural carbohydrate levels significantly rise, even in well-fertilized plants (Körner *et al.*, 1995; Farrar and Williams, 1991). These increases in carbohydrate accumulation are excessive because the plants do not fully mobilize the stored carbohydrate reserves during periods of high demand. In leaves exposed to elevated CO_2, more carbohydrates are accumulated during the day than are mobilized in the following night period (Wong, 1990; Körner *et al.*, 1995). Even at today's CO_2 level, high accumulation of carbohydrate is commonly observed, leading to speculation that plants currently maintain excessive photosynthetic capacity (Gifford and Evans, 1981; Stitt, 1991; Stitt and Schulze, 1994).

In addition to sink capacity, sucrose synthesis and export through the phloem appear to be important control points for carbohydrate utilization, and may be the major explanation for excessive carbohydrate accumulation at CO_2 levels above the current ambient (Körner *et al.*, 1995). Plants that have enhanced carbohydrate export capacity, either because they have higher sucrose synthesis capacity or because they are more efficient phloem loaders, exhibit less carbohydrate accumulation and often greater responsiveness to elevated CO_2 (Körner *et al.*, 1995; Micallef and Sharkey, 1996; Micallef *et al.*, 1995). Control over export is an important point on which evolution may have acted to ensure adequate carbon storage in leaves for night and unfavorable periods. Mature leaves are poor sinks and therefore must rely on stored reserves during nonphotosynthetic periods (Turgeon, 1984; ap Rees, 1984). Should local reserves become exhausted, senescence may soon follow (Lloyd, 1980; Gan and Amasino, 1997). Establishing a regulatory setpoint that ensures adequate carbohydrate reserves in leaves could involve control at either phloem loading or sucrose synthesis, and would likely result in more conservative setpoints if plants are adapted to low CO_2 conditions. If CO_2 levels were to rise above adaptive setpoints for carbohydrate regulation, then excessive leaf carbohydrate accumulation could result. Although high carbohydrate levels inhibit transcription of photosynthetic enzymes (Stitt, 1991; Van Oosten *et al.*, 1995), this process does not appear effective enough to reduce photosynthesis to where carbon supplies become limiting. Consequently, leaf carbohydrate levels remain high. In many cases, carbohydrates accumulate to a point where leaf damage is evident (Ehret and Jolliffe, 1985; Stitt, 1991, 1994). Thus, strategies that may be adaptive in low CO_2 may promote carbohydrate toxicity in CO_2-enriched conditions.

IV. Testing the Hypothesis of Low CO_2 Adaptation

We are well aware that plants in natural environments are highly variable in their pattern of carbohydrate use. Adaptation to low CO_2 does not imply that plants are conservative in absolute terms, but are conservative in comparison to what would be optimal if their habitat has been enriched with CO_2. To test the hypothesis, variation in carbohydrate use that is already present in natural populations could be exploited to examine fitness traits in low, normal, and high CO_2 conditions. For example, plants with different allocation to storage pools, differences in phloem loading (Körner *et al.*, 1995), or differences in sink potential could be screened across a range of CO_2 levels to determine relative performance and survivability. Along these lines, Ward and Strain (1997) recently screened *Arabidopsis thaliana* genotypes from different elevations for reproductive responsiveness to CO_2 variation between 200 and 700 μmol mol^{-1}. They found all genotypes were highly responsive between 200 and 350 μmol mol^{-1}, but poorly responsive to CO_2 enrichment from 350 to 700 μmol mol^{-1}. Genetic manipulation of the processes controlling carbohydrate acquisition and allocation could also be employed to examine the value of single genes in varying CO_2 conditions. Some of this work has already been conducted with transgenic crop plants exposed to high CO_2 (Micallef *et al.*, 1995; Stitt, 1994). To be most telling, however, these experiments should include treatments that allow for interactions between stress and low CO_2. Extreme conditions are strong selection agents in natural environments, and the fitness of traits affecting acquisition and use of carbohydrates may be most apparent during carbohydrate-depleting extremes.

V. Summary

Responses of plants to CO_2 enrichment have commonly been interpreted in terms of imbalances between sources and sinks, or between nutrient supply and carbon availability. Clearly, increasing CO_2 perturbs the relationship between carbon and nutrient dynamics to produce a condition where control over growth is more dependent on nutrients than carbon. Even at today's level of CO_2, carbon can have low control over biomass productivity because CO_2 enrichment above the current ambient often produces negligible growth responses. As discussed elsewhere in this volume, the ultimate factor affecting productivity may be the capacity of the soil to supply mineral nutrients, and failure to observe growth responses to CO_2 enrichment may reflect finite rates of nutrient supply. As proposed here, an alternative possibility for failure to respond to CO_2 enrichment is that plants may have evolved strategies which confer fitness to much lower levels of atmospheric CO_2, and the nutrient limitations now expressed may be a result of low flexibility in the expression of traits that enhance survival of carbon deficiency. Resolving the significance of these two possibilities is important because if nutrient supply is the primary determinant of CO_2 response, then vegetation responses to CO_2 enrichment could be set regardless of the allocation patterns expressed by plants. Alternatively, if plants are adapted to lower CO_2 levels, then evolutionary responses that correct imbalances and reestablish a condition of shared control between finite resources could lead to substantial enhancement of CO_2 responsiveness. For example, instead of heavily investing in photosynthesis, plants may instead allocate internal reserves to nutrient acquisition. By considering what CO_2 level plants are adapted to, we may be in a better position to understand and, therefore, predict long-term vegetation responses to CO_2 enrichment.

Acknowledgments

We thank Sam Puvendren, Susan Dexter, and Dr. Tammy Sage for assistance in the low CO_2 experiments discussed in this chapter. Research in this report was supported by National Science and Engineering Council of Canada grant 91-37100-6619 to R.F.S.

References

Ackerly, D. D., and Bazzaz, F. A. (1995). Plant growth and reproduction along CO_2 gradients: Non-linear responses and implications for community change. *Global Change Biol.* **1,** 199–207.

Allen, L. H., Bisbal, E. C., Boote, K. J., and Jones, P. H. (1991). Soybean dry matter allocation under subambient and superambient levels of carbon dioxide. *Agron. J.* **83,** 875–883.

ap Rees, T. (1984). Sucrose metabolism. *In* "Storage Carbohydrates of Vascular Plants" (D. H. Lewis, ed.), pp. 53–73. Cambridge University Press, Cambridge.

Baker, J. T., Allen, L. H., Boote, K. J., Jones, P., and Jones, J. W. (1990). Rice photosynthesis and evapotranspiration in subambient, ambient, and superambient carbon dioxide concentrations. *Agron. J.* **82,** 834–840.

Barnola, J. M., Raynaud, D., Dorotkevich, Y. S., and Lorius, C. D. (1987). Vostok ice core provides 160,000 year record of atmospheric CO_2. *Nature* **329,** 408–418.

Bazzaz, F. A. (1990). The response of natural ecosystems to the rising global CO_2 levels. *Annu. Rev. Ecol. Syst.* **21,** 167–196.

Beck, E. (1994). The morphogenic response of plants to soil nitrogen: Adaptive regulation of biomass distribution and nitrogen metabolism by phytohormones. *In* "Flux Control in Biological Systems" (E. D. Schulze, ed.), pp. 119–151. Academic Press, San Diego.

Bloom, A. J., Chapin, F. S. III, and Mooney, H. A. (1985). Resource limitation in plants—an economic analogy. *Annu. Rev. Ecol. Syst.* **16,** 363–392.

Brooks, A., and Farquhar, G. D. (1985). Effect of temperature on the CO_2/O_2 specificity of ribulose-1,5-bisphosphate carboxylase/oxygenase and the rate of respiration in the light. *Planta* **165,** 397–406.

Chapin III, F. S. (1980). The mineral nutrition of wild plants. *Annu. Rev. Ecol. Syst.* **21,** 423–447.

Chapin III, F. S. (1988). Ecological aspects of plant mineral nutrition. *Adv. Plant Nutrition* **3,** 161–191.

Chapin III, F. S. (1991). Integrated responses of plants to stress: A centralized system of physiological responses. *BioScience* **41,** 29–36.

Chapin III, F. S., Schulze, E. D., and Mooney, H. A. (1990). The ecology and economics of storage in plants. *Annu. Rev. Ecol. Syst.* **21,** 423–47.

Chaves, M. M., and Pereira, J. S. (1992). Water stress, CO_2 and climate change. *J. Exp. Bot.* **43,** 1131–1139.

Cook, R. E. (1985). Growth and development in clonal plant populations. *In* "Population Biology and Evolution of Clonal Organisms" (J. B. C. Jackson, L. W. Buss, and R. E. Cook, eds.), pp. 259–296. Yale University Press, New Haven.

Cowling, S. A., and Sage, R. F. (1998). Interactive effects of low atmospheric CO_2 and elevated temperature on growth, photosynthesis and respiration in *Phaseolus vulgaris. Plant Cell Environ.* **21,** 427–435.

Diaz, S., Grime, J. P., Harris, J., McPherson, E. (1993). Evidence of a feedback mechanism limiting plant response to elevated carbon dioxide. *Nature* **364,** 616–617.

Diemer, M. W. (1994). Mid-season gas exchange of an alpine grassland under elevated CO_2. *Oecologia* **98,** 429–435.

Dippery, J. K., Tissue, D. T., Thomas, R. B., and Strain, B. R. (1995). Effects of low and elevated CO_2 on C_3 and C_4 annuals: I. Growth and biomass allocation. *Oecologia* **101,** 13–20.

Ehret, D. L., and Jolliffe, P. A. (1985). Leaf injury to bean plants grown in carbon dioxide enriched atmospheres. *Can. J. Bot.* **63,** 2015–2020.

Farquhar, G. D., and von Caemmerer, S. (1982). Modelling of photosynthetic response to environmental conditions. *In* "Physiological Plant Ecology II" (O. L. Lange, P. S. Nobel, C. B. Osmond, and H. Ziegler, eds.), pp. 549–587. Springer Verlag, Berlin.

Farrar, J. F., and Williams, M. L. (1991). The effects of increased atmospheric carbon dioxide and temperature on carbon partitioning, source-sink relations and respiration. *Plant Cell Environ.* **14,** 819–830.

Field, C. B., Chapin III, F. S., Matson, P. A., and Mooney, H. A. (1992). Responses of terrestrial ecosystems to the changing atmosphere: A resource-based approach. *Annu. Rev. Ecol. Syst.* **23,** 201–235.

Gan, S., and Amasino, R. M. (1997). Making sense of senescence: Molecular genetic regulation and manipulation of leaf senescence. *Plant Physiol.* **113,** 313–319.

Gifford, R. M., and Evans, L. T. (1981). Photosynthesis, carbon partitioning and yield. *Annu. Rev. Plant Physiol.* **32,** 485–509.

Givinish, T. J. (1988). Adaptation to sun and shade: A whole-plant perspective. *Aust. J. Plant Physiol.* **15,** 63–92.

Grime, J. P. (1988). The C-S-R model of primary plant strategies—origins, implications and tests. *In* "Plant Evolutionary Biology" (L. D. Gottlieb and S. K. Jain, eds.), pp. 371–393. Chapman and Hall, London.

Grime, J. P., and Campbell, B. D. (1991). Growth rate, habitat productivity, and plant strategy as predictors of stress response. *In* "Response of Plants to Multiple Stresses" (H. A. Mooney, W. E. Winner, E. J. Pell, and E. Chu, eds.), pp. 143–159. Academic Press, San Diego.

Jasper, J. P., and Hayes, J. M. (1990). A carbon isotope record of CO_2 levels during the late Quaternary. *Nature* **347,** 462–464.

Johnson, H. B., Polley, H. W., and Mayeux, H. S. (1993). Increasing CO_2 and plant–plant interactions: Effects on natural vegetation. *Vegetatio* **104/105,** 157–170.

Jouzel, J., Barkov, N. I., Barnola, J. M., Bender, M., Chappellaz, J., Genthon, C., Kotlyakov, V. M., Lipenkov, V., Lorius, C., Petit, J. R., Raynaud, D., Raisbeck, G., Ritz, C., Sowers, T., Stievenard, M., Yiou, F., and Yiou, P. (1993). Extending the Vostok ice-core record of paleoclimate to the penultimate glacial period. *Nature* **364,** 407–412.

Kingsolver, J. G. (1996). Physiological sensitivity and evolutionary responses to climate change. *In* "Carbon Dioxide, Populations, and Communities" (C. Körner and F. A. Bazzaz, eds.), pp. 3–12. Academic Press, San Diego.

Koch, G. W., and Mooney, H. A. (1996). Responses of terrestrial ecosystems to elevated CO_2: a synthesis and summary. *In* "Carbon Dioxide and Terrestrial Ecosystems" (G. W. Koch and H. A. Mooney, eds.), pp. 415–429. Academic Press, San Diego.

Körner, C. (1995). Biodiversity and CO_2: Global change under way. *GAIA* **4,** 234–242.

Körner, C., Pelaez-Riedl, S., and Van Bel, J. E. (1995). CO_2 responsiveness of plants: A possible link to phloem loading. *Plant Cell Environ.* **18,** 595–600.

Körner, C., Diemer, M., Schäppi, B., and Zimmermann, L. (1996). Response of alpine vegetation to elevated CO_2. *In* "Carbon Dioxide and Terrestrial Ecosystems" (G. W. Koch and H. A. Mooney, eds.), pp. 177–196. Academic Press, San Diego.

Kutzbach, J. E., and Ruddiman, W. F. (1993). Simulated climatic changes: Results of the COHMAP climate-model experiments. *In* "Global Climates Since the Last Glacial Maximum" (H. E. Wright, J. E. Kutzbach, T. Webb, W. F. Ruddiman, F. A. Street-Perrott, and P. J. Bartlein, eds.), pp. 24–93. University of Minnesota Press, Minneapolis.

Leadley, P. W., and Körner, C. (1996). Effects of elevated CO_2 on plant species dominance in a highly diverse calcareous grassland. *In* "Carbon Dioxide, Populations, and Communities" (C. Körner and F. A. Bazzaz, eds.), pp. 51–81. Academic Press, San Diego.

Lloyd, E. S. (1980). Effects of leaf age and senescence on the carbon distribution of *Lolium. J. Exp. Bot.* **31,** 1067–1079.

Luo, Y., Chen, J. L., Reynolds, J. F., Field, C. B., and Mooney, H. A. (1997). Disproportional increases in photosynthesis and plant biomass in a California grassland exposed to elevated CO_2: A simulation analysis. *Functional Ecol.* **11,** 696–704.

Micallef, B. J., and Sharkey, T. D. (1996). Genetic and physiological characterization of *Flaveria linearis* plants having a reduced activity of cytosolic fructose-1,6-bisphosphatase. *Plant Cell Environ.* **19,** 1–9.

Micallef, B. J., Haskins, K. A., Vanderveer, P. J., Roh, K., Shewmaker, C. K., and Sharkey, T. D. (1995). Altered photosynthesis, flowering, and fruiting in transgenic tomato plants that have an increased capacity for sucrose synthesis. *Planta* **196,** 327–334.

Mitton, J. B., and Grant, M. C. (1996). Genetic variation and the natural history of quaking aspen. *Bioscience* **46,** 25–31.

Mooney, H. A., and Chiariello, N. R. (1984). The study of plant function—the plant as a balanced system. *In* "Perspectives on Plant Population Ecology" (R. Dirzo and J. Sarukhan, eds.), pp. 305–323. Sinauer Associates, Sunderland, MA.

Neftel, A., Oeschger, H., Staffelback, T., and Stauffer, B. (1988). CO_2 record in the Byrd ice core 50,000–5,000 years B.P. *Nature* **331,** 609–611.

Osmond, C. B., Austin, M. P., Berry, J. A., Billings, W. D., Boyer, J. S., Dacey, W. H., Nobel, P. S., Smith, S. D., and Winner, W. E. (1987). Stress physiology and the distribution of plants. *BioScience* **37,** 38–48.

Polley, H. W., Johnson, H. B., and Johnson, H. S. (1992). Growth and gas exchange of oats (*Avena sativa*) and wild mustard (*Brassica Kaber*) at subambient CO_2 concentrations. *Int. J. Plant Sci.* **15,** 453–461.

Polley, H. W., Johnson, H. B., Marino, B. D., and Mayeux, H. S. (1993a). Increase in C_3 plant water-use-efficiency and biomass over glacial to present CO_2 concentrations. *Nature* **361,** 61–63.

Polley, H. W., Johnson, H. B., Mayeux, H. S., and Malone, S. R. (1993b). Physiology and growth of wheat across a subambient carbon dioxide gradient. *Ann. Bot.* **71,** 347–356.

Raymo, M. E. (1992). Global climate change: A three million year perspective. *In* "Start of a Glacial" (G. J. Kukla and E. Went, eds.), pp. 207–224. Springer Verlag, Berlin.

Rogers, H. H., Runion, G. B., and Krupa, S. V. (1994). Plant responses to atmospheric CO_2 enrichment with emphasis on roots and the rhizosphere. *Environ. Pollut.* **83,** 155–189.

Sage, R. F. (1990). A model describing the regulation of ribulose-1,5-bisphosphate carboxylase, electron transport, and triose phosphate use in response to light intensity and CO_2 in C_3 plants. *Plant Physiol.* **94,** 1728–1734.

Sage, R. F. (1995). Was low atmospheric CO_2 during the Pleistocene a limiting factor for the origin of agriculture? *Global Change Biol.* **1,** 93–106.

Sage, R. F., and Pearcy, R. W. (1987). The nitrogen use efficiency of C_3 and C_4 plants. *Plant Physiol.* **84,** 954–958.

Sage, R. F., and Reid, C. D. (1992). Photosynthetic acclimation to sub-ambient CO_2 (20 Pa) in the C_3 annual *Phaseolus vulgaris* L. *Photosynthetica* **27,** 605–617.

Sage, R. F., and Reid, C. D. (1994). Photosynthetic response mechanisms to environmental change in C_3 plants. *In* "Plant Environment Interactions" (R. E. Wilkinson, ed.), pp. 413–499. Marcel Dekker, New York.

Sage, R. F., Ustin, S. L., and Manning, S. J. (1989). Boron toxicity in the rare serpentine plant, *Streptanthus morrisonnii. Environ. Pollut.* **61,** 77–93.

Schäppi, B. (1996). Growth dynamics and population development in an alpine grassland under elevated CO_2. *Oecologia* **106,** 93–99.

Schäppi, B., and Körner, C. (1996). Growth responses of an alpine grassland to elevated CO_2. *Oecologia* **105,** 43–52.

Schulze, E. D. (1994). The regulation of plant transpiration: Interactions of feedforward, feedback, and futile cycles. *In* "Flux Control in Biological Systems" (E. D. Schulze, ed.), pp. 203–235. Academic Press, San Diego.

Schulze, E. D., Robichaux, R. H., Grace, J., Rundel, P. W., and Ehleringer, J. R. (1987). Plant water balance. *BioScience* **37,** 30–37.

Silander, Jr., J. A. (1985). Microevolution in clonal plants. *In* "Population Biology and Evolution of Clonal Organisms" (J. B. C. Jackson, L. W. Buss, and R. E. Cook, eds.), pp. 107–152. Yale University Press, New Haven.

Stitt, M. (1991). Rising CO_2 levels and their potential significance for carbon flow in photosynthetic cells. *Plant Cell Environ.* **14,** 741–762.

Stitt, M. (1994). Flux control at the level of the pathway: Studies with mutants and transgenic plants having a decreased activity of enzymes involved in photosynthesis partitioning. *In* "Flux Control in Biological Systems" (E. D. Schulze, ed.), pp. 13–36. Academic Press, San Diego.

Stitt, M., and Schulze, E. D. (1994). Plant growth, storage, and resource allocation: From flux control in a metabolic chain to the whole-plant level. *In* "Flux Control in Biological Systems" (E. D. Schulze, ed.), pp. 57–118. Academic Press, San Diego.

Thomas, S. C., and Jasienski, M. (1996). Genetic variability and the nature of microevolutionary responses to elevated CO_2. *In* "Carbon Dioxide, Populations, and Communities" (C. Körner and F. A. Bazzaz, eds.), pp. 51–81. Academic Press, San Diego.

Tilman, D. (1993). Carbon dioxide limitation and potential direct effects of its accumulation on plant communities. *In* "Biotic Interactions and Global Change" (P. M. Kareiva, J. G. Kingsolver, and R. B. Huey, eds.), pp. 333–346. Sinauer Associates, Sunderland, MA.

Tilman, D., and Wedin, D. (1991). Plant traits and resource reduction for five grasses growing on a nitrogen gradient. *Ecology* **72,** 685–700.

Tissue, D. T., Griffin, K. L., Thomas, R. B., and Strain, B. R. (1995). Effects of low and elevated CO_2 on C_3 and C_4 annuals: II. photosynthesis and leaf biochemistry. *Oecologia* **101,** 21–28.

Trabalka, J. R. (1985). Atmospheric carbon dioxide and the global carbon cycle, DOE/ER-0239. United States Department of Energy, Washington, DC.

Turgeon, R. (1984). Efflux of sucrose from minor veins of tobacco leaves. *Planta* **161,** 120–128.

Van Oosten, J. J., Wilkins, D., and Besford, R. T. (1995). Acclimation of tomato to different carbon dioxide concentrations. Relationships between biochemistry and gas exchange during leaf development. *New Phytol.* **130,** 357–367.

Vasek, F. C. (1980). Creosote bush: Long-lived clones in the Mojave Desert. *Am. J. Bot.* **67,** 246–255.

von Caemmerer, S., and Farquhar, G. D. (1981). Some relationships between the biochemistry of photosynthesis and the gas exchange of leaves. *Planta* **153,** 376–387.

Ward, J. K., and Strain, B. R. (1997). Effects of low and elevated CO_2 partial pressure on growth and reproduction of *Arabidopsis thaliana* from different elevations. *Plant Cell Environ.* **20,** 254–260.

Wong, S. (1990). Elevated atmospheric partial pressure of CO_2 and plant growth: II. Non-structural carbohydrate content in cotton plants and its effect on growth parameters. *Photosyn. Res.* **23,** 171–180.

12

Scaling against Environmental and Biological Variability: General Principles and A Case Study

Yiqi Luo

I. Introduction

Research on plant and ecosystem responses to CO_2 is primarily driven by large-scale issues. Modeling studies of global carbon (C) cycling have long suggested that the global C budget cannot be balanced without storage of C in terrestrial ecosystems (Bacastow and Keeling, 1973). Recently, evidence has been presented to support the idea that terrestrial ecosystems play a critical role in modulating the C balance of the earth system. For example, Tans *et al.* (1990) analyzed spatial and temporal distributions of atmospheric CO_2 concentration (C_a) and concluded that a large amount of C is absorbed by terrestrial ecosystems. Other research has found that temporal variations and latitudinal gradients of $^{13}C/^{12}C$ ratios in the atmosphere result, at least partly, from C isotope discrimination in plants, again indicating the possibility that terrestrial ecosystems sequester C (Tans *et al.*, 1993; Lloyd and Farquhar, 1994; Ciais *et al.*, 1995). In addition to research findings regarding the global terrestrial C sink, signing of the Climate Convention has forced nations to assess their contributions to C sources and sinks and to evaluate the processes that control CO_2 accumulation in the atmosphere. The great need to quantify global terrestrial C sink and to assess regional C sources and sinks provides a major challenge in scaling studies.

Scaling has long been a scientific issue. Scale has been explicitly considered in achieving representative data from vegetation sampling (Curtis and McIntosh, 1950), allometric studies of animals (Schmidt-Nielsen, 1984) and plants (West *et al.*, 1997), and in the biomechanical studies of morphology

(McMahon, 1975). Defining appropriate scales at which underlying processes occur has been a useful way of revealing critical causes of aggregation of plant distributions (Greig-Smith, 1983). Scaling of animal sizes has been found to be critical in determining physiology and habitat. Scaling-up of ecophysiological measurements to predict large-scale energy and material fluxes has recently become a major scientific activity (Ehleringer and Field, 1993), largely attributable to our inability to make direct measurements at the scales that require solution.

This chapter presents a case study on scaling of leaf photosynthesis to predictions of global terrestrial C influx stimulated by rising C_a. The nature and approaches of scaling studies are discussed in an attempt to establish criteria for scaleability of parameters. These criteria are then used to review recent results on scaling photosynthesis from leaf to globe. This review establishes supportive evidence for parameter scaling and also identifies sources of uncertainty. Four supplementary studies are also presented to examine rigorously some of the uncertain sources involved in extrapolating photosynthetic responses from small to large scales. Based on the current evidence, it is concluded that separating photosynthetic responses to rising atmospheric CO_2 into two components, that is, photosynthetic sensitivity and acclimation, facilitates scaling-up studies. A baseline prediction can be provided by using the sensitivity component to estimate global terrestrial C influx stimulated by a yearly increase in C_a.

II. The Nature and Approaches of Scaling-Up Studies

Scaling studies in nature use knowledge at one scale to address issues at either higher or lower scales. When issues at higher scales are addressed using knowledge obtained from a lower scale, it is a scaling-up study. The reverse is a scaling-down study. In CO_2 research, we primarily use knowledge from experimental studies at leaf, plant, and small ecosystem scales to address issues at regional and global scales. Thus, the scaling-up study is the most commonly used approach.

The need for scaling-up studies is generated by the desire to solve problems at large scales. The issues of global environmental change and/or regional assessment force us to address large-scale, long-term issues (e.g., global terrestrial C sink, regional C sinks and sources, and the landscape impact of natural fire). Although considerable efforts have been invested in making larger scale measurements using remote sensing, micrometeorological techniques, and aircraft, either the measurement scales have been several orders of magnitude smaller than the scales under consideration or the measured data have not been directly applicable to the problems (Norman, 1993). On the other hand, leaf and plant physiology as well as small-scale ecosystem structure and function have been extensively studied in response to biological and environmental variables. Extensive knowledge bases have been established. Thus, the challenge of extrapolating our knowledge of leaf and plant physiology as well as small ecosystem function is critical to our ability to predict regional and global changes (Fig. 1).

Because of the difficulties in making direct measurements at the regional and global scales, any scaling operation must rely on linkages of biological objects to certain conceptual frameworks (Allen *et al.*, 1993) within which a set of logical rules has to be established for scaling. In the study of biogeochemical cycling in terrestrial ecosystems, scaling operations are primarily conducted on energy and material fluxes (e.g., C and N) and rarely involve transformation of types (e.g., from food supply to number of animals). Although different types of species, soils, and/or weather patterns are involved in energy and material fluxes, these types only vary rates of fluxes leading to heterogeneity of environments and interspecific variation. Indeed, environmental heterogeneity and interspecific variation are the two general issues in extrapolating small-scale knowledge to large-scale changes (Fig. 1). Any methods that can diminish or reduce the effects

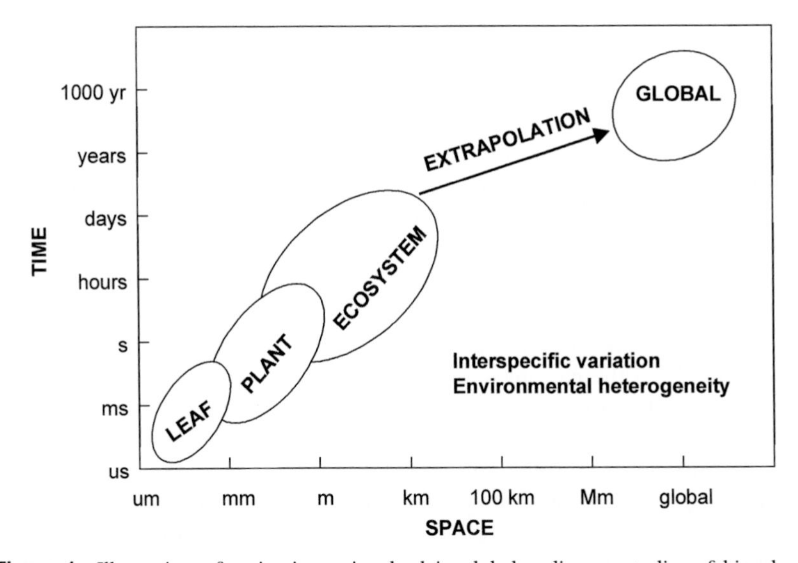

Figure 1 Illustration of major issues involved in global scaling-up studies of biospheric processes. Global change research is usually designed to address large-scale and long-term questions. However, we cannot make any measurements at such large scales. By extrapolating small-scale measurements (leaf, plant, or small ecosystem) to predict large-scale changes, we encounter two general problems; interspecific variation and environmental heterogeneity. Any method that can eliminate or diminish the effects of these variables on the parameters in question will reduce uncertainties in predicting large-scale changes.

of these two general variables on the parameters in question can reduce uncertainties in predicting large-scale changes. Therefore, scaling-up studies must search for methods to reduce the uncertainties associated with interspecific variation and environmental heterogeneity.

Several schemes have recently been used to address this need in scaling-up studies, including summation, averaging, and aggregation (Jarvis, 1995). Those schemes have been used in regional and global modeling that divides geographical maps of world vegetation and soils into grids with relatively uniform environmental conditions and species composition. Each grid is then represented by mean values of ecosystem processes in computation of regional and global estimates. This practice is hereafter referred to as the *grid-based approach.* The Vegetation/Ecosystem Modeling and Analysis Project (VEMAP), for example, has used this approach by dividing the U.S. land area into 0.5-deg grids (VEMAP Members, 1995). In each grid, a typical soil and vegetation type were chosen to represent the mean properties of soil and vegetation. Representative values were then used to parameterize various models including, for example, the photosynthetic model by Farquhar *et al.* (1980), the stomatal conductance model by Jarvis and McNaughton (1986), and models of nutrient and water feedbacks. This approach helps reduce spatial variability. Difficulties in model parameterization and vegetation delineation, however, still make such quantification unsatisfactory (Schimel, 1995).

An alternative approach in scaling-up studies is to search for scaleable parameters that characterize intrinsic properties of a system in question. This method is hereafter referred to as the *process-oriented approach.* Field (1991) has argued for a functional convergence hypothesis in which evolution has shaped plants such that some physiological processes reflect the availability of all the resources required for plant growth and thus become scaleable predictors of environmental conditions and resource availability. The validity of this convergence hypothesis has been examined in the context of the photosynthesis-nitrogen relationship at the leaf level, nitrogen-light at the canopy level, and light use efficiency at the ecosystem scale. Although it varies in plot-scale measurements, Field (1991) argued that light use efficiency (defined as the ratio of dry matter production to the integrated energy absorbed) in combination with remote sensing data is a reasonable predictor of net primary productivity at large scales. Another example of searching for scaleable parameter is quantum yield of CO_2 uptake—a physiological parameter—that is used to predict global distributions of C_3 and C_4 plants (Ehleringer *et al.,* 1997). Both theoretical and experimental studies have indicated that the quantum yield of C_3 plants is independent of growth environments as well as species and varies with temperature and CO_2 concentration. At the current ambient CO_2 concentration, the quantum yield of C_3 plants is higher than that of C_4 plants in low temperature and lower than that of C_4 plants in higher temperature

(Ehleringer and Bjorkman, 1977). The C_3/C_4 crossover temperatures, where one photosynthetic pathway becomes superior to the other, have been used to delineate the boundary of C_4 species distribution in global terrestrial ecosystems (Ehleringer, 1978). The scaleability of the quantum yield to prediction of global C_3/C_4 distributions is particularly high for monocotyledonous C_4 species but a little complicated by other environmental variables for dicotyledonous C_4 species (Ehleringer *et al.*, 1997).

This process-oriented approach of identifying scaleable parameters has also been used by Luo and Mooney (1996) in scaling-up photosynthesis from leaf to globe. They have examined leaf photosynthetic sensitivity to an increase in C_a (normalized photosynthetic response to a small increase in C_a) and found that the sensitivity is independent of interspecific variation and growth environments for C_3 plants and only varies with measurement temperature. This independence establishes the potential to simplify global extrapolation of leaf-level studies and to provide a baseline prediction of an increment in global C influx stimulated by rising C_a (Luo *et al.*, 1996; Luo, 1997). The scaleability of photosynthesis from the leaf to the globe will be explored in more detail in the next two sections.

A natural ecosystem is a complex assemblage of many processes. The process-oriented approach focuses on identifying scaleable parameters within this complex assemblage in an attempt to reduce uncertainties in scaling-up studies. We also have to recognize that some ecosystem processes are not scaleable. For example, photosynthetic acclimation involves changes in many biochemical processes and morphological properties. These changes are often species and environment specific. As a result, the search for a scaleable parameter to represent the divergent processes of photosynthetic acclimation may not be realistic. In this case, a mean value may be the best representation we can ever have in scaling-up studies. Precision in the use of these unscaleable processes can only be improved as a result of extensive measurements. The grid-based approach in combination with functional types and other aggregation methods will be the most effective in scaling-up studies involving these processes. It follows that the process-oriented and grid-based approaches may have to be used interactively in most cases. While the process-oriented approach is useful in improving precision in predicting large-scale change, the grid-based approach will be helpful in examining the system in more realistic circumstances.

III. Scaling Photosynthesis from Leaf to Globe: A Two-Component Model

A. Sensitivity and Acclimation

To facilitate scaling-up studies of photosynthesis, Luo *et al.* (1996) have suggested separation of photosynthetic response to rising C_a into two general components: sensitivity and acclimation. Sensitivity of leaf photosynthesis to C_a is determined by competition between carboxylation and oxygenation of ribulose-bisphosphate (RuBP). Both carboxylation (i.e., photosynthesis) and oxygenation (i.e., photorespiration) are catalyzed by ribulose-1,5-bisphosphate carboxylase/oxygenase (Rubisco) (Andrews and Lorimer, 1987). Increased carbon dioxide concentration competes with oxygen and decreases the oxygenase activity of Rubisco (Farquhar *et al.*, 1980; Lawlor, 1993), leading to an increased ratio of carboxylation to oxygenation. Leaf photosynthetic sensitivity to CO_2 has been extensively studied (see a comprehensive review by Farquhar and von Cammerer, 1982) and found to be similar among species (Allen *et al.*, 1987; Sharkey, 1988), to be insensitive to growth environments (Farquhar and von Cammerer, 1982), and to vary with temperature and CO_2 and O_2 concentration (Polglase and Wang, 1992; Kirschbaum, 1994).

Acclimation describes the changes in photosynthetic capacity induced by elevated CO_2 (Stitt, 1991; Luo *et al.*, 1994). Studies have shown, for example, that photosynthetic capacity in elevated CO_2 increases for plant species *Glycine max* (Campbell *et al.*, 1988) and decreases for *Gossypium hirsutum* (Wong, 1990) and *Lolium perenne* (Ryle *et al.*, 1992). Phytosynthetic acclimation may result from redistribution of nitrogen among various photosynthetic enzymes (Sage, 1990), adjustments in source-sink relationships (Stitt, 1991), and/or increased mesophyll growth (Vu *et al.*, 1989; Sims *et al.*, 1998).

The photosynthetic sensitivity and acclimation components are embedded in the Farquhar *et al.* (1980) model that assumes that photosynthesis is limited either by the light-driven regeneration of RuBP or by Rubisco:

$$P_1 = J\frac{C_i - \Gamma}{4.5\ C_i + 10.5\ \Gamma}, \tag{1}$$

$$P_2 = V_{cmax}\frac{C_i - \Gamma}{C_i + K}, \tag{2}$$

where J is the electron transport rate (μmol m^{-2} s^{-1}) and varies with light and the maximum electron transport rate J_{max}, V_{cmax} is the maximum RuBP carboxylase activity (μmol m^{-2} s^{-1}), C_i is the intercellular CO_2 concentration (μmol mol^{-1}), Γ is the CO_2 compensation point (μmol mol^{-1}) without dark respiration, and K is an enzymatic kinetics, equaling to

$$K = \frac{K_C}{1 + \frac{O}{K_O}}, \tag{3}$$

where K_C and K_O are the Michaelis–Menten constants for CO_2 (μmol mol^{-1}) and oxygen (mmol mol^{-1}), respectively and O is the partial pressure of oxygen at the site of carboxylation (= 0.21 mol mol^{-1}).

The terms $(C_i - \Gamma)/(4.5C_i + 10.5\Gamma)$ and $(C_i - \Gamma)/(C_i + K)$ in Eqs. (1) and (2) describe the photosynthetic sensitivity to CO_2, which is the enzymatic reactions of Rubisco to oxygen and CO_2 caused by a small variation in CO_2 concentration. The parameters V_{cmax} and J_{max} quantify the amount of available Rubisco enzymes and electrons, respectively. They indicate the capacity of plant photosynthesis and have been found to vary by 30-fold among 109 species surveyed by Wullschleger (1993). These two parameters can be altered by growth in different environments, with varying levels of CO_2, nutrients, light, and temperature. The variation of J_{max} and V_{cmax} caused by different growth environments is photosynthetic acclimation.

To conduct a scaling analysis of leaf-level results, Luo and Mooney (1996) mathematically defined photosynthetic sensitivity to a small increase in atmospheric C_a as:

$$\mathcal{L} = \frac{1}{P}\frac{dP}{dC_a}, \tag{4}$$

where $\mathcal{L}$ is a leaf-level function (ppm^{-1}) denoting the normalized leaf photosynthetic response to one unit C_a change. With an assumption that C_i is proportional to C_a as

$$C_i = \alpha \cdot C_a, \qquad \text{for } 0 < \alpha < 1, \tag{5}$$

the corresponding $\mathcal{L}$ function is

$$\mathcal{L}_1 = \frac{15\,\alpha\,\Gamma}{(\alpha\,C_a - \Gamma)(4.5\,\alpha\,C_a + 10.5\,\Gamma)}, \tag{6}$$

$$\mathcal{L}_2 = \frac{\alpha(K + \Gamma)}{(\alpha\,C_a - \Gamma)(\alpha\,C_a + K)}. \tag{7}$$

Variables V_{cmax} and J are eliminated from Eqs. (6) and (7) because $\mathcal{L}_1$ is a measure of relative response. The elimination of V_{cmax} and J_{max} indicates that sensitivity is independent of species, growth environment of nutrients, light, temperature, and water stress. Because the physiological process of electron transport is less sensitive to CO_2 concentration than is carboxylation, $\mathcal{L}_1$ and $\mathcal{L}_2$ define the lower and upper limits of the $\mathcal{L}$ function, respectively.

If photosynthetic capacity, which is quantified by the variables V_{cmax} and J_{max}, varies with C_a, photosynthetic acclimation occurs. In this case, V_{cmax} and J_{max} are a function of C_a. Equations (6) and (7) would have to be modified as follows:

$$\mathscr{L}_1' = \mathscr{L}_1 + \frac{1}{J}\frac{dJ}{dC_a} \quad (8)$$

$$\mathscr{L}_2' = \mathscr{L}_2 + \frac{1}{V_{cmax}}\frac{dV_{cmax}}{dC_a}. \quad (9)$$

The first terms in the Eqs. (8) and (9) are the photosynthetic sensitivity as in Eqs. (6) and (7). The second terms in Eqs. (8) and (9) describe photosynthetic acclimation. Thus, photosynthetic response to CO_2 is separated into two components: sensitivity and acclimation.

B. Properties of Photosynthetic Sensitivity and Acclimation

A series of studies has been conducted using both theoretical and experimental approaches to examine sensitivity and acclimation in the context of scaling photosynthesis from the leaf to the globe. Major conclusions are summarized as follows:

1. Photosynthetic sensitivity as indicated by the $\mathscr{L}$ function in Eqs. (6) and (7) is independent of diverse growth environments and interspecific variation. This conclusion is supported by both theoretical analysis (Luo and Mooney, 1996) and experimental data (Luo *et al.*, 1996). Sensitivity is a function of C_a and measurement temperature.

2. Photosynthetic acclimation, as indicated in the second terms of the Eqs. (8) and (9), is species and environment specific. Diversity of acclimative changes, however, can be predicted from changes in leaf N concentration (Luo *et al.*, 1994) and is determined by a balance between biochemical down-regulation and morphological up-regulation (Luo *et al.*, 1998).

3. Photosynthetic sensitivity and acclimation are independent of each other and additive in determining photosynthetic response to CO_2 (Luo *et al.*, 1996). Acclimation may result in either an increase or a decrease in photosynthetic capacity and rate. Sensitivity, however, always leads to an increase in photosynthetic rate as C_a increases.

4. Extrapolation of the $\mathscr{L}$ function results in an estimate of the marginal increase in global terrestrial ecosystem C influx caused by a marginal increase in C_a. The estimate is 0.26 Gt (10^{15} g) C yr^{-1} in 1993 caused by a 1.5 ppm increase in C_a and declines with time, primarily due to decreased sensitivity of photosynthesis to C_a in its high range.

5. Uncertainty exists in the estimated increment of global terrestrial C influx and originates primarily from three sources: (1) the sensitivity component itself, (2) acclimative changes at various scales from leaf to biome shifting, and (3) composition of C_4 plants in the earth system combined with the present flux of global photosynthetic C fixation (Luo and Mooney, 1995; Luo *et al.*, 1996). To improve estimation of global C influx and storage induced by rising C_a, each of these sources of uncertainty

deserves careful examination. The next section provides four supplementary studies in an attempt to examine some of the sources.

IV. Supplementary Studies

A. Oscillation of Sensitivity: Light versus Enzyme Limitation

There are two theoretical curves for the $\mathscr{L}$ function that are derived from two equations in the Farquhar *et al.* (1980) photosynthetic model. The two equations represent photosynthesis limited by either RuBP regeneration or Rubisco. The biochemical basis for the two photosynthetic equations has been discussed in terms of limiting factors of photosynthetic C fixation (Farquhar and von Cammerer, 1982). Switching between the two phases of photosynthetic reactions has also been demonstrated by measured photosynthetic responses to a variety of light or CO_2 levels. As a consequence of the switching of photosynthesis between the two phases, it is expected that experimental data reflecting photosynthetic sensitivity will oscillate between the two curves of the $\mathscr{L}$ function, resulting in a 30% uncertainty relative to its mean value (Luo *et al.*, 1996). Because this 30% uncertainty is significant in global extrapolation, the variation has been examined using data from a mesocosm experiment conducted by a group of scientists at the Desert Research Institute (DRI), Reno, Nevada.

This research utilized DRI's EcoCELLs open-flow, mass-balance mesocosms that continuously operate for the measurement of C, water, and energy fluxes (Griffin *et al.*, 1996). Sunflowers (*Helianthus annus*) were grown in two EcoCELLs, one with ambient CO_2 (399 ppm) and the other with elevated CO_2 (746 ppm). The experiment started with seeds on 5 July 1997 and ended with a full canopy cover on 29 August 1997. C fluxes were continuously measured using Li-Cor 6262 (Li-Cor Inc., Lincoln, NE). Photosynthetic C fixation was small relative to soil respiration during the seedling stage in the first 35 d and gradually increased to 50 μmol m^{-2} ground area s^{-1} in the late growing stage (Luo *et al.*, 1999). The C flux data used to test the oscillation were collected on 24 August 1997, 54 d after the experiment began and after the canopies were fully closed. Leaf area index (LAI) reached approximately four in both treatments. Little leaf photosynthetic acclimation was detected during the experimental period (Sims *et al.*, 1999).

The C fluxes from the two EcoCELLs over a time course from dawn to noon on 24 August 1997 were used to calculate the photosynthetic sensitivity to CO_2 using the formula $(F_e - F_a)/F_a$, where F_e and F_a are the measured C fluxes at elevated and ambient CO_2, respectively. The measured data in the EcoCELLs using Li-Cor 6262 represent net ecosystem productivity (NEP), equaling the photosynthetic C influx minus plant and soil respiration. Canopy photosynthesis in this study was estimated from NEP plus daytime plant and soil respiration. The latter was estimated from nighttime plant and soil respiration corrected by a temperature function using Q_{10} equal to 2 for the difference between night (13°C) and day (28°C) time temperature.

Data plotted in Fig. 2 show that the photosynthetic sensitivity resulted in an increase of canopy C fixation by 25 to 55% as the light level increased from virtually zero at dawn to 1300 μmol m^{-2} s^{-1} at noon. These results were generally consistent with the theoretical analysis that photosynthetic sensitivity is low when light limits photosynthesis and high when Rubisco limits photosynthesis during periods of high light intensity. The theoretical sensitivity values oscillate between 0.29 and 0.70 with light. The lower values of measured sensitivity, compared to the theoretical ones, were probably caused by estimation errors in daytime respiration. As discussed by Luo *et al.* (1996), if actual respiration is less than the estimate, then measured values of the $\mathscr{L}$ function will be greater than those plotted in Fig. 2, resulting in a closer match between the measured and theoretical values and vice versa. Despite the slightly lower values in the measured sensitivity, this data

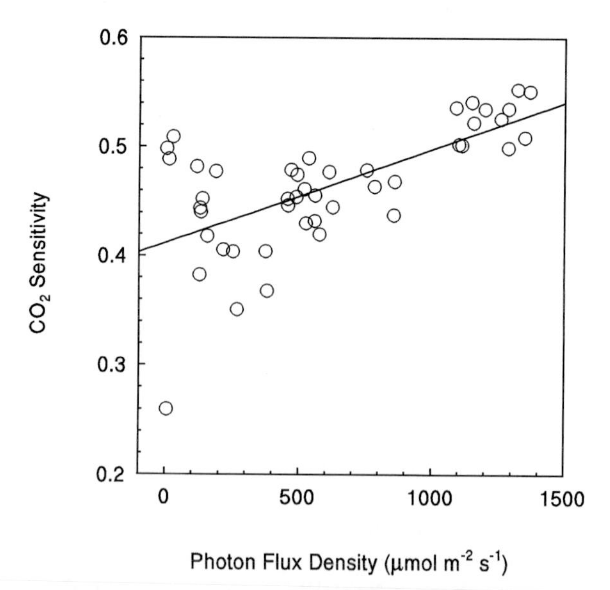

Figure 2 CO_2 sensitivity of ecosystem photosynthesis as a function of light availability illustrated by sunflower (*Helianthus annus*) plants grown at two CO_2 levels (399 versus 746 ppm, respectively). The plant canopy was fully closed, and measurements were made every 15 min continuously for 24 h. Data from sunrise to noon on 24 August 1997 were used to calculate CO_2 sensitivity at each measurement point. Measurements indicated similar LAIs between the two canopies and no photosynthetic acclimation. Sensitivity is largely attributable to biochemical stimulations. Data suggested that photosynthetic stimulation by CO_2 was low in the morning when the light level was low and high at noon when the light level was high, consistent with theoretical analysis.

set suggested that photosynthetic sensitivity does oscillate between the two theoretical curves of the $\mathcal{L}$ function (Luo and Mooney, 1996). Consequently, the uncertainty associated with the two curves will be considered inherent in natural ecosystems. The mean value between the two curves is likely the best representation of photosynthetic sensitivity in the natural world.

B. Stomatal Conductance and Photosynthetic Sensitivity

Photosynthetic sensitivity to CO_2, by definition, is the change in enzymatic reactions in the chloroplast within leaves. The passage of CO_2 from the air to the reaction sites in the chloroplast is controlled by stomata which have the potential to influence sensitivity. The mechanism by which sensitivity is actually affected by stomatal conductance has not been carefully examined. A theoretical analysis indicates that when the C_i/C_a ratio associated with variation in stomatal conductance is altered by 0.10 from 0.70, photosynthetic sensitivity varies by 7% (Luo and Mooney, 1996). The supplementary study described later evaluates the effects of stomatal conductance on the $\mathcal{L}$ function using data from an experimental study(Griffin and Luo, 1999).

Seeds of *Glycine max* were planted in 5-liter pots filled with potting soil on 12 April 1995. Five pots were placed in each of two EcoCELLs (Griffin *et al.*, 1996) at the Desert Research Institute, Reno, Nevada. CO_2 concentration was maintained at 360 ppm in one EcoCELL and at 700 ppm in the other. All pots were watered twice weekly with deionized water. One week after planting all pots were thinned to a single plant, and 2 weeks after planting all pots were given 500 ml of half-strength Hoagland's solution. Growth temperature was maintained at 28°/20°C (day/night), and relative humidity was kept at a constant 50% during the experiment. Solar radiation typically exceeded 1300 μmol m^{-1} s^{-1} on sunny days. Measurements of leaf level CO_2 and H_2O fluxes were made with Li-6400 (Li-Cor., Lincoln, Nebraska). All measurements were made at a constant air temperature of 25°C, light intensity of 1500 μmol m^{-2} s^{-1}, and relative humidity of 50%. The photosynthetic response to ambient and intercellular CO_2 concentration was measured on expanded and expanding leaves by varying the CO_2 concentration from 100 to 1500 ppm in 150 roughly equal steps.

Growth CO_2 concentration and leaf ages resulted in substantial differences in stomatal conductance and C_i/C_a ratio. The latter ratio varied from 0.4 to 0.95. The $\mathcal{L}$ values were calculated using A/C_i (assimilation/intercellular CO_2 concentration) and A/C_a (assimilation/ambient CO_2 concentration) response curves. When A/C_i response is used to derive $\mathcal{L}$, the $\mathcal{L}$ function becomes:

$$\mathcal{L}_1 = \frac{15\,\Gamma}{(C_i - \Gamma)(4.5\,C_i + 10.5\,\Gamma)}, \tag{10}$$

$$\mathcal{L}_2 = \frac{K + \Gamma}{(C_i - \Gamma)(C_i + K)}. \tag{11}$$

Equations (10) and (11) differ from Eqs. (6) and (7) by a factor of the C_i/C_a ratio. To compare the effects of stomatal conductance on the $\mathscr{L}$ function, the $\mathscr{L}$ values derived from A/C_a response curves times the C_i/C_a ratio were plotted against the $\mathscr{L}$ values derived from A/C_i response curves in Fig. 3. Data are scattered close to the 1 : 1 line, indicating that variation in the C_i/C_a ratio associated with stomatal conductance may not induce a large change in the $\mathscr{L}$ function. This result is consistent with the theoretical predictions (Luo and Mooney, 1996).

C. Temperature and Photosynthetic Sensitivity

Temperature has a dual effect on photosynthesis. First, growth in certain temperature regimes may induce change in photosynthetic capacity (i.e., V_{cmax} and J_{max}). Temperature-induced change in photosynthetic capacity will not affect photosynthetic sensitivity, however, as indicated by both theoretical analysis (Luo and Mooney, 1996) and validation using experimental data (Luo *et al.*, 1996). The second effect caused by fluctuation in temperature in both diurnal and annual time courses is the regulation of enzymatic activities of Rubisco (i.e., instantaneous effect of temperature). The instantaneous effects of temperature and the implication for global

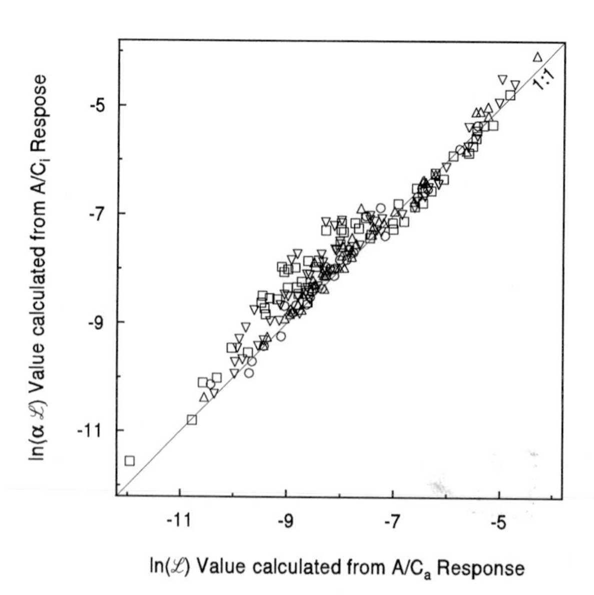

Figure 3 Correlation of $\mathscr{L}$ values derived from photosynthetic response to intercellular CO_2 versus $\mathscr{L}$ values derived from photosynthetic response to ambient CO_2. ○ = expanding leaves of plants grown at 350 ppm; △ = expanding leaves of plants grown at 700 ppm; □ = expanded leaves of plants grown at 350 ppm; ▽ = expanded leaves of plants grown at 700 ppm.

terrestrial C fluxes and storage have been examined by Kirshbaum (1994) as well as Polglase and Wang (1992). However, Luo and Mooney (1996) did not explicitly incorporate the instantaneous effect of temperature into their model but made the spatial extrapolation of the $\mathcal{L}$ function to the global scale with the assumption that effective global mean temperature in relation to terrestrial photosynthetic machinery is 20°C. The validity of the extrapolation of the $\mathcal{L}$ function and the assumption on effective temperature have yet to be examined.

I have recently examined the instantaneous effects of temperature on the extrapolation of the $\mathcal{L}$ function using a spatially explicit Terrestrial Uptake and Release of Carbon (TURC) model (Ruimy *et al.*, 1996). TURC is a diagnostic model for the estimation of continental gross primary productivity (GPP) and net primary productivity (NEP). The model uses a remotely sensed vegetation index to estimate the fraction of solar radiation absorbed by canopies combined with radiation use efficiency to predict GPP. The $\mathcal{L}$ function is integrated into the TURC model, which computes the increment in GPP in grid cells. The $\mathcal{L}$ function varies with temperature through parameters K_O, K_C, and Γ in Eqs. (3), (6), and (7), following Harley *et al.* (1992) and Badger and Collatz (1977) as

$$K_O = \exp[9.959 - \frac{14.51}{0.00831\ (273.2 + T)}], \tag{12}$$

$$K_C = \exp[35.79 - \frac{80.47}{0.00831(273.2 + T)}], \tag{13}$$

$$\Gamma = 42\ \exp[\frac{9.46\ (T - 25)}{273.2 + T}], \tag{14}$$

where T is the temperature in °C. When these temperature functions were integrated into Eqs. (6) and (7), estimated $\mathcal{L}$ values and GPP increments varied by season and location. In the Northern Hemisphere in January, when temperature and photosynthetic activity are both low, the estimated $\mathcal{L}$ value was approximately 0.0006 for both the lower and upper limits. At the same time of the year in the Southern Hemisphere where temperature and photosynthetic activity are both high, the estimated $\mathcal{L}$ values were approximately 0.0019 for the lower limit and 0.0032 for the upper limit at the current ambient CO_2. This variation in the $\mathcal{L}$ values with the temperature makes the extrapolation of the $\mathcal{L}$ function more complicated than that originally proposed by Luo and Mooney (1996). On the other hand, data presented in the paper by Luo *et al.* (1996) support that the $\mathcal{L}$ function is independent of interspecific variation and growth environments of light, nutrients, water, temperature, and CO_2 concentration. This property still helps simplify the global extrapolation of the $\mathcal{L}$ function.

I have also examined the effects of the instantaneous temperature on the GPP increments. The increments in each grid cell were added to calculate the global terrestrial GPP increment stimulated by the increase in C_a. The global GPP increment in 1987 caused by a 1.5-ppm increase in atmospheric CO_2 was estimated to range from 0.154 to 0.342 Gt C. This is the equivalent of a 0.116 and 0.249% increase over the base value of GPP (133 Gt C yr^{-1}). That increment is equivalent to the estimate with the $\mathscr{L}$ function estmate at 20.16°C. The temperature of 20°C used in the original calculation by Luo and Mooney (1996) is very close to the estimation with the spatially explicit model TURC. In short, the instantaneous effect of temperature complicates the global extrapolation of the $\mathscr{L}$ function. However, the aggregated estimate using the $\mathscr{L}$ function at a global effective temperature of 20°C is close to the disaggregated estimate with a spatially explicit model.

D. Ecosystem N Dynamics and Photosynthetic Acclimation

Photosynthetic acclimation potentially adds a high degree of uncertainty when extrapolating leaf photosynthesis to a global scale. Acclimation has been studied extensively and found to vary greatly in relation to both species and growth environments (Gunderson and Wullschleger, 1994; Luo *et al.*, 1994). This species- and environment-specific variation in acclimation makes the extrapolation across scales very difficult. On the other hand, the relationship between leaf photosynthesis and N concentration has been found to be similar across diverse biomes ranging from tropical and temperate forests to alpine tundra and deserts (Reich *et al.*, 1997). As a result, the N–photosynthesis relationship may assist extrapolation of leaf photosynthetic acclimation to the global scale. In addition, Luo *et al.* (1994) have found that leaf photosynthetic acclimation is well predicted by CO_2-induced changes in leaf N concentration. Thus, it is critical to understand how leaf N and photosynthetic acclimation are affected by ecosystem N availability.

The relationship between ecosystem N dynamics and photosynthetic acclimation has been studied using both modeling and experimental approaches. Models with strong links between nutrient cycling and plant production generally predict long-term photosynthetic down-regulation because of nutrient constraints (VEMAP Members, 1995; Ryan *et al.*, 1996; Mooney *et al.*, 1999). An elegant analysis on C and N interactions was conducted by Comins and McMurtrie (1993) using the G'DAY model. They developed photosynthetic and N-recycling constraint curves in determining the semiequilibrium, long-term ecosystem responses to doubling CO_2. The G'DAY model predicts an instantaneous increase of 27% in photosynthesis after doubling CO_2. This is based principally on increased photosynthesis per unit N. As the additional amount of photosynthetically fixed C is used for plant growth and/or stored in soil as organic matter, ecosystem N limitation gradually develops. As a consequence, leaf C:N ratio increases and leaf photosynthesis and ecosystem productivity are strongly limited by ecosystem N availability. Thus, a long-term increase of only 5% is predicted when growth is limited by both photosynthetic and N-cycling constraints (Comins and McMurtrie, 1993). Similarly, other models also predict substantial down-regulation in elevated CO_2. The extent to which photosynthesis is down-regulated depends on assumptions on plant and soil C:N ratios, litter decomposition, net mineralization, and N fixation and losses (Kirschbaum *et al.*, 1994; McMurtrie and Comins, 1996; Rastetter *et al.*, 1997).

Although early experimental studies in growth chambers and greenhouses usually indicated photosynthetic down-regulation mediated by nitrogen supply, predicted strong down-regulation in photosynthesis has not been confirmed by most field experiments across ecosystems with diverse soil N availability. A Global Change and Terrestrial Ecosystems (GCTE) review conducted by Mooney *et al.* (1999) qualitatively assessed 27 ecosystem experiments (14 in herbaceous and 13 in woody systems) using either open-top chambers (OTC) or free-air CO_2 enrichment (FACE) facilities. Of these, 7 exhibited no down-regulation in photosynthesis; 19 had some; and only 1 experienced complete down-regulation. A quantitative review by Curtis and Wang (1998) using a meta-analysis method found no significant photosynthetic down-regulation in 24 woody species. Drake *et al.* (1997) reviewed 42 studies, some of which have been ongoing for 4–10 yr, and found that a significant reduction in photosynthetic capacity as a result of growth in elevated CO_2 is the exception rather than the rule when there is no rooting-volume limitation.

This discrepancy between the model predictions and experimental results prompted a study addressing N demand and supply mechanisms in ecosystems, instead of nitrogen constraints on photosynthesis, in response to elevated CO_2 (Luo and Reynolds, 1999). A terrestrial C sequestration (TCS) model was developed to simulate photosynthetic C production; allocation of C into leaves, wood, and roots; litterfall from plant parts; microbial decomposition of litter; and soil organic C formation into slow and passive C pools. The TCS model adopted a C influx module developed by Luo *et al.* (1996) and Luo and Mooney (1996) as well as plant and soil dynamic modules similar to CENTURY (Parton *et al.*, 1987; Schimel *et al.*, 1994). Processes that represent the C and N interactions in the model include photosynthetic dependency on leaf N concentration, adjustments in plant C:N ratio, litter decomposition, net mineralization (gross mineralization − immobilization), rhizosphere expansion, and N fixation. The model was parameterized primarily using Integrated Forest Studies (IFS) field data from the Duke Forest (Binkley and Johnson, 1991) and was used to examine both the amount of N required (N demand) to balance additional C influx and the mechanisms of N supply that could potentially be involved in meeting N demand in response to elevated CO_2.

Study results indicated that a large amount of N is required to balance additional C influx in elevated CO_2. For example, when the CO_2 concentration was abruptly increased from the ambient level to 560 ppm in 1996 in the Duke Forest FACE experiment, measurements indicated a 40–60% increase in leaf photosynthesis for both the dominant and other species (Ellsworth *et al.,* 1995; R. Thomas and E. DeLucia, personal communication). Using these measurements as a guide, we set the TCS model to simulate an approximately 50% increase in ecosystem photosynthesis with assumptions of no acclimation for 113 yr (Fig. 4A). The model simulated respiratory C release from both plant parts and soil pools as well as net ecosystem productivity (NEP = ecosystem C influx − efflux). NEP was predicted to increase to 265 g C m^{-2} yr^{-1} in the first year of the CO_2

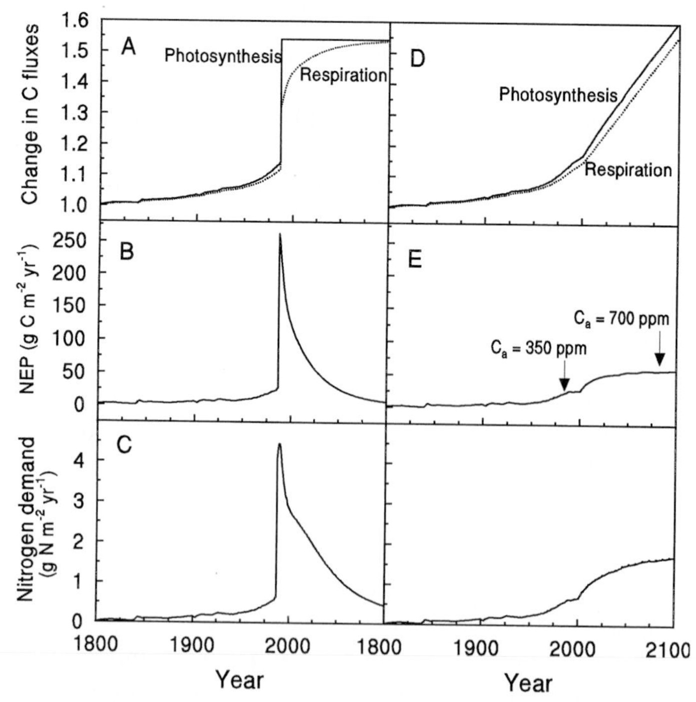

Figure 4 Ecosystem C and N dynamics in response to a step vs. gradual increase in atmospheric CO_2. In the simulation, C:N ratio in the model is set at 40 for leaf; 150 for woody stem and root; 50 for fine root; 15, 20, and 10 for organic matter in active, slow, and passive soil C pools, respectively, according to Comins and McMurtrie (1993). Other assumptions include (1) no photosynthetic acclimation, (2) no adjustments in other ecosystem C processes, and (3) a constant step increase in CO_2 concentration for 113 yr.

experiment, assuming no N limitation, and then to decline gradually (Fig. 4B). An NEP of 265 g C m^{-2} yr^{-1} would require an additional 5.3 g N m^{-2} yr^{-1} to balance C:N relationships (Fig. 4C). Using the model to simulate a gradual increase in C_a (as occurs in the natural world), ecosystem photosynthesis and respiration both increased gradually with a time lag in respiratory C release (Fig. 4D). In response to a gradual C_a increase, NEP was predicted to be 27 g C m^{-2} yr^{-1} in 1987 (C_a = 350 ppm) and 58 g C m^{-2} yr^{-1} in 2085 (C_a = 700 ppm) (Fig. 4E). The N demand was predicted to be 0.7 and 1.43 g N m^{-2} yr^{-1} in 1987 and 2085, respectively (Fig. 4F).

We also used the model to explore three N supply mechanisms that might meet N demand in response to rising CO_2: (1) adjustment in the C:N ratio in live biomass, (2) net mineralization, and (3) rhizosphere expansion (Fig. 5). The simulations indicated that when an ecosystem only

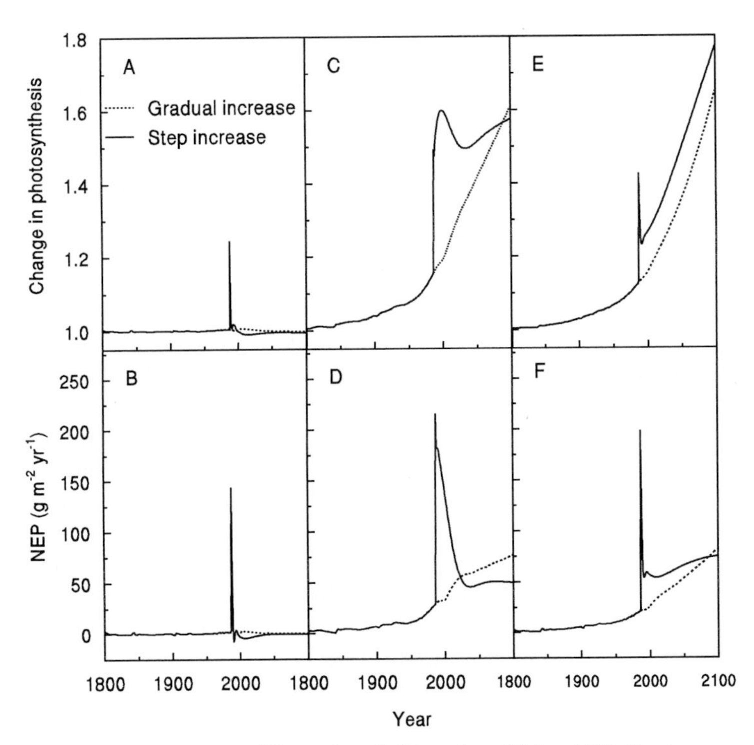

Figure 5 Three mechanisms of N supply to balance the additional C influx in response to elevated CO_2 in regulating photosynthetic acclimation and NEP. Panels A and B show the consequences of plants adjusting the C:N ratio in leaves, fine roots, and wood to use the scarce N resource efficiently. Panels C and D show the rising CO_2 effect of increased fine root growth that results in more efficient N uptake, thereby stimulating N mineralization. Panels E and F show the effects of increased rooting depth and root length density, resulting in more soil exploration.

adjusts the C:N ratio in live biomass without changes in other parameters, the whole ecosystem is not responsive to an increase in CO_2. Photosynthetic C influx and NEP were also found to be strongly limited by N in both gradual and step increases of CO_2 (Figs. 5A and B). Photosynthetic down-regulation completely eliminated CO_2 stimulation of ecosystem C influx.

In simulation studies of CO_2-induced change in net mineralization and rhizosphere expansion, we set parameter values so that photosynthetic response to a gradual CO_2 increase would only be slightly limited by N supply. Using that set of parameters, we used the model to predict responses of photosynthesis and NEP to the step increase in CO_2. When the model simulated the stimulation of net N mineralization by increasing the rooting system in elevated CO_2, ecosystem photosynthesis declined slightly in the first few years, then bounced back and reached a peak in year 13 (Fig. 5C). Afterward, photosynthesis declined and went up again. NEP reached 210 g C m^{-2} yr^{-1} in the first year of the step CO_2 experiment and then declined (Fig. 5D). When the increased rooting system expanded the rhizosphere in elevated CO_2, ecosystem photosynthesis was down-regulated to the lowest level in year 3 of the step CO_2 experiment and then steadily increased to the end of the simulation experiment (Fig. 5E). NEP was 200 g C m^{-2} yr^{-1} in the first year of the step experiment and then quickly declined to 60 g C m^{-2} yr^{-1} (Fig. 5F).

In short, ecosystem C and N interactions are very complicated. Current modeling studies on nutrient constraints to C production in elevated CO_2 may be strongly hindered by our limited understanding of the fundamental aspects of ecosystem N dynamics, resulting in a substantial discrepancy between model predictions and experimental results. The latter indicate minimal photosynthetic down-regulation in elevated CO_2 across diverse ecosystems, suggesting that acclimation may not be a dominant factor in estimating global photosynthetic C influx in response to elevated CO_2. However, the theoretical basis for this lack of strong regulation of photosynthesis in intact ecosystems is not clear yet. If the step experiments with an abrupt increase in CO_2 concentration do not induce much N limitation, it is even less likely that N will induce photosynthetic down-regulation in response to a gradual increase in C_a. In the latter case, N demand is much less than in the step increase in CO_2 (Fig. 5).

V. Summary

Scaling-up studies of leaf photosynthesis have been generated by our great need to quantify global terrestrial C sink and assess regional C sources and sinks. However, at the regional and global scales, we are unable to make direct measurements of C production and have to rely almost entirely on extrapolating knowledge derived from leaf-level studies. When we extrapolate leaf physiology to predict global change, we encounter two general issues: environmental variability and biological diversity. Any methods that can diminish or eliminate the effects of these two general variables on the parameters in question can reduce uncertainties in predicting large-scale changes. Therefore, the focus of scaling-up studies should be on methods to reduce uncertainties associated with environmental variability and biological diversity.

While schemes using summation, averaging, and aggregation are frequently used in scaling-up studies, identifying scaleable parameters is probably the most effective approach to reducing uncertainties. The scaleable parameters should characterize intrinsic properties of a system and have minimal variability associated with environmental heterogeneity and genetic diversity. A successful example of using this approach is the extrapolation of quantum yield of CO_2 uptake to predict global distributions of C_3 and C_4 plants. Similarly, leaf photosynthetic sensitivity to rising C_a can be used to predict the increment of global GPP. Leaf photosynthetic sensitivity is independent of growth environments and interspecific diversity and is also a function of instantaneous temperature and CO_2 concentration. Extrapolating the sensitivity from leaf to globe results in an estimate of a marginal increment of 0.26 Gt C yr^{-1} in global terrestrial C influx in 1993 caused by a yearly increase in atmospheric CO_2. While this method greatly improves the precision of global predictions, uncertainties in the estimated increment may be further reduced by rigorously examining three sources: (1) the sensitivity component, (2) photosynthetic acclimation at various scales from leaf to biome shifting, and (3) composition of C_4 plants in the earth system and global GPP.

This chapter presented four supplementary studies examining (1) whether photosynthetic sensitivity oscillates between light- versus enzyme-limited processes, leading to a 30% approximate uncertainty in the global estimation; (2) how stomatal conductance affects the global estimation; (3) how the global estimation is affected by temperature variation; and (4) whether ecosystem N availability provides any predictability to the acclimation. These supplementary studies led to the following conclusions: (1) Sensitivity intrinsically oscillates in natural ecosystems as light availability fluctuates; (2) stomatal conductance has a negligible effect on sensitivity and thus on the global estimation of additional C influx; (3) fluctuation in instantaneous temperature results in spatial and temporal variation in the $\mathscr{L}$ function and hinders simple extrapolation of the function from the leaf to the globe (note that a study with a spatially explicit model indicated that global extrapolation of the $\mathscr{L}$ function using an average temperature of 20°C provided a reliable estimate of the increment in C influx); and (4) photosynthetic acclimation may be regulated by N availability, but this regulation has not been found to be significant in field studies across diverse ecosystems. Accordingly, acclimation may not be a major component in estimating terrestrial C influx and storage in response to rising C_a. In general, separating photosynthetic responses to rising C_a into sensitivity and acclimation components facilitates scaling-up studies. Scaling up the sensitivity component provides a baseline prediction of additional global C influx stimulated by a yearly increase in C_a.

Acknowledgments

The author thanks the group of scientists including Drs. W. Cheng, J. Coleman, D. Johnson, and D. Sims in the Biological Science Center, Desert Research Institute, for the data set in Fig. 2 and D. Hui for preparing that figure. Preparation of this manuscript was financially supported by DOE grant DE-FG05-95ER62083 and TECO grant NSF-IBN-9524036.

References

Allen, T. F. H., and Hoejstra, T. W. (1992). "Toward a Unified Ecology." Columbia, New York.

Allen, Jr., L. H., Boote, K. L., Jones, J. W., Jones, P. H., Valle, R. R., Acock, B., Rogers, H. H., and Dalhman, R. C. (1987). Response of vegetation to rising carbon dioxide: Photosynthesis, biomass, and seed yield of soybean. *Global Biogeochem. Cycles* **1,** 1–14.

Allen, T. F. H., King, A. W., Milne, B. T., Johnson, A., and Turner, S. (1993). The problem of scaling in ecology. *Evol. Trends Plants* **7,** 3–8.

Andrews, T. J., and Lorimer, G. H. (1987). Rubisco: Structure, mechanisms, and prospects for improvement. *In* "The Biochemistry of Plants, A Comprehensive Treatise" (P. K. Stumpf and E. E. Conn, eds.), pp. 131–218. Academic, San Diego.

Bacastow, R., and Keeling, C. D. (1973). Atmospheric carbon dioxide and radiocarbon in the natural carbon cycle: II. Changes from A.D. 1700 to 2070 as deduced from a geochemical model. *In "Carbon and the Biosphere, Proceedings of the 24th Brookhaven Symposium in Biology"* (G. M. Woodwell and E. V. Pecan, eds.), AEC Symp. Ser., Vol. 30, pp. 86–135. Atomic Energy Commission, Upton, NY.

Badger, M. R., and Collatz, G. J. (1977). Studies on the kinetic mechanism of ribulose-1,5-bisphosphate carboxylase and oxygenase reactions, with particular reference to the effect of temperature on kinetic parameters. *Carnegie Institution of Washington Yearbook* **76,** 355–361.

Ball, J. T., Woodrow, I. E., and Berry, J. A. (1987). A model predicting stomatal conductance and its contribution to the control of photosynthesis under different environment conditions. *In "Progress in Photosynthesis Research"* (I. Giggins, ed.), Vol IV, pp. 221–224. Martinus Nijhoff Publishers, The Netherlands.

Binkley, D., and Johnson, D. W. (1991). Southern pines. *In* "Atmospheric Deposition and Forest Nutrient Cycling: A Synthesis of the Integrated Forest Study" (D. W. Johnson and S. E. Lindberg, eds.), pp. 534–543. Springer Verlag, New York.

Campbell, W. J., Allen, Jr., L. H., and Bowes, G. (1988). Effects of CO_2 concentration on Rubisco activity, amount and photosynthesis in soybean leaves. *Plant Physiol.* **88,** 1310–1316.

Ciais, P., Tans, P. P., Trolier, M., White, J. W. C., and Francey, R. J. (1995). A large Northern Hemisphere terrestrial CO_2 sink indicated by the $^{13}C/^{12}C$ ratio of atmospheric CO_2. *Science* **269,** 1098–1102.

Comins, H. N., and McMurtrie, R. E. (1993). Long-term response of nutrient-limited forests to CO_2 enrichment: Equilibrium behavior of plant–soil models. *Ecol. Appl.* **3,** 666–681.

Curtis, J. T., and McIntosh, R. P. (1950). The interrelation of certain analytic and synthetic phytosocialogical characters. *Ecology* **31,** 434–455.

Curtis, P. S., and Wang, X. (1998). A meta-analysis of elevated CO_2 effects on woody plant mass, form, and physiology. *Oecologia* **113,** 299–313.

Drake, B. G., Gonzàlez-Meler, M. A., and Long, S. P. (1997). More efficient plants: A consequence of rising atmospheric CO_2? *Annu. Rev. Plant Physiol. Plant Mol. Biol.* **48,** 607–637.

Ehleringer, J. R. (1978). Implications of quantum yield differences to distributions of C_3 and C_4 grasses. *Oecologia* **31,** 255–267.

Ehleringer, J. R., and Björkman, O. (1977). Quantum yield for CO_2 uptake in C_3 and C_4 plants: Dependence on temperature, CO_2 and O_2 concentrations. *Plant Physiol.* **59,** 86–90.

Ehleringer, J. R., and Field, C. B. (eds.). (1993). "Scaling Physiological Processes: Leaf to Globe." Academic Press, San Diego.

Ehleringer, J. R., Cerling, T. E., Helliker, B. R. (1997). C_4 photosynthesis, atmospheric CO_2, and climate. *Oecologia* **112,** 285–299.

Ellsworth, D. S., Oren, R., Huang, C., Phillips, N., and Hendrey, G. R. (1995). Leaf and canopy responses to elevated CO_2 in a pine forest under free-air CO_2 enrichment. *Oecologia* **104,** 139–146.

Farquhar, G. D., and von Caemmerer, S. (1982). Modeling of photosynthetic response to environment. *In* "Encyclopedia of Plant Physiology, New Series, Vol. 12B. Physiological Plant Ecology II" (O. L. Lange, P. S. Nobel, C. B. Osmond, and H. Ziegler, eds.), pp. 549–587. Springer Verlag, New York.

Farquhar, G. D., von Caemmerer, S., and Berry, J. A. (1980). A biochemical model of photosynthetic CO_2 assimilation in leaves of C_3 species. *Planta* **149,** 79–90.

Field, C. B. (1991). Ecological scaling of carbon gain to stress and resource availability. *In* "Response of Plants to Multiple Stresses" (H. A. Mooney, W. E. Winner, and E. J. Pell, eds.), pp. 35–65. Academic Press, San Diego.

Griffin, K. L., and Luo, Y. (in press). An experimental study on the sensitivity and acclimation of *Glycine max* (L.) Merr. leaf photosynthesis to CO_2 partial pressure, A direct test of the leaf-level function. *Environmental and Experimental Botany.*

Griffin, K. L., Ross, P. D., Sims, D. A., Luo, Y., Seemann, J. R., Fox, C. A., and Ball, J. T. (1996). EcoCELLs: Tools for mesocosm scale measurements of gas exchange. *Plant Cell Environ.* **19,** 1210–1221.

Greig-Smith, P. (1983). "Quantitative Plant Ecology." Blackwell, London.

Gunderson, C. A., and Wullschleger, S. D. (1994). Photosynthetic acclimation in trees to rising atmospheric CO_2: A broader perspective. *Photosynth. Res.* **39,** 369–388.

Harley, P. C., Thomas, R. B., Reynolds, J. F., and Strain, B. R. (1992). Modeling photosynthesis of cotton grown in elevated CO_2. *Plant Cell Environ.* **15,** 271–282.

Jarvis, P. G. (1995). Scaling processes and problems. *Plant Cell Environ.* **18,** 1079–1089.

Jarvis, P. G., and McNaughton, K-G. (1986). Stomatal control of transpiration: Scaling from leaf to region. *Advances in Ecological Research* **15,** 1–47.

Kirschbaum, M. U. F. (1994). The sensitivity of C_3 photosynthesis to increasing CO_2 concentration: A theoretical analysis on its dependence on temperature and background CO_2 concentration. *Plant Cell Environ.* **17,** 747–754.

Kirschbaum, M. U. F., King, D. A., Comins, H. N., McMurtrie, R. E., Medlyn, B. E., Pongracic, S., Murty, D., Keith, H., Raison, R. J., Khanna, P. K., and Sheriff, D. W. (1994). Modelling forest response to increasing CO_2 concentration under nutrient-limited conditions. *Plant Cell Environ.* **17,** 1081–1099.

Lawlor, D. W. (1993). "Photosynthesis: Molecular, Physiological and Environmental Process," 2nd ed. Longman, White Plains, NY.

Lloyd, J., and Farquhar, G. D. (1994). [13]C discrimination during CO_2 assimilation by the terrestrial. *Oecologia* **99,** 201–215.

Luo, Y. (1997). Quantifying the global terrestrial carbon influx and sink associated with an increase in atmospheric CO_2 concentration. *In* "Proceedings: Thirteen Annual Pacific Climate (PACLIM) Workshop," Asiolomar, CA, April 14–17, 1996.

Luo, Y., Field, C. B., and Mooney, H. A. (1994). Predicting responses of photosynthesis and root fraction to elevated CO_2: Interaction among carbon, nitrogen and growth. *Plant Cell Environ.* **17,** 1195–1204.

Luo, Y., Griffin, K. L., and Sims, D. A. (1998). Nonlinearity of photosynthetic responses to growth in rising atmospheric CO_2: An experimental and modeling study. *Global Change Biol.* **4,** 173–183.

Luo, Y., Hui, D., Cheng, W., Coleman, J. S., Johnson, D. W., and Sims, D. A. (in review). Canopy quantum yield in a mesocosm study. *Oecologie.*

Luo, Y., and Mooney, H. A. (1995). Long-term studies on carbon influx into global terrestrial ecosystems: Issues and approaches. *J. Biogeogr.* **22,** 797–803.

Luo, Y., and Mooney, H. A. (1996). Stimulation of global photosynthetic carbon influx by an increase in atmospheric carbon dioxide concentration. *In* "Carbon Dioxide and Terrestrial Ecosystems" (G. W. Koch and H. A. Mooney, eds.), pp. 381–397. Academic Press, San Diego.

Luo, Y., and Reynolds, J. F. (in press). Validity of extrapolating field CO_2 experiments to predict carbon sequestration in natural ecosystems. *Ecology.*

Luo, Y., Sims, D., Thomas, R., Tissue, D., and Ball, J. T. (1996). Sensitivity of leaf photosynthesis to CO_2 concentration is an invariant function for C_3 plants: A test with experimental data and global applications. *Global Biogeochem. Cycles* **10,** 209–222.

McMahon, T. A. (1975). The mechanical design of trees. *Sci. Am.* **233,** 93–102.

McMurtrie, R. E., and Comins, H. N. (1996). The temporal response of forest ecosystems to doubled atmospheric CO_2 concentration. *Global Change Biol.* **2,** 49–57.

Mooney, H. A., Canadell, J., Chapin, F. S., Ehleringer, J., Körner, C., McMurtrie, R., Parton, W. J., Pielka, L., and Schulze, E.-D. (in press). Ecosystem physiology responses to global change. *In* "Implications of Global Change for Natural and managed Ecosystems: A Synthesis of GCTE and Related Research" (B. H. Walker, W. L. Steffen, J. Canadell, and J. S. I. Ingram, eds.) Cambridge University Press, Cambridge.

Norman, J. M. (1993). Scaling processes between leaf and canopy levels. *In* "Scaling Physiological Processes: Leaf to Globe" (J. R. Ehleringer and C. B. Fields, eds.), pp. 41–76. Academic Press, San Diego.

Parton, W. J., Schimel, D. S., Cole, C. V., and Ojima, D. S. (1987). Analysis of factors controlling soil organic matter levels in Great Plains grasslands. *Soil Sci. Soc. Am. J.* **51,** 1173–1179.

Polglase, P. J., and Wang, Y. P. (1992). Potential CO_2-enhanced carbon storage by the terrestrial biosphere. *Aust. J. Bot.* **40,** 641–656.

Rastetter, E. B., Ågren, G. I., and Shaver, G. R. (1997). Responses of N-limited ecosystems to increased CO_2: A balanced-nutrition, coupled-element-cycles model. *Ecol. Appl.* **7,** 444–460.

Reich, P. B., Walters, M. B., and Ellsworth, D. S. (1997). From tropics to tundra: Global convergence in plant functioning. *Proc. Natl. Acad. Sci.* **94,** 13,730–13,734.

Ruimy, A., Derard, G., and Saugier, B. (1996). TURC: A diagnostic model of continental gross primary productivity and net primary productivity. *Global Biogeochem. Cycles* **10,** 269–286.

Ryan, M. G., Hunt, E. R., McMurtrie, R. E., Ågren, G. I., Aber, J. D., Friend, A. D., Rastetter, E. B., Pulliam, W. M., Raison, R. J., and Linder, S. (1996). Comparing models of ecosystem function for temperate conifer forests. I. Model description and validation. *In* "Global Change: Effects on Coniferous Forests and Grasslands" (A. I. Breymeyer, D. O. Hall, J. M. Melillo, and G. I. Ågren, eds.), pp. 313–362. John Wiley and Sons Ltd., Chichester.

Ryle, G. J. A., Powell, C. E., and Powell, V. (1992). Effect of elevated CO_2 on the photosynthesis, respiration, and growth of perennial ryegrass. *J. Exp. Bot.* **43,** 811–818.

Sage, R. F. (1990). A model describing the regulation of ribulose-1,5-bisphosphate carboxylase, electron transport and triose-phosphate use in response to light intensity and CO_2 in C_3. *Plant Physiol.* **94,** 1728–1734.

Schimel, D. S., Braswell, B. H., Holland, E. A., McKeown, R., Ojima, D. S., Painter, T. H., Parton, W. J., and Townsend, A. R. (1994). Climatic, edaphic, and biotic controls over storage and turnover of carbon in soils. *Global Biogeochem. Cycles* **8,** 279–293.

Schimel, D. S. (1995). Terrestrial ecosystems and the carbon cycle. *Global Change Biol.* **1,** 77–91.

Schmidt-Nielsen, K. (1984). "Scaling: Why Is Animal Size So Important?" Cambridge, London.

Sharkey, T. D. (1988). Estimating the rate of photorespiration in leaves. *Physiol. Plant.* **73,** 147–152.

Sims, D. A., Cheng, W., Luo, Y., and Seemann, J. R. (in press). Photosynthetic acclimation to elevated CO_2 in a sunflower canopy. *J. Exp. Bot.*

Sims, D. A., Seemann, J., and Luo, Y. (1998). Elevated CO_2 concentration has independent effects on expansion rates and thickness of soybean leaves across light and nitrogen gradients. *J. Exp. Bot.* **49,** 583–591.

Stitt, M. (1991). Rising CO_2 levels and their potential significance for carbon flow in photosynthetic cell. *Plant Cell Environ.* **14,** 741–762.

Tans, P. P., Fung, I. Y., and Takahashi, T. (1990). Observational constraints on the global atmospheric CO_2 budget. *Science* **247,** 1431–1438.

Tans, P., Berry, J. A., and Keeling, R. F. (1993). Oceanic $^{13}C/^{12}C$ observations: A new window on ocean CO_2 uptake. *Global Biogeochem. Cycles* **7,** 353–368.

VEMAP Members. (1995). Vegetation/ecosystem modeling and analusis project: Comparing biogeography and biogeochemistry models in a continental-scale study of terrestrial ecosystem responses to climate change and CO_2 doubling. *Global Biogeochem. Cycles* **4,** 407–437.

Vu, J. C. V., Allen, L. H., Jr., and Bowes, G. (1989). Leaf ultrastructure, carbohydrates and protein of soybeans grown under CO_2 enrichment. *Environ. Exp. Bot.* **29,** 141–147.

West, G. B., Brown, J. H., and Enquist, B. J. (1997). A general model for the origin of allometric scaling laws in biology. *Science* **276,** 122–126.

Wong, S. C. (1990). Elevated atmospheric partial pressure of CO_2 and plant growth, II, Non-structural carbohydrate content in cotton plants and its effect on growth parameters. *Photosynth. Res.* **23,** 171–180.

Wullschleger, S. D. (1993). Biochemical limitations to carbon assimilation in C_3 plants—A retrospective analysis of the A/C_i curves from 109 species. *J. Exp. Bot.* **44,** 907–920.

13

Nutrients: Dynamics and Limitations

Göran I. Ågren, Gaius R. Shaver, and Edward B. Rastetter

I. Introduction

Carbon is, leaving oxygen aside, the quantitatively most important element in living organisms. An increasing availability of this element through an increasing atmospheric carbon dioxide concentration is therefore expected to have a direct impact on, at least some, living organisms. We discuss in this chapter the constraints on this impact arising from the interactions between carbon and the other, for living organisms, essential elements, which we will exemplify with nitrogen. These constraints are basically of two kinds: (1) the extent to which different components in an ecosystem can adjust the element ratios, ecosystem stoichiometry, that is, how flexible are plant and soil C : N ratios; and (2) the extent to which an increasing availability of one element can change the availability of other elements, that is, if an increase in carbon dioxide concentration can increase the retention of nitrogen in the ecosystem.

We analyze this question by considering simple ecosystems consisting of only a plant subsystem and a soil subsystem and between which flows of elements take place. The extent of exchange of elements with the surroundings of the ecosystem and to which extent this exchange can be modified by internal ecosystem processes is an important aspect.

II. Ecosystem Stoichiometry

Stoichiometry places constraints on the amount of carbon stored in ecosystems. From the perspective of carbon–nitrogen interactions, there are three ways in which the amount of carbon in the ecosystem can change (Rastetter *et al.,* 1992): (1) Change the total amount of nitrogen in the ecosystem while maintaining C:N ratios. The increasing atmospheric carbon dioxide provides the primary producers with more carbon but this does not necessarily have consequences for availability of other elements. The enhanced nitrogen deposition in many areas may not have any counterpart for other elements. We will consider this question further in the next section. (2) Change the C:N ratios of the ecosystem components. Plants growing under high nutrient availability tend to have higher nutrient concentrations than those grown under low nutrient availability and soils differ depending on parent material and vegetation (Vitousek and Howarth, 1991). (3) Change the distribution of nitrogen among ecosystem components. Because soils generally have lower C:nutrient ratios than plants, moving elements other than carbon from soils to plants leads to lower ecosystem C:element ratios.

A. Plant Element Ratios

The fact that plants increase their nutrient concentration with increasing nutrient availability is so well known that we do not discuss it further here. Effects of increasing carbon availability have, however, not been studied as extensively and results are more controversial with respect to how carbon:element ratios will be affected. A recent review by McGuire *et al.* (1995) showed that, on the average, the leaf nitrogen decreased by 21% in response to elevated carbon dioxide but a decrease in stem and fine root nitrogen concentration could not be clearly shown. Note also that all the studies reviewed by McGuire *et al.* are for small trees and observations on stem nitrogen changes might not be representative of what happens in larger trees. This point is important, because McMurtrie and Comins (1996) showed in the G'DAY model that long-term changes in ecosystem carbon store as a result of an increasing carbon dioxide concentration are sensitive to nitrogen concentrations in long-lived woody biomasses.

The most obvious effects of increasing carbon dioxide should be on plant growth rate and there are many reviews written on this subject (e.g., Eamus and Jarvis, 1989; Ceulemans and Mousseau, 1994). To understand the consequences of interactions between carbon dioxide and nutrients, we need concepts that allow us to define what we mean by interactions and methods that permit us to measure these. However, before going into these questions, we would like to make an observation: *Plant ecologists and physiologists in general do not seem to appreciate the importance of the interactions between carbon dioxide and nutrients.* We base this observation on the attitude revealed by the description of methods used in studies of carbon dioxide effects on plant properties. Poorter (1993) reviewed a large number of studies of carbon dioxide effects on plant growth rate. From that review we have taken 55 papers (all of the papers accessible in the university library of the senior author) dealing with C_3 plants and checked how nutrition was handled. The results are

- 11 (20%) do not mention nutrition at all,
- 31 (56%) mention that nutrients have been added in various quantities, but not the effects on the nutritional status, and
- 13 (24%) have measured plant nutrient status.

It is worth observing that the uptake of nutrients is not always measured even in studies with titles like "Effects of carbon dioxide and nitrogen . . .". Moreover, in none of the studies in Poorter's review was plant nutrient status followed during the course of the experiment; only final nutrient content or concentration is given. This attitude should be contrasted with the very strong recommendations given by Ingestad and Ågren (1992, 1995) on the necessity of strict control of nutrition in plant growth studies. The only published study, to our knowledge, where control over both carbon dioxide and nutrients has been maintained is by Pettersson *et al.* (1993) with birch seedlings.

A suitable tool for analyzing integrated plant responses, when nitrogen is the growth-limiting element, is the nitrogen productivity (Ågren, 1985a, 1985b, 1988, 1994, 1996; Ågren and Bosatta, 1996a). The idea behind nitrogen productivity is that plant growth can be no more rapid than allowed by the enzymatic activity. One limitation on the enzymatic activity is the amount of enzymes in the plant. As a surrogate for the amount of enzymes, it is assumed that the total amount of plant nitrogen can serve (Ågren, 1985a, 1985b). This idea is used in the nitrogen productivity (P_N), which connects the growth rate of a plant (dW/dt) to its nitrogen amount (N) according to

$$\frac{dW}{dt} = P_N(\mathrm{N} - c_{\mathrm{N,min}}W), \tag{1}$$

where $c_{\mathrm{N,min}}$ is the minimum nitrogen concentration in the plant necessary for growth. The nitrogen productivity, P_N, is independent of nitrogen but varies with other growth-controlling factors. A more convenient form of Eq. (1) is obtained by expressing it in terms of the relative growth rate and the plant nitrogen concentration:

$$R_W = \frac{1}{W}\frac{dW}{dt} = P_N(c_N - c_{\mathrm{N,min}}). \tag{2}$$

The relation between relative growth rate and plant nitrogen concentration is thus a straight line with the nitrogen productivity as slope (Fig. 1). The nitrogen productivity, P_N, integrates the effect of all other factors controlling growth and as long as these factors remain constant the relative growth

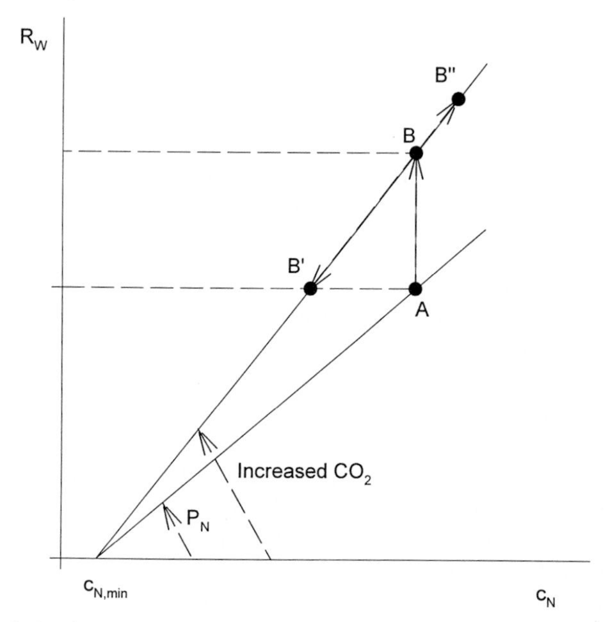

Figure 1 Relation between relative growth rate, R_W, and plant nitrogen concentration, c_N, at two different levels of carbon dioxide concentration. See text for further explanation.

rate can, in response to changes in nitrogen supply, only move up and down along the line defined by P_N and $c_{N,min}$. With all other factors constant and the nitrogen supply such that a constant nitrogen concentration is maintained in the plant, the plant will grow at an exponential rate

$$W(t) = W_0 e^{P_N c_N t}. \tag{3}$$

Changes in factors other than nitrogen supply can now be studied through their impact on the nitrogen productivity. An increase in carbon dioxide concentration will, at least instantaneously, mean an increase in the photosynthetic rate. This will supply the growth enzymes with more substrate and an increased nitrogen productivity is expected. Since a change in the nitrogen productivity affects the growth rate in an exponential manner, even small changes in the nitrogen productivity result in large effects on the plant size. For example, an increase in relative growth rate [$P_N(c_N - c_{N,min})$] from 15 to 17% d^{-1} will over a 35-d period (a typical experimental period) result in a doubling of the plant size. The plant performance is therefore sensitive to changes in the nitrogen productivity and to changes in the plant nitrogen concentration.

An illustration of what might happen to a plant that experiences a change in carbon dioxide availability is given in Fig. 1. A plant growing with a fixed exponential rate (constant R_W and c_N), point A, is exposed to an increase in carbon dioxide concentration. The short-term response will be an increase in nitrogen productivity as discussed earlier but the response in plant nitrogen concentration is slower; the plant moves to point B. What happens in a longer time perspective depends now on the nitrogen supply. If the nitrogen supply, in relative terms, is unchanged (resource limitation) the relative growth rate of the plant must be necessity adjust to this supply rate (Ingestad and Ågren, 1992) and the plant moves to point B′. The result is a plant growing at the same relative growth rate as before the change in the carbon dioxide concentration, but with a lower nitrogen concentration; the plant is, however, larger. This seems to be a common observation in carbon dioxide experiments. There are, however, other possibilities. The increased availability of carbon might mean that the plant can invest relatively more in roots, which also sometimes is observed. If the nitrogen uptake by the plant were limited by its investment in uptake capacity we would now see an increase in relative uptake rate and in the extreme, the relative growth rate of the plant might even exceed that seen immediately following the increase in carbon dioxide, point B″. It seems appropriate to define the strength of the interaction by how much the plant performance deviates from point B′. Plants at B′ have not changed their uptake of other elements in response to the change in carbon dioxide, that is, no interaction. Any deviation (up or down) from B′ along the nitrogen productivity line is a sign of (positive or negative) interaction. It is impossible to predict what happens under changing carbon dioxide unless we also understand the simultaneous response in nutrient uptake; observed changes in nutrient concentrations alone are not sufficient to determine the existence of interactions. The assumption that $c_{N,min}$ is unaffected by the carbon dioxide is in accordance with empirical evidence as shown in Fig. 2; effects on $c_{N,min}$ would also have only minor consequences. This figure also shows the magnitude of response to expect.

We would also like to point out problems with using nutrient use efficiencies (NUEs) in this context. Figure 3 compares two, initially identical, plants but where in one of the plants the amount of the limiting nutrient is increased in a step. Otherwise the plants are left without a supply of nutrients. From Eq. (1) it is easy to see that these two plants will now grow in a linear fashion. NUE, calculated as the inverse of the nutrient concentration as is usually done (however, see Grubb, 1989), will initially be lower in the plant that got the shot of nutrients. However, because of its more rapid growth rate its nutrient concentration will decrease most rapidly and at some point in time it will have the lower nutrient concentration and hence the higher NUE. The point here is that NUE is not an intrinsic plant property but a consequence of the interaction between plant growth and nutrient uptake and as such varies with the time of observation. Further

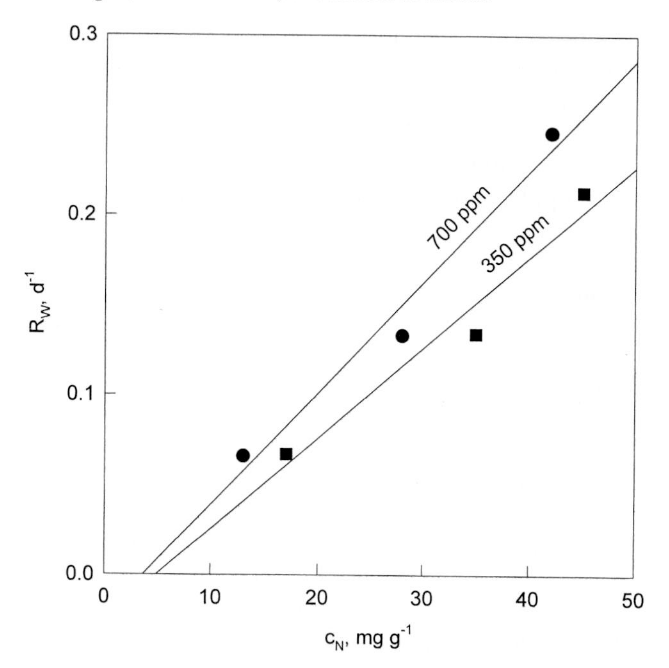

Figure 2 Relation between relative growth rate, R_W, and plant nitrogen concentration, c_N, at two different levels of carbon dioxide concentration, 350 and 700 ppm, in an experiment with birch seedlings. The small number of samples for each treatment makes statistical analyses less meaningful but the experiences with this kind of experiments show that the results are highly reproducible and stable. Redrawn from Pettersson et al. (1993) with permission.

illustrations of this kind of interaction are given by Wikström (1994) and Ågren and Bosatta (1996a).

B. Soil Element Ratios

The effects of carbon dioxide on plants will propagate to the soil through changes in plant litter. An obvious change is, of course, that an increase in litter amount will increase soil carbon. Less obvious is the consequences this will have for the partitioning of the ecosystem nitrogen. With a fixed total amount of nitrogen available, the increased ecosystem carbon will lead to a dilution of the nitrogen (Rastetter *et al.*, 1992; Comins and McMurtrie, 1993; Kirschbaum *et al.*, 1994). Model calculations indicate that this dilution should lead to a net transfer of nitrogen from the vegetation to the soil (Rastetter *et al.*, 1992). A temperature increase accompanying the increase in carbon dioxide can, however, reverse the sign of this transfer.

Another factor to be taken into account is the effects of changing litter chemistry. There is first an effect of the decreased nutrient concentration in the live plant components (Fig. 1) that may or may not be transmitted

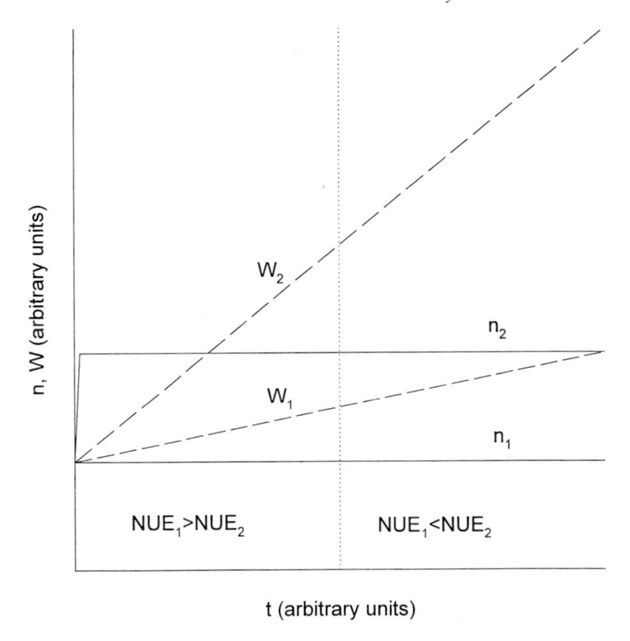

Figure 3 Development of plant weights (W_1 and W_2, dashed lines) for two different amounts of a limiting nutrient (n_1 and n_2, solid lines) and the border (dotted line) where the nutrient use efficiencies for the two treatments change order.

to the litter. Second, carbon dioxide experiments indicate also that the carbon chemistry of the plant changes with carbon dioxide. Initial decomposition rate constants have been estimated from C:N and lignin:N ratios (Aber and Melillo, 1982; Melillo *et al.*, 1982) and could potentially be used together with an assumption of a simple exponential decay to estimate effects on steady-state stores of soil carbon and nitrogen. However, extrapolating such initial values seems dangerous. An alternative is to use the continuous quality theory (Bosatta and Ågren, 1991, 1994; Ågren and Bosatta, 1996a), which accounts for changing properties of a litter during the decomposition process.

Ågren and Bosatta (1996b) showed that the quality, q, of a litter during its decomposition could be defined as a weighted mean of its chemical composition

$$q = 1.25c_{as} + 1.00c_{ex} + 0.65c_{ai}, \tag{4}$$

where c_{as} is the fraction of acid solubles, c_{ex} the fraction of extractables (polar and nonpolar), and c_{ai} the fraction of acid insolubles. During decomposition, these fractions change in such a way that the quality decreases from its intial value q_0. The fraction of a litter remaining at a given time is a function of the ratio of the current quality and the initial quality

(Ågren and Bosatta, 1996a). With a constant rate of litter input of initial nitrogen : carbon ratio of r_0 the steady-state stores of soil carbon, C_{SS}, and nitrogen, N_{SS}, can be calculated by summing the remainders of an infinite number of litter cohorts

$$C_{SS} = Aq_0^{-7} \tag{5}$$

$$N_{SS} = \left[\frac{f_N}{f_C}(1 - B) + Br_0\right]C_{SS}, \tag{6}$$

where A and B (= 0.32) are numerical factors without interest here and f_N/f_C is the nitrogen : carbon ratio of the decomposers. In contrast to some other ways of defining litter quality, our definition in Eq. (4) is done independently of the nitrogen concentration, r_0, of the litter.

We will use two studies of carbon dioxide effects on plant chemistry that give the information necessary to apply Eqs. (4)–(6) to predict steady-state consequences of changes in plant chemistry (Norby *et al.*, 1986; Melillo *et al.*, 1995). The implications of these two studies interpreted by Eqs. (4)–(6) are given in Table I. The only effect on steady-state carbon stores is through the changes in litter quality. Nitrogen stores are also completely dominated by changes in litter quality; litter nitrogen concentrations are so much lower than decomposer nitrogen concentrations that with time the decomposer properties will dominate. There are some species differences with no effect of growth CO_2 concentration on carbon store with white oak leaves but an ~15% increase with the sugar maple and red oak leaves. The only effect on nitrogen store is that accompanying the increased carbon store. The fate of the nitrogen added to the soil is discussed in the next section.

We could also have applied the analysis to the initial specific decomposition rate of the litters. Quantitatively, the differences between litters from different CO_2 environments would have been the same as those for the carbon stores; the high-quality litters disappear, of course, more rapidly. Our predicted effects on decomposition rates are much less than those derived on the basis of C:N and lignin:N ratios (Melillo *et al.*, 1995).

III. Open versus Closed Systems

Although carbon is available in basically unlimited amounts to plants and uptake is only limited by the effort put into acquiring it, other elements are only available in limited quantities and normally held in a tight cycle within the ecosystem. However, the cycle is not perfectly tight but small (compared to the internal turnover) amounts are lost to or acquired from

Table I Steady-State Stores of Soil Carbon and Nitrogen Obtained with Different Litters and from Plants Grown in Different CO_2 Concentrations

Litter	Carbon dioxide concentration (ppm)	c_{as}	c_{ex}	c_{ai}	r_0 (mg g^{-1})	q_0	C_{SS}[a]	N_{SS}[a]	N_{SS}/C_{SS}
White oak leaves[b]	362	0.505	0.449	0.045	7.9	1.110	1	0.114	0.114
	690	0.482	0.488	0.029	6.1	1.109	1.000	0.113	0.113
Red maple leaves[c]	350	0.350	0.536	0.115	12.8	1.047	1	0.117	0.117
	700	0.300	0.565	0.136	4.3	1.027	1.167	0.128	0.112
Sugar maple leaves[c]	350	0.354	0.537	0.109	18.0	1.050	1	0.120	0.120
	700	0.287	0.598	0.115	5.1	1.031	1.136	0.127	0.112

[a] A value of 0.16 is assumed for f_N/f_C. The parameter A has been chosen such that $C_{ss} = 1$ for each litter at ambient carbon dioxide concentration. [b] Data from Norby *et al.* (1986). [c] Data from Melillo *et al.* (1995).

the surroundings. This has important consequences for ecosystem behavior as first pointed out by Rastetter and Shaver (1992).

In a closed system, any nitrogen that is added to the soil comes from the vegetation and should be accompanied by a decrease in carbon storage in the vegetation. This conclusion is trivial, and because of the lower C:N ratio in soils such a transfer of nitrogen leads to a net ecosystem decrease in carbon storage. However, if we open the system such that nitrogen can flow in and out of the system and if we also allow the microbial community to respond to the level of inorganic nitrogen (a variable f_N in the terminology of the previous section), the system behavior is less trivial (Rastetter *et al.*, 1997).

First of all, the increased carbon availability for the decomposers should push them to operate at lower nitrogen concentrations (f_N decreases). As a consequence, soil nitrogen storage decreases [Eq. (6)]. On the other hand, increasing carbon availability relative to nitrogen should make the decomposers more competitive for nitrogen, which should tend to maintain their nitrogen status. We cannot presently predict the net outcome of these two counteracting forces but its consequences for total ecosystem carbon storage are large. Rastetter *et al.* (1997) calculated for a closed system an increase of 2% in ecosystem carbon storage in a world with doubled CO_2 when decomposers are inefficient in retaining inorganic nitrogen, compared to a minimal change when decomposers are maximally efficient. In addition, the temporal behavior of ecosystems to disturbances differs substantially.

When an ecosystem is opened for exchange of nitrogen with the surroundings, the importance of the competition between decomposers and vegetation is less drastic because both can now favorably compete with the loss mechanisms. In the long run, in an open system the efficiency of decomposers to retain inorganic nitrogen matters little although again responses over periods up to 100 yr in forest can differ. Increases in CO_2 can now have a much larger impact. With an opened nitrogen cycle, Rastetter *et al.* (1997) predicted an 18% increase in ecosystem carbon, that is, a ninefold difference compared to the closed system. Thus, in the closed system decomposer retention efficiency matters for carbon storage but more important is whether the system is open or closed with respect to nitrogen.

IV. Discussion

The increasing atmospheric carbon dioxide concentration means an increasing availability of carbon for the biosphere. The availability of other essential resources, notably mineral nutrients, cannot be expected to increase in proportion, although mechanisms not discussed here such as nitrogen fixation and solubilization of phosphate through exudation of organic acids could be enhanced. We have pointed here at some consequences of changes in the relative availability of elements. Whether these consequences should be called interactions or not is, of course, a matter of taste, but it seems that the term is normally used so loosely that any consequence will be termed an interaction. A more precise use of the term will require some form of model.

Feeding plants more carbon will decrease their nitrogen concentration, which normally is associated with an increase in their root fraction. Such behavior is observed in some experiments with carbon dioxide, but not always (Ceulemans and Mousseau, 1994; Roumet *et al.*, 1996). What one should expect in a given situation depends entirely on the possibilities the plant has to increase its nutrient uptake to balance the increase in carbon availability; strict control over or monitoring of the nutrient uptake in carbon dioxide experiments is therefore imperative. Changes in the relative uptake of elements could also make other elements than nitrogen limiting.

The increased flux of carbon to an energy (carbon) limited decomposer community is probably not of a sufficient magnitude to have any qualitative effects on the steady-state properties of the soil organic matter such as the quality of the organic matter or the carbon : nitrogen ratio. The transient effect of changes in carbon : nitrogen ratios and in litter quality are both such they will prolong the immobilization phase of nitrogen in fresh litters. However, the rate of change in the increasing atmospheric carbon dioxide is probably slow enough that this will not be noticed.

A problem in evaluating effects on soil organic matter is that most information is available for leaf litters although these may contribute a minor part to the store of soil organic matter (Ågren and Bosatta, 1987). It is also important that experiments to produce litters be conducted under conditions that resemble natural ones. The large change in lignin concentration [(c_{ai} in Eq. (4)] in the white oak experiment (Table I) is not found in other studies (e.g., Chu *et al.*, 1996) and could be an effect of light conditions in the nursery. The large change in nitrogen concentration between the low and high carbon dioxide treatments for the other litters has a large impact and warrants additional studies.

Ecosystems are completely open with respect to carbon because the atmosphere can be regarded as an infinite sink and source. This is not the case for other elements where ecosystems are almost closed. The understanding of the controls of the openness is still lacking although its importance is evident (e.g., Hedin *et al.*, 1995; Townsend *et al.*, 1996).

We have suggested here some means of evaluating effects of an increasing atmospheric carbon dioxide concentration on the element balance in the vegetation and soil compartments of ecosystems. However, before such evaluations can yield quantitatively reliable predictions of changes in ecosystem element stores, let alone of the globe, a much firmer empirical basis is required, both in terms of its quality as well as its extent.

References

Aber, J. D., and Melillo, J. M. (1982). Nitrogen immobilization in decaying leaf litter as a function of initial nitrogen and lignin content. *Can. J. Bot.* **60,** 2263–2269.

Ågren, G. I. (1985a). Theory for growth of plants derived from the nitrogen productivity concept. *Physiol. Plant.* **64,** 17–28.

Ågren, G. I. (1985b). Limits to plant production. *J. Theoret. Biol.* **113,** 89–92.

Ågren, G. I. (1988). The ideal nutrient productivities and nutrient proportions. *Plant Cell Environ.* **11,** 613–620.

Ågren, G. I. (1994). The interaction between CO_2 and plant nutrition: Comments to a paper by Coleman, McConnaughay and Bazzaz. *Oecologia* **98,** 239–240.

Ågren, G. I. (1996). Nitrogen productivity or photosynthesis minus respiration to calculate plant growth? *Oikos* **76,** 529–535.

Ågren, G. I., and Bosatta, E. (1987). Theoretical analysis of the long term dynamics of carbon and nitrogen in soils. *Ecology* **68,** 1181–1189.

Ågren, G. I., and Bosatta, E. (1996a). "Theoretical Ecosystem Ecology—Understanding Element Cycles." Cambridge University Press, Cambridge.

Ågren, G. I., and Bosatta, E. (1996b). Quality: A bridge between theory and experiment in soil organic matter studies. *Oikos* **76,** 522–528.

Bosatta, E., and Ågren, G. I. (1991). Dynamics of carbon and nitrogen in the organic matter of the soil: A generic theory. *Am. Naturalist* **138,** 227–245.

Bosatta, E., and Ågren, G. I. (1994). Theoretical analysis of microbial biomass dynamics in soils. *Soil Biol. Biochem.* **26,** 143–148.

Ceulemans, R., and Mousseau, M. (1994). Effects of elevated atmospheric CO_2 on woody plants. *New Phytol.* **127,** 425–446.

Chu, C. C., Field, C. B., and Mooney, H. A. (1996). Effects of CO_2 and nutrient enrichment on tissue quality of two California annuals. *Oecologia* **107,** 433–440.

Comins, H. N., and McMurtrie, R. E. (1993). Long-term response of nutrient-limited forest to CO_2 enrichment: Equilibrium behavior of plant-soil models. *Ecol. Appl.* **3,** 666–681.

Eamus, D., and Jarvis, P. G. (1989). Direct effects of increase in the global atmospheric CO_2 concentration on natural and commercial temperate and trees and forest. *Adv. Ecol. Res.* **19,** 1–55.

Grubb, P. J. (1989). The role of mineral nutrients in the tropics: A plant ecologist's view. *In* "Mineral Nutrients in Tropical Forest and Savannah Ecosystems" (J. Proctor, ed.), pp. 417–439. Blackwell Scientific Publications, Oxford.

Hedin, L. O., Armesto, J. J., and Johnson, A. H. (1995). Patterns of nutrient loss from unpolluted, old-growth temperate forests: Evaluation of biogeochemical theory. *Ecology* **76,** 492–509.

Ingestad, T., and Ågren, G. I. (1992). Theories and methods on plant nutrition and growth. *Physiol. Plant.* **84,** 177–184.

Ingestad, T., and Ågren, G. I. (1995). Plant nutrition and growth: Basic principles. *Plant Soil* **168/169,** 15–20.

Kirschbaum, M. U. F., King, D. A., Comins, H. N., McMurtrie, R. E., Medlyn, B. E., Pongracic, S., Murty, D., Keith, H., Raison, R. J., Khanna, P. K., and Sheriff, D. W. (1994). Modelling forest response to increasing CO_2 concentration under nutrient-limited conditions. *Plant Cell Environ.* **17,** 1081–1089.

McGuire, A. D., Melillo, J. M., and Joyce, L. A. (1995). The role of nitrogen in the response of forest net primary production to elevated atmospheric carbon dioxide. *Annu. Rev. Ecol. Syst.* **26,** 473–503.

McMurtrie, R. E., and Comins, H. N. (1996). The temporal response of forest ecosystems to doubled atmospheric CO_2 concentration. *Global Change Biol.* **2,** 49–57.

Melillo, J. M., Aber, J. D., and Muratore, J. F. (1982). Nitrogen and lignin control of hardwood leaf litter decomposition dynamics. *Ecology* **63,** 621–626.

Melillo, J. M., Kicklighter, D. W., McGuire, A. D., Peterjohn, W. T., and Newkirk, K. M. (1995). Global change and its effects on soil organic carbon stocks. *In* "Role of Nonliving Organic Matter in the Earth's Carbon Cycle" (R. G. Zepp and C. Sonntag, eds.), pp. 175–189. John Wiley & Sons, Chichester.

Norby, R. J., Pastor, J., and Melillo, J. M. (1986). Carbon–nitrogen interactions in CO_2-enriched white oak: Physiological and long-term perspectives. *Tree Physiol.* **2,** 233–241.

Pettersson, R., McDonald, A. J. S., and Stadenberg, I. (1993). Response of small birch plants (*Betula pendula* Roth.) to elevated CO_2 and nitrogen supply. *Plant Cell Environ.* **16,** 1115–1121.

Poorter, H. (1993). Interspecific variation in the growth response of plants to an elevated ambient CO_2 concentration. *Vegetatio* **104/105,** 77–97.

Rastetter, E. B., and Shaver, G. R. (1992). A model of multiple-element limitations for acclimating vegetation. *Ecology* **73,** 1157–1174.

Rastetter, E. B., McKane, R. B., Shaver, G. R., and Melillo, J. M. (1992). Changes in C storage by terrestrial ecosystems: How C–N interactions restrict responses to CO_2 and temperature. *Water Air Soil Pollut.* **64,** 327–344.

Rastetter, E. B., Ågren, G. I., and Shaver, G. R. (1997). Responses to increased CO_2 concentration in N-limited ecosystems: Application of a balanced-nutrition, coupled-element-cycles model. *Ecol. Appl.* **7,** 444–460.

Roumet, C., Bel, M. P., Sonie, L., Jardon, F., and Roy, J. (1996). Growth response of grasses to elevated CO_2: A physiological plurispecific analysis. *New Phytol.* **133,** 595–603.

Townsend, A. R., Braswell, B. H., Holland, E. A., and Penner, J. E. (1996). Spatial and temporal patterns in terrestrial carbon storage due to deposition of fossil fuel nitrogen. *Ecol. Appl.* **6,** 806–814.

Vitousek, P. M., and Howarth, R. H. (1991). Nitrogen limitation on land and in the sea: How can it occur? *Biogeochemistry* **13,** 87–115.

Wikström, J. F. (1994). A theoretical explanation of the Piper–Steenbjerg effect. *Plant Cell Environ.* **17,** 1053–1060.

14

Ecosystem Modeling of the CO_2 Response of Forests on Sites Limited by Nitrogen and Water

Ross E. McMurtrie and Roderick C. Dewar

I. Introduction

Controlled experiments have consistently shown that, under optimal growing conditions, plant growth is greatly increased by elevated CO_2 (Luxmoore *et al.*, 1993; Idso and Idso, 1994; Mooney *et al.*, 1998)—the *CO_2 fertilization effect* (CFE). However, there is considerable uncertainty about the magnitude of the CFE under natural conditions where growth is limited by water and/or nutrients. This uncertainty is illustrated by conflicting conclusions of the Intergovernmental Panel on Climate Change (IPCC): Houghton *et al.* (1995) concluded that "when the availability of water and nutrients is taken into account, the [CFE] is likely to be reduced . . . by around a half"; in contrast, Kirschbaum *et al.* (1996a, 1996b) concluded that the CFE should be diminished under nutrient limitation but enhanced under water limitation.

The CFE under nutrient limitation has been extensively studied using both experiments and models. In short-term, controlled-environment experiments, a large relative CFE is commonly observed even under severe nutrient limitation (Eamus and Jarvis, 1989; Idso and Idso, 1994; Körner, 1996; Long *et al.*, 1996). In the longer term, however, there is little evidence of a CFE in nutrient-limited natural environments. For example, in a long-term field experiment at an infertile wetland site in the Alaskan tundra, only a transient CFE was detected (Tissue and Oechel, 1987). In contrast, a large CFE was sustained over 6 yr in a nutrient-rich temperate wetland in Maryland (Drake *et al.*, 1996; Körner, 1996). These contrasting experimental results have led to much debate over the role of nutrients (and other environmental factors) in moderating short-term effects of CO_2 enrichment (Long, 1991; Mooney *et al.*, 1991; McMurtrie and Wang, 1993; Ceulemans and Mousseau, 1994; Gifford, 1994; Long *et al.*, 1996).

Modeling studies have verified that, under nutrient limitation, the long-term CFE may be less than the short-term CFE (Rastetter *et al.*, 1991; Comins and McMurtrie, 1993; Hudson *et al.*, 1994; Kirschbaum *et al.*, 1994; McMurtrie and Comins, 1996). This decrease in the CFE with time can be understood in terms of negative soil feedbacks: Increased growth at elevated CO_2 leads to enhanced nutrient immobilization in biomass and soils, and hence to reduced availability of soil nutrients to support further growth (e.g., Diaz *et al.*, 1993).

Less is known about the CFE under water limitation, because few experiments have been conducted involving manipulations of both CO_2 and soil moisture. However, the few published studies suggest that the short-term CFE is amplified under water limitation (Allen, 1990; Morison, 1993; Idso and Idso, 1994; for an exception see Guehl *et al.*, 1994), as observed, for example, in an experiment involving free-air CO_2 enrichment (FACE) of wheat grown with a plentiful water supply and 50% reduced water supply (Kimball *et al.*, 1995), and in field experiments on Kansas tallgrass prairie (Owensby *et al.*, 1996) and Californian Mediterranean grassland (Field *et al.*, 1996). A physiological explanation of this amplification is suggested by the consistent experimental finding that water use efficiency (WUE), defined as carbon uptake per unit water loss, increases at elevated CO_2 (Morison, 1993; Polley *et al.*, 1993; Picon *et al.*, 1996), as observed, for example, in both the Alaskan and Maryland experiments referred to earlier (Mooney *et al.*, 1991). Experiments indicate that leaf-scale WUE (defined as net photosynthesis/transpiration) increases approximately linearly with CO_2 concentration (Morison, 1993; Long *et al.*, 1996). Associated with this increase in WUE is a commonly observed decrease in stomatal conductance at elevated CO_2 (Morison, 1993; Field *et al.*, 1995). However, for some species, including many conifers, stomatal conductance is insensitive to CO_2 (e.g., Morison, 1985; Bunce, 1992; Beerling *et al.*, 1996; Curtis, 1996). Plants with no stomatal closure at elevated CO_2 may nevertheless show large increases in WUE because of the CO_2 response of photosynthesis (Shugart *et al.*, 1986).

Several forest models have been used to simulate the CFE under water limitation. One comprehensive study involved applying five growth/water balance models to control and irrigated stands of *Pinus radiata* growing near Canberra, Australia (Ryan *et al.*, 1996a, 1996b). For four models (BIOMASS, BIOME-BGC, HYBRID, PnET-CN) the simulated CFE on the water-limited, control stand was more than double that on the irrigated stand (Ryan *et al.*, 1996b). Only one model (CENTURY) predicted a larger CFE for the irrigated stand. In a second study, where three models (BIOME-BGC,

CENTURY, TEM) were applied to a range of biomes, the CFE was negatively correlated with rainfall (Mooney *et al.*, 1998; Pan *et al.*, 1998). The model prediction of an enhanced CFE under water-stressed conditions is consistent with experimental observations (Gifford, 1992; Kimball *et al.*, 1995; Kirschbaum *et al.*, 1996b).

The evidence mentioned from both experiments and models suggests that the CFE is enhanced under water limitation, but diminished under long-term nutrient limitation. However, it is difficult to extrapolate from this evidence to predict the combined effects of water and nutrient limitations because of inadequately understood feedbacks within and between the water and nutrient cycles (Field *et al.*, 1995). For example, under elevated CO_2, increased WUE may improve soil water and/or nutrient availability in the short term, leading to enhanced net carbon production and foliage allocation and, hence, increased leaf area; however, increased leaf area may then lead to more rapid exhaustion of soil water and nutrient reserves, negating the short-term benefits of improved WUE.

The objective of this paper is to analyze the long-term CFE when both water and nitrogen (N) are limiting, by using a model to integrate the interactions between the C, N, and water cycles. Our approach is to evaluate the CFE using a "quasi-equilibrium" analysis of the G'DAY forest ecosystem model. That analysis, first described by McMurtrie *et al.* (1992), exploits the fact that forest ecosystems consist of many pools with a wide range of turnover times. Thus, on a given timescale, the system may be divided into "fast" and "slow" turnover pools, where the former are in approximate (or quasi-) equilibrium and the latter are effectively constant. Quasi-equilibrium implies that the total flux of C into the fast pools is balanced by the total efflux of C (and similarly for N). The C and N flux balance of the fast pools leads to a powerful simplification of the model.

The quasi-equilibrium analysis of G'DAY may be depicted graphically in terms of two relationships between NPP and foliage N:C ratio: the *photosynthetic constraint* to production (describing the quasi-equilibrium C balance of the fast pools), and the *N availability constraint* to production (describing the quasi-equilibrium N balance of the fast pools). NPP on the given timescale is then represented by the intersection of the two constraint curves. The photosynthetic constraint increases with foliar N concentration for most forests, which have suboptimal tissue N concentrations. The shape of the N-availability constraint depends on how input and output processes scale with leaf N concentration; it usually has a negative slope (Comins and McMurtrie, 1993; McMurtrie and Comins, 1996). Thus, the two constraint curves usually have opposite slopes, as illustrated in Figs. 1 and 2. Elevated CO_2 raises the photosynthetic constraint curve, producing the well-documented short-term increase in production, but the longer term response depends also on the shape of the N availability constraint curve

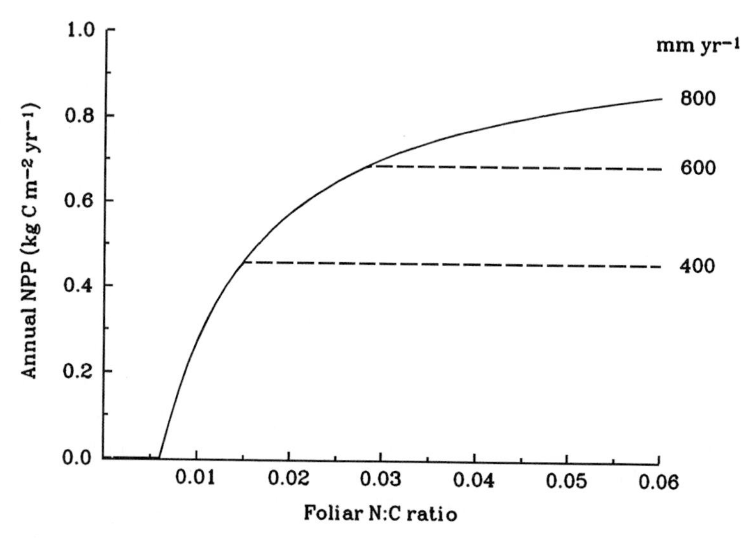

Figure 1 Dependence of photosynthetic constraint curve on effective rainfall (P) over the range 400–800 mm yr^{-1}. Parameters as in Table I.

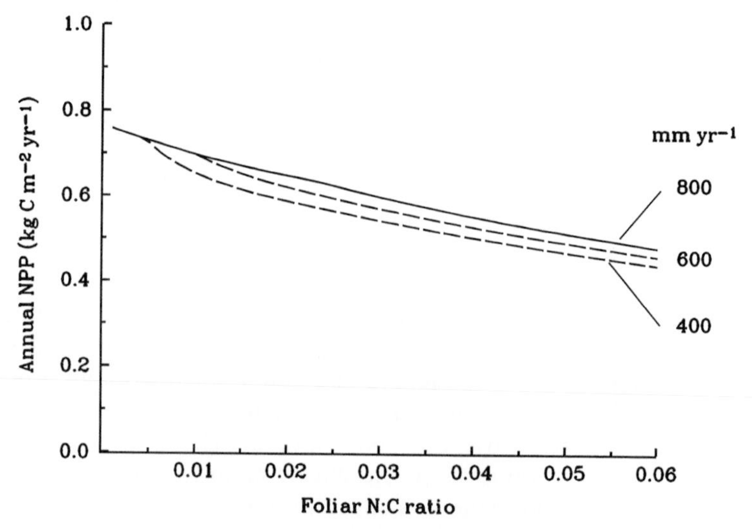

Figure 2 Dependence of long-term N availability constraint curve on effective rainfall (P) over the range 400–800 mm yr^{-1}. Parameters as in Table I.

Table I Definition of Symbols and Parameter Values for *P. radiata*[a]

Symbol	Definition	Value and units
C_p	C content of passive SOM pool	3 kg C m^{-2} yr^{-1}
$[CO_2]_a$	Atmospheric CO_2 concentration	350, 700 ppm
c_O	Water use efficiency coefficient	2.8×10^{-6} $[c_O]$
D	Air saturation vapor pressure deficit	0.856 kPa
E	Annual water-limited transpiration rate	mm yr^{-1}
F	Foliage C	kg C m^{-2}
f_W	Water dependence of k_i	v
f_T	Air temperature dependence of k_i	v
G	NPP	kg C m^{-2} yr^{-1}
G_L,	Light-limited NPP	kg C m^{-2} yr^{-1}
G_L*	Quasi-equilibrium value of G_L	kg C m^{-2} yr^{-1}
G_W	Water-limited NPP	kg C m^{-2} yr^{-1}
k	Canopy light extinction coefficient	0.5
k_i	SOM decomposition rate (i = s, p)	yr^{-1}
$k_{0,\,i}$	Decomposition coefficient (i = s, p)	yr^{-1}
$\mathcal{L}$	Fractional lignin content of litter	0.25
N_D	Atmospheric N deposition rate	4×10^{-4} kg m^{-2} yr^{-1}
N_F	N fixation rate	4×10^{-4} kg m^{-2} yr^{-1}
N_L	Rate of N loss in gaseous emission and leaching	kg N m^{-2} yr^{-1}
N_{min}	N mineralization rate	kg N m^{-2} yr^{-1}
N_{Ri}	N release rate from decay (i = s, p, w)	kg N m^{-2} yr^{-1}
N_{Si}	N sequestration rate (i = s, p, w)	kg N m^{-2} yr^{-1}
P	Annual effective rainfall input	400, 800 mm yr^{-1}
q	Water use efficiency	kg C (kg H2O)$^{-1}$
ε	Light utilization coefficient	kg C GJ^{-1} PAR
η_i	C allocation fractions (i = f, w, r)	0.1, 0.7, 0.2
ϕ_{abs}	Canopy-absorbed PAR	GJ PAR m^{-2} yr^{-1}
ϕ_0	Incident PAR	3 GJ m^{-2} yr^{-1}
ν_i	N:C ratio (i = f, r, w, p, po)	v, v, 0.0015, 0.05, 0.0
Ω_{pf}, Ω_{pr}	Fraction of leaf and root litterfall sequestered in passive SOM	v, v
ζ	Fraction of N_{min} lost in gaseous emission	0.05
σ	Specific leaf area	11 m^2 (kg C)$^{-1}$
τ_f	Average leaf lifespan	3.3 yr

[a] C carbon; $[c_O]$, kg C (kg H_2O)$^{-1}$ kPa (ppm CO_2)$^{-1}$; N, nitrogen; NPP, net primary production; PAR, photosynthetically active radiation; SOM, soil organic matter; V, variable. Subscripts: *f*, foliage; *w*, newly formed wood; r, fine roots; s, slow SOM; p, passive SOM; po, newly formed passive SOM. Values of other parameters as in Kirschbaum *et al.* (1994).

(Fig. 6 in Comins and McMurtrie's, 1993). McMurtrie and Comins (1996) studied the CFE on various timescales by examining how the corresponding constraint curves and their intersection shift at elevated CO_2. We extend this approach by examining how water limitation influences both the photosynthetic and N availability constraint curves. This approach thus provides a theoretical framework for analyzing the CFE under multiple resource limitations, through their combined effect on both plant and soil processes. This chapter describes how water balance can be incorporated in Comins and McMurtrie's (1993) quasi-equilibrium analysis.

This chapter begins with an outline of how the G'DAY model has been modified to incorporate water limitations, and derives the photosynthetic and N availability constraints to production (Section II). The quasi-equilibrium analysis and its graphical representation are then used to evaluate the CFE in relation to nutrient and water limitation in various combinations (Section III). We then discuss the extent to which the modeled CFE is due to direct effects of CO_2 on plant production versus indirect effects on litter and soil decomposition (Section IV).

II. Modifications to Incorporate Water Limitation in G'DAY

G'DAY is a forest ecosystem model that simulates the mechanisms of C and N cycling in the plant and soil. The plant is represented by three biomass pools (foliage, stem, fine root), and the soil by four litter pools and three soil organic matter (SOM) pools (called active, slow, and passive SOM) (Parton *et al.*, 1987). The C-cycling model describes photosynthesis, allocation and senescence, and the decomposition of litter and SOM. The N-cycling model describes atmospheric N deposition, biological fixation, plant N uptake, allocation and retranslocation from senescing tissue, N release and immobilization by decomposing litter and SOM, and gaseous N emission from soil. Because the model is fully described elsewhere (Comins and McMurtrie, 1993), we focus on modifications to incorporate water balance in G'DAY. The aim here is to derive the photosynthetic and N availability constraints for a water-limited site.

A. Light- and Water-Limited Net Primary Productivity

Our approach to incorporating the effects of water limitation into G'DAY, on an annual timescale, is based on Dewar's (1997) model of daily light and water use, itself derived from the RESCAP crop model (Monteith *et al.*, 1989). Annual net primary productivity (G) is assumed to be either light limited or water limited:

$$G = \min(G_L, G_W), \tag{1}$$

where G_L and G_W are light- and water-limited NPP, respectively. Light-limited NPP is assumed to be proportional to absorbed radiation (ϕ_{abs}) (Monteith, 1977):

$$G_L = \varepsilon([CO_2]_a, \nu_f)\, \phi_{abs}, \tag{2a}$$

where ε is a light utilization coefficient that depends on ambient CO_2

concentration ($[CO_2]_a$) and foliage N : C ratio (ν_f), as described by Kirschbaum *et al.* (1994) for *P. radiata* growing near Canberra, Australia. Absorbed radiation is given by Beer's law:

$$\phi_{abs} = \phi_0 \, [1 - \exp(-k \, \sigma \, F)], \tag{2b}$$

where ϕ_0 is incident radiation, k is the light extinction coefficient, σ is specific leaf area, and F is foliar carbon. (See Table I for a list of symbols.). When growth is light limited, water use is equal to G_L divided by the water use efficiency (WUE) (q).

Water-limited NPP (G_W) is evaluated as the maximum rate of soil moisture extraction by roots (E) multiplied by the WUE. Dewar (1997) assumed that E is proportional to the product of root carbon and available soil moisture; soil moisture was determined from the daily balance of rainfall, canopy interception (proportional to canopy leaf area), water use, drainage, and runoff. For the purposes of the quasi-equilibrium analysis, we adopt a simpler approach. We assume that a fixed fraction of incident rainfall is evaporated through canopy interception (ignoring any dependence on canopy leaf area), and therefore adopt the annual effective rainfall input (P) as a fixed parameter. We then assume that, when growth is water limited, there is no drainage or runoff, and we ignore soil evaporation and understorey evapotranspiration, so that E is equal to P in quasi-equilibrium; for the nonirrigated *P. radiata* stands for which G'DAY was parameterized by Kirschbaum *et al.* (1994), annual runoff and drainage amounted to approximately 5% of annual rainfall (Myers and Talsma, 1992), and soil evaporation and understorey transpiration were predicted by simulation studies to be small components of annual water balance (R. E. McMurtrie, unpublished results). Therefore, annual water-limited NPP is simply:

$$G_W = q \, P. \tag{3a}$$

We assume that WUE is proportional to the atmospheric CO_2 concentration ($[CO_2]_a$) (Morison, 1993), and inversely proportional to the daylight atmospheric saturation vapor pressure deficit (D):

$$q = \frac{c_0 \, [CO_2]_a}{D}, \tag{3b}$$

where c_0 is a constant coefficient. The assumption that $q \propto 1/D$ is a reasonable approximation for rough canopies that are well coupled to the atmosphere, such as forests (Dewar, 1997).

B. Photosynthetic Constraint to Production

The photosynthetic constraint to production is the relationship between G and the foliage N : C ratio when foliage carbon (F, a fast pool) is in quasi-equilibrium (Comins and McMurtrie, 1993), that is, when annual leaf production equals leaf fall. Expressing leaf production as a constant proportion (η_f) of NPP, and assuming a fixed average leaf life span (τ_f), gives

$$\eta_f\, G = F/\tau_f. \tag{4}$$

If growth is light limited, the photosynthetic constraint is obtained by eliminating F from Eqs. (2) and (4) to obtain an implicit equation relating G_L at canopy closure to foliar N:C and $[CO_2]_a$:

$$G_L^* = \varepsilon\, \phi_0\, [1 - \exp(-k\, \sigma\, \eta_f\, \tau_f\, G_L^*)], \tag{5}$$

where the asterisk indicates that G_L has been evaluated at equilibrium. If growth is water limited, the photosynthetic constraint is independent of the foliage N:C ratio, and is given by Eq. (3) for G_W.

The full photosynthetic constraint to production, given by the minimum of G_L^* and G_W, is illustrated in Fig. 1 for *P. radiata* growing near Canberra, Australia, in an environment with mean annual soil and daytime air temperatures of 10 and 14°C, respectively (Kirschbaum *et al.,* 1994), for $[CO_2]_a$ of 350 ppm, and for effective annual rainfall (P) ranging from 400 to 800 mm yr^{-1}. At high rainfall, growth is always light limited, and the photosynthetic constraint curve has a large positive initial slope, which levels off smoothly at high foliar N:C. At lower rainfall, the photosynthetic constraint is truncated above a certain foliar N:C ratio because of water limitation. As rainfall declines further, the water-limited plateau of the photosynthetic constraint curve becomes lower.

The photosynthetic constraint curve shown in Fig. 1 is derived using a simple water balance model based on estimated annual water use [Eq. (3)]. We have also evaluated the photosynthetic constraint using a more complex water balance model with rainfall events occurring at regular time intervals, and with E evaluated from Dewar's (1997) full water balance model where soil water extraction is proportional to root mass. In running that model over a single drying cycle between successive rain events, we must consider three possibilities: (1) that production is light limited throughout the drying cycle, (2) that production is water limited throughout the drying cycle, and (3) that production is light limited for an initial period after rainfall, but later becomes water limited. When these three possibilities are analyzed, the photosynthetic constraint has a shape similar to Fig. 1; the sole difference is that the sharp truncation from light to water limitation (Fig. 1) is replaced by a smooth curve corresponding to the third possibility where growth is partially limited by both light and water. Because the simple and complex models give qualitatively similar behavior, we choose to present only results from the simpler version.

C. N Availability Constraint to Production

Rainfall also affects rates of decomposition (k_i) of soil and litter pools (i = pool label), which are temperature and moisture dependent:

$$k_i = k_{0,i} f_T f_W, \tag{6a}$$

where f_T and f_W denote the air-temperature and moisture dependencies, and $k_{0,i}$ is a constant for each pool. The functions f_T and f_W are derived from Parton *et al.* (1987). The function f_W is an empirical index of soil moisture status, defined in terms of the ratio of annual effective rainfall (P) to potential water use ($\varepsilon\, \phi_0/q$), where ϕ_0 is the incident radiation:

$$f_W = \min\left(\frac{q\,P}{\varepsilon\,\phi_0}, 1\right). \tag{6b}$$

These decomposition rates affect the N availability constraint, which is derived by assuming N fluxes into equilibrated pools are equal to fluxes out. We consider the timescale of the long-term equilibrium, when all pools except woody biomass and passive SOM are equilibrated. This leads to the following N balance equation:

$$N_D + N_F + N_{Rp} + N_{Rw} = N_{Sp} + N_{Sw} + N_L. \tag{7}$$

The terms on the left-hand side of this equation represent the N fluxes into the fast pools; N_D and N_F are N inputs from atmospheric deposition and fixation, respectively; N_{Rp} and N_{Rw} are the rates of N release due to the decomposition of passive SOM and woody litter, respectively. The terms on the right-hand side represent N fluxes out of the fast pools; N_{Sp} and N_{Sw} are the rates of N immobilization in newly formed passive SOM and wood, respectively; N_L represents losses due to gaseous emission or leaching.

Following Comins and McMurtrie (1993), N_{Sw} is given by the product of the N : C ratio of newly formed wood (ν_W) and woody C production ($\eta_W G$); similarly, N_{Sp} is the product of the N : C ratio of newly formed passive SOM (ν_{po}) and the rate of C sequestration in passive SOM (expressed in terms of the fractions of annual leaf and root litterfall C that are eventually sequestered in passive SOM, Ω_{pf} and Ω_{pr}, respectively). The sequestration fractions Ω_{pf} and Ω_{pr} were derived by Comins and McMurtrie (1993) for the CENTURY soil carbon model (Parton *et al.*, 1987). These fractions are functions of litter quality, characterized by the lignin : N ratio (Melillo *et al.*, 1982), such that C sequestration in SOM increases when litter quality decreases; typical values of Ω_{pf} are 0.009 (when the lignin fraction is $\mathcal{L}$ = 0.1 and the leaf N : C ratio is ν_f = 0.02) and 0.013 (when $\mathcal{L}$ = 0.3 and ν_f = 0.02).

The rate of N release from decomposing passive SOM is $N_{Rp} = k_p C_p \nu_p$, where k_p, C_p, and ν_p are, respectively, the SOM decomposition rate [given by Eq. (6)], C content, and N : C ratio. Assuming that N losses are proportional to the rate of N mineralization ($N_L = \xi\, N_{min}$), and ignoring woody litter decay ($N_{Rw} = 0$) and N retranslocation prior to root and leaf senescence, Eq. (7) then leads to the N availability constraint to production (Comins and McMurtrie, 1993):

$$G = \frac{N_D + N_F + k_p C_p \nu_p}{\eta_W \nu_W + (\eta_f \Omega_{pf} + \eta_r \Omega_{pr}) \nu_{po} + \frac{\xi}{1 - \xi} (\eta_f \nu_f + \eta_W \nu_W + \eta_r \nu_r)}, \quad (8)$$

which describes a second relationship between G and the foliage N:C ratio (ν_f).

Figure 2 shows how the N availability constraint depends on effective rainfall over the range from 400 to 800 mm yr^{-1}. The curve has a negative slope because increasing the foliar N:C ratio enhances gaseous N emission losses, leaving less N available to support growth. Increasing rainfall raises the N availability constraint, reflecting the response of the soil moisture index (f_W) and its effect on the passive SOM decomposition rate (k_p).

The long-term equilibrium NPP is given by the intersection of the photosynthetic and N availability constraint curves, where both the C and N quasi-equilibrium conditions on the fast pools are satisfied.

III. Modeled Responses to CO_2 in Relation to Water and N Limitation

McMurtrie and Comins (1993) describes how G'DAY responds to a step change of $[CO_2]_a$ from 350 to 700 ppm. Immediately following the increase of $[CO_2]_a$, there is a *short-term* transient increase of NPP, which lasts little more than a decade, and this is followed by a reduced equilibrium response. The constraint curve analysis described earlier provides an estimate of NPP at the long-term equilibrium when all pools except woody biomass and passive SOM are equilibrated (Comins and McMurtrie, 1993).

Figure 3 shows the effect of CO_2 enrichment on both constraint curves for *P. radiata* (using parameter values specified in Table I), for various combinations of N and rainfall inputs. For a site with high rainfall and moderate N inputs, equilibrium NPP (given by the intersection of the two constraint curves) is light limited at both 350 and 700 ppm (Fig. 3A, points A and C, respectively). The light-limited segments of the photosynthetic constraint curves are equivalent to those derived by McMurtrie and Comins (1996) who ignored water dependence; our light-limited curves are not identical to their curves because we have changed values of four parameters, the allocation coefficients (η_f, η_W, η_r) and leaf life span (τ_f). The N availability constraint curve is raised at elevated $[CO_2]_a$; this response reflects the effect of improved soil moisture status on the decomposition rate of passive SOM (k_p); Eq. (6b) predicts that the soil moisture index (if less than 1 initially) is increased at elevated $[CO_2]_a$ because WUE increases more than

ε. Combining the effects of doubled $[CO_2]_a$ on both constraint curves leads to a 10.5% increase in the equilibrium NPP (from 0.621 to 0.686 kg C m^{-2} yr^{-1}).

For a site with low rainfall and moderate N inputs (Fig. 3B), equilibrium NPP is water limited at 350 ppm (point A) but light limited at 700 ppm (point C). The net effect of doubled $[CO_2]_a$ is a 48.8% increase in the equilibrium NPP (from 0.458 to 0.681 kg C m^{-2} yr^{-1}). In this case, the equilibrium NPP at 700 ppm is almost the same as at high rainfall (Fig. 3A); the larger percentage increase at low rainfall (Fig. 3B) is due to the fact that doubling $[CO_2]_a$ alleviates water limitation at 350 ppm.

For a site with low N inputs and either high rainfall (Fig. 3C) or low rainfall (Fig. 3D), long-term equilibrium NPP is light limited at both 350 and 700 ppm (points A and C, respectively). The net effect of doubled $[CO_2]_a$ is to increase the equilibrium NPP at high rainfall by 4.6% (from 0.406 to 0.424 kg C m^{-2} yr^{-1}) and at low rainfall by 18.1% (from 0.359 to 0.424 kg C m^{-2} yr^{-1}). The lower percentage CFE at these two sites, relative to the respective sites with moderate N inputs, reflects the effect of N inputs on the N availability constraint curve.

IV. Discussion

A. Long-Term CFE

Previous ecosystem model simulations have incorporated interactions between C and N dynamics (Schimel 1990; Rastetter *et al.*, 1991, 1992; Comins and McMurtrie, 1993; Melillo *et al.*, 1993; Hudson *et al.*, 1994; Medlyn and Dewar, 1996), and between C, N, and water dynamics (Thornley and Cannell, 1996), to demonstrate that the CFE is diminished under N limitation. Our study extends these analyses by examining the CFE in relation to both N and rainfall inputs, using a relatively simple ecosystem model that incorporates some of the major interactions between C, N, and water dynamics. The quasi-equilibrium analysis complements simulation-based approaches by providing a direct, graphical insight into the roles of plant and soil processes in determining the long-term CFE. Our results support the commonly held proposition that the CFE is amplified under water limitation, but is reduced under long-term nutrient limitation. With moderate N inputs, the amplification of the CFE by water limitation is caused mainly by improved WUE (see further analysis later), which directly affects plant production and indirectly affects soil N availability through improved soil moisture (Fig. 3B), leading to a switch from water- to light-limited growth at elevated CO_2. The constraining effect of N availability is illustrated by the predicted responses at low N inputs (Figs. 3C and D),

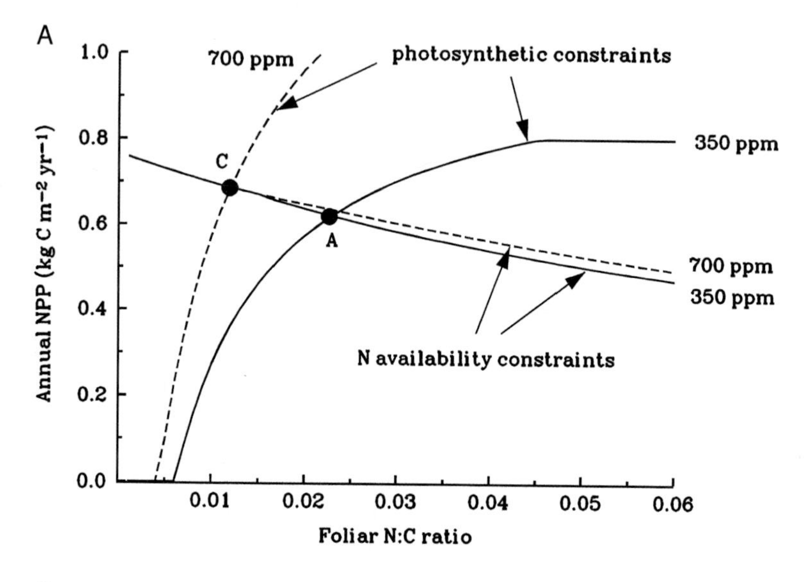

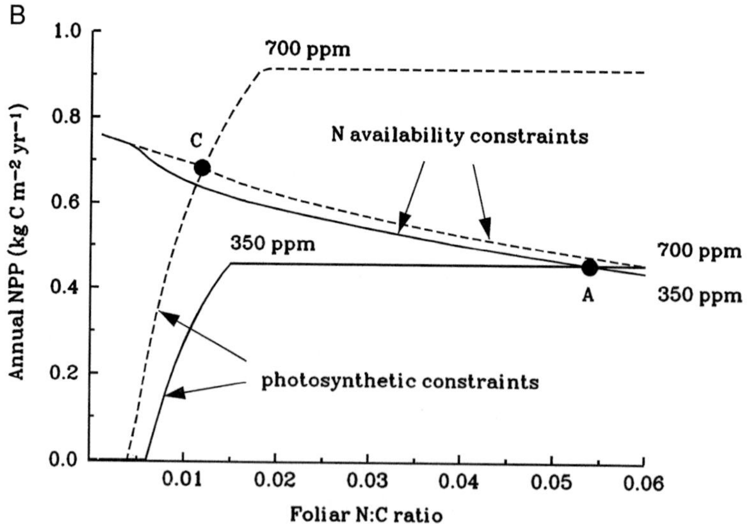

Figure 3 Graphical analysis of long-term CO_2 response of NPP (indicated by the transition from point A to point C), with (A) moderate N inputs ($N_D + N_F = 0.0008$ kg N m^{-2} yr^{-1}), high rainfall ($P = 700$ mm yr^{-1}), (B) moderate N inputs, low rainfall ($P = 400$ mm yr^{-1}), (C) low N inputs ($N_D + N_F = 0.0004$ kg N m^{-2} yr^{-1}), high rainfall, (D) low N inputs, low rainfall. Other parameters as in Table I. Also in (D), the medium-term CO_2 response of NPP is indicated by the transition from point A to point B.

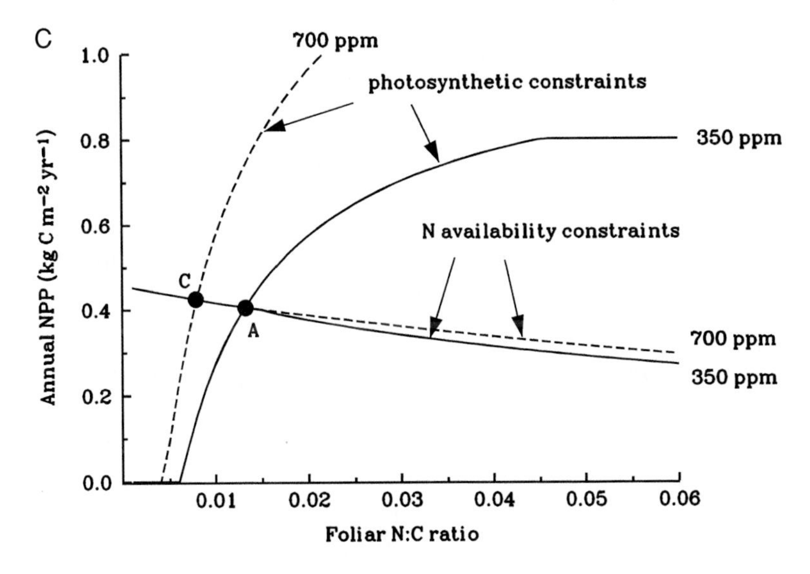

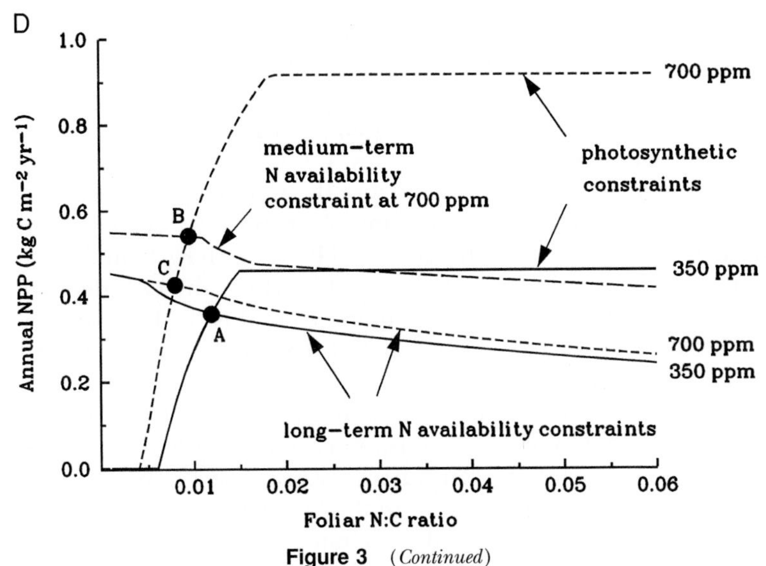

Figure 3 (*Continued*)

for which growth is light limited at both 350 and 700 ppm, even at a low rainfall site.

Thus, the magnitude of the long-term CFE depends on whether growth, represented by Eq. (1), is light or water limited at current $[CO_2]_a$. If growth is water limited at current $[CO_2]_a$, then CO_2 enrichment is likely to alleviate the water limitation and lead to a large CFE (+49% in Fig. 3B). If growth is light limited at current $[CO_2]_a$, then a smaller relative CFE is predicted (+11, 5, and 18% in Figs. 3A, C, and D, respectively). This conclusion may provide a useful, general guide in identifying ecosystems whose long-term relative CFE is likely to be large; if an ecosystem at current $[CO_2]_a$ experiences significant soil water deficit that limits its productivity, then a large, long-term, relative CFE may be expected. (Keep in mind, however, that the absolute CFE on a site that is water limited and, hence, unproductive might be smaller than on a non-water-limited site, even though its relative CFE is larger.)

B. Physiological Factors Contributing to the CFE

Elevated $[CO_2]_a$ has two main effects in the preceding model. First, it increases rates of carbon production (either through increased ε or increased WUE, depending on whether growth is light or water limited, respectively); second, it increases SOM decomposition (through increased WUE/ε). The contribution of each effect to the long-term CFE can be quantified using the graphical analysis shown in Fig. 3, by evaluating the constraint curves with only that effect incorporated into the model.

For example, in Fig. 3A, where growth is light limited at current $[CO_2]_a$, the overall CO_2 response is determined by the vertical shifts of the photosynthetic constraint (due to increased ε) and of the N availability constraint (due to increased WUE/ε). In Fig. 3A, the contribution of increased ε alone can be approximately estimated as the difference between the NPP values where the N availability constraint at 350 ppm intersects the photosynthetic constraints at 350 ppm (giving NPP = 0.621 kg C m^{-2} yr^{-1}) and 700 ppm (NPP = 0.686 kg C m^{-2} yr^{-1}). Therefore, the CO_2 dependence of ε alone leads to a 10.5% long-term increase of NPP. The contribution of the increase in WUE/ε beyond that of ε alone can be estimated from the difference between the NPP values where the photosynthetic constraint at 700 ppm intersects the N availability constraints at 350 and 700 ppm; in this case, the two N availability constraints coincide where they intersect the 700 ppm photosynthetic constraint (because $f_W = 1$ in both cases), and there is no additional contribution to the long-term NPP response. Thus, in Fig. 3A, with high rainfall and moderate N inputs, the overall CFE of 10.5% in response to doubled CO_2 can be attributed entirely to increased ε and its effect on carbon production.

A similar analysis can be performed for each of the cases considered in Fig. 3 (see Table II). This analysis reveals that at high rainfall (cases A and C), the overall CFE, which is relatively small, is entirely due to the effect of increased ε on carbon production—because, at high rainfall, the N availability constraint curves are insensitive to WUE/ε over a wide range of leaf N:C ratios (Fig. 2). At low rainfall, both the photosynthetic and N

Table II Percentage Increase in Long-Term Equilibrium NPP in Response to Doubled $[CO_2]_a$ (i.e., the Relative CO_2 Fertilization Effect, CFE) for Various Levels of N Input and Rainfall[a]

Case	% CFE due to effects on C production alone	Additional % CFE due to effects on SOM decay	Total % CFE
Case A: moderate N inputs high rainfall	10.5 (ε)	0	10.5
Case B: moderate N inputs low rainfall	40.2 (WUE & ε)	8.6 (WUE/ε)	48.8
Case C: low N inputs high rainfall	4.6 (ε)	0	4.6
Case D: low N inputs low rainfall	9.2 (ε)	8.9 (WUE/ε)	18.1

[a] Corresponding to the four cases considered in Fig. 3. The overall CFE is separated into the contribution from effects of high CO_2 on the photosynthetic constraint (carbon production) and the additional effects of high CO_2 on the N availability constraint (soil organic matter decay); the physiological factor mediating each effect is indicated in parentheses. C, carbon; SOM, soil organic matter.

availability constraint curves are sensitive to CO_2 (Figs. 3B and D), but there is a fundamental difference depending on the level of N inputs. If N inputs are high, growth is water limited at current CO_2 and light limited at high CO_2 (Fig. 3B). Consequently, the large percentage CFE associated with the shift of the photosynthetic constraint (+40.2%, Table II) is due primarily to increased WUE (though there is a small effect associated with increased ε). There is a smaller additional CFE (of +8.6%) due to the effect of increased WUE/ε on the N availability constraint. On the other hand, if both rainfall and N inputs are low (Fig. 3D), then growth is light limited at both CO_2 levels, and the overall CFE of +18.1% is due to both increased ε (+9.2%) affecting the photosynthetic constraint, and increased WUE/ε (+8.9%) affecting the N availability constraint, the contributions from plant and soil processes being of similar magnitude.

The results shown in Table II can be summarized as follows: (1) The CFE is largest in case B (low rainfall, moderate N inputs) primarily because water limitations to carbon production are alleviated by improved WUE; (2) the CFE is relatively small in cases A and C (high rainfall), when the response is entirely due to increased ε, affecting carbon production; (3) the CFE is intermediate in case D (low rainfall, low N inputs), when the response is due to increases in both ε and WUE, affecting carbon production and SOM decomposition in approximately equal measure.

C. Medium versus Long-Term CFE

The preceding analysis has focused on the long-term (century-timescale) response to high CO_2, achieved when all pools except wood and passive SOM are at equilibrium. A similar analysis can be applied to determine the medium-term CO_2 response achieved when all pools except wood, passive SOM, and slow SOM are at equilibrium (which occurs on a timescale of one to two decades after CO_2 is doubled) (McMurtrie and Comins, 1996; Comins, 1997). The results of this analysis are illustrated in Fig. 3D for the case with low N inputs and low rainfall. According to Fig. 3D, NPP increases from 0.359 kg C m^{-2} yr^{-1} when $[CO_2]_a = 350$ ppm (point A) to 0.539 kg C m^{-2} yr^{-1} (+50.2%) at the medium-term equilibrium at 700 ppm (point B), and then declines to 0.424 kg C m^{-2} yr^{-1} (+18.1%) at the long-term equilibrium (point C). The decline in NPP between the medium- and long-term equilibria at high CO_2 occurs because improved WUE at high CO_2 leads to enhanced decomposition, and hence to a gradual loss of slow SOM, leading to a long-term reduction in soil N availability.

D. Model Uncertainties

There are several uncertainties in the preceding model of ecosystem response to CO_2 enrichment. One uncertainty concerns the validity of our assumption that plant WUE is proportional to CO_2. This assumption is supported by gas exchange data from high CO_2 experiments, in many of which enhanced WUE is associated with reduced stomatal conductance (Morison, 1993; Long *et al.*, 1996). However, for many plants stomatal conductance does not decline at high CO_2, and it is relevant to examine how the WUE of these plants depends on CO_2. This question is explored in a preliminary way in Fig. 4, which illustrates the relationship between leaf net photosynthesis (A, μmol m^{-2} s^{-1}) and intercellular CO_2 concentration ($[CO_2]_i$) for *P. radiata*, the *demand function* (from McMurtrie and Wang, 1993). The realized value of A is given to a first approximation (ignoring boundary layer effects) by the intersection of the demand function and the *supply function* for the rate of CO_2 transport from the atmosphere, $A = g_S([CO_2]_a - [CO_2]_i)$, where g_S is stomatal conductance to CO_2. In Fig. 4, intersection points are shown for $[CO_2]_a$ equal to 350 ppm (point A), and for $[CO_2]_a$ equal to 700 ppm in the two cases in which stomatal conductance (g_S) is (1) independent of $[CO_2]_a$ and equal to 0.21 mol m^{-2} s^{-1} (point B) and (2) reduced as $[CO_2]_a$ increases from 350 and 700 ppm such that the ratio $[CO_2]_i/[CO_2]_a$ remains unchanged (point C). In case (1), WUE, which is proportional to the ratio A/g_S, increases by 69% when $[CO_2]_a$ increases from 350 to 700 ppm. In case (2), CO_2 enrichment leads to a 100% increase in WUE, consistent with Eq. (3b). These two calculations confirm that large improvements in WUE can occur at high CO_2 even when stomatal conductance does not decline.

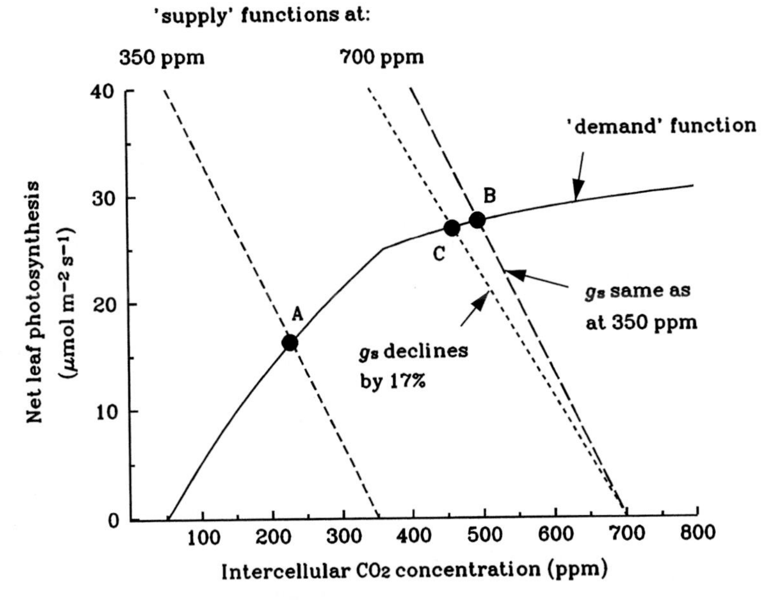

Figure 4 Rates of net leaf photosynthesis (demand) and CO_2 diffusion from the atmosphere (supply) as functions of intercellular $[CO_2]$ (cf. McMurtrie and Wang, 1993). The operating point is given by the intersection of the two functions. Three operating points are shown: for $[CO_2]_a = 350$ ppm (point A), and for $[CO_2]_a = 700$ ppm with unaltered stomatal conductance (point B) and with stomatal conductance reduced by 17% (point C); the latter case results in twice the intercellular $[CO_2]$ at point A and thus a doubling of the WUE.

A second uncertainty concerns whether reduced leaf transpiration at high CO_2 will be counteracted by feedbacks in the water cycle that are not currently represented in the model (Field *et al.*, 1995). For example, any reduction in tree water use is likely to lead to changes in foliage allocation and retention. Foliage allocation may increase in response to enhanced soil moisture and N availability (Medlyn and Dewar, 1996), tending to counter the effects of reduced leaf transpiration. In our calculations we assumed fixed allocation fractions. More crucially, the empirical soil moisture index (f_W) is independent of leaf area, so that in any case the present model cannot capture feedbacks from leaf area to soil moisture; it would be more realistic to relate f_W more directly to soil moisture status and its dependence on tree water use (e.g., Dewar, 1997). We have also neglected the positive effect of stomatal closure at high CO_2 on leaf temperature, and thus overestimated the reduction in leaf transpiration.

A third uncertainty concerns the effect of incorporating soil evaporation, understorey evapotranspiration, runoff, and drainage in the water balance model. Morison (1993) addresses this issue by evaluating canopy water use efficiency, defined as NPP divided by transpiration plus soil evaporation. He argues that if soil evaporation represents a large fraction of water balance, then canopy WUE will be less responsive to high CO_2 than WUE defined by Eq. (3). His argument was supported by experimental evidence that doubling of CO_2 leads to a 100% increase of leaf-scale WUE, but smaller increases in canopy WUE. In our analysis we have assumed that soil evaporation, understorey evapotranspiration, runoff, and drainage are zero in water-limited systems, but not if water is nonlimiting. Thus, G'DAY would predict a doubling of NPP at 700 ppm only for sites that are water limited at both 350 and 700 ppm. Of the four cases considered in Fig. 3, none is water limited at 700 ppm. Thus, all four cases give NPP responses much less than the increase in leaf-scale WUE.

A fourth uncertainty concerns the effect of elevated CO_2 on belowground versus aboveground allocation, and consequences for both nutrient and water acquisition. There is considerable experimental evidence of enhanced belowground carbon allocation under CO_2 enrichment (e.g., Mooney *et al.*, 1998). There is also theoretical evidence for enhanced root allocation (Medlyn and Dewar, 1996). If root production is enhanced at high CO_2, then both nutrient and water uptake may be enhanced, tending to alleviate both nutrient and water stress. However, increased belowground carbon allocation could lead to enhanced exudation of carbon substrate into the rhizosphere and enhanced root turnover, which could stimulate the activity of microbial decomposers, but could also have subtle effects on rates of nutrient immobilization in soils (e.g., Körner, 1996; McMurtrie and Comins, 1996), and hence on productivity. In particular, stimulation of microbial activity at high CO_2, leading to enhanced N fixation, could counter the above negative soil feedback on the CFE under nutrient limitation (Gifford, 1994).

A fifth uncertainty concerns our assumption that plant and soil N : C ratios are unaltered at high CO_2; the overall CFE, and the relative contributions of plant and soil processes to it, are likely to differ considerably if these parameters change at high CO_2 (Kirschbaum *et al.*, 1994; McMurtrie and Comins, 1996). These and other uncertainties may be addressed using the graphical analysis given earlier. Because of its basis on C and N conservation, the quasi-equilibrium analysis is not model specific and therefore provides a general framework for examining the contribution of various plant and soil processes to the CFE on different timescales.

E. The Missing CO_2 Sink

In Table II, our result that under water limitation the long-term CFE is amplified by a factor of 4.6 at moderate N inputs (cf. cases A and B) and by a factor of 3.9 at low N inputs (cf. Cases c and d) raises the possibility that the CFE for water-limited forests may represent a significant component of the so-called "missing" CO_2 sink (Schimel, 1995). On the one hand, increases in WUE with rising CO_2 may enhance soil decomposition and promote soil C loss. On the other hand, improved WUE may lead to enhanced growth rates on water-limited sites (and, in woody ecosystems, to a shift of N from soil pools with low C:N ratio to woody biomass with high C:N ratio). Because of the difference between the C:N ratios of soil and wood, a widespread net shift of N from soil to wood could potentially lead to a large increase in carbon storage in terrestrial ecosystems (Rastetter *et al.*, 1992). The contribution of water-limited vegetation to the global carbon budget needs further investigation. Keep in mind, however, that many water-limited ecosystems are relatively unproductive, and hence make little contribution to the annual global carbon budget; thus, even though the relative CFE might be larger under water-limited than under non-water-limited conditions, the absolute stimulation of production might be smaller. Another reason for caution in extrapolating our results to larger spatial scales relates to our assumption that soil evaporation is negligible; this assumption is not valid for many water-limited ecosystems.

V. Summary

Our equilibrium-based analysis of the G'DAY forest ecosystem model suggests that the CO_2 fertilization effect (CFE) is likely to be amplified under water limitation, but reduced under long-term nitrogen limitation. The modeled, long-term, relative CFE is largest (49% increase of NPP with a doubling of CO_2) for a water-limited site with moderate N inputs, and is least (+4.6%) for a well-watered site with low N inputs. In our simulations the amplification of the CFE by water limitation is primarily due to enhanced water use efficiency, which has a large, direct effect on carbon uptake, and a smaller, indirect effect due to stimulation of soil decomposition.

Acknowledgments

We acknowledge the support of the dedicated grants scheme of the National Greenhouse Advisory Committee and the Australian Research Council.

References

Allen, L. H. (1990). Plant responses to rising carbon dioxide and potential interactions with air pollutants. *J. Environ. Quality* **19,** 15–34.

Beerling, D. J., Heath, J., Woodward, F. I., and Mansfield, T. A. (1996). Drought–CO_2 interactions in trees: Observations and mechanisms. *New Phytol.* **134,** 235–242.

Bunce, J. A. (1992). Stomatal conductance, photosynthesis, and respiration of temperate deciduous tree seedlings grown outdoors at an elevated concentration of carbon dioxide. *Plant Cell Environ.* **15,** 541–549.

Ceulemans, R., and Mousseau, M. (1994). Effects of elevated atmospheric CO_2 on woody plants. *New Phytol.* **127,** 425–446.

Comins, H. N. (1997). Analysis of nutrient-cycling dynamics, for predicting sustainability and CO_2-response of nutrient-limited forest ecosystems. *Ecol. Mod.* (in press).

Comins, H. N., and McMurtrie, R. E. (1993). Long-term response of nutrient limited forests to CO_2 enrichment: Equilibrium behavior of plant–soil models. *Ecol. Appl.* **3,** 666–681.

Curtis, P. S. (1996). A meta-analysis of leaf gas exchange and nitrogen in trees grown under elevated carbon dioxide. *Plant Cell Environ.* **19,** 127–137.

Dewar, R. C. (1997). A model of light and water use elevated for *Pinus radiata. Tree Physiol.* **17,** 259–265.

Diaz, S., Grime, J. P., Harris, J., and McPherson, E. (1993). Evidence of a feedback mechanism limiting plant response to elevated carbon dioxide. *Nature* **364,** 616–617.

Drake, B. G., Peresta, G., Beugeling, E., and Matamala, R. (1996). Long-term elevated CO_2 exposure in a Chesapeake Bay wetland: Ecosystem gas exchange, primary production, and tissue nitrogen. *In* "Carbon Dioxide and Terrestrial Ecosystems" (G. W. Koch and H. A. Mooney, eds.), pp. 197–214. Academic Press, San Diego.

Eamus, D., and Jarvis, P. G. (1989). The direct effects of increase in the global atmospheric CO_2 concentration on natural and commercial temperate trees and forests. *Adv. Ecol. Res.* **19,** 1–55.

Field, C. B., Jackson, R. B., and Mooney, H. A. (1995). Stomatal responses to CO_2: Implications from the plant to the global scale. *Plant Cell Environ.* **18,** 1214–1225.

Field, C. B., Chapin, F. S. III, Chiariello, N. R., Holland, E. A., and Mooney, H. A. (1996). The Jasper Ridge CO_2 experiment: Design and motivation. *In* "Carbon Dioxide and Terrestrial Ecosystems" (G. W. Koch and H. A. Mooney, eds.), pp. 112–145. Academic Press, San Diego.

Gifford, R. M. (1992). Interaction of carbon dioxide with growth-limiting environmental factors in vegetation productivity: Implications for the global carbon cycle. *Adv. Bioclimatol.* **1,** 24–58.

Gifford, R. M. (1994). The global carbon cycle: A viewpoint on the missing sink. *Aust. J. Plant Physiol.* **21,** 1–15.

Guehl, J. M., Picon, C., Aussenac, G., and Gross, P. (1994). Interactive effects of elevated CO_2 and soil drought on growth and transpiration efficiency and its determinants in two European forest tree species. *Tree Physiol.* **14,** 707–724.

Houghton, J. T., Meira Filho, L. G., Bruce, J., Hoesung Lee, Callander, B. A., Haites, E., Harris, N., and Maskell, K. (1995). "Climate Change 1994. Radiative Forcing of Climate Change and an Evaluation of the IPCC IS92 Emission Scenarios." Intergovernmental Panel on Climate Change, Cambridge University Press, Cambridge, UK.

Hudson, R. J. M., Gherini, S. A., and Goldstein, R. A. (1994). Modeling the global carbon cycle: Nitrogen fertilisation of the terrestrial biosphere and the "missing" CO_2 sink. *Glob. Biogeochem. Cycles* **8,** 307–333.

Idso, K. E., and Idso, S. B. (1994). Plant responses to atmospheric CO_2-enrichment in the face of environmental constraints: A review of the last 10 years' research. *Agric. For. Meteorol.* **69,** 153–203.

Kimball, B. A., Pinter, P. J., Garcia, R. L., LaMorte, R. L., Wall, G. W., Hunsaker, D. J., Wechsung, G., Wechsung, F., and Kartschall, T. (1995). Productivity and water use of wheat under free-air CO_2 enrichment. *Global Change Biol.* **1,** 425–445.

Kirschbaum, M. U. F., King, D. A., Comins, H. N., McMurtrie, R. E., Raison, R. J., Pongracic, S., Murty, D., Keith, H., Medlyn, B. E., Khanna, P. K., and Sheriff, D. W. (1994). Modelling forest response to increasing CO_2 concentration under nutrient-limited conditions. *Plant Cell Environ.* **17,** 1081–1099.

Kirschbaum, M. U. F., Bullock, P., Evans, J. R., Goulding, K., Jarvis, P. G., Noble, I. R., Rounsevell, M., and Sharkey, T. D. (1996a). Ecophysiological, ecological and soil processes in terrestrial ecosystems: A primer on general concepts and relationships. *In* "Climate Change 1995. Impacts, Adaptations, and Mitigation of Climate Change: Scientific-Technical Analyses" (R. T. Watson, M. C. Zinyowera, and R. H. Moss, eds.), pp. 57–74. IPCC Second Assessment Report, Cambridge University Press, Cambridge, UK.

Kirschbaum, M. U. F., Fischlin, A., Cannell, M. G. R., Cruz, R. V. O., Galinski, W., and Cramer, W. P. (1996b). Climate change impacts on forests. *In* "Climate Change 1995. Impacts, Adaptations, and Mitigation of Climate Change: Scientific-Technical Analyses" (R. T. Watson, M. C. Zinyowera, and R. H. Moss, eds.), pp. 95–129. IPCC Second Assessment Report, Cambridge University Press, Cambridge, UK.

Körner, Ch. (1996). The response of complex multispecies systems to elevated CO_2. *In* "Global Change and Terrestrial Ecosystems" (B. Walker and W. Steffen, eds.), pp. 20–42. International Geosphere-Biosphere Programme Book Series, Cambridge University Press, Cambridge.

Long, S. P. (1991). Modification of the response of photosynthetic productivity to rising temperature by atmospheric CO_2 concentrations: Has its importance been underestimated? *Plant Cell Environ.* **14,** 315–332.

Long, S. P., Osborne, C. P., and Humphries, S. W. (1996). Photosynthesis, rising atmospheric carbon dioxide concentration and climate change. *In* "Global Change: Effects on Coniferous Forests and Grasslands" (A. I. Breymeyer, D. O. Hall, J. M. Melillo, and G. I. Ågren, eds.), Scientific Committee on Problems of the Environment, Vol. 56, pp. 121–159. John Wiley and Sons Ltd., Chichester, UK.

Luxmoore, R. J., Wullschleger, S. D., and Hanson, P. J. (1993). Forest responses to CO_2 enrichment and climate warming. *Water Air Soil Pollut.* **70,** 309–323.

McMurtrie, R. E., and Comins, H. N. (1996). The temporal response of forest ecosystems to doubled atmospheric CO_2 concentration. *Global Change Biol.* **2,** 49–57.

McMurtrie, R. E., and Wang, Y.-P. (1993). Mathematical models of the photosynthetic response of tree stands to rising CO_2 concentrations and temperatures. *Plant Cell Environ.* **16,** 1–13.

McMurtrie, R. E., Comins, H. N., Kirschbaum, M. U. F., and Wang, Y.-P. (1992). Modifying existing forest growth models to take account of direct effects of elevated CO_2. *Aust. J. Bot.* **40,** 657–677.

Medlyn, B. E., and Dewar, R. C. (1996). A model of the long-term response of carbon allocation and productivity of forests to increased CO_2 concentration and nitrogen deposition. *Global Change Biol.* **2,** 367–376.

Melillo, J. M., Aber, J. D., and Muratore, J. F. (1982). Nitrogen and lignin control of hardwood leaf litter decomposition dynamics. *Ecology* **63,** 621–626.

Melillo, J. M., McGuire, A. D., Kicklighter, D. W., Moore, B., Vorosmarty, C. J., and Schloss, A. L. (1993). Global climate change and terrestrial net primary production. *Nature* **363,** 234–240.

Monteith, J. L. (1977). Climate and the efficiency of crop production in Britain. *Phil. Trans. R. Soc. London* **B281,** 277–294.

Monteith, J. L., Huda, A. K. S., and Midya, D. (1989). RESCAP: A resource capture model for sorghum and pearl millet. *In* "Modelling the Growth and Development of Sorghum and Pearl Miller" (S. M. Virmani, H. L. S. Tandon and G. Alagarswamy, eds.), ICRISAT Research Bulletin 12, pp. 30–34. Patancheru, India.

Mooney, H. A., Drake, B. G., Luxmoore, R. J., Oechel, W. C., and Pitelka, L. F. (1991). Predicting ecosystem responses to elevated CO_2 concentrations. *BioScience* **41,** 96–104.

Mooney, H. A., Canadell, J., Chapin, F. S., Ehleringer, J., Körner, McMurtrie, R., Parton, W. J., Pitelka, L., and Schulze, E.-D. (1998). Ecosystem physiology responses to global change. *In* "Implications of Global Change for Natural and Managed Ecosystems: A Synthesis of GCTE and Related Research" (B. H. Walker, W. L. Steffen, J. Canadell, and J. S. I. Ingram, eds.), Cambridge University Press, Cambridge, UK.

Morison, J. I. L. (1985). Sensitivity of stomata and water use efficiency to high CO_2. *Plant Cell Environ.* **8,** 467–474.

Morison, J. I. L. (1993). Responses of plants to CO_2 under water limited conditions. *Vegetatio* **104/105,** 193–209.

Myers, B. J., and Talsma, T. (1992). Site water balance and tree water status in irrigated and fertilised stands of *Pinus radiata*. *For. Ecol. Manage.* **52,** 17–42.

Owensby, C. E., Ham, J. M., Knapp, A., Rice, C. W., Coyne, P. I., and Auen, L. M. (1996). Ecosystem-level responses of tallgrass prairie to elevated CO_2. *In* "Carbon Dioxide and Terrestrial Ecosystems" (G. W. Koch and H. A. Mooney, eds.), pp. 147–162. Academic Press, San Diego.

Pan, Y., Melillo, J. M., McGuire, A. D., Kicklighter, D. W., Pitelka, L. F., Hibbard, K., Pierce, L. L., Running, S. W., Ojima, D. S., Parton, W. J., Schimel, D. S., and other VEMAP Members. (1997). Modeled responses of terrestrial ecosystems to elevated atmospheric CO_2: A comparison of simulation studies among the biogeochemistry models of the vegetation/ecosystem modeling and analysis project (VEMAP). *Oecologia* **114,** 389–404.

Parton, W. J., Schimel, D. S., Cole, C. V., and Ojima, D. S. (1987). Analysis of factors controlling soil organic matter levels in Great Plains grasslands. *Soil Sci. Soc. Am. J.* **51,** 1173–1179.

Picon, C., Guehl, J. M., and Aussenac, G. (1996). Growth dynamics, transpiration and water-use efficiency in *Quercus robur* plants submitted to elevated CO_2 and drought. *Ann. Sci. For.* **53,** 431–446.

Polley, J. W., Johnson, H. B., Marino, B. D., and Mayeux, H. S. (1993). Increase in C_3 plant water-use efficiency and biomass over Glacial to present CO_2 concentrations. *Nature* **361,** 61–63.

Rastetter, E. B., Ryan, M. G., Shaver, G. R., Melillo, J. M., Nadelhoffer, K. J., Hobbie, J. E., and Aber, J. D. (1991). A general biogeochemical model describing the responses of the C and N cycles in terrestrial ecosystems to changes in CO_2, climate and N deposition. *Tree Physiol.* **9,** 101–126.

Rastetter, E. B., McKane, R. B., Shaver, G. R., and Melillo, J. M. (1992). Changes in C storage by terrestrial ecosystems: How C–N interactions restrict responses to CO_2 and temperature. *Water Air Soil Pollut.* **64,** 327–344.

Ryan, M. G., Hunt, E. R., McMurtrie, R. E., Ågren, G. I., Aber, J. D., Friend, A. D., Rastetter, E. B., Pulliam, W. M., Raison, R. J., and Linder, S. (1996a). Comparing models of ecosystem function for temperate conifer forests. I. Model description and validation. *In* "Global Change: Effects on Coniferous Forests and Grasslands" (A. I. Breymeyer, D. O. Hall, J. M. Melillo, and G. I. Ågren, eds.), Scientific Committee on Problems of the Environment, Vol. 56, pp. 313–362. John Wiley and Sons Ltd., Chichester, UK.

Ryan, M. G., McMurtrie, R. E., Ågren, G. I., Hunt, E. R., Aber, J. D., Friend, A. D., Rastetter, E. B., and Pulliam, W. M. (1996b). Comparing models of ecosystem function for temperate conifer forests. II. Simulations of the effect of climate change. *In* "Global Change: Effects on Coniferous Forests and Grasslands" (A. I. Breymeyer, D. O. Hall, J. M. Melillo, and G. I. Ågren, eds.), Scientific Committee on Problems of the Environment, Vol. 56, pp. 363–387. John Wiley and Sons Ltd., Chichester, UK.

Schimel, D. S. (1990). Biogeochemical feedbacks in the earth system. *In* "Global Warming: The Greenhouse Report" (J. Leggett, ed.), pp. 68–82. Oxford University Press, Oxford.

Schimel, D. S. (1995). Terrestrial ecosystems and the carbon cycle. *Global Change Biol.* **1,** 77–91.

Shugart, J. J., Antonovsky, M. J., Jarvis, P. G., and Sandford, A. P. (1986). CO_2, Climatic change and forest ecosystems. *In* "The Greenhouse Effect, Climate Change and Ecosystems" (B. Bolin, B. R. Döös, J. Jäger, and R. A. Warrick, eds.), Scientific Committee on Problems of the Environment, Vol. 56, pp. 475–521. John Wiley and Sons Ltd., Chichester, UK.

Thornley, J. H. M., and Cannell, M. G. R. (1996). Temperate forest responses to carbon dioxide, temperature and nitrogen: A model analysis. *Plant Cell Environ.* **19,** 1331–1348.

Tissue, D. T., and Oechel, W. C. (1987). Response of *Eriophorum vaginatum* to elevated CO_2 and temperature in the Alaskan tussock tundra. *Ecology* **68,** 401–410.

III

Synthesis and Summary

15

Diverse Controls on Carbon Storage under Elevated CO_2: Toward a Synthesis

Christopher B. Field

Results of recent experiments and models provide a solid foundation for understanding some critical aspects of ecosystem responses to rising atmospheric CO_2. That foundation is more solid for some ecosystems than for others. It is generally more complete for agricultural systems and substantially less solid for long-term responses of unmanaged ecosystems, especially for changes that involve rising CO_2 in combination with other anthropogenic forcings. For a number of processes with large leverage over both the sign and magnitude of ecosystem CO_2 responses, the foundation ranges from rickety, to flimsy, to nonexistent. In some ways, the history of recent CO_2 research is a clear success, combining creative individual researchers, dedicated international attention, and responsive funding agencies. But in others, it is a history of working with blinders, which allowed the community to focus intensively on some components of the problem while ignoring others. These blinders were useful, even essential, in the early stages of CO_2 research. Now, it is critical that the blinders be set aside and that we consider the full range of ecosystem CO_2 responses. In the coming years, we will need to address carbon residence times, species changes, hydrologic discharge, climate feedbacks, and disturbance with the intensity currently devoted to photosynthesis, plant respiration, and growth.

I. Impacts of Rising Atmospheric CO_2

Most of the existing literature on plant and ecosystem responses to elevated CO_2 emphasizes a narrow slice of the relevant mechanisms and response variables. Studies on the input side of the carbon balance, photosynthesis and net primary production (NPP), have been the dominant themes of CO_2 research. Although these are clearly related to the primary mechanisms through which elevated CO_2 alters ecosystem processes, they are just as clearly only part of the story. The identities and roles of the other actors depend on the perspective of the inquiry.

Ecosystems provide a number of critical goods and services (Costanza *et al.*, 1997; Daily, 1997; Ehrlich and Mooney, 1983). For agricultural ecosystems, the value of these is clear. It is based on the economic value of the crops produced, correcting for inputs from costs of management. For ecosystems that are managed less intensively, the goods and services are more diverse, ranging from harvestable wood products to watershed protection to pest control. The value of these is difficult to calculate, but potentially immense. One recent estimate places the current economic value of ecosystem services in excess of the output of all the world's economies (Costanza *et al.*, 1997).

A broad range of ecosystem services is potentially CO_2 sensitive. Net primary production and carbon storage are but two examples. Other examples that are much less studied include freshwater resources (Hatton *et al.*, 1992; Idso and Brazel, 1984), climate (Betts *et al.*, 1997; Sellers *et al.*, 1996), the abundance and distribution of valuable species and genetic resources, and aesthetics or a sense of place. None of these services can be accurately quantified in isolation, and the greatest value of one or more may emerge through their impacts on other services. For example, changes in biome boundaries in response to elevated CO_2 could have large impacts on NPP and carbon storage. Subtle changes, like altered flammability or sensitivity to a pest or pathogen, could have major impacts on ecosystem services ranging from carbon storage, to maintenance of soil fertility, to sensitivity to biological invasion.

Given the potentially large value of ecosystem services and the diversity of ecosystem CO_2 responses, it is appropriate to ask whether the current distribution of research activities addresses the range and relative importance of the scientific issues. The answer to this question cannot be definitive at this time, partly because the values of the ecosystem services are still uncertain, and partly because the magnitude of the CO_2 effects on the services cannot be determined without targeted research. Still, the connections between CO_2 effects and ecosystem services are clear enough that they should inform syntheses of current and future research.

The diversity of ecosystem goods and services that are potentially CO_2 sensitive argues for a broad approach to designing and interpreting research. Sufficient breadth to capture all of the major processes should also be a priority for the research efforts focused on individual goods or services. This is not a plea for every project to consider every process related to CO_2 responses. But it is critical for the teams and individuals working on these issues to define the problems broadly, to appreciate the relevance of processes that may be outside the expertise or interest of any particular research group, and to encourage the community to support a range of coordinated activities. To illustrate the need for breadth, this paper addresses some of the processes that control ecosystem carbon storage, one of the most discussed, but not one of the best studied, CO_2 responses. In the following paragraphs, I use simple models to explore the role of a number of potentially high-impact processes that merit increased future attention. My focus on these processes is intended to highlight rather than survey, just as the focus on carbon storage is intended as an example of the diversity of factors that impinge on a major ecosystem good or service.

II. NPP and Carbon Storage

The value of the ecosystem services connected with NPP and carbon storage depends on perspective. The economic value of the world's crops is roughly \$1.4 trillion yr^{-1} (Naylor and Ehrlich 1997) or about 8% of global gross domestic product (GDP). If a CO_2 doubling led to a 30% increase in the NPP and crop yield, and if economic income scaled with crop yield, then income from agriculture could increase by more than \$400 billion yr^{-1}. Though optimistic, these estimates still capture only one aspect of a multidimensional issue. For the millions of people currently seriously malnourished, a CO_2-dependent increase in crop yield could spell the end of hunger. On the other hand, problems of access and distribution could concentrate the increased NPP on regions and socioeconomic groups already well nourished and suffering more from the consequences of crop surpluses than shortages.

The value of carbon storage is one of the main topics under discussion in the negotiations related to the Framework Convention on Climate Change. Transferable carbon credits, or payments for forgoing emissions, represent an explicit valuation. If the world's governments agree to emissions credits in the range of \$50–120 ton^{-1}, then the value of current terrestrial sink of approximately 1.6 Pg yr^{-1} (Schimel *et al.*, 1995) is approximately \$80–190 billion yr^{-1}. Another way to look at the value of the current terrestrial carbon storage is that the world's economies could produce the same consequences for the atmosphere through cutting carbon emissions between 20 and 25%. If GDP were proportional to carbon emissions, then cutting carbon emissions by this amount would imply forgoing about \$4 trillion of yearly economic activity. With either approach to evaluating terrestrial carbon storage, the value is great enough to place a high priority on a complete understanding.

Numerous simulation models address the details of the relationship between the dynamics of NPP and C storage, including effects of temperature, moisture, etc. The fundamentals are, however, quite simple and can be usefully explored with a very simple model. In fact, a simple model can have the profound advantage of distinguishing the aspects of the responses that are fundamental to all ecosystems from those that reflect interactions among sometimes subtle details. Here, I consider the behavior of a simple model of an ecosystem with annual plants. This model is based on three key mechanisms that come as close as anything in ecology to general laws (Fig. 1). First, decomposition is first order with respect to the quantity of carbon in the decomposing pool. Doubling the size of a pool doubles the amount of carbon released in any time period. Second, photosynthesis and growth are often sensitive to nutrient limitation. Decreasing nutrient supply reduces NPP. Third, carbon can be stored in ecosystems in pools with turnover lines ranging from less than a year to several centuries.

Carbon storage (cumulative: g m^{-2}) or net ecosystem production (NEP, the rate of storage: g m^{-2} yr^{-1}) in response to elevated CO_2 is critically dependent on the relative dynamics of production and decomposition. When NPP rises, either as a step or gradually, it tends to get ahead of the equilibrium with carbon losses, resulting in positive NEP (Fig. 2). For a given rate of NPP increase, the NEP is greater when the decomposing pool has a longer turnover time (Thompson *et al.*, 1996). As long as NPP

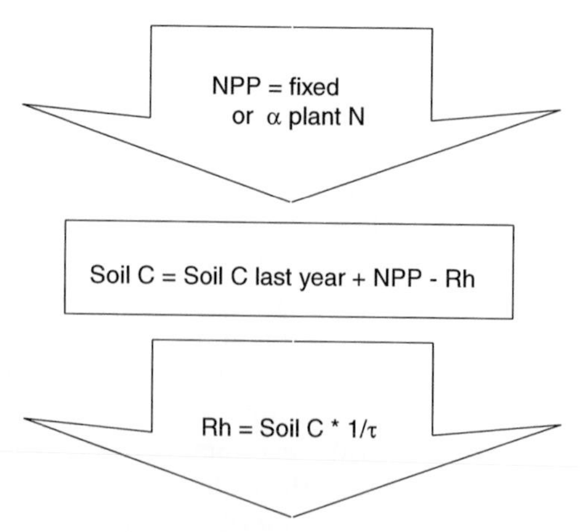

Figure 1 Schematic of the simple carbon model used for these simulations. The model assumes an annual plant community, with 100% of the NPP transferred to the soil pools at the end of each year. Heterotrophic respiration (Rh) is first order with respect to soil C. It is given by the ratio of soil C to the C residence time (τ). NEP is soil and plant C in year n minus soil and plant C in year $n - 1$.

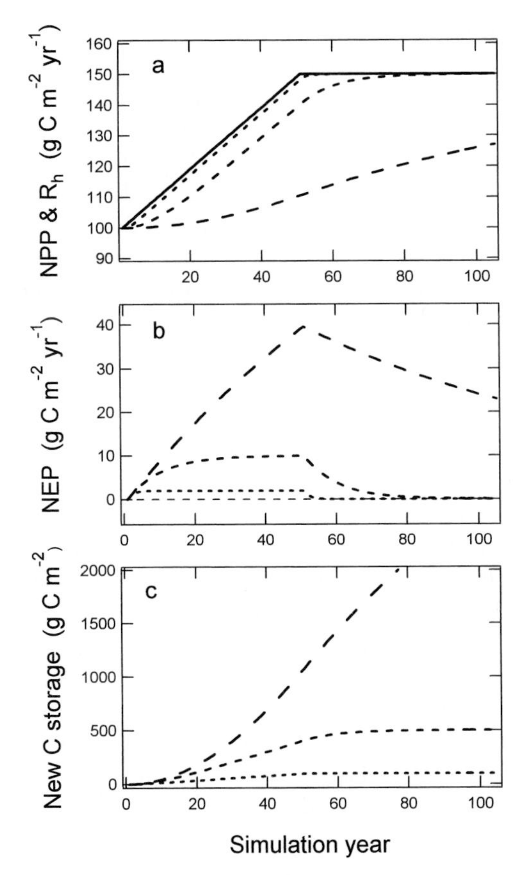

Figure 2 C dynamics resulting from gradual changes in NPP. For simulations with (a) NPP increasing by 50% over 50 yr (solid line) and heterotrophic respiration for ecosystems with C residence times of 1 yr (short dashes), 10 yr (medium dashes), and 100 yr (long dashes). (b) NEP (the rate of carbon storage) for simulated ecosystems with C residence times of 1 yr (short dashes), 10 yr (medium dashes), and 100 yr (long dashes). (c) cumulative C storage (the time integral of NEP) for ecosystems with C residence times of 1 yr (short dashes), 10 yr (medium dashes), and 100 yr (long dashes).

continues to increase, the positive NEP can persist. But if NPP stabilizes, even if it is at the maximum reached directly after a CO_2 doubling, carbon losses eventually reach the same level, and NEP or new storage drops to zero. After the NEP drops to zero, the ecosystem will still contain more carbon than it did initially, but the carbon content will not continue to increase. If carbon is stored in a pool with very slow turnover, the amount of carbon stored may be large, and the time required for the storage rate to decay to zero can be many decades. But the first-order nature of decomposition insures that the NEP or storage rate eventually drops to zero, except in unusual circumstances where organic matter is effectively removed from the local carbon cycle. Examples of this removal include freezing in permafrost or trapping in anoxic layers deep in a bog.

A step increase in NPP that is sustained indefinitely does not typically imply a sustained carbon sink (Fig. 3). Because heterotrophic respiration tends to scale with the total quantity of respirable material, increases in the pool of soil organic matter lead to increased respiration, until the outflows balance the inflows and accumulation ceases. Positive NEP with increased storage in litter and soil organic matter may continue for decades, but not indefinitely. This result is clear from many simulation studies. At least in essence, it is almost certainly correct, given the simple processes involved.

Yet, this response is not clear in results from experimental studies. Why? The explanation has at least three components. First, for reasonable turnover rates of the carbon pools, the decay of the NEP should occur over one to several decades. Few if any of the experimental projects have operated long enough for increases in the size of the respiring pools to depress NEP significantly. Second, because the annual NEP is typically a small fraction of the total ecosystem carbon pool, it is likely to be detectable from pool data only when integrated over several years of observations (Hungate *et al.*, 1996b). An integration that includes both an initial period of high NEP and a later period of falling NEP will yield a single average value between the initial and the final rates. Only with a second integration interval should it be possible to detect declining NEP. Third, interannual

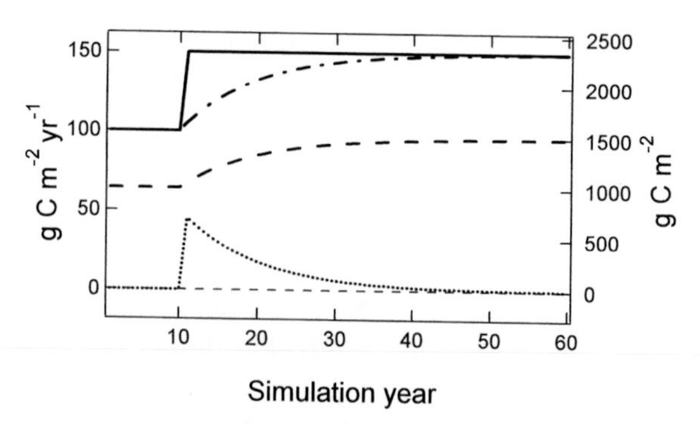

Figure 3 C dynamics associated with a step change in NPP, where some treatment (like a CO_2 doubling) leads to a sustained 50% increase in NPP, beginning in year 10, for an ecosystem with a C residence time of 10 yr. NPP (solid line), heterotrophic respiration (dash-dot line), and NEP (dotted line) are plotted relative to the left axis. Total soil C (dashed line) is plotted relative to the right axis.

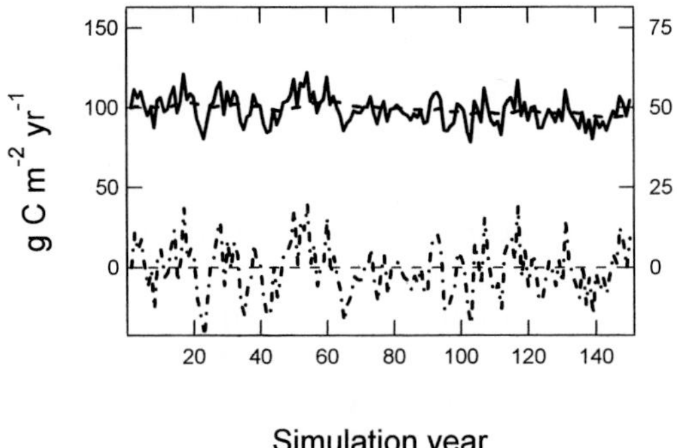

Figure 4 Dynamics of NEP (dash-dot line; right axis) and heterotrophic respiration (dashed line; left axis) resulting from interannual variation in NPP (solid line; left axis). The NPP variation is random, with a maximum value of 20%.

variation in NPP can produce large excursions in NEP, which can potentially obscure progressive changes (Fig. 4).

III. Carbon Turnover Dynamics

If elevated CO_2 does not lead to changes in the distribution of carbon among pools, the cumulative change in carbon storage at equilibrium, when NEP falls to zero following a persistent increase in NPP, is proportional to the magnitude of the NPP response. This is a simple consequence of the first-order nature of respiration. A 5% NPP increase leads to a 5% cumulative increase in C storage, and a 50% NPP increase leads to a 50% cumulative increase in C storage. While this relationship highlights the role of the NPP response in carbon storage, it also identifies the role of initial carbon stocks, which are controlled by residence time. When the residence time is long, initial storage is a large multiple of NPP, and additional storage from an NPP increase is large. When residence time is short, cumulative storage per NPP increase is also small.

In real ecosystems, carbon residence times span a broad range, from days or weeks for easily metabolizable cell contents to centuries or millenia for organic matter that is physically protected by close association with clay mineral structures. In the CENTURY model of Parton and colleagues (Parton *et al.*, 1987, 1993), this diversity of residence times is represented in a number of pools of soil organic matter, litter, and living biomass. The PnET model of Aber and Federer (1992) represents this diversity by tracking individual cohorts of SOM and assigning time-varying decomposition coefficients. ^{14}C provides powerful tools for verifying these residence times

(Trumbore *et al.*, 1996), especially when coupled with techniques for partitioning the soil organic matter into fractions of differing decomposability. The ratio of total ecosystem carbon to NPP is an estimate of the average age of carbon in all pools, but this estimate is often quite misleading as a basis for estimating additional C storage from the NPP increase. A large pool of very old carbon can dramatically extend the mean age without contributing significantly to carbon dynamics on the timescale of a few centuries (Thompson and Randerson, 1999).

If the ecosystem-scale response to elevated CO_2 involves changes in the distribution of carbon among pools with different turnover times, the implications for carbon storage can be profound. A partitioning of all or even most of an NPP increase to a pool with short residence time results in minimal storage, while directing extra carbon to a pool with long residence time leads to large amounts of storage (Fig. 5). The term controlling the incremental storage is the partitioning of the extra NPP. Since this is only a fraction of the total NPP, changes are intrinsically difficult to observe and may be clear only after extended periods. One potentially confusing aspect of this control on C storage is that many changes in partitioning will not affect C storage in the first few years following a step increase in NPP. Distribution of extra NPP to a pool with a residence time of 10 instead of 100 yr cuts incremental C storage 10-fold, but is probably not visible until near the end of 10 yr.

Observational evidence on the partitioning of new NPP is limited. Hungate *et al.* (1997b) inferred partitioning of extra carbon to short-residence pools in annual grassland. Hendrick and Pregitzer (1996) report increased allocation to root turnover, providing circumstantial evidence for increased allocation to short-lived, readily decomposable components. Of course, the partitioning of incremental NPP to woody stems or other long-lived tissues resistant to decomposition could result in large amounts of storage per unit of NPP increase.

IV. Nutrient Limitation

The role of nutrient limitation in modulating the CO_2 responses of NPP and carbon storage is one of the most widely misunderstood and poorly studied issues in ecosystem ecology. This is a topic where proponents of both positions often pass in the dark, with each group basing their analysis on concepts and conditions so far from the mind-set of the other that communication is minimal. The terms are similar, but the time frames and mechanisms are largely nonoverlapping. Experimental evidence for suppressed responses of NPP to elevated CO_2 under nutrient limitation is mixed. Some studies report proportionally smaller NPP responses when

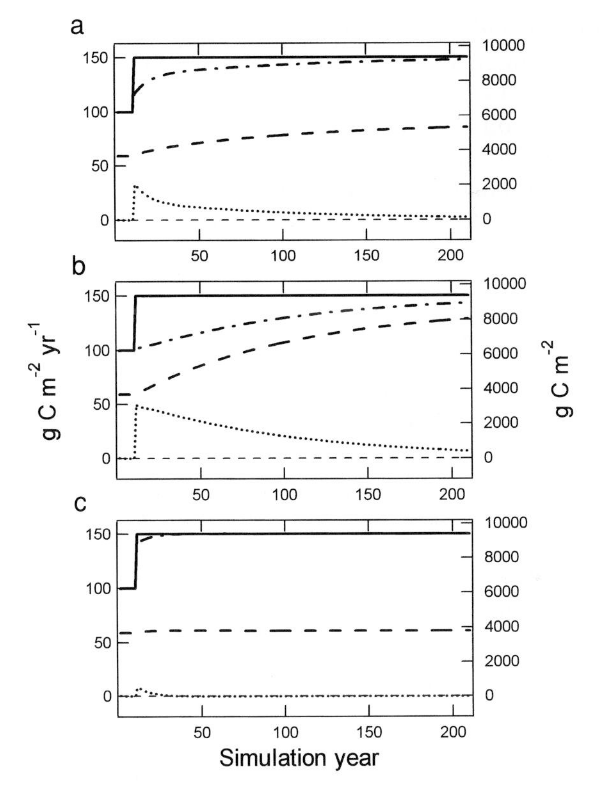

Figure 5 C dynamics associated with a sustained 50% increase in NPP, beginning in year 10, with a range of C allocations among pools with different residence times. Prior to the NPP increase, 1/3 of the NPP is allocated to each of three pools, with C residence times of 1, 10, and 100 yr. In (a), each pool continues to receive 1/3 of the NPP. In (b), all of the additional NPP goes to the pool with the 100-yr residence time. In (c), 83% of the additional NPP goes to the pool with a residence time of 1 yr and 17% goes to the pool with a residence time of 10 yr. For all panels, the traces are NPP (solid line), heterotrophic respiration (dash-dot line), NEP (dotted line) (all plotted relative to the right axis), and total soil C (dashed line, plotted relative to the right axis).

nutrients are limiting (Larigauderie *et al.*, 1988) but others do not (Idso and Idso, 1994). Lloyd and Farquhar (1996) use physiologically based criteria to argue that NPP responses to elevated CO_2 should increase with nutrient limitation. Yet, in many biogeochemistry models, including all three in VEMAP 1 (VEMAP Members, 1995), the long-term responses of NPP and carbon storage to elevated CO_2 are strongly constrained by N. The mechanism driving this response is a progressive decrease in nitrogen availability resulting from an increased pool of organic nitrogen immobilized in biomass and soil organic matter (Fig. 6). Unless elevated CO_2 is accompanied by increased N inputs, decreased N losses, or increased partitioning of N to high C:N pools within the ecosystem, decreased N availability in response to increased N storage is essentially unavoidable.

The model-based conclusion that increased N storage leads to decreased N availability is fundamentally different from an empirical test of the response of NPP to elevated CO_2 under limiting versus abundant nutrients.

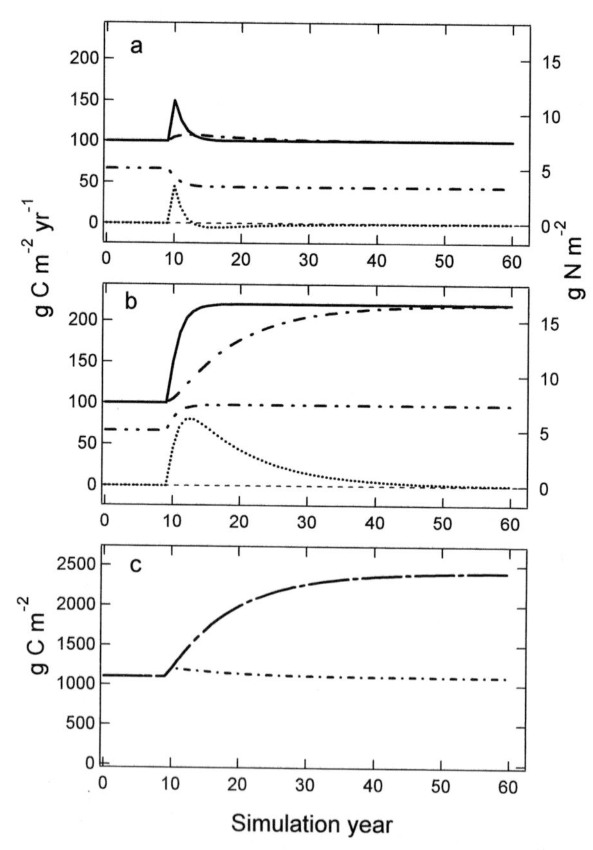

Figure 6 C dynamics associated with a step change in NPP, with N limitation or N additions. In (a), NPP increases by 50% in year 10, when the plant C:N ratio increases by 50%. Because the total N in plants and soil is fixed, any transfer of N to the soil reduces N availability for the plant and constrains NPP. In (b), the plant C:N ratio increases at year 10, but N fixation or deposition adds N in an amount equal to 40% of the plant N every year. For both (a) and (b), the traces are NPP (solid line), heterotrophic respiration (dash-dot line), NEP (dotted line) (all plotted relative to the left axis), and total plant N (dash-double-dot line, plotted relative to the right axis). Panel (c) shows the total ecosystem C (plant plus soil) for the case with N limitation (short dash-dot line) and N additions (long dash-dot line).

The essence of the mechanism represented in the biogeochemistry models is that, regardless of initial availability, the availability of N and other nutrients decreases under elevated CO_2 as a consequence of increased carbon storage. It is the progressive decrease in nutrient availability that constrains the responses of NPP and carbon storage, not the initial levels.

A number of experimental studies have asked whether elevated CO_2 alters N availability. The results of these studies are mixed, with some evidence for decreases due to decreased decomposability (Díaz *et al.*, 1993), increases due to increased carbon substrate (Zak *et al.*, 1993), increases due to increased soil moisture (Hungate *et al.*, 1997a), and plant species effects (Hungate *et al.*, 1996a). Although interesting and important, none of these results is directly relevant to the issue of progressive nutrient limitation from increased carbon storage. These experimental studies on changes in nutrient availability under elevated CO_2 address the early stages of the CO_2 response, before ecosystem carbon stocks can change substantially. The biogeochemistry models that predict progressive nutrient limitation under elevated CO_2 are largely insensitive to changes in litter decomposability (Mooney *et al.*, 1998), reflecting the relative importance of the long-, medium-, and short-term controls on nutrient availability.

The primary controls on N availability on the time frame of decades to millennia are inputs, losses, and immobilization in the large pools in the ecosystem, especially soil organic matter. Current models are far from satisfactory in their treatment of these primary controls on long-term nitrogen limitation. Among the biogeochemistry models that treat the N cycle at all, some assume that the total stock of ecosystem N is constant (e.g., McMurtrie and Comins, 1996). Others simulate inputs and losses, but often with simple assumptions about inputs. For example, N inputs are sometimes assumed constant (e.g., Running and Hunt, 1993) or proportional to some aspect of the water budget (e.g., Parton *et al.*, 1993). The comprehensive treatments of N inputs through deposition (Townsend *et al.*, 1996) have not been coupled with a thorough treatment of CO_2 responses.

Nitrogen inputs through biological fixation are not treated thoroughly in any of the major biogeochemistry models. This is a potentially serious gap, given the hypothesis that symbiotic N fixers are frequently energy limited (Vitousek and Howarth, 1991) and, therefore, candidates for enhanced ecological success under elevated CO_2. This possibility is consistent with model simulations (Vitousek and Field, 1999) and with observations suggesting that average responses of NPP to a CO_2 doubling may be greater in legumes than in nonlegumes (Poorter *et al.*, 1996). It is also consistent with observations documenting increases in the relative abundance of N fixers in multispecies communities grown under elevated CO_2 (Hebeisen *et al.*, 1997; Lüscher *et al.*, 1996) and with evidence for N transfer from fixers to nonfixers (Zanetti *et al.*, 1997).

The sensitivity of N losses to elevated CO_2 is poorly studied. Under experimental conditions with abundant nutrients, doubled CO_2 led to increased microbial N immobilization and decreased emissions of NO following the onset of the rainy season (Hungate *et al.*, 1997c). But under other conditions, elevated CO_2 led to increased denitrification and N loss (Smart *et al.*, 1997). Some models assume that N losses decrease under elevated CO_2 because plants can use additional carbon to improve their ability to compete for nutrients (Raich *et al.*, 1991).

If NPP is dependent on plant N and if increased inputs, decreased losses, or N redistribution lead to increased plant N under elevated CO_2, the nutrient effect could go in the opposite direction, with progressive increases in carbon storage (Fig. 6). In a real ecosystem, the C stocks would certainly not continue to increase indefinitely. Another factor or factors, for example, light, another nutrient, or water, would place an upper bound on the N stimulation of the CO_2 response. Yet the possible role of mechanisms that could decrease N limitation under elevated CO_2 clearly warrants further exploration.

V. Disturbance

Disturbance is a giant lever in the global carbon balance. Changes in the frequency or intensity of the disturbances that reset ecosystem carbon stocks and population dynamics could lead to dramatic increases or decreases in terrestrial carbon storage and other ecosystem goods and services. Yet disturbance is completely untreated in most of the experiments and models designed to explore the future of the carbon cycle.

Consider a forest situation in which elevated CO_2 leads to a 25% increase in NPP, a 25% decrease in the time to canopy closure, and a 25% increase in wood accumulation. If these responses or other changes also produce a 25% decrease in the mean interval between fires, the average C storage over a mix of plots in all stages of regrowth remains unchanged, with all of the increased C inputs appearing as increased losses through fire. On the other hand, factors that extend the disturbance cycle under elevated CO_2, from improved water relations to decreased sensitivity to a pest, could increase terrestrial carbon storage.

When the probability of disturbance increases with biomass, as with fire cycles in many regions (McKenzie *et al.*, 1996), disturbance may largely eliminate CO_2-stimulated changes in carbon storage. Where fires are suppressed, additional C storage from elevated CO_2 becomes more likely, though the benefits of the added storage may be offset by the extra costs of fire fighting. Where fires are rare, outbreaks of pests and pathogens might or might not respond to biomass accumulation. For disturbances that are independent of biomass, for example hurricanes, elevated CO_2 is unlikely to alter either frequency or intensity through its effects on the plants. Climate change may, however, increase hurricane frequency and/or intensity (Knutson *et al.*, 1998), perhaps offsetting direct effects of CO_2.

VI. Ecological Dynamics

Most experimental studies with elevated CO_2 extend over less than a single generation of the dominant plants. Yet, a broad range of ecosystem services is potentially sensitive to species composition. From NPP and carbon storage, through runoff, soil protection, flammability, and aesthetics, species changes could be major actors, with their roles enhanced in some cases by other global changes.

In principle, changes in dominance under elevated CO_2 could enhance or degrade ecosystem services. At present, we have relatively little foundation for predicting either the schedule or the direction of species changes under elevated CO_2. Changes in the frequency of disturbance could be a major control on the life forms represented (Wardle *et al.*, 1997). The availability and abundance of potential invaders could be critically important, especially if fragmentation or large distances impact the colonization ability of local natives.

General patterns of changes in dominance in response to elevated CO_2 are largely unknown, as a consequence of the limited duration of the observations. At least two potentially general patterns are, however, beginning to emerge. First, in grassland or savanna ecosystems, elevated CO_2 tends to favor species that depend on late season moisture reserves. This class includes C_4 perennial grasses in Kansas (Owensby *et al.*, 1996), shrubs in Texas (Polley *et al.*, 1994), and late season C_3 annuals in California (Chiariello and Field, 1996; Field *et al.*, 1996). At least in grasslands, CO_2 effects on the water budget appear to boost the dominance of the species active during the period when drought risk is greatest. Second, elevated CO_2 may consistently enhance the relative success of symbiotic N fixers, especially in settings where N fixation is energy limited. This pattern has been observed in Swiss pasture ecosystems (Hebeisen *et al.*, 1997), and California annual grasslands (N. R. Chiariello, unpublished data). Other examples of changes in species dominance or competitive success under elevated CO_2 are difficult to generalize, because they appear to depend on a range of features of the individual species involved (Bazzaz, 1990).

Few experiments have tracked multispecies populations through several generations under elevated CO_2. If differential reproduction follows differential growth and competitive success, it is possible that elevated CO_2 can lead to major changes in ecosystem structure and function. Increased areas dominated by woody plants and increased abundance of symbiotic N fixers are two possibilities with the potential for high impact on the carbon cycle. Other species changes could have profound effects on other ecosystem services, ranging from the loss of economically important species to biological invasions that degrade grazing, water, or recreational services.

Changes in dominance at other trophic levels may also be important. The responses of microbial community structure can be profound, including altered abundance of protozoans (Lussenhop *et al.*, 1998; Treonis and Lussenhop, 1997) and changes in the relative success of mycorrhizal versus nonmycorrhizal fungi (Rillig *et al.*, 1997). Altered structure of the soil microbial community could impact both the partitioning of nutrients between plants and microbes and the decomposability of litter and soil organic matter. CO_2-dependent changes in the success of pathogens (Malmström and Field, 1997) and herbivores (Lincoln *et al.*, 1993; Lindroth, 1996) could precipitate further restructuring of the plant community.

The current generation of biogeography models (e.g., Neilson, 1995; Prentice *et al.*, 1992; Woodward *et al.*, 1995) and the next generation of dynamic global vegetation models are a fantastic advance in the community's ability to represent species dynamics and integrate these with biogeochemistry. The indication that shifts in biome boundaries can be important in the responses of NPP and carbon storage to elevated CO_2 (VEMAP Members, 1995) is a solid start. Sophisticated gap-scale models are already providing insights into possible changes in within-ecosystem community composition under elevated CO_2 (Bolker *et al.*, 1995). Next steps will include integrating these approaches and combining them with strategies for representing human impacts on ecological dynamics.

VII. Conclusions

Most of the current knowledge base on plant and ecosystem responses to elevated CO_2 concerns the input side of the carbon balance, especially photosynthesis and plant growth or NPP. While these quantities are relevant to carbon storage and a range of other ecosystem services, they are an insufficient foundation for a comprehensive understanding. A number of processes have the potential for large leverage on the long-term impacts of elevated CO_2, possibly changing the sign as well as the magnitude of the response of a range of critical ecosystem services.

Two classes of actions can improve the ability of the research community to deal with these high-leverage processes. First, it is important to consider ecosystem-scale CO_2 responses in the context of the full range of potentially sensitive ecosystem goods and services. The past focus on carbon balance has been very productive, but it has failed to encourage critical thinking about other issues, ranging from economically valuable species to runoff to recreational values. Especially during the next few decades, changes in these other goods and services may be as important or even more important than carbon storage.

Second, it is essential to develop an improved set of experiments, models, and time-series observations that provide improved access to the long-term components of ecosystem CO_2 responses. For experiments, three kinds of strategies should be pursued. First, we should take advantage of model systems that exhibit the full range of dynamics in a reasonable time period. Powerful models can be synthetic ecosystems with tailored communities and soils, or they can be natural communities with short-lived dominants. Second, increased emphasis on tracers, including stable isotopes, can provide critical insights into small changes in large pools (Hungate *et al.*, 1996b). Practical carbon tracers for ambient CO_2 treatments should be a top priority. Third, there is no alternative to running an experiment long enough for the full range of dynamics to emerge. Just as the community has benefited from long-term fertilizer plots, we need long-term CO_2 plots, operated for at least several generations of the dominant plants. Among the issues that will need to be addressed before establishing such an experiment, access (by invaders, herbivores, pests, pathogens), management (by harvesting, spraying, etc.), and control of disturbance (by fire, hurricanes, etc.) are critical.

Improved models for dealing with the full range of ecosystem CO_2 responses will need to emphasize the interactions among biogeochemistry, ecological dynamics, and disturbance. The treatment of nutrient inputs and outputs is a key area, as is a strategy for representing the limits to N fixation. Modelers and experimentalists need to collaborate to develop a comprehensive picture of the generalities in ecosystem CO_2 responses, so the models effectively capture differences among ecosystems, climate zones, and resource environments.

Even a major investment in experiments and models will still require many years before they lead to comprehensive understanding of the long-term components of CO_2 responses. Observations on other sites, including CO_2 springs, long-term fertilizer plots, and areas with fortuitous transitions (e.g., C_3 to C_4, etc.) can provide access to key processes, often in many more sites than it is practical to manipulate.

Finally, elevated CO_2 will occur in the context of diverse global changes. In some cases, it may be appropriate to infer interactions between responses to elevated CO_2 and other global changes. In general, however, we need to begin thinking about a comprehensive strategy for blurring the boundary between CO_2 research and the broader framework of global change research.

Acknowledgments

Thanks are due the U.S. National Science Foundation (BSR 90-20134, DEB 97-27059) for supporting work with elevated CO_2 on Jasper Ridge and to the U.S. Department of Energy for support of studies to evaluate biogeochemistry models. The foundations for the ideas presented here came from discussions with Jim Randerson, Matt Thompson, Nona Chiariello, Yiqi Luo, Dave Schimel, Bruce Hungate, Fakhri Bazzaz, and Peter Vitousek.

References

Aber, J. D., and Federer, C. A. (1992). A generalized, lumped parameter model of photosynthesis, evapotranspiration and net primary production in temperate and boreal forest ecosystems. *Oecologia* **92,** 463–474.

Bazzaz, F. A. (1990). The response of natural ecosystems to the rising global CO_2 levels. *Annu. Rev. Ecol. System.* **21,** 167–196.

Betts, R. A., Cox, P. M., Lee, S. E., and Woodward, F. I. (1997). Contrasting physiological and structural vegetation feedbacks in climate change simulations. *Nature* **387,** 796–799.

Bolker, B. M., Pacala, S. W., Bazzaz, F. A., Canham, C. D., and Levin, S. A. (1995). Species diversity and ecosystem response to carbon dioxide fertilization: Conclusions from a temperate forest model. *Global Change Biol.* **1,** 373–381.

Chiariello, N. R., and Field, C. B. (1996). Annual grassland responses to elevated CO_2 in long-term community microcosms. *In* "Community, Population and Evolutionary Responses to Elevated Carbon Dioxide Concentration" (C. K. and F. A. Bazzaz, eds.), pp. 139–157. Academic Press, San Diego.

Costanza, R., d'Arge, R., de Groot, R., Farber, S., Grasso, M., Hannon, B., Limburg, K., Naeem, S., O'Neill, R. V., Paruelo, J., Raskin, R. G., Sutton, P., and van den Belt, M. (1997). The value of the world's ecosystem services and natural capital. *Nature* **387,** 253–260.

Daily, G. C. (eds.). (1997). "Nature's Services." Island Press, Washington, DC.

Díaz, S., Grime, J. P., Harris, J., and McPherson, E. (1993). Evidence of a feedback mechanism limiting plant response to elevated carbon dioxide. *Nature* **364,** 616–617.

Ehrlich, P. R., and Mooney, H. A. (1983). Extinction, substitution, and ecosystem services. *BioScience* **33,** 248–254.

Field, C. B., Chapin, F. S., III, Chiariello, N. R., Holland, E. A., and Mooney, H. A. (1996). The Jasper Ridge CO_2 experiment: Design and motivation. *In* "Carbon Dioxide and Terrestrial Ecosystems" (G. W. Koch and H. A. Mooney, eds.), pp. 121–145. Academic Press, San Diego.

Hatton, T. J., Walker, J., Dawes, W. R., and Dunin, F. X. (1992). Simulations of hydroecological responses to elevated CO_2 at the catchment scale. *Aust. J. Bot.* **40,** 679–696.

Hebeisen, T., Luscher, A., Zanetti, S., Fischer, B., Hartwig, U. A., Frehner, M., Hendrey, G. R., Blum, H., and Noseberger, J. (1997). Growth responses of *Trifolium repens* L. and *Lolium perenne* L. as monocultures and bi-species mixtures to free air CO_2 enrichment and management. *Global Change Biol.* **3,** 149–161.

Hendrick, R. L., and Pregitzer, K. S. (1996). Temporal and depth-related patterns of fine root dynamics in northern hardwood Forests. *J. Ecol.* **84,** 167–177.

Hungate, B. A., Canadell, J., and Chapin III, F. S. (1996a). Plant species mediate changes in soil microbial N in response to elevated CO_2. *Ecology* **77,** 2505–2516.

Hungate, B. A., Jackson, R. B., Field, C. B., and Chapin III, F. S. (1996b). Detecting changes in soil carbon in annual grasslands under CO_2 enrichment. *Plant Soil* **187,** 135–145.

Hungate, B. A., Chapin III, F. S., Zhong, H. L., Holland, E. A., and Field, C. B. (1997a). Stimulation of grassland nitrogen cycling under carbon dioxide enrichment. *Oecologia* **109,** 149–153.

Hungate, B. A., Holland, E. A., Jackson, R. B., Chapin III, F. S., Mooney, H. A., and Field, C. B. (1997b). The fate of carbon in grasslands under carbon dioxide enrichment. *Nature* **388,** 576–579.

Hungate, B. A., Lund, C. P., Pearson, H. L., and Chapin III, F. S. (1997c). Elevated CO_2 and nutrient addition alter soil N cycling and trace gas fluxes with early season wet-up in a California annual grassland. *Biogeochemistry* **37,** 89–109.

Idso, K. E., and Idso, S. B. (1994). Plant responses to atmospheric CO_2 enrichment in the face of environmental constraints: A review of the past 10 years' research. *Agric. For. Meteorol.* **69,** 153–202.

Idso, S. B., and Brazel, A. J. (1984). Rising atmospheric carbon dioxide may increase streamflow. *Nature* **312,** 51–53.

Knutson, T. R., Tuleya, R. E., and Kurihara, Y. (1998). Simulated increase of hurricane intensities in a CO_2 warmed climate. *Science* **279,** 1018–1021.

Larigauderie, A., Hilbert, D. W., and Oechel, W. C. (1988). Effect of CO_2 enrichment and nitrogen availability on resource acquisition and resource allocation in a grass, *Bromus mollis. Oecologia* **77,** 544–549.

Lincoln, D. E., Fajer, E. D., and Johnson, R. H. (1993). Plant–insect herbivore interactions in elevated CO_2 environments. *Trends Ecol. Evol.* **8,** 64–68.

Lindroth, R. L. (1996). Consequences of elevated atmospheric CO_2 for forest insects. *In* "Carbon, Dioxide, Populations, and Communities" (C. Körner and F. A. Bazzaz, eds.), pp. 347–361. Academic Press, San Diego.

Lloyd, J., and Farquhar, G. D. (1996). The CO_2 dependence of photosynthesis, plant growth responses to elevated atmospheric CO_2 concentrations and their interaction with soil nutrient status. I. General principals and forest ecosystems. *Funct. Ecol.* **10,** 4–33.

Lüscher, A., Hebeisen, T., Zanetti, S., Hartwig, U. A., Blum, H., Hendrey, G. R., and Nösberger, J. (1996). Differences between legumes and nonlegumes of permanent grassland in their responses to free-air carbon dioxide enrichment: Its effect on competition in a multispecies mixture. *In* "Carbon Dioxide, Populations, and Communities" (C. Körner and F. A. Bazzaz, eds.), pp. 287–300. Academic Press, San Diego.

Lussenhop, J., Treonis, A., Curtis, P. S., Teeri, J. A., and Vogel, C. S. (1998). Response of soil biota to elevated atmospheric CO_2 in poplar model systems. *Oecologia* **113,** 247–251.

Malmström, C. M., and Field, C. B. (1997). Virus-induced differences in the response of oat plants to elevated carbon dioxide. *Plant Cell Environ.* **20,** 178–189.

McKenzie, D., Peterson, D. L., and Alvarado, E. (1996). Extrapolation problems in modeling fire effects at large spatial scales: A review. *Int. J. Wildland Fire* **6,** 165–176.

McMurtrie, R. E., and Comins, H. N. (1996). The temporal response of forest ecosystems to doubled atmospheric CO_2 concentration. *Global Change Biol.* **2,** 49–59.

Mooney, H. A., Canadell, J., Chapin III, F. S., Ehleringer, J., Körner, C., McMurtrie, R., Parton, W. J., Pitelka, L., and Schulze, E.-D. (1998). Ecosystem physiology responses to global change. *In* "Implication of Global Change for Natural and Managed Ecosystems: A Synthesis of GCTE and Related Research" (B. H. Walker, W. L. Steffen, J. Canadell, and J. S. I. Ingram, eds.). Cambridge University Press, Cambridge.

Naylor, R., and Ehrlich, P. R. (1997). Natural pest control services and agriculture. *In* "Nature's Services" (G. C. Daily, ed.), pp. 151–174. Island Press, Washington, DC.

Neilson, R. P. (1995). A model for predicting continental-scale vegetation distribution and water balance. *Ecol. Appl.* **5,** 362–385.

Owensby, C. E., Ham, J. M., Knapp, A., Rice, C. W., Coyne, P. I., and Auen, L. M. (1996). Ecosystem-level responses of tallgrass prairie to elevated CO_2. *In* "Carbon Dioxide and Terrestrial Ecosystems" (G. W. Koch and H. A. Mooney, eds.), pp. 147–162. Academic Press, San Diego.

Parton, W. J., Schimel, D. S., Cole, C. V., and Ojima, D. S. (1987). Analysis of factors controlling soil organic matter levels in Great Plains grasslands. *Soil Sci. Soc. Am. J.* **51,** 1173–1179.

Parton, W. J., Scurlock, J. M. O., Ojima, D. S., Gilmanov, T. G., Scholes, R. S., Schimel, D. S., Kirchner, T., Menaut, J.-C., Seastedt, T., Garcia Moya, E., Kamnalrut, A., and Kinyamario, J. L. (1993). Observations and modeling of biomass and soil organic matter for the grassland biome worldwide. *Global Biogeochem. Cycles* **7,** 785–809.

Polley, H. W., Johnson, H. B., and Mayeux, H. S. (1994). Increasing CO_2: Comparative responses of the C_4 grass *Schizachyrium* and grassland invader *Prosopis. Ecology* **75,** 976–988.

Poorter, H., Roumet, C., and Campbell, B. D. (1996). Interspecific variation in the growth response of plants to elevated CO_2: A search for functional types. *In* "Carbon Dioxide, Populations, and Communities" (C. Körner and F. A. Bazzaz, eds.), pp. 375–412. Academic Press, San Diego.

Prentice, I. C., Cramer, W., Harrison, S. P., Leemans, R., Monserud, R. A., and Solomon, A. M. (1992). A global biome model based on plant physiology and dominance, soil properties and climate. *J. Biogeog.* **19,** 117–134.

Raich, J. W., Rastetter, E. B., Melillo, J. M., Kicklighter, D. W., Steudler, P. A., Peterson, B. J., Grace, A. L., Moore III, B., and Vörösmarty, C. J. (1991). Potential net primary production in South America. *Ecol. Appl.* **1,** 399–429.

Rillig, M. C., Allen, M. F., Klironomos, J. N., Chiariello, N. R., and Field, C. B. (1997). Plant species-specific changes in root inhabiting fungi in a California annual grassland: Responses to elevated CO_2 and nutrients. *Oecologia* **113,** 252–259.

Running, S. W., and Hunt, E. R. (1993). Generalization of a forest ecosystem process model for other biomes, BIOME-BGC, and an application for global-scale models. *In* "Scaling Physiological Processes: Leaf to Globe" (J. R. Ehleringer and C. B. Field, eds.), pp. 141–158. Academic Press, San Diego.

Schimel, D., Enting, I. G., Heimann, M., Wigley, T. M. L., Raynaud, D., Alves, D., and Siegenthaler, U. (1995). CO_2 and the carbon cycle. *In* "Climate Change 1994: Radiative Forcing of Climate Change and an Evaluation of the IPCC IS92 Emission Scenarios" (J. T. Houghton, L. G. Meira Filho, J. Bruce, H. S. Lee, B. A. Callander, E. Haites, N. Harris, and K. Maskell, eds.), pp. 33–71. Cambridge University Press, Cambridge.

Sellers, P. J., Bounoua, L., Collatz, G. J., Randall, D. A., Dazlich, D. A., Los, S., Berry, J. A., Fung, I., Tucker, C. J., Field, C. B., and Jenson, T. G. (1996). A comparison of the radiative and physiological effects of doubled CO_2 on the global climate. *Science* **271,** 1402–1405.

Smart, D. R., Ritchie, K., Stark, J. M., and Bugbee, B. (1997). Evidence that elevated CO_2 levels can indirectly increase rhizosphere denitrifier activity. *Appl. Environ. Microbiol.* **63,** 4621–4624.

Thompson, M. V., and Randerson, J. T. (1999). Impulse response functions of terrestrial carbon cycle models: Method and application. *Global Change Biol.* (in press).

Thompson, M. V., Randerson, J. T., Malmström, C. M., and Field, C. B. (1996). Change in net primary production and heterotrophic respiration: How much is necessary to sustain the terrestrial carbon sink? *Global Biogeochem. Cycles* **10,** 711–726.

Townsend, A. R., Braswell, B. H., Holland, E. A., and Penner, J. E. (1996). Spatial and temporal patterns in terrestrial carbon storage due to deposition of anthropogenic nitrogen. *Ecol. Appl.* **6,** 806–814.

Treonis, A. M., and Lussenhop, J. F. (1997). Rapid response of soil protozoa to elevated CO_2. *Biol. Fertil. Soils* **25,** 60–62.

Trumbore, S. E., Chadwick, O. A., and Amundson, R. (1996). Rapid exchange between soil carbon and atmospheric carbon dioxide driven by temperature change. *Science* **272,** 393–398.

VEMAP Members. (1995). Vegetation/ecosystem modeling and analysis project: Comparing biogeography and biogeochemistry models in a continental-scale study of terrestrial ecosystem responses to climate change and CO_2 doubling. *Global Biogeochem. Cycles* **9,** 407–438.

Vitousek, P. M., and Field, C. B. (1999). Ecosystem constraints to symbiotic nitrogen fixers: A simple model and its implications. *Biogeochemistry* (in press).

Vitousek, P. M., and Howarth, R. W. (1991). Nitrogen limitation on land and in the sea: How can it occur? *Biogeochemistry* **13,** 87–115.

Wardle, D. A., Zackrisson, O., Hörnberg, G., and Gallet, C. (1997). The influence of island area on ecosystem properties. *Science* **277,** 1296–1299.

Woodward, F. I., Smith, T. M., and Emanuel, W. R. (1995). A global land primary productivity and phytogeography model. *Global Biogeochem. Cycles* **9,** 471–491.

Zak, D. R., Pregitzer, K. S., Curtis, P. S., Terri, J. A., Fogel, R., and Randlett, D. L. (1993). Elevated atmospheric carbon dioxide and feedback between carbon and nitrogen cycles. *Plant Soil* **151,** 105–117.

Zanetti, S., Hartwig, U. A., Van Kessel, C., Lüscher, A., Hebeisen, T., Frehner, M., Fischer, B. U., Hendrey, G. R., Blum, H., and Nosberger, A. (1997). Does nitrogen nutrition restrict the CO_2 response of fertile grassland lacking legumes? *Oecologia* **112,** 17–25.

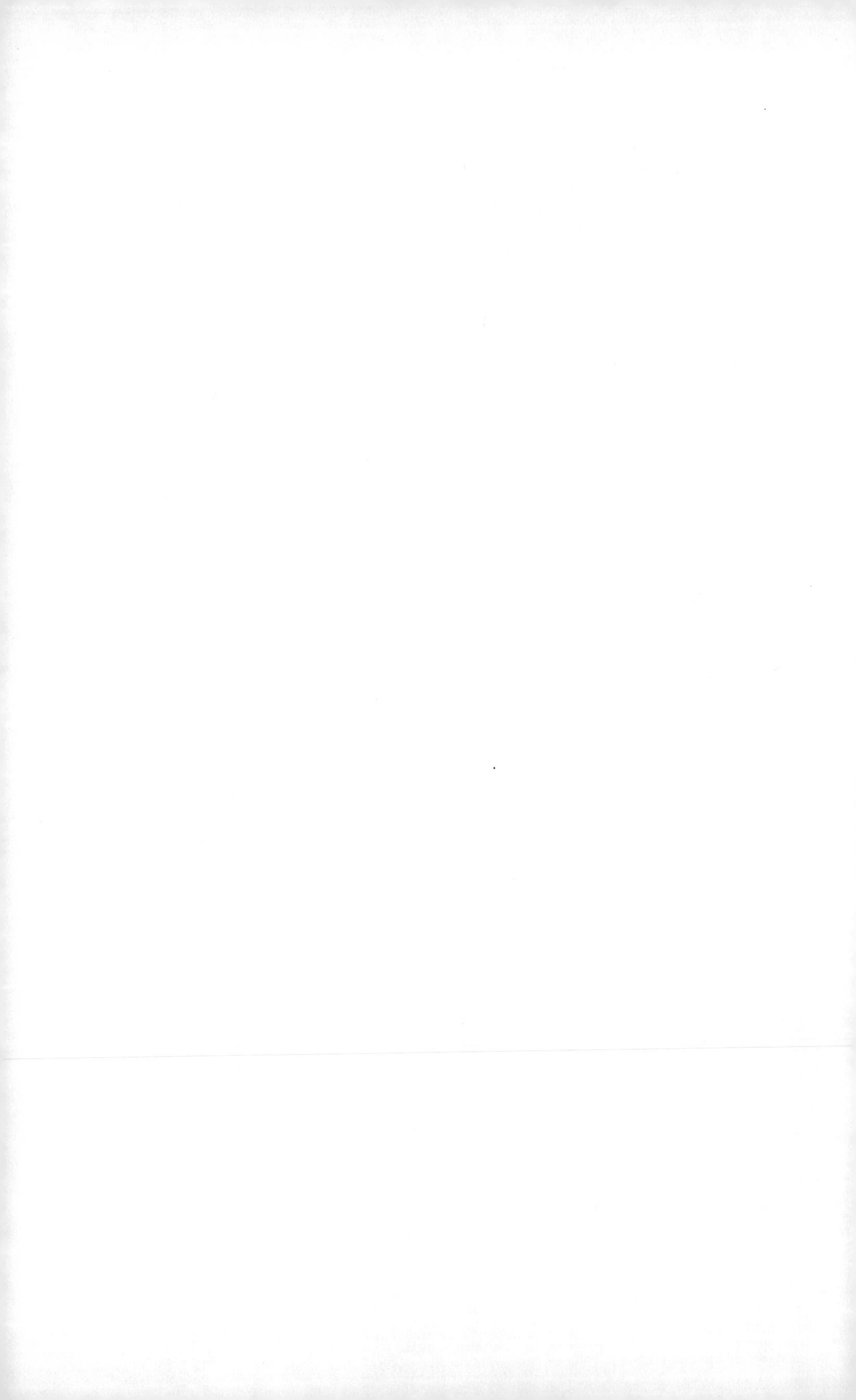

16

Interactive Effects of Carbon Dioxide and Environmental Stress on Plants and Ecosystems: A Synthesis

Yiqi Luo, Josep Canadell, and Harold A. Mooney

I. Introduction

There is now a clear understanding of the multiple-driver nature of global change, and the need to address complex and nonlinear responses as we try to predict the future function and structure of the world's biomes. Furthermore, as we prepare to meet the demands of the Framework Convention on Climate Change, and particularly of the Kyoto Protocol, a sound understanding of the carbon cycle at the ecosystem, regional, and global level is urgently needed. In this respect, there is a necessity for improving our knowledge on the temporal and spatial dynamics of sources and sinks of carbon. It is even more important to understand the controls of source–sink dynamics and how they may change in the future as global change keeps progressing (e.g., increase of atmospheric CO_2, N deposition, air temperature).

The effects of carbon dioxide on plants and ecosystems have been extensively studied in the past two decades. These studies have provided great insights into potential changes in plant and ecosystem functions and structure in the next century when atmospheric CO_2 increases to twice the current CO_2 concentration. How the CO_2 effects on plants and ecosystems are regulated by stresses has not been carefully examined. In the real world, rising atmospheric CO_2 concentration is always interacting with other environmental and biological stresses in determining actual changes in material and energy fluxes in ecosystems. It is critical to evaluate the interactive responses of plants and ecosystems to rising atmospheric CO_2 and environmental stress. This chapter summarizes and synthesizes major knowns and unknowns presented in the chapters compiled in this book. Built upon the synthesis, we propose future research needs in order to improve our predictive understanding of the interactive effects of rising atmospheric CO_2 and environmental stress.

II. Interactive Effects of Carbon Dioxide and Stresses on Plants and Ecosystems

Research on CO_2 and stress interactions is needed to address one question in two ways. One is whether or not elevated CO_2 ameliorates or exacerbates environmental stresses. The other is how environmental stress moderates the direct effect of elevated CO_2 on plants and ecosystems. Results from plant-level studies have suggested that elevated CO_2 is likely to ameliorate mild drought, salinity, UV-B, and ozone stresses, to exacerbate nutrient stress, and to interact with temperature in a complex fashion. Accordingly, direct effects of elevated CO_2 on plants are likely to be amplified under mild drought and salinity stresses but dampened by nutrient stress. At the ecosystem scale, we have very limited evidence to suggest one way or another on impacts of CO_2 and stress interactions because major feedback mechanisms have not been evaluated using multifactorial experiments (Table I).

A. Water and CO_2

Extensive experimental data have generally supported a conclusion that rising atmospheric CO_2 directly reduces stomatal conductance and then transpiration per unit leaf area (Chapter 1). The magnitude of reduction in stomatal conductance varies with species and growth environments, by 36% on average for 11 crop and herb species and 23% for 23 tree species when growth CO_2 increases from ambient to twice ambient levels. Stomatal conductance is generally reduced more for plants grown in growth chambers than in fields utilizing open-top chambers (OTC) or free-air CO_2 enrichment (FACE) facilities. Reduced stomatal conductance in elevated CO_2 is almost always associated with a decrease in water loss via leaf transpiration and an increase in leaf water potential and expansive growth (Chapter 1).

Translation of reduced stomatal conductance and leaf transpiration to plant and canopy levels is complicated by numerous factors, including leaf area growth (Chapter 1), root growth, canopy structure and closure, canopy water interception and loss, soil surface evaporation, and species replacement (Chapter 2) (Table I). Elevated CO_2 generally results in larger leaves and higher leaf area growth (Chapter 1) and more root growth (Chapter 8) than ambient CO_2 does. Increased leaf growth counteracts the reduced stomatal conductance in determining water loss, whereas increased root growth explores more soil water resource. Both lead to more water consumption. Interactions of these physical and biological processes are site specific, leading to diverse responses of ecosystem hydrological cycling to elevated CO_2. Ecosystem-level measurements indicated that soil water content increased in agricultural forb ecosystems and in annual and perennial grasslands (Chapters 2 and 10). Results from FACE studies, however, did not indicate much change in soil water content in the elevated CO_2 plots compared to that in the ambient CO_2 plots (Oren *et al.*, 1998) (Table I). Even if the ecosystem water consumption is similar between the two CO_2 treatments, gross primary productivity is expected to increase in elevated CO_2 due to increased water use efficiency (Chapter 1). Whether or not net primary productivity will consequently increase in elevated CO_2 depends on feedback processes of carbon allocation, carbon loss via respiration, leaf and root turnover, and carbon use efficiency associated with changes in nonstructural carbohydrate storage and leaf and root mass per unit area (Luo *et al.*, 1997).

Chapter 2 pointed out that plot-level studies using OTCs and FACE facilities may not capture hydrological processes that operate at landscape, regional, and continent scales. These processes include CO_2-induced change in regional precipitation, within-continent water cycling between the biosphere and the atmosphere, and planetary boundary layer. Experimental studies by manipulating atmospheric CO_2 at the landscape or larger scales are beyond our technical capability. It may be a viable, alternative approach to analyze long-term watershed hydrological data. Chapter 2 analyzed 40-yr watershed hydrological data during the 1956–1996 period from the Hubbard Brook Experimental Forest in the White Mountain National Forest, New Hampshire, and concluded that watershed evapotranspiration may have slowed with rising atmospheric CO_2 in only one of five forested watersheds. That approach deserves more exploration in addressing large-scale, long-term impacts resulting from CO_2 and water interactions.

B. Temperature and CO_2

Temperature affects numerous physiological and ecological processes at several hierarchical levels and thus interacts with CO_2 in a complex fashion (Chapters 3 and 4). At the biochemical level, temperature regulates membrane permeability, enzyme kinetics, synthesis, and stability. For example, temperature differentially affects carboxylation and oxygenation kinetics of ribulose-1,5-bisphosphate carboxylase/oxygenase (Rubisco) and then

Table I Plant and Ecosystem Responses to Environmental Stress That Interacts with Elevated CO_2[a]

	Plant response		Ecosystem response	
Stress	Direct mechanism	Experimental results	Feedback mechanisms	Experimental results
Water	Stomatal conductance	Reduced stomatal conductance Increased WUE and leaf expansive growth	Leaf area growth Canopy closure Interception/evaporation	Extended growing seasons in annual and perennial grasslands Little change in Duke and Arizona FACE
Temperature	Numerous biochemical and physiological processes	Increased CO_2 stimulation of photosynthesis with temperature Little leaf temperature increases observed in CO_2 experiments	Nutrient mineralization Migration/colonization Water balance C_3/C_4 competition	Accelerated nutrient cycling and increased respiratory C loss as indicated from soil warming experiments Decreased primary productivity
Salinity	Reduced water uptake	No effect on salt accumulation and turgor pressure Increased growth of salt-affected plants	Soil salinity Community structure Groundwater recharge	Increased productivity in salt marsh Altered community structure (more C_3 plants)

UV-B	Plant growth Secondary metabolites	No interactive effects on photosynthesis, productivity Change in allocation and morphogenesis	Decreased litter quality Photodegradation of litter Changes in community of decomposers	Changed herbivory by UV-B but not by elevated CO_2 in the tundra Increased mycorrhizal infection, lignin content Reduced decomposition in litter bags
Ozone	Stomatal conductance Antioxidant synthesis	Declined O_3 damage symptoms Increased carbohydrate and metabolite pools Decreased Rubisco	Plant protein content Soil nutrient availability	
Nutrients	Carbon and nutrient coupling in plant tissues	Reduced tissue N concentration Increased root growth Little downward regulation in photosynthesis across field studies	Altered C:N ratio Short-circuit C cycling Increased N supply	Variable changes in N mineralization Increased soil exploration Increased N fixation Increased carbon input into soil

[a] Direct interactions of elevated CO_2 with stress first take place through plant processes (direct mechanisms). The direct effects at the plant level are translated to ecosystem responses through a variety of feedback mechanisms. Also listed is evidence related from experimental studies on plant and ecosystem responses to the stress and CO_2 interactions. Note that results at the ecosystem level are based on a limited set of field experiments. Even fewer of them were in factorial design to study specifically the CO_2 and stress interactions.

controls the responsiveness of photosynthesis to elevated CO_2. As a consequence, the magnitude of the stimulation of C_3 plant photosynthesis and growth by elevated CO_2 tends to increase with temperature (Chapter 3) (Table I). At the plant level, temperature affects water relations, development, carbon partitioning, morphology, phenology, and reproduction. High temperature stress, for example, impairs reproductive development, exacerbating downward regulation of photosynthesis and growth limitation in CO_2-enriched plants.

At the community scale, temperature governs the success of migration and colonization of different plant species, inducing changes in species distribution and vegetation movement. Chapter 3 provided case studies in Arctic and alpine regions to illustrate that temperature and CO_2 interactive effects may be primarily exhibited by potentially adverse factors that result from a prolonged winter. A warmer winter is a major component of current climatic change. Winter warming that is greater than summer warming may cause remarkable changes in species distribution. But other factors, including a high degree of polymorphism, gene migration between adjacent populations in contrasting microsites, a genetic memory in the seed bank and in hybrid population, and longevity, contribute to the capacity of plant populations to withstand the impacts of climate change. It is a challenge for modelers to incorporate these processes when predicting the impacts of both rising atmospheric CO_2 and temperature on vegetation distribution in the arctic and alpine regions while manipulative studies of these processes are yet beyond experimental reality.

At the ecosystem scale, temperature alters primary productivity, water balance, nutrient availability, and fire (Table I). However, few studies, either experimental or simulation based, have addressed the combined effects of elevated CO_2 and temperature stress on ecosystem processes. Chapter 4 synthesized our knowledge on effects of temperature and CO_2 on these key ecosystem processes and made substantial inferences about potential ecosystem responses to temperature and CO_2 interactions. Elevated CO_2 is predicted to stimulate primary production most in water-limited ecosystems due to a substantial enhancement of plant water use efficiency though this may be offset by higher leaf area production (Chapters 1 and 2). In contrast, higher temperature could have markedly negative effects on plant production in water-limited ecosystems due to higher evapotranspiration rates and/or more temperature stress. In cold ecosystems, warming might stimulate ecosystem production by increasing nutrient availability. Overall ecosystem responses will probably vary widely across different ecosystem types because elevated CO_2 and global warming seem to have opposite effects on primary production, ecosystem water dynamics, and nutrient availability. Thus, regional predictions will be difficult with any level of precision.

C. Salinity, UV-B, Ozone, and CO_2

The total area of saline, sodic, or alkaline land counts up to 7% of the world land area, approximately 950 million hectares. Most of the salt-affected land is man-made, resulting from irrigation and clearing of perennial vegetation. Understanding how elevated CO_2 might mitigate the salinity problem has important applications for world food production. Chapter 5 provided a comprehensive overview of this issue. Salinity inhibits plant growth primarily through reduced water uptake from saline soil solution and excessive amount of salts in live cells causing damage. Elevated CO_2 enhances growth of plants in saline soil due to reduced water uptake (Table I). Possible mechanisms involved in CO_2 amelioration of salt-affected plants include decreased salt accumulation in leaves, increased turgor, or increased carbon supply to growing tissues. Results from numerous experimental studies have provided little evidence for increased turgor and decreased salt accumulation in elevated CO_2 in comparison to that in ambient CO_2. Thus, increased carbohydrate supply to growing tissues may be the primary mechanism of increased growth of salt-affected plants. Although elevated CO_2 may have ameliorating effects on salt-affected plants, its impact on soil salinity is much less clear. Chapter 5 discussed a variety of mechanisms at the ecosystem scale, including increased canopy closure, ecosystem water loss, CO_2-induced global warming, and increased groundwater recharge (Table I). The only evidence available from the field experiment in a salt marsh in Chesapeake Bay is that community structures shifted to have more C_3 plants and fewer C_4 plants in elevated CO_2.

Recent field experiments have provided evidence that an elevated UV-B level does not affect photosynthesis but reduces plant elongation growth, and alters production of plant secondary metabolites, for example, flavonoids, tannins, and lignin, leading to changes in primary productivity (Chapter 6). Thus, significant interaction between CO_2 and UV-B is generally manifested by changes in biomass allocation and feedback processes through, for example, litter decomposition (Table I). Plant morphogenesis, including height, shoot and leaf length, leaf thickness and area, and auxiliary branching, is often altered under elevated UV-B. But limited data are available concerning the influence of CO_2 and UV-B interaction on those characteristics. Increased production of secondary metabolites, which are complex polyphenolics, influences plant–animal, plant–microorganism interactions, and litter decomposition (Table I). Both elevated CO_2 and enhanced UV-B reduce the quality of plant tissues and litter, leading to reduced insect herbivory and litter decomposition.

Chapter 7 suggested that a decline in O_3 sensitivity due to elevated CO_2 is probably an integration of multiple processes, including reduction in stomatal conductance, possible anatomical changes in the leaf, and flexibility to induce antioxidants when needed. When plants are grown in elevated CO_2, O_3-induced foliage symptoms are reduced. Biochemical mechanisms for the CO_2 and O_3 interaction can be complicated. Plants grown in elevated CO_2 result in increased carbohydrate and decreased content of enzymes, such as Rubisco. Data presented in Chapter 7 indicated that plants grown in elevated CO_2 have greater flexibility to shift carbohydrates to increase pools of antioxidants as needed. On the other hand, the O_3 target is protein, like Rubisco. Reduction of Rubisco concentrations in elevated CO_2 could have an adverse interaction with O_3 because each Rubisco molecule has a higher chance of being damaged by O_3 in elevated than in ambient CO_2. Thus, responses of plants to elevated CO_2 and O_3 interactions are influenced substantially by nutrient availability. If nutrients are not limiting, as in agricultural situations, CO_2 could have an ameliorating effect on O_3 toxicity. In natural ecosystems, where nutrients are more limiting, elevated CO_2 may render the plant less flexible to synthesizing protective compounds and as a result offer less protection from O_3.

D. Nutrients and CO_2

The interactive effects of nutrients and elevated CO_2 on plants have been extensively studied. It is a consistent conclusion across numerous experiments that growth in elevated CO_2 almost always results in lower nutrient concentration in plant tissues than in ambient CO_2. Despite the decrease in tissue nutrient concentration, three recent reviews, conducted by Curtis and Wang (1998), Drake *et al.* (1997), and Mooney *et al.* (1999), have found little reduction in photosynthetic capacity for plants grown in elevated CO_2 relative to that in ambient CO_2 (also see Chapter 12). In addition, factorial experiments with multiple levels of nitrogen and CO_2 concentration have suggested that plants responded to elevated CO_2 even at low nutrient supply levels (Johnson *et al.*, 1996). Responsiveness of plants to elevated CO_2 usually increases with nutrient supply levels. Thus, it is critical to understand how ecosystems regulate nutrient availability and then influence CO_2 effects on plant and ecosystem carbon processes (Fig. 1). Several chapters in this book are devoted to various aspects of ecosystem nutrient dynamics.

There is no doubt that elevated CO_2 generally stimulates root growth (Chapter 8), which extends the potential to increase carbon deposition into rhizosphere and nutrient uptake by plants (Chapters 8–10, 12–15) (Fig. 1). Elevated CO_2 tends to increase rhizodeposition either through increased root growth and turnover, or increased deposition per unit of root mass (e.g., root exudation), or both. Although root exudation is critical for predicting soil carbon and nutrient dynamics in ecosystems, virtually no report is available on the amount of root exudates in elevated CO_2. Continuous ^{14}C-labeling studies have shown that plants grown in elevated

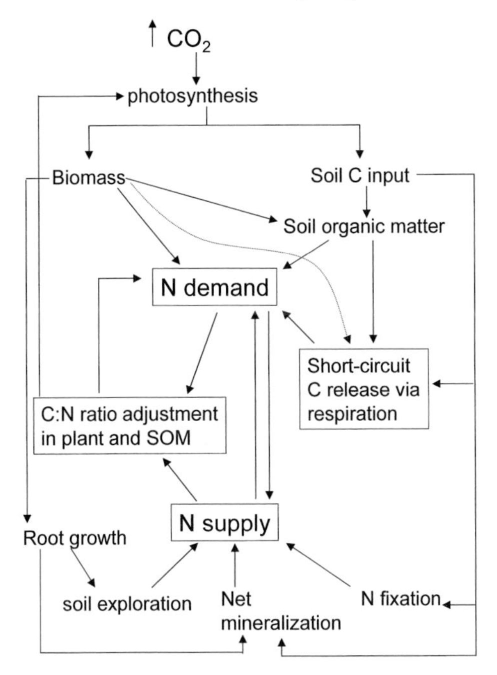

Figure 1 A conceptual model of ecosystem carbon and nitrogen interactions as affected by elevated CO_2. Elevated CO_2 generates additional nitrogen demand primarily through increased plant biomass growth and soil carbon storage. To meet the additional nitrogen demand, three general mechanisms are used: adjustment of C:N ratios in plant biomass and soil organic matter (SOM), short-circuit dissipation of additional carbon, and additional nitrogen supply through soil exploration, net mineralization, and nitrogen fixation.

CO_2 allocated more carbon to total rhizosphere respiration, suggesting a substantial amount of short-term carbon influx into the rhizosphere through fast pathways such as exudation (Chapter 9). Root growth and turnover have also been found to increase considerably in elevated CO_2 (Chapter 8), leading to increased medium-term carbon influx to the rhizosphere (Fig. 1).

Increased carbon input to soil in elevated CO_2 has been reported to increase, decrease, and have no effect on nitrogen availability to plants. These idiosyncratic results may be partly explained by differences in the timescales of experimental measurements and plant species involved in the experiments and partly explained by different measurement techniques (Chapters 9 and 10). Elevated CO_2 tends to increase rhizosphere symbiants such as mycorrihizae and rhizobia across several types of associations. N fixation has generally been found to increase in elevated CO_2. Inputs of nitrogen through N fixation may not be quantitatively significant in the short term, but merit attention for their effects on nitrogen stocks in the long term (Chapter 10) (Fig. 1). Whether or not the structure of rhizosphere communities will be changed in elevated CO_2 is not clear yet. Most of the data, however, are from growth chamber experiments under highly disturbed and highly controlled conditions. Rhizosphere dynamics urgently need to be studied in field experiments (Chapter 9).

Chapter 10 offered compelling experimental data as well as comprehensive conceptual models to support a notion that soil nitrogen cycling is not only altered directly by increased carbon input to soil but also indirectly by increased soil moisture. Soil moisture content has been found to be significantly altered in several ecosystems, including annual and perennial grasslands, agricultural fields, and microcosm experiments in elevated CO_2. CO_2-induced change in soil water content resulted in changes in soil microbial activities and nitrogen transformation (Chapter 10). Because ecosystem hydrological cycling depends on leaf area growth (Chapters 1 and 2), the indirect effects of elevated CO_2 through changed soil moisture content on nitrogen cycling are likely to vary among ecosystems. Elevated CO_2 may cause the largest reduction in evapotranspiration in ecosystems where aboveground growth responses to elevated CO_2 are the smallest, leading to a large change in N cycling.

Long-term impacts of various mechanisms of nitrogen supply as well as C : N ratio adjustment in plant and soil organic matter (SOM) on ecosystem carbon and nitrogen dynamics were quantitatively evaluated in Chapters 12–14. Examined also were implications of three-way interactions among carbon dioxide, nitrogen, and water for ecosystem productivity and carbon sequestration (Chapter 14). Soil respiratory carbon release has often been found to increase substantially (Chapter 9), leading to short-circuit carbon cycling (Fig. 1). The short-circuit carbon release reduces nitrogen demand in elevated CO_2 and thus may mitigate nitrogen stress in natural ecosystems. It may also be partly responsible for little photosynthetic down-regulation observed from most of the OTC and FACE experiments in natural ecosystems. The impact of the short-circuit mechanism, however, has not been evaluated using either experimental or modeling approaches.

III. Evolutionary, Scaling, and Modeling Studies of CO_2 and Stress Interactions

Rising CO_2 concentration in the atmosphere is a long-term, large-scale phenomenon. It not only affects short-term physiological and ecological processes such as those described in Chapters 1–10, but also regulates evolutionary courses as well as large spatial scale processes. Several chapters in this book offer approaches to place physiological and ecological research in broad evolutionary and scaling perspectives.

Most of the research addressing biological responses to variation in atmospheric CO_2 has focused on physiological responses of plants to doubling of the present CO_2 level. But much less has been considered on evolutionary responses of plants to the long-term CO_2 change in the atmosphere. Chapter 11 explored the latter issues and argued that the world's flora may be adapted to the preindustrial CO_2 level under which plants are selected for stress-tolerant mechanisms in order to survive under carbohydrate deficiency. These mechanisms include low growth potential, conservative allocation patterns, and storage investment. At optimal environmental conditions, plant performance modestly decreases with declining CO_2. As CO_2 levels decline, the inhibitory range of environmental conditions broadens substantially. Many conditions that are currently nonstressful may have been stressful during past episodes of low CO_2, and conditions that are now moderately stressful may have been lethal at low CO_2. Thus, strong selection pressure may have favored stress-tolerant mechanisms at low CO_2. The hypothetical evolutionary mechanism provides an alternative explanation of imbalances between sources and sinks or between nutrient supply and carbon availability often observed in experiments. If evolutionary processes do correct the imbalances and reestablish a shared control among multiple resources, a gradual increase of atmospheric CO_2 over a time frame of a century could lead to substantial enhancement of CO_2 responsiveness. Chapter 11 also urged incorporation of stress and low CO_2 interactions in experimental studies of plant evolutionary responses to rising atmospheric CO_2.

Scaling has become one of the major scientific activities in global change research partly because plant and ecosystem studies are primarily driven by large-scale issues and partly because we are unable to make direct measurements at regional and global scales (Chapters 2 and 12). Our predictions of large-scale terrestrial carbon sinks and sources rely on the scaling up of our knowledge from leaf, plant, and small ecosystem studies. Conventional scaling-up schemes include summation, averaging, and aggregation in association with gridded geographical information systems of world vegetation, soil, and climate conditions. Chapter 12 argued that a scaling-up study fundamentally has to cope with two general factors: environmental variability and biological diversity. The challenge in scaling-up studies is to reduce uncertainties associated with these two general factors. In addition to the conventional approaches, Chapter 12 also suggested that scaling studies can be accomplished by identifying scaleable parameters. Such a parameter characterizes intrinsic properties of a system in question and thus reduces extrapolation uncertainties caused by environmental and biological variability. For example, quantum yield of CO_2 uptake, that is, a physiological parameter, has been successfully used to delineate relative distribution of C_3 and C_4 plants in terrestrial ecosystems over the globe. In addition, identification of scaleable parameters may provide a unique approach to deal with the complexity of CO_2 and stress interactions. An invariant leaf-level function, for instance, that describes photosynthetic sensitivity to a small increment in atmospheric CO_2 is powerful in extrapolating leaf-level studies to predict marginal increments in carbon influx and storage caused by rising atmospheric CO_2.

Modeling has been an essential tool in studying plant and ecosystem responses to stress and CO_2 interactions because of a critical need to synthesize and extrapolate plot-level measurements to predict long-term, large-scale ecosystem responses to global change. Various modeling approaches have been developed in the literature. Most of the models have been built on the fundamental basis of stoichiometry that carbon dynamics in terrestrial ecosystems are strongly coupled with nitrogen cycling (Chapter 13). Increased input of carbon resulting from rising atmospheric CO_2 should result in changes in the total amount of nitrogen in the ecosystem while maintaining C:N ratios, or changes in the C:N ratios of ecosystem components, or changes in the distribution of nitrogen among ecosystem components. This basic stoichiometrical relationship leads to the theory of nitrogen productivity that the relative plant growth rate is determined by plant nitrogen concentration. When the concept is applied to the whole ecosystem net primary productivity, it is crucial to consider whether the ecosystem is open or closed in terms of nitrogen input. Because the availability of mineral nutrients cannot be expected to increase in proportion to the carbon increase, changes in the relative availability of elements are predicted to cause changes in root fraction, C:N ratio of litter and soil organic matter, and immobilization of nitrogen (Fig. 1).

It is a common practice to utilize the minimum limitation notion or the multiplicity concept to simulate multiple stresses in interacting with rising atmospheric CO_2. Using the former notion, Chapter 14 extended the G'DAY (General Decomposition And Yield) model to predict interactive effects of nitrogen and water with elevated CO_2 on forest ecosystems. The model predicted that the CO_2 fertilization effect is likely to be amplified under water limitation, but reduced under long-term nitrogen limitation. The amplification of the CO_2 fertilization effect by water limitation is primarily due to enhanced water use efficiency, which has a large, direct effect on carbon uptake, and smaller, indirect effect due to stimulation of soil decomposition.

IV. Future Research Needs

A. Experiments

The primary goal of experimentation is to identify and quantify the mechanisms underlying ecosystem responses. The challenge here is to provide relevant ecophysiological data to develop and test models, and to incorporate ecosystem and biosphere metabolic processes that are not currently well understood.

Abundant research on plant physiological responses to various environmental factors has taken place during the last two decades. However, if one of the main goals of our research is to predict ecosystem or larger-scale responses to global change, there is a critical need for research on processes relevant at the ecosystem and landscape levels. In this respect, great progress has been made during the past 6 yr in understanding ecosystem-level responses to elevated CO_2, particularly in grassland systems (Mooney *et al.*, 1999). However, as we begin recognize the multifactorial nature of global change, it becomes obvious that single-factor experiments will provide limited information to predict the consequences of several simultaneously occurring environmental changes. Therefore, multifactorial experiments, in which interactions among factors can be tested, are critically important. Among many of the global changes that are already occurring, special attention should be paid to air and soil warming, nitrogen deposition, and changes in water availability in addition to elevated CO_2, ozone, and UV-B. They are all major controls of net ecosystem exchange and, therefore, of the ecosystem and global carbon cycle.

There are still important technical difficulties associated with setting up such multifactorial experiments in the field, particularly for the interaction between elevated CO_2 and warming in large stature systems (e.g., forests). Although various techniques are available to increase atmospheric CO_2 at the stand level and each has a unique strength, the free-air carbon dioxide enrichment (FACE) experiments are the least environmentally disturbing technique and, therefore, the preferred one. Natural CO_2 springs also offer excellent opportunities for studying undisturbed systems exposed to long-term elevated CO_2.

There is no one technology for warming experiments that can be clearly considered the best, although convective heating should be preferred over other types of heating. In addition, natural experiments using thermal gradients can provide us, as in the case of natural CO_2 springs, with unique opportunities to study long-term adaptations and community dynamics, which cannot be studied in short-term studies (<5 yr). Soil/vegetation transplant experiments can equally provide valuable information.

A list of recommendations follows for whole ecosystem experiments to better address responses to global change (see also Canadell *et al.*, 1999; Schulze *et al.*, 1999):

1. Experiments with either elevated CO_2, or warming, or both should investigate the interactions with other factors such as nitrogen deposition and water availability. If possible, it is desirable to include ozone, salinity, and UV-B.
2. As we learn more about the existence of thresholds and nonlinearity of plant and ecosystem responses, multifactorial treatments with more than two levels in each factor (e.g., ambient concentration, 500 ppm, 700 ppm in CO_2 treatments) are recommended.
3. Experiments should run long enough to allow, if possible, at least one life cycle of the dominant species. Such an experiment will allow inter- and intraspecific competition and feedback processes (e.g., litter decomposition and nutrient availability changes) to manifest. This recommendation may be feasible only for herbaceous systems.
4. The study of mature systems is encouraged over young, expanding systems. The latter is more prone to show unsustainable large responses that are likely to be quite different from those revealed from the mature systems.
5. Field experiments that have coupled plant–soil systems and undisturbed soils are recommended in order to characterize more realistic belowground processes (e.g., rhizosphere dynamics).

Finally, a major effort should be directed at setting up experiments in biomes that have been studied insufficiently or not at all but are critical for understanding the global carbon cycle (Fig. 2). These include such major ecosystems as boreal forests, savanna, and both humid and dry tropical forests. However, even with a major scientific guided effort (both for funding and personnel), it may not be possible to study enough sites to cover the basic ecosystem types. This makes the use of model ecosystems (e.g., annual grasslands) an essential tool in understanding the basic operating mechanisms involved in response to global changes.

B. Modeling

Whole ecosystem experiments will be closely linked to the development and operation of dynamic ecosystem models. This linkage will help to guide the interpretation of the results, sharpen the focus on understanding the mechanisms underlying the observed responses, and ensure the broader applicability of the results to other systems.

At the regional and global levels, ecosystem and biospheric models have shown to be very valuable. However, they do not account for important landscape processes such as disturbances (e.g., fire) and biome shifts that

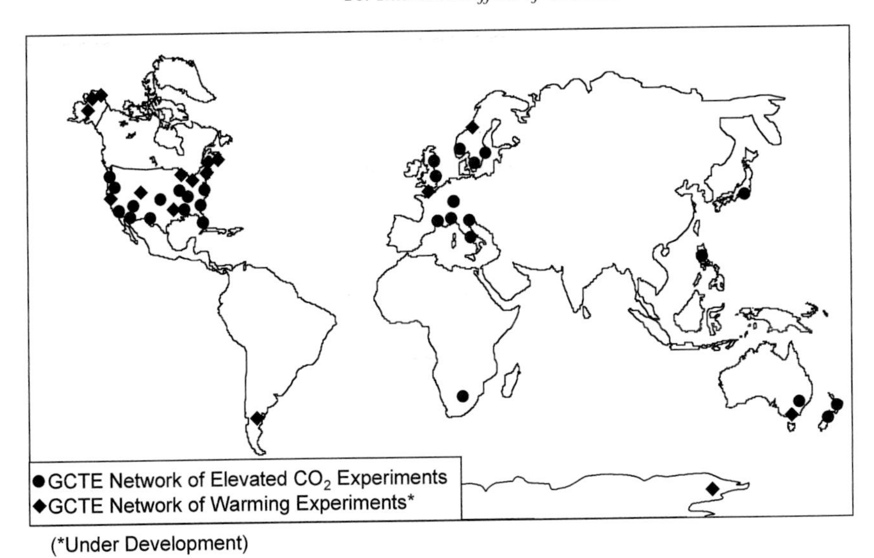

Figure 2 Global distribution of GCTE (global change and terrestrial ecosystem) networks of elevated CO_2 experiments and warming experiments. This global map is also indicative of where future efforts should be placed in biomes that have not been studied or have been less studied.

are known to occur under climate change. To account for these important landscape processes, which have major impacts on the regional and global carbon cycle, new modeling approaches are needed. One exemplary approach is to couple vegetation movement with biogeochemical cycling. Such models account for disturbances and biome shifts as well as changes in ecosystem functioning and they also quantify changes in carbon pools in a more realistic way.

V. Conclusions

Global change is a multifactorial process, involving not only rising atmospheric CO_2 concentration but also increases in nitrogen deposition, global warming, ozone, UV-B radiation, and salinity in association with variations in precipitation. Research on the effects of elevated CO_2 on plants and ecosystems in the past two decades has provided a great foundation for developing a predictive ability with regard to terrestrial responses to global change. However, our understanding of stress and CO_2 interactions in determining actual changes in plant and ecosystem structure and function is extremely limited. This book, while providing the most up-to-date knowledge and thinking on multistress interactions, is designed to stimulate future research on that matter.

References

Canadell, J., Noble, I., and Ingram, J. (in press). "Global Change and Terrestrial Ecosystems. Implementation Plan." IGBP Report 47, Stockholm.

Curtis, P. S., and Wang, X. (1998). A meta-analysis of elevated CO_2 effects on woody plant mass, form, and physiology. *Oecologia* **113,** 299–313.

Drake, B. G., Gonzàlez-Meler, M. A., and Long, S. P. (1997). More efficient plants: A consequence of rising atmospheric CO_2? *Annu. Rev. Plant Physiol. Plant Mol. Biol.* **48,** 607–637.

Johnson, D. W., Henderson, P. H., Ball, J. T., and Walker, R. F. (1996). Effects of CO_2 and N on growth and N dynamics in ponderosa pine: Results from the first two growing seasons. *In* "Carbon Dioxide and Terrestrial Ecosystems." (G. W. Koch and H. A. Mooney eds.), pp. 23–40. Academic Press, San Diego.

Luo, Y., Chen, J. L., Reynolds, J. F., Field, C. B., and Mooney, H. A. (1997). Disproportional increases in photosynthesis and plant biomass in a California grassland exposed to elevated CO_2: A simulation analysis. *Funct. Ecol.* **11,** 697–704.

Mooney, H., Canadell, J., Chapin, F. S., Ehleringer, J., Körner, Ch., McMurtrie, R., Parton, W., Pitelka, L., and Schulze, D.-E. (1999). Ecosystem physiology responses to global change. *In* "The Terrestrial Biosphere and Global Change. Implications for Natural and Managed Ecosystems" (B. H. Walker, W. L. Steffen, J. Canadell, J. S. I. Ingram, eds.), pp. 141–189. Cambridge University Press, London (in press).

Oren, R., Ewers, B. E., Todd, P., Phillips, N., and Katul, G. (1998). Water balance delineates the soil layer in which moisture affects canopy conductance. *Ecological Applications* **8,** 990–1002.

Schulze, D.-E., Canadell, J., Scholes, B., Ehleringer, J., Hunt, T., Sutherst, B., Chapin III, F. S., and Steffen, W. (in press). "The study of ecosystems in the context of global change. *In* "The Terrestrial Biosphere and Global Change. Implications for Natural and Managed Ecosystems" (B. H. Walker, W. L. Steffen, J. Canadell, J. S. I. Ingram, eds.), pp. 19–44. Cambridge University Press, London.

Index

Physiological Ecology

A Series of Monographs, Texts, and Treatises

T. T. KOZLOWSKI. Growth and Development of Trees, Volumes I and II, 1971
D. HILLEL. Soil and Water: Physical Principles and Processes, 1971
V. B. YOUNGER and C. M. McKELL (Eds.). The Biology and Utilization of Grasses, 1972
J. B. MUDD and T. T. KOZLOWSKI (Eds.). Responses of Plants to Air Pollution, 1975
R. DAUBENMIRE. Plant Geography, 1978
J. LEVITT. Responses of Plants to Environmental Stresses, Second Edition
Volume I: Chilling, Freezing, and High Temperature Stresses, 1980
Volume II: Water, Radiation, Salt, and Other Stresses, 1980
J. A. LARSEN (Ed.). The Boreal Ecosystem, 1980
S. A. GAUTHREAUX, JR. (Ed.). Animal Migration, Orientation, and Navigation, 1981
F. J. VERNBERG and W. B. VERNBERG (Eds.). Functional Adaptations of Marine Organisms, 1981
R. D. DURBIN (Ed.). Toxins in Plant Disease, 1981
C. P. LYMAN, J. S. WILLIS, A. MALAN, and L. C. H. WANG. Hibernation and Torpor in Mammals and Birds, 1982
T. T. KOZLOWSKI (Ed.). Flooding and Plant Growth, 1984
E. L. RICE. Allelopathy, Second Edition, 1984
M. L. CODY (Ed.). Habitat Selection in Birds, 1985

R. J. HAYNES, K. C. CAMERON, K. M. GOH, and R. R. SHERLOCK (Eds.). Mineral Nitrogen in the Plant–Soil System, 1986

T. T. KOZLOWSKI, P. J. KRAMER, and S. G. PALLARDY. The Physiological Ecology of Woody Plants, 1991

H. A. MOONEY, W. E. WINNER, and E. J. PELL (Eds.). Response of Plants to Multiple Stresses, 1991

F. S. CHAPIN III, R. L. JEFFERIES, J. F. REYNOLDS, G. R. SHAVER, and J. SVOBODA (Eds.). Arctic Ecosystems in a Changing Climate: An Ecophysiological Perspective, 1991

T. D. SHARKEY, E. A. HOLLAND, and H. A. MOONEY (Eds.). Trace Gas Emissions by Plants, 1991

U. SEELIGER (Ed.). Coastal Plant Communities of Latin America, 1992

JAMES R. EHLERINGER and CHRISTOPHER B. FIELD (Eds.). Scaling Physiological Processes: Leaf to Globe, 1993

JAMES R. EHLERINGER, ANTHONY E. HALL, and GRAHAM D. FARQUHAR (Eds.). Stable Isotopes and Plant Carbon–Water Relations, 1993

E.-D. SCHULZE (Ed.). Flux Control in Biological Systems, 1993

MARTYN M. CALDWELL and ROBERT W. PEARCY (Eds.). Exploitation of Environmental Heterogeneity by Plants: Ecophysiological Processes Above- and Belowground, 1994

WILLIAM K. SMITH and THOMAS M. HINCKLEY (Eds.). Resource Physiology of Conifers: Acquisition, Allocation, and Utilization, 1995

WILLIAM K. SMITH and THOMAS M. HINCKLEY (Eds.). Ecophysiology of Coniferous Forests, 1995

MARGARET D. LOWMAN and NALINI M. NADKHARNI (Eds.). Forest Canopies, 1995

BARBARA L. GARTNER (Ed.). Plant Stems: Physiology and Functional Morphology, 1995

GEORGE W. KOCH and HAROLD A. MOONEY (Eds.). Carbon Dioxide and Terrestrial Ecosystems, 1996

CHRISTIAN KÖRNER and FAKHRI A. BAZZAZ (Eds.). Carbon Dioxide, Populations, and Communities, 1996

THEODORE T. KOZLOWSKI and STEPHEN G. PALLARDY. Growth Control in Woody Plants, 1997

J. J. LANDSBERG and S. T. GOWER. Application of Physiological Ecology to Forest Management, 1997

FAKHRI A. BAZZAZ and JOHN GRACE (Eds.). Plant Resource Allocation, 1997

LOUISE E. JACKSON (Ed.). Ecology in Agriculture, 1997

ROWAN F. SAGE and RUSSELL K. MONSON (Eds.). C_4 Plant Biology, 1999

YIQI LUO and HAROLD A. MOONEY (Eds.). Carbon Dioxide and Environmental Stress, 1999